Aerosol Science

Aerosol Science

Technology and Applications

IAN COLBECK

School of Biological Sciences,
University of Essex, UK

MIHALIS LAZARIDIS

Department of Environmental Engineering,
Technical University of Crete, Greece

WILEY

This edition first published 2014

Registered office
John Wiley & Sons Ltd, The Atrium, Southern Gate, Chichester, West Sussex, PO19 8SQ, United Kingdom

For details of our global editorial offices, for customer services and for information about how to apply for permission to reuse the copyright material in this book please see our website at www.wiley.com.

Library of Congress Cataloging-in-Publication Data

Aerosol science : technology and applications / edited by Ian Colbeck, Mihalis Lazaridis.
pages cm
Includes index.
ISBN 978-1-119-97792-6 (cloth)
1. Aerosols–Industrial applications. 2. Aerosols–Environmental aspects. I. Colbeck, I. (Ian), editor of compilation. II. Lazaridis, Mihalis, editor of compilation
TP244.A3A336 2014
660′.294515–dc23

2013028905

A catalogue record for this book is available from the British Library.

ISBN: 978-1-119-97792-6

Set in 10/12pt Sabon by Laserwords Private Limited, Chennai, India
Printed and bound in Malaysia by Vivar Printing Sdn Bhd

1 2014

Contents

List of Contributors

Wolfram Birmili, Leibniz Institute for Tropospheric Research, Germany

Da-Ren Chen, Department of Mechanical and Nuclear Engineering, Virginia Commonwealth University, USA

Mansoo Choi, Division of WCU Multiscale Mechanical Design, School of Mechanical and Aerospace Engineering, Seoul National University, South Korea

Ian Colbeck, School of Biological Sciences, University of Essex, UK

Sagnik Dey, Centre for Atmospheric Sciences, IIT Delhi, India

Yannis Drossinos, European Commission, Joint Research Centre, Italy

Sarah Dunnett, Department of Aeronautical and Automotive Engineering, Loughborough University, UK

Pat Goodman, School of Physics, Environmental Health Sciences Institute, Dublin Institute of Technology, Ireland

Otto Hänninen, Department of Environmental Sciences, University of Eastern Finland, Finland

Marie Harang, Institute of Pharmaceutical Sciences, King's College London, UK

Katsumi Hirose, Department of Materials and Life Sciences, Faculty of Science and Technology, Sophia University, Japan

Victoria Hutter, Department of Pharmacy, University of Hertfordshire, UK

Maria Kanakidou, Environmental Chemical Processes Laboratory, Department of Chemistry, University of Crete, Greece

Jana S. Kesavan, Edgewood Chemical Biological Center, Aberdeen Proving Ground, USA

Markku Kulmala, Department of Physics, University of Helsinki, Finland

Ka man Lai, Department of Biology, Hong Kong Baptist University, China

Mihalis Lazaridis, Department of Environmental Engineering, Technical University of Crete, Greece

Katrianne Lehtipalo, Department of Physics, University of Helsinki, Finland

Christel Leyronas, INRA, UR407 de Pathologie Végétale, France

Cindy E. Morris, INRA, UR407 de Pathologie Végétale, France

Darragh Murnane, Department of Pharmacy, University of Hertfordshire, UK

Zaheer Ahmad Nasir, Healthy Infrastructure Research Centre, Department of Civil, Environmental & Geomatic Engineering, University College London, UK

Philippe C. Nicot, INRA, UR407 de Pathologie Végétale, France

John Ogren, NOAA ESRL GMD, USA

Peter V. Pikhitsa, Division of WCU Multiscale Mechanical Design, School of Mechanical and Aerospace Engineering, Seoul National University, South Korea

David Y. H. Pui, Mechanical Engineering Department, University of Minnesota, USA

Jean-Philippe Putaud, European Commission, Joint Research Centre, Italy

L. Reijnders, IBED, University of Amsterdam, The Netherlands

Jose-Luis Sagripanti, Edgewood Chemical Biological Center, Aberdeen Proving Ground, USA

Mikko Sipilä, Department of Physics, University of Helsinki, Finland

Jonathan Symonds, Cambustion, UK

Jonathon Taylor, Healthy Infrastructure Research Centre, Department of Civil, Environmental & Geomatic Engineering, University College London, UK

Sachchida Nand Tripathi, Department of Civil Engineering, IIT Kanpur, India

Alfred Wiedensohler, Leibniz Institute for Tropospheric Research, Germany

Preface

An aerosol is a stable suspension of solid and liquid particles in a gas. Aerosols are ubiquitous throughout the environment and are very important to public health. It is important that we understand their dynamics so that we can quantify their effects on humans. Airborne particulate matter is a complex mixture of many different chemical species, originating from a variety of sources. Particulate matter can act as a transport medium for several chemical compounds, as well as for biological materials absorbed or adsorbed upon them. The field of aerosol science and technology has advanced significantly over the past 20 years, with ultrafine particles gaining particular interest, not only for their health properties but also for their industrial applications.

Particles in the atmosphere have important effects on both air pollution and climate. There is also increasing concern that terrorist attacks could result in the contamination of the atmosphere with chemical, biological or radiological materials. However, aerosols have been assessed by the Intergovernmental Panel on Climate Change (IPCC) as having the largest radiative forcing uncertainty. Geoengineering has been proposed as a feasible way of mitigating climate change, with many of the suggested approaches involving injecting aerosols into the atmosphere.

This book reviews the technological applications of aerosol science, together with the current scientific status of aerosol modelling and measurements. It presents the fundamental properties of aerosols and introduces some aspects of aerosol dynamics. Topics such as satellite aerosol remote sensing, the effect of aerosols on climate change and atmospheric nucleation are also included. So too are applications related to the health implications of aerosols and, therefore, to topics related to human exposure and infection control. Quantification of the health impact of aerosols plays a crucial role in environmental protection. Today, the pharmaceutical industry is under increasing pressure to realise the full potential of the lungs for local and systematic treatment of diseases. This has resulted in novel aerosol-delivery devices capable of producing particles of defined characteristics for improved delivery. Nanostructured material synthesis, the safety of emerging nanomaterials and filtration are important engineering topics in aerosol science.

The book also gives significant attention to specific aerosol sources such as vehicle emissions and bioaerosols. It discusses the importance of radioactive aerosols as tracers of atmospheric processes following Fukushima. Identification of the significance of specific aerosol sources is important to human exposure assessment and has been identified as an area with significant knowledge gaps.

20 March 2013 | Mihalis Lazaridis and Ian Colbeck

1

Introduction

Mihalis Lazaridis[1] and Ian Colbeck[2]
[1]Department of Environmental Engineering, Technical University of Crete, Greece
[2]School of Biological Sciences, University of Essex, UK

1.1 Introduction

An aerosol is defined as a suspension of liquid or solid in a gas. Aerosols are often discussed as being either 'desirable' or 'undesirable'. The former include those specifically generated for medicinal purposes and those intentionally generated for their useful properties (e.g. nanotechnology, ceramic powders); the latter are often associated with potential harmful effects on human health (e.g. pollution). For centuries, people thought that there were only bad aerosols. Early writers indicated a general connection between lung diseases and aerosol inhalation. In 1700, Bernardo Ramazzini, an Italian physician, described the effect of dust on the respiratory organs, including descriptions of numerous cases of fatal dust diseases (Franco and Franco, 2001).

Aerosols are at the core of environmental problems, such as global warming, photochemical smog, stratospheric ozone depletion and poor air quality. Recognition of the effects of aerosols on climate can be traced back to 44 BC, when an eruption from Mount Etna was linked to cool summers and poor harvests. People have been aware of the occupational health hazard of exposure to aerosols for many centuries. It is only relatively recently that there has been increased awareness of the possible health effects of vehicular pollution, and in particular submicron particles.

The existence of particles in the atmosphere is referred to in the very early literature (see Husar, 2000; Calvo *et al.*, 2012). In the 1800s, geologists studied atmospheric dust in connection with soil formation, and later that century meteorologists recognised the ability

Aerosol Science: Technology and Applications, First Edition. Edited by Ian Colbeck and Mihalis Lazaridis.

of atmospheric particles to influence rain formation, as well as their impact on both visible and thermal radiation (Husar, 2000).

The environmental impact of the long-range transport of atmospheric particles has also been widely discussed (Stohl and Akimoto, 2004). Around 1600, Sir Francis Bacon reported that the Gasgogners of southern France had filed a complaint to the King of England claiming that smoke from seaweed burning had affected the wine flowers and ruined the harvest. During the eighteenth century, forest fires in Russia and Finland resulted in a regional haze over Central Europe. Even then, Wargentin (1767) and Gadolin (1767) (quoted in Husar, 2000) indicated that it would be possible to map the path of the smoke based on the locations of the fires and its appearance at different locations. Danckelman (1884) mentions that hazes and smoke from burnings in the African savannah have been observed in various regions of Europe since Roman times.

The possibility of atmospheric particles forming from gaseous chemical reactions was pointed out by Rafinesque (1819). In his paper entitled 'Thoughts on Atmospheric Dust', he makes a number of pertinent observations: '*Whenever the sun shines in a dark room, its beams display a crowd of lucid dusty molecules of various shapes, which were before invisible as the air in which they swim, but did exist nevertheless. These form the atmospheric dust; existing every where in the lower strata of our atmosphere*'; '*The size of the particles is very unequal, and their shape dissimilar*'.

In spite of the widespread occurrence of aerosols in nature and their day-to-day creation in many spheres of human activity, it is only in comparatively recent times that a scientific study has been made of their properties and behaviour. During the late nineteenth and early twentieth centuries, many scientists working in various fields became interested in problems that would now be considered aerosol-related. The results were fairly often either byproducts of basic research, related to other fields or just plain observations that roused curiosity. Several of the great classical physicists and mathematicians were attracted by the peculiar properties of particulate clouds and undertook research on various aspects of aerosol science, which have since become associated with their names, for example Stokes, Aitken and Rayleigh.

Whatever the usage, the fundamental rules governing the behaviour of aerosols remain the same. Rightly or wrongly, the terms 'aerosols' and 'particles' are often freely interchanged in the literature. Aerosols range in size range from 0.001 μm ($0.001\,\mu m = 10^{-9}\,m = 1\,nm = 10\,\text{Å}$) to 100 μm ($10^{-4}$ m), so the particle sizes span several orders of magnitude, ranging from almost macroscopic down to near molecular sizes. All aerosol properties depend on particle size, some very strongly. The smallest aerosols approach the size of large gas molecules and have many of the same properties; the largest are visible grains that have properties described by Newtonian physics.

Figure 1.1 shows the relative size of an aerosol particle (diameter 0.1 μm) compared with a molecule (diameter 0.3 nm, average spacing 3 nm, mean free path 70 nm (defined as the average distance travelled by a molecule between successive collisions)).

There are various types of aerosol, which are classified according to physical form and method of generation. The commonly used terms are 'dust', 'fume', 'smoke', 'fog' and 'mist'. Virtually all the major texts on aerosol science contain definitions of the various categories. For example, for Whytlaw-Gray and Patterson (1932):

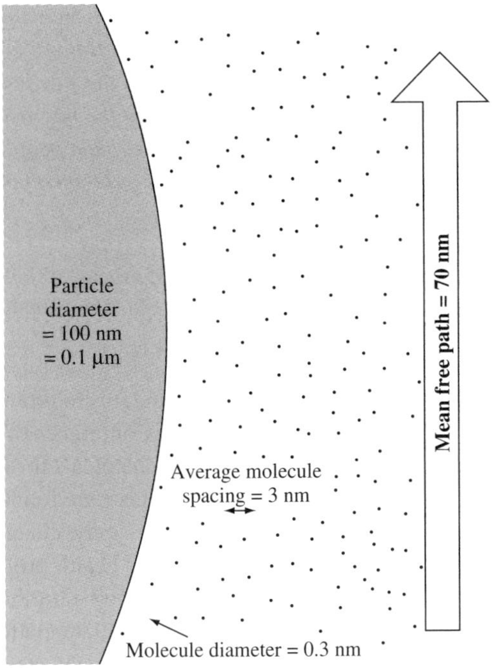

Figure 1.1 *Relative size of an aerosol particle (diameter 0.1 μm) compared with a molecule (diameter 0.3 nm).*

> ***Dust:*** 'Dusts result from natural and mechanical processes of disintegration and dispersion.'
> ***Smoke:*** 'If suspended material is the result of combustion or of destructive distillation it is commonly called smoke.'

while more recently, for Kulkarni, Baron and Willeke (2011):

> ***Dust:*** 'Solid particles formed by crushing or other mechanical action resulting in physical disintegration of a parent material. These particles have irregular shapes and are larger than about 0.5 μm.'
> ***Smoke:*** 'A solid or liquid aerosol, the result of incomplete combustion or condensation of supersaturated vapour. Most smoke particles are submicrometer in size.'

It is clear right from the early literature that dust and smoke are not defined in terms of particle size but in terms of their formation mechanism.

The actual meanings of 'smoke' and 'dust' have recently been the subject of an appeal at the New South Wales Court of Appeal (East West Airlines Ltd v. Turner, 2010). The New South Wales Dust Diseases Tribunal had previously found in favour of a flight attendant who inhaled smoke in an aircraft. The initial trial judge concluded that '*In ordinary common parlance, dust encompasses smoke or ash. Dust may need to be distinguished from gas, fume or vapour. The distinction would be that dust comprises particulate matter.*

Smoke comprises particulate matter and, accordingly, is more comfortably described as dust rather than gas, fume or vapour. I do not consider that there is a distinction between smoke and dust such that smoke cannot be dust. When the particulate matter settled, it would, to most people, be recognised as dust. If, through the microscope or other aid, one could see the particulate matter without the smoky haze, most people would recognise the particulate matter as dust. The dictionary definitions would encompass smoke as dust'.

The Court of Appeal stated:

> ... *His Honour did not find that, as a matter of general principle, 'smoke' was a 'dust'* ... *This was not a decision as to a point of law but a factual determination. There was ample evidence before his Honour to justify that conclusion.*

Various governments worldwide have instigated standards to protect workers from toxic substances in workplaces. For example, the American Conference of Governmental Industrial Hygienists (ACGIH) publishes a list of over 600 chemicals for which 'threshold limit values' have been established. Approximately 300 of these are found in workplaces in the form of aerosols. Aerosol science is thus central to the study, characterisation and monitoring of atmospheric environments. Aerosols can cause health problems when deposited on the skin, but generally the most sensitive route of entry into the body is through the respiratory system. Knowledge of the deposition of particulate matter in the human respiratory system is important for dose assessment and the risk analysis of airborne pollutants. The deposition process is controlled by physical characteristics of the inhaled particles and by the physiological factors of the individuals involved. Of the physical factors, particle size and size distribution are among the most important. The same physical properties that govern aerosols in the atmosphere apply within the lungs.

Aerosols in the atmosphere are either primary or secondary in nature. Primary aerosols are atmospheric particles that are emitted or injected directly into the atmosphere, whereas secondary aerosols are atmospheric particles formed by *in situ* aggregation or nucleation from gas-phase molecules (gas to particle conversion). Particles in the atmosphere consist of a mixture of solid particles, liquid droplets and liquid components contained within the solid particles. Particles are variable in relation to their concentration and their physicochemical and morphological characteristics. Particles can be products of combustion, suspensions of soil materials, suspensions of sea spray or secondary formations from chemical reactions in the atmosphere (Figure 1.2).

Aerosols have diverse effects ranging from those on human health to those on visibility and the climate. They are also very important in public health and understanding of their dynamics is essential to the quantification of their effects. Human exposure to aerosols occurs both outdoors and indoors. They are also important in numerous technological applications, such as the delivery of drugs to the lungs, delivery of fuels for combustion and the production of nanomaterials.

The World Health Organization's Global Burden of Disease WHO GBD project concluded that that 3.2 million people die prematurely every year from cancer, heart disease and other illnesses that are attributable to particulate air pollution; 65% of these deaths occur in Asia. Brauer *et al.* (2012) have reported that 99% of the population in South and East Asia lives in areas where the WHO Air Quality Guideline (annual average of

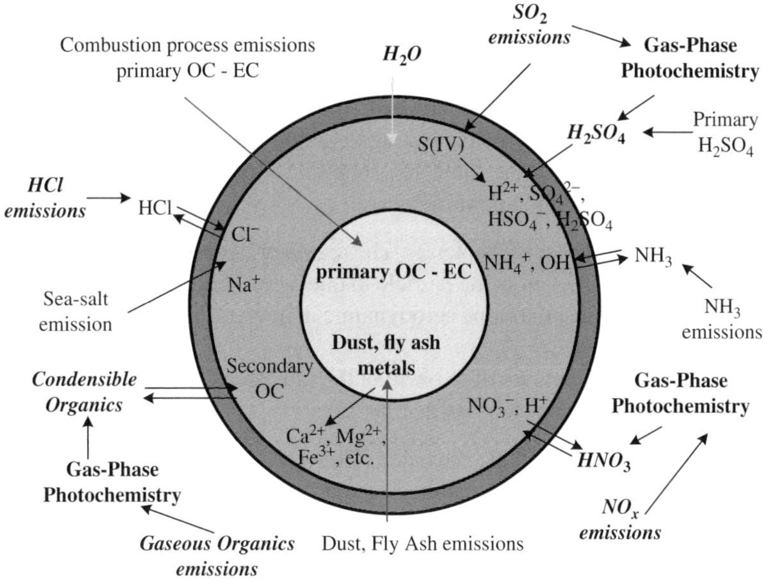

Figure 1.2 *Schematic representation of the chemical reactions and processes associated with the chemical composition of particulate matter.*

10 μg/m^3) for $PM_{2.5}$ is exceeded. Particulate matter pollution was also ranked ninth of all the risk factors in terms of years lost due to disability by Brauer *et al.* (2012).

Aerosols have the potential to change the global radiation balance. A 2007 report by the Intergovernmental Panel on Climate Change (IPCC) estimated the effect of aerosols on the climate since the start of the industrial era to be around 20% of that of greenhouse gases. Aerosols are thought to be responsible for a negative forcing and therefore to have mitigated some of the expected global warming over this period (Kulmala, Riipinen and Kerminen, 2012).

1.2 Size and Shape

Particle size is the most important descriptor for the prediction of aerosol behaviour. When its particles are all the same in size, an aerosol is termed 'monodisperse'. This is extremely rare in nature. Generally, particles vary in size, and this is called 'polydisperse'. When its particles are chemically identical, an aerosol is called 'homogeneous'. Particle shapes can be divided into three general classes:

- **Isometric**: The particle's three dimensions are roughly equal, for example spherical particles.
- **Platelets**: The particle has two long dimensions and a third small one, for example leaves and discs.
- **Fibres**: The particle has one long dimension and two much smaller ones, for example needles and asbestos.

Most of our knowledge regarding aerosol behaviour relates to isometric particles. Concern over the health hazards of fibres has prompted their study.

When particles are spherical, their radius or diameter can be used to describe their size. Since most particles are not spherical, however, other parameters must be used. Often the diameter is defined in terms of particle setting velocity. All particles with similar settling velocities are considered to be the same size, regardless of their actual size, composition or shape. The two most common definitions are:

- **Aerodynamic diameter** (see Chapter 2): The diameter of a unit-density sphere with the same aerodynamic properties as the particle in question. This means that particles of any shape or density will have the same aerodynamic diameter if their settling velocity is the same.
- **Stokes diameter**: The diameter of a sphere of the same density as the particle in question that has the same settling velocity as that particle.

Stokes diameter and aerodynamic diameter differ only in that Stokes diameter includes the particle density whereas aerodynamic diameter does not. Equivalent diameter is also commonly used. When particle size is measured by a specific technique, the measurement usually corresponds to a specific physical property; if electrically induced motion is used then a mobility equivalent diameter is reported.

1.3 Size Distribution

Determination of the aerosol size distribution is one of the most important aspects in the measurement and modelling of aerosol dynamics. The diameter of an ambient particle can be determined by various means, including light-scattering measurements, characterisation of the aerodynamic resistance of the particle and measurement of its electrical mobility or settling velocity. It is necessary to refer to an equivalent diameter independent of the measurement method and therefore the Stokes and aerodynamic equivalent diameter have been introduced. The aerodynamic diameter is defined as the diameter of a spherical particle with equal settling velocity as the particle under consideration but with material density of 1 g/cm^3 (Hinds, 1999).

Particles can be categorised according to their size based on (i) their observed modal distribution (Hinds, 1999), (ii) the 50% cut-off diameter or (iii) dosimetric variables related to human exposure. In the latter case, the most common divisions are $PM_{2.5}$ and PM_{10}. PM_{10} is defined as airborne particulate matter passing through a sampling inlet with a 50% efficiency cut-off at 10 μm aerodynamic diameter that transmits particles below this size (European Commission, 2008); $PM_{2.5}$ is similarly defined. The division in particle size is related to the possibility of $PM_{2.5}$ particles penetrating to the lower parts of the human respiratory tract. In the modal distribution, several subcategories can be observed: nucleation mode, Aitken mode, accumulation mode, ultrafine particles and fine and coarse particles. These terms are discussed in Chapter 2.

An important region of the size distribution is the ultrafine part of the nuclei mode. Understanding of the physics and chemistry of very small clusters containing a few hundreds of molecules represents a theoretical and experimental challenge. Figure 1.3 depicts some typical aerosol size ranges and their related properties.

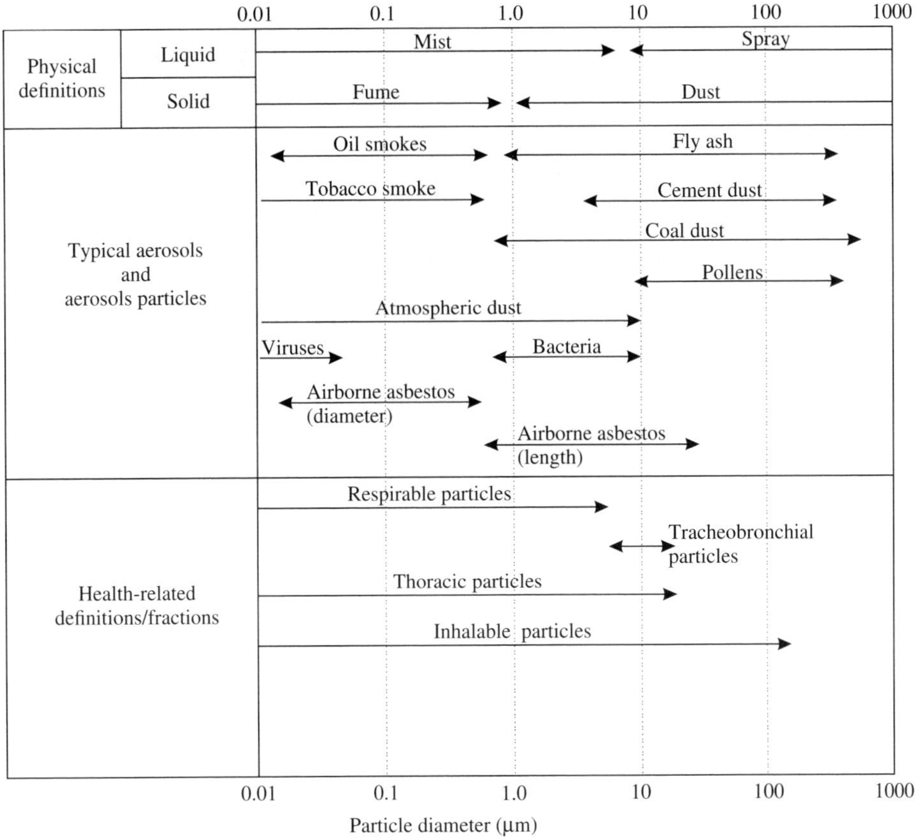

Figure 1.3 *Particle size range for aerosols.*

The various aerosol modes are associated with different sources and mechanisms of formation and with different chemical characteristics, as depicted in Figure 1.1 for the number and volume distributions. Examples of aerosols in the Aitken mode include soot, sulfuric acid and crystal bio-organic particles; in the accumulation mode, ammonium sulfate, marine organics and biomass smoke; and in the coarse mode, dust, sea salt and pollen.

Figure 1.4 shows an example of the aerosol size distribution and morphology obtained from electron microscopy. Number and volume size distribution are depicted together with the chemical composition by size for a number of aerosol types.

The logarithmic canonical distribution of particle mass is used to describe aerosol dynamics. The multilognormal model is widely used to describe aerosol size distribution (Seinfeld and Pandis, 2006; Lazaridis, 2011). The multilognormal distribution is mathematically expressed as:

$$\frac{\mathrm{d}N}{\mathrm{d}(\log(D_p))} = \sum_{i=1}^{n} \frac{N_i}{\sqrt{2 \cdot \pi} \cdot \log(\sigma_{g,i})} \cdot \exp\left[-\frac{\left(\log\left(D_p\right) - \log(\overline{D}_{pg,i})\right)^2}{2 \cdot \log^2(\sigma_{g,i})} \right] \tag{1.1}$$

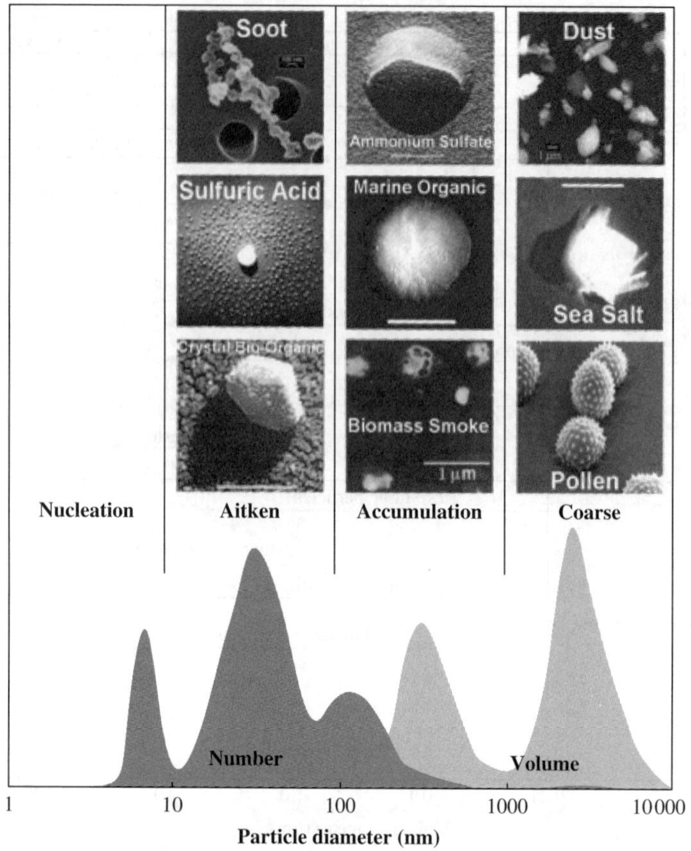

Figure 1.4 *Aerosol size distribution and morphology for various aerosol types. Reproduced with permission from Heintzenberg* et al. *(2003). Copyright © 2003, Springer Science + Business Media.*

where n is the number of modes, N_i the number concentration in each mode, D_p the aerosol diameter, $D_{pg,i}$ the geometric mean diameter in each mode and $\sigma_{g,i}$ the geometric standard deviation.

Aerosol behaviour in the atmosphere is controlled by internal and external processes. Internal processes act within the system boundaries, while external processes processes act across boundaries (Whitby and McMurry, 1997). Internal processes include coagulation, condensation, evaporation, adsorption/desorption, heterogeneous chemistry and nucleation mechanisms (Figure 1.5). External processes involve convection, diffusion and the effect of external forces such as thermophoresis (Hinds, 1999).

Figure 1.6 presents some typical atmospheric aerosol distributions by number and volume. The volume distribution has different features to the number distributions; it is usually bimodal, with a minimum ~1 μm (the dividing limit between coarse and fine particles). The arithmetic distribution has a maximum at the ultrafine mode (nucleation mode), whereas the

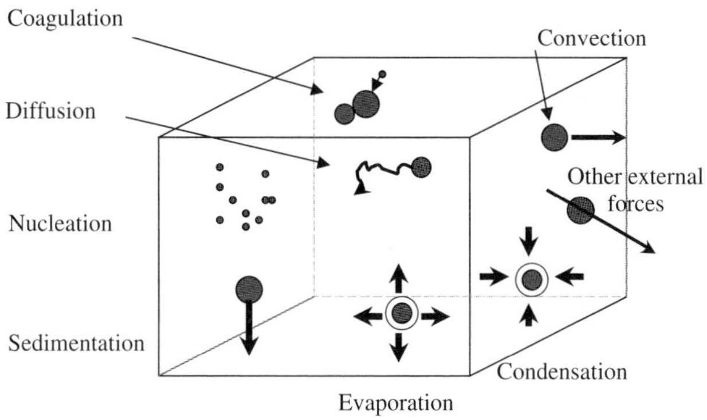

Figure 1.5 *Internal and external processes that control aerosol behaviour.*

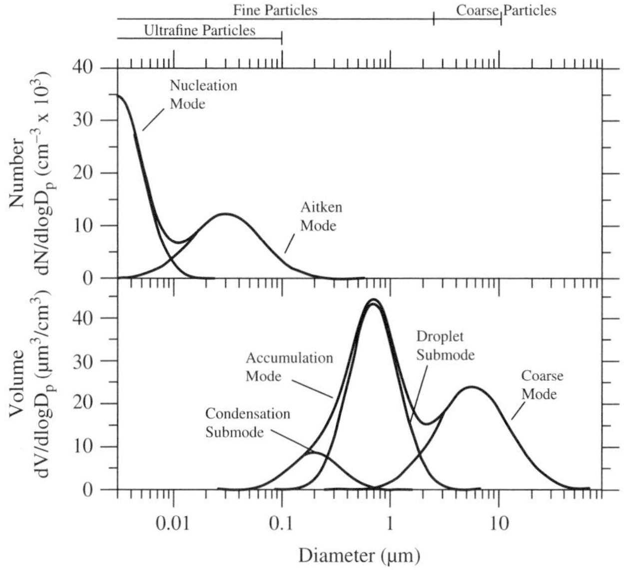

Figure 1.6 *Typical ambient aerosol distributions by number and volume. (Adapted from Seinfeld and Pandis, 2006.)*

volume distribution presents two logarithmic distributions, one at the accumulation mode and one at the coarse mode. It should be remembered that 1 million 1 μm particles have the same volume as a single 100 μm.

The separation of fine and coarse particles is a determined factor, since particles in these two regions are different with respect to their source, chemical composition, processes for removal from the atmosphere, optical properties and effects on human health (Hinds, 1999; Lazaridis, 2011).

1.4 Chemical Composition

The composition of an atmospheric aerosol is determined from its source, which can include the emission of primary and secondary particles produced in the atmosphere. The main components of an aerosol include sulfate, nitrate, ammonium, chloride, elemental carbon, organic carbon, water, chloride and crustal material (Seinfeld and Pandis, 2006).

Crustal material, biogenic matter and sea salt make up the majority of natural aerosols. Anthropogenic aerosols comprise primary emitted soot (elemental carbon) and secondary formed carbonaceous material (organic carbon) and inorganic matter (nitrates, sulfates, ammonium and water).

Figure 1.7 shows the distribution of particles and the physicochemical processes associated with different particle sizes.

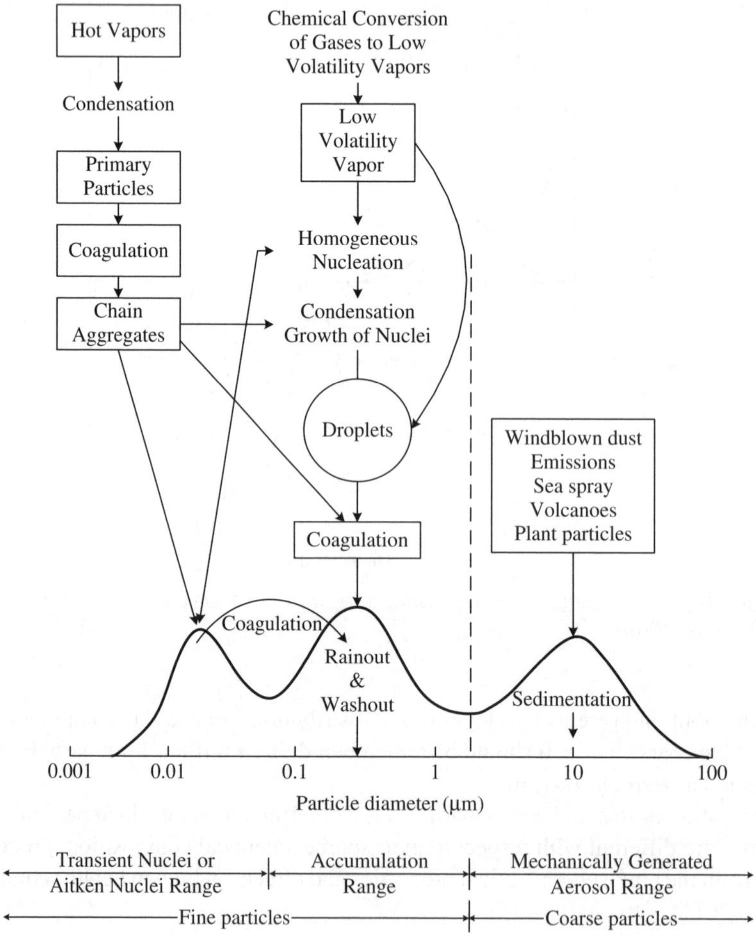

Figure 1.7 Physicochemical processes related to aerosol particle size.

An important part of secondary aerosol particles in the atmosphere is composed of secondary formed organic matter (Turpin and Huntzicker, 1991) produced from the oxidation of organic compounds. Partitioning of gas-particle organic compounds in the atmosphere is important in determining their association with fine particulate matter (Seinfeld and Pandis, 2006; Lazaridis, 2011). The number of different chemical forms of organic matter and the absence of direct chemical analysis mean that fractional aerosol yields, fractional aerosol coefficients and adsorption/absorption methodologies for describing the incorporation of organic matter in the aerosol phase are mainly experimentally determined. An important pathway for secondary organic particle formation arises from biogenic hydrocarbons. There are very large quantities of globally emitted biogenic hydrocarbons that are highly reactive (Hoffmann *et al.*, 1997).

Bioaerosols include all airborne particles of biological origin; that is, bacteria, fungi, fungal spores, viruses and pollen, as well as their fragments, including various antigens. Aerodynamic diameters can range from about 0.5 to 100 μm (Nevalainen *et al.*, 1991; Cox and Wathes, 1995). Airborne microorganisms become nonviable and fragmented over time, due to desiccation. Indoor air contains a complex mixture of (i) bioaerosols such as fungi, bacteria and allergens and (ii) nonbiological particles such as dust, tobacco smoke, cooking-generated particles, motor vehicle exhaust particles and particles from thermal power plants. Exposure to several of these biological entities, as well as to microbial fragments such as cell-wall fragments and flagella and microbial metabolites such as endotoxin, mycotoxins and volatile organic compounds (VOCs), can result in adverse health effects. In particular, an increase in asthma attacks and bronchial hyper-reactivity has been correlated to increased bioaerosol levels. Bioaerosols are usually measured in standard colony forming units per volume (CFU/cubic metre counts). An in-depth consideration of bioaerosols can be found in Chapter 16 and in a recent review by Després *et al.* (2012).

1.5 Measurements and Sampling

According to Kerker (1997), the first recorded use of laboratory-generated aerosols was by Leonardo da Vinci (1452–1519), who wanted to account for the blue colour of the sky. Centuries later, Tyndall (1869) noted that if a beam of light was passed through a suspension and viewed at an angle against a dark background, the presence of particles was revealed by the scattered light. Tyndall's legacy to aerosol science was great (Gentry and Lin, 1996), including in particular his proposal of a connection between the light scattered by an aerosol during the early stages of its formation, when the particles were small, and the colour of the sky and the polarisation of light. Tyndall assumed that all small particles behaved in this manner and considered the light of the sky a specific instance of a general physical phenomenon. This work and a theoretical treatment by Rayleigh (1871) gave the scattering of blue light by very small particles and the preferential transmission of red light, so strikingly exemplified by the vivid colours of sunset, a ready explanation. At first Rayleigh believed the blue sky was caused by the presence of fine particles such as those Tyndall had experimented with, but sometime later he revised this notion, noting that particles as such were not necessary and that the blue sky '*can be explained by diffraction from the molecules of air themselves*'.

In the past, exploding wires were used to generate aerosols. Although scientific interest in the phenomenon didn't truly begin until the 1920s, the first paper on exploding wires was read before the Royal Society in December 1773 by Nairne (1774). He used an exploding wire to prove that the current in all parts of a series circuit is the same. Some 40 years later, Singer (1815) and Singer and Crosse (1815) reported on more experiments involving exploding wires. Faraday (1857) demonstrated how exploding wires could be uses to produce a metal film or mirror. He was the first scientist to systematically use the exploding wire technique to generate aerosols. He was also first to characterise the aerosols and to develop techniques that allowed certain of their optical properties to be examined (Gentry, 1995).

The requirement to measure aerosols in a range of fields has increased dramatically over the last 2 decades. As a result, there are now a large number of instruments on the market, ranging from small portable devices for personnel exposure monitoring to research-laboratory-based instrumentation. Selection of an instrument depends upon the aims of your research and on determining compliance with standards, quantifying trends and identifying hotspots. In other words, you must decide on (i) what you want to measure (which metric: number, mass, volume and size distribution or concentration), (ii) whether measurement response time critical, (iii) how long you will sample for and (iv) whether you need to collect a sample.

Any sample should be representative of its environment, taking into account timing, location and particle size distribution. As will be discussed in Chapter 3, the sampling system can influence the transmitted sample. Particles do not behave in the same way as gas molecules when dispersed in air. They deposit under gravity, impact on bends due to particle inertia, are deposited on internal surfaces by molecular and turbulent diffusion and are affected by thermal, electrostatic and acoustic forces.

Generally, particulate sampling devices are divided into two types: those that collect a sample on a substrate and those that conduct *in situ* real-time measurements. With the former, one most ensure that the substrate is compatible with subsequent analysis: for gravimetric analysis, the substrate should be weight-stable; for microscopy, the filter should be transparent to radiation (optical or electron); for biological aerosol, recovery of organisms from the filter should be possible; and for chemical analysis, the substrates should have low levels of the compound under analysis or be capable of incineration. With the latter, either extractive or external sensing techniques can be used. Extractive methods require the aerosol to be brought into the instrument (e.g. optical particle counters), whereas external sensing methods are noninvasive.

In summary, a wide variety of techniques and instruments are available by which to measure and characterise aerosols. Each has advantages and disadvantages in terms of size range, concentration range, measurement resolution, speed of response and so on. New instruments are always being developed.

References

Brauer, M., Amann, M., Burnett, R.T. *et al.* (2012) Exposure assessment for estimation of the global burden of disease attributable to outdoor air pollution. *Environmental Science and Technology*, **45**, 652–660.

Calvo, A.I., Alves, C., Castro, A. *et al.* (2012) Research on aerosol sources and chemical composition: past, current and emerging issues. *Atmospheric Research*, **120–121**, 1–28.

Cox, C.S. and Wathes, C.M. (1995) Bioaerosols in the environment, in *Bioaerosols Handbook* (eds C.S. Cox and C.M. Wathes), Lewis Publishers, Boca Raton, FL, pp. 11–14.

Danckelman, V. (1884) Die Bevoelkungsverhaeltnisse des suedwstlichen Africas. *Meteorologische Zeitschrift*, **8**, 301–311.

Després, V.R., Huffman, J.A., Burrows, S.M. *et al.* (2012) Primary biological aerosol particles in the atmosphere: a review. *Tellus Series B*, **64**, 15598. doi: 10.3402/tellusb.v64i0.15598

East West Airlines Ltd v. Turner (2010). NSWCA 53; BC201001873 - 01 Apr 2010 – Supreme Court of New South Wales, Court of Appeal.

European Commission (2008) Directive 2008/50/EC of the European Parliament and of the Council of 21 May 2008 on ambient air quality and cleaner air for Europe. *Official Journal of the European Union*, **L 152**, 1–44.

Faraday, M. (1857) Experimental relations of gold (and other metals) to light. *Philosophical Transactions of the Royal Society of London*, **147**, 145–181.

Franco, G. and Franco, F. (2001) Bernardino Ramazzini: the father of occupational medicine. *American Journal of Public Health*, **91**, 1382.

Gadolin, J. (1767). Bedenken von Sonnenrauch. Abhandlung Der Königlichen Schwedischen Akademie der Wissenschaften, Abhandlungen für die Monate April, Mai, Juni, 1767.

Gentry, J.W. (1995) The aerosol science contributions of Michael Faraday. *Journal of Aerosol Science*, **26**, 341–349.

Gentry, J.W. and Lin, J.C. (1996) The legacy of John Tyndall in aerosol science. *Journal of Aerosol Science*, **27**, S503–S504.

Heintzenberg, J., Raes, F., Schwartz, S. *et al.* (2003) Tropospheric aerosols, in *Atmospheric Chemistry in a Changing World* (eds G. Brasseur, R. Prinn and A.P. Pszenny), Springer, Berlin, Heidelberg, pp. 125–156.

Hinds, W.C. (1999) *Aerosol Technology: Properties, Behavior, and Measurement of Airborne Particles*, 2nd edn, John Wiley & Sons, Inc., New York.

Hoffmann, T., Odum, J.R., Bowman, F. *et al.* (1997) Formation of organic aerosols from the oxidation of biogenic hydrocarbons. *Journal of Atmospheric Chemistry*, **26**, 189–222.

Husar, R.B. (2000). Atmospheric aerosol science before 1900, in *History of Aerosol Science*, (eds Preining, O. and Davis E.J.), Verlag der Oesterreichischen Akademie der Wissenschaften, Wien, pp. 25–36.

Kerker, M. (1997) Light scattering instrumentation for aerosol studies: an historical overview. *Aerosol Science and Technology*, **27**, 522–540.

Kulkarni, P., Baron, P.A. and Willeke, K. (2011) *Aerosol Measurement: Principles, Techniques, and Applications*, 3rd edn, John Wiley & Sons, Inc., Hoboken, NJ.

Kulmala, M., Riipinen, I. and Kerminen, V.M. (2012) Aerosols and climate change, in *From the Earth's Core to Outer Space*, (ed Haapala I.), Springer, Berlin, Heidelberg, pp. 219–226.

Lazaridis, M. (2011) *First Principles of Meteorology and Air Pollution*, Springer Science + Business Media, Dordrecht.

Nairne, E. (1774) Electrical experiments. *Philosophical Transactions of the Royal Society of London*, **64**, 79–89.

Nevalainen, A., Pasanen, A.L., Niininen, M. *et al.* (1991) The indoor air quality in Finnish homes with mold problems. *Environment International*, **17**, 299–302.

Rafinesque, C. (1819) Thoughts on atmospheric dust. *American Journal of Science*, **1**, 397–400.

Rayleigh, L. (1871) On the scattering of light by small particles. *Philosophical Magazine*, **41**, 446–454.

Singer, G.J. (1815) Some account of the electrical experiments of M. De Nelis, of Malines in the Netherlands. *Philosophical Magazine*, **46**, 259–264.

Seinfeld, J.H. and Pandis, S.N. (2006) *Atmospheric Chemistry and Physics: From Air Pollution to Climate Change*, 2nd edn, John Wiley & Sons, Inc., New York.

Singer, G.J. and Crosse, A. (1815) Account of some electrical experiments of M. De Nelis, of Malines in the Netherlands: with an extension of them. *Philosophical Magazine*, **46**, 161–166.

Stohl, A. and Akimoto, H. (2004) *Intercontinental Transport of Air Pollution*, Springer, Berlin.

Turpin, B.J. and Huntzicker, J.J. (1991) Secondary formation of organic aerosol in the Los Angeles Basin: a descriptive analysis of organic and elemental carbon concentrations. *Atmospheric Environment. Part A. General Topics*, **25**, 207–215.

Tyndall, J. (1869) On the blue colour of the sky, the polarisation of skylight, and the polarisation of light by cloudy matter generally. *Proceedings of the Royal Society of London*, **17**, 223–233.

Wargentin, P. (1767). Anmerkungen ueber Sonnenrauch. Abhandlung Der Königlichen Schwedischen Akademie der Wissenschaften, Abhandlungen für die Monate April, Mai, Juni, 1767.

Whitby, E.R. and McMurry, P.H. (1997) Modal aerosol dynamics modeling. *Aerosol Science and Technology*, **27**, 673–688.

Whytlaw-Gray, R. and Patterson, H.S. (1932) *Smoke: A Study of Aerial Disperse Systems*, Edward Arnold and Company, London.

2

Aerosol Dynamics

Mihalis Lazaridis[1] *and Yannis Drossinos*[2]
[1]*Department of Environmental Engineering, Technical University of Crete, Greece*
[2]*European Commission, Joint Research Centre, Italy*

2.1 Introduction

Airborne particulate matter (PM) contains various chemical components and ranges in size from few nanometres to several hundred micrometres (Hinds, 1999). It is apparent that PM is not a single pollutant, and its mass includes a mixture of numerous pollutants distributed differently at different sizes. Particle size is an essential parameter that determines the chemical composition, optical properties, deposition of particles and inhalation in the human respiratory tract (RT) (Hinds, 1999; Friedlander, 2000; Seinfeld and Pandis, 2006; Lazaridis, 2011). Particle size is specified by the particle diameter, d_p, which is most commonly expressed in micrometres. Particles represent a very small fraction, less than 0.0001%, of the total aerosol mass or volume (Drossinos and Housiadas, 2006). The gas phase mainly influences the particle flow through hydrodynamic forces.

Particles may be classified into a number of categories; based on their size, they can be categorized according to (i) their observed modal distribution, (ii) the 50% cut-off diameter of the measurement instrument or (iii) dosimetric variables that are related to human exposure to atmospheric concentrations. However, these categories are not rigorously defined and they are usually application-specific.

In category (i), several subcategories can be identified:

- **Nucleation mode**: Particles with diameter $< 0.1\ \mu m$, which are formed by nucleation processes. The lower size limit of this category is not very well defined, but it is close to 3 nm.

Aerosol Science: Technology and Applications, First Edition. Edited by Ian Colbeck and Mihalis Lazaridis.

- **Aitken mode**: Particles with diameter 0.01 μm < d_p < 100 nm. They originate from vapour nucleation or the growth of preexisting particles as a result of condensation.
- **Accumulation mode**: Particles with diameter 0.1 μm < d_p < 1 up to 3 μm. The upper limit coincides with a relative minimum of the total particle volume distribution. Particles in this mode are formed either by coagulation of smaller particles or by condensation of vapour constituents. The number of particles in this category does not increase with condensational growth. Furthermore, the removal mechanisms of particles in this category are very slow and as a result there is an accumulation of particles.
- **Ultrafine particles**: Particles in the Aitken and nucleation modes.
- **Fine fraction**: d_p < 2 μm.
- **Coarse fraction**: d_p > 2 μm.

Particles in the atmosphere have a distribution of sizes; lognormal distributions are commonly used to describe these distributions (Hinds, 1999; Lazaridis, 2011). Figure 2.1 presents typical atmospheric aerosol distributions by number, surface area and volume (for spherical particles).

A logarithmic normal distribution is used to represent the distribution of particle mass/number/surface area (Hinds, 1999).

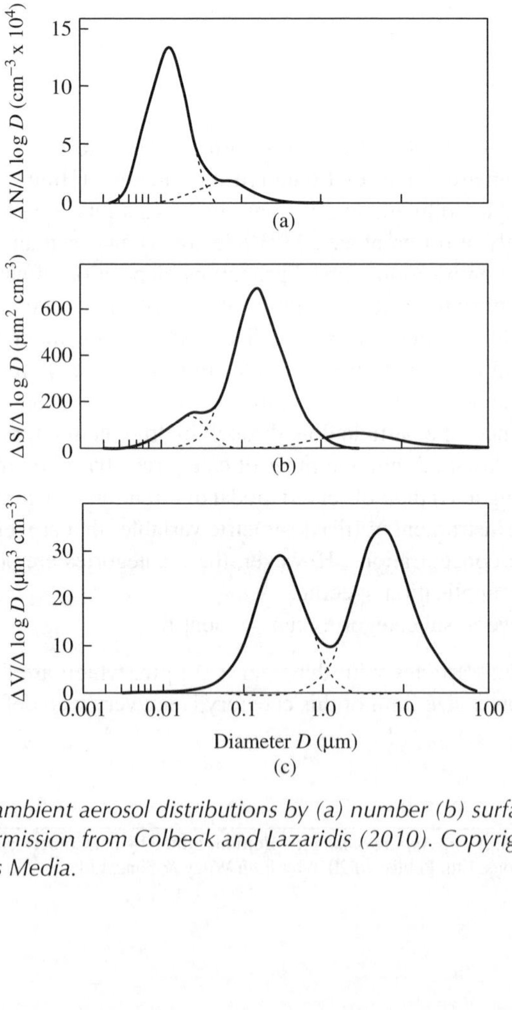

Figure 2.1 *Typical ambient aerosol distributions by (a) number (b) surface area (c) volume. Reproduced with permission from Colbeck and Lazaridis (2010). Copyright © 2010, Springer Science and Business Media.*

The frequency function of a unimodal logarithmic normal distribution can be expressed as:

$$df(d_p) = \frac{1}{\sqrt{2\pi}\ln\sigma_g}\exp\left(-\frac{(\ln d_p - \ln d_g)^2}{2\ln{\sigma_g}^2}\right)d\ln d_p \tag{2.1}$$

and that of a bimodal as:

$$df(d_p) = \frac{\alpha}{\sqrt{2\pi}\ln(\sigma_{g,F})}\exp\left(-\frac{(\ln d_p - \ln d_{gF})^2}{2\ln{\sigma_{g,F}}^2}\right)d\ln d_p + \frac{1-\alpha}{\sqrt{2\pi}\ln(\sigma_{g,C})}\exp\left(-\frac{(\ln d_p - \ln d_{gC})^2}{2\ln{\sigma_{g,C}}^2}\right)d\ln d_p \tag{2.2}$$

where α is the fraction of fine particles, $(1-\alpha)$ the fraction of coarse particles and d_g the mean geometric diameter. The coefficients F and C refer to fine and coarse particles, respectively. The geometric standard deviation σ_g of the distribution given by:

$$\ln\sigma_g = \left(\frac{\sum n_i(\ln d_i - \ln d_g)^2}{N-1}\right)^{1/2} \tag{2.3}$$

where N is the total number of particles and d_g the geometric mean diameter is given by:

$$\ln(d_g) = \frac{\sum n_i \ln d_i}{N} \tag{2.4}$$

There are different distributions that characterize specific particle properties, such as their number, surface area, volume and mass. The number distribution describes the particle number at different sizes, whereas the mass distribution describes the particle mass at different particle sizes. The distributions, if taken to unimodal lognormals, are characterised by a geometric mean diameter and geometric standard deviation.

This chapter presents a general overview of the dynamics of atmospheric aerosols, including the aerosol general dynamic equation (GDE), and of relevant physical processes such as agglomeration, coagulation, gas-to-particle conversion, deposition and resuspension.

2.2 General Dynamic Equation

The variation in space and time of the particle size distribution is described by the GDE, a population-balance equation. The particle size distribution within a fixed-volume element is influenced by processes within the volume (internal processes) and processes that transport particles across the volume boundaries (external processes) (Friedlander, 2000). Internal processes include coagulation, agglomeration, fragmentation and gas-to-particle conversion. External processes include transport across boundaries due to gas flow, particle diffusion, particle motion induced by concentration or temperature gradients and sedimentation (Drossinos and Housiadas, 2006). The GDE is a nonlinear, integrodifferential equation subject to different initial and boundary conditions.

2.2.1 Discrete Particle Size Distribution

Smoluchowski (1916) derived the first equation describing the effect of particle coagulation on the discrete particle size distribution resulting from Brownian motion and motion induced by a laminar shear. The equation refers to internal processes. For a discrete particle size distribution, as a result of coagulation between particles, particles are both removed from and added to size bins. If two particles of masses m_1 and m_2 collide and subsequently coagulate, the mass of the particle formed is $m_3 = m_1 + m_2$. If $K(m_1, m_2)\, n_1\, n_2$ is the coagulation rate between particles of masses m_1 and m_2 then $dn_1/dt = -K(m_1, m_2)\, n_1\, n_2$, $dn_2/dt = -K(m_1, m_2)\, n_1\, n_2$ and $dn_3/dt = K(m_1, m_2)\, n_1\, n_2$. There is a net loss of one particle per coagulation event, but the total mass is conserved. Generalising these equations we obtain:

$$\frac{dn_k}{dt} = \frac{1}{2} \sum_{i+j=k} K(m_i, m_j) n_i n_j - n_k \sum_{i=1}^{\infty} K(m_i, m_k) n_i \tag{2.5}$$

where $i+j=k$ means that the summation is taken over those size grid points for which $m_k = m_i + m_j$. The factor ½ avoids over-counting. Here m_k includes all particles in size bin k. The first term on the right-hand side (RHS) of Equation 2.5 represents the gain in bin k due to coagulation between smaller particles, whereas the second term represents the loss of particles from size bin k due to coagulation with particles in all size bins (including coagulation events between two particles both of which are in bin k). The theory of particle coagulation is reviewed extensively in Fuchs (1964), Friedlander (2000), Williams and Loyalka (1991) and Seinfeld and Pandis (2006).

If the collision frequency function $K(m_i, m_j)$ is constant and independent of particle size, and denoted by K, the equation for the discrete size distribution can be solved analytically to get (Drossinos and Housiadas, 2006):

$$n_k(t) = \frac{N_o}{\left(1 + \frac{t}{\tau}\right)^{k+1}} \left(\frac{t}{\tau}\right)^{k-1} \tag{2.6}$$

where N_o is the initial total number of particles per unit volume and τ is the characteristic agglomeration time scale ($\tau = 2/[K_o N_o]$). For Brownian coagulation of identical particles of diameter d_p and diffusion coefficient D, in a fluid of dynamic viscosity μ at absolute temperature T, the (constant) collision frequency evaluates to:

$$K_o = 8\pi D d_p = 8 k_B T/(3\mu).$$

A generalisation of the Smoluchowski equation includes other internal processes. Accounting for condensation and evaporation, the time-dependent equation for the discrete size distribution becomes (Friedlander, 2000):

$$\frac{dn_k}{dt} = \frac{1}{2} \sum_{i+j=k} K(m_i, m_j) n_i n_j - n_k \sum_{i=1}^{\infty} K(m_i, m_k) n_i + \alpha_{k+1} s_{k+1} n_{k+1} - \alpha_k s_k n_k \tag{2.7}$$

The last two terms on the RHS model condensation and evaporation. The term α_k is the evaporative flux and s_k is the effective surface area of evaporation of a k-mer. It is assumed that evaporation occurs via the loss of single molecules.

2.2.2 Continuous Particle Size Distribution

The continuous distribution is introduced for particle volumes much larger than the molecular volume. In this case, the discrete distribution $n_k(t)$ can be replaced by the continuous distribution $n(\upsilon; \boldsymbol{r}, t)$. The variable $\boldsymbol{r}$ refers to the location of the distribution in space and υ to the particle volume. The variation of $n(\upsilon; \boldsymbol{r}, t)$ with time can be expressed as (Friedlander, 2000):

$$\frac{\partial n}{\partial t} + \nabla \cdot (n\ \mathbf{v}_p) = \left.\frac{\partial n}{\partial t}\right|_{g-p} + \left.\frac{\partial n}{\partial t}\right|_{coag} \tag{2.8}$$

where $\mathbf{v}_p$ is the average particle velocity. The first term on the RHS corresponds to gas-to-particle conversion and the second to particle coagulation. For transport mechanisms that can be considered to act independently (a reasonable assumption for aerosol particles) and for particles of negligible inertia, the average particle velocity becomes:

$$\mathbf{v}_p = \mathbf{u} - \mathrm{D}\nabla \ln n + \mathbf{v}^{\mathrm{ext}}$$

where $\boldsymbol{u}$ is the fluid velocity and $\mathbf{v}^{\mathrm{ext}}$ the sum of all other transport velocities; for example, thermophoretic, electrostatic, gravitational and so on.

Two main methods have been elaborated to model aerosol dynamics using a detailed aerosol size distribution: the sectional method and the moment method. The numerical techniques usually do not take into account spatial inhomogeneities and assume that the aerosol is spatially well mixed. Therefore, the numerical methods apply equally well to Equation 2.7 and to its continuous generalisation Equation 2.8. The main objective is to solve the aerosol GDE using a comprehensive method to treat the complexity of the aerosol size distribution dynamics (Seinfeld and Pandis, 2006). However, since the inclusion of detailed aerosol dynamic models in mesoscale or regional modelling is a difficult and computationally intensive task, various simplifications have been introduced through the omission of specific terms of the GDE.

In the sectional method, the size distribution is divided into several size bins (sections), logarithmically spaced. A common assumption is that all the particles in each section have the same chemical composition (internally mixed assumption).

In the moment method, the moments of the aerosol size distribution are expressed in terms of the distribution parameters. The most important moments of the aerosol size distribution refer to the determination of the total number concentration, the geometric mean diameter and the average surface area and volume per particle (Friedlander, 2000).

2.3 Nucleation: New Particle Formation

Nucleation is the initial stage of a first-order phase transition that takes place in various energetically metastable or unstable systems (Debenedetti, 1996). Homogeneous nucleation refers to new particle formation in the absence of preexisting particles. It has several applications in fields ranging from atmospheric science to nanoparticle formation in engine emissions, or combustion processes in general.

In the atmosphere, where various condensable vapours exist in low concentrations, binary (two-component) or multicomponent nucleation is the predominant particle formation mechanism (Seinfeld and Pandis, 2006; Lazaridis, 2011). Even though homogeneous nucleation is not an important mechanism for the determination of the aerosol mass size distribution, it provides a source of numerous newly formed particles that shape the number size distribution.

2.3.1 Classical Nucleation Theory

The classical nucleation theory (CNT) was developed by Becker and Döring (1935) and Zeldovich (1942) for isothermal nucleation. It is based on the phenomenological concept of a droplet that is viewed as a group of molecules which interact strongly among themselves and weakly with the rest of the system. According to the classical theory, the nucleating cluster is treated with equilibrium thermodynamics as a macroscopic droplet whose free energy of formation depends crucially on the bulk surface tension. The nucleation rate depends exponentially on the reversible work of cluster formation, since nucleation is an activated process.

The kinetics by which small clusters of the new phase gain or lose molecules is based on ideas developed in chemical kinetics. It is assumed that clusters grow or shrink via the gain or loss of single molecules, an approximation that is reasonable for the condensation at low pressures of nonassociated vapours. However, the classical theory, being a phenomenological approach, lacks a sound microscopic foundation.

Homogeneous unary (single-component) nucleation occurs in a supersaturated vapour. The saturation ratio of a chemical species A in air at temperature T is defined as:

$$S = \frac{p_A}{p_A^s(T)} \tag{2.9}$$

where p_A is the partial pressure of the condensable gaseous species A and $p_A^s(T)$ is the saturation pressure of A, which is in equilibrium with the liquid phase at temperature T (over a planar vapour–liquid interface).

The theory of nucleation is based on the solution of a system of equations that describes the concentration variation of clusters with the addition or subtraction of molecules (Debenedetti, 1996). It is assumed that clusters impact with air molecules at a rate that is equilibrated thermally at time periods that are small compared with that necessary for the addition or subtraction of a molecule. This denotes that the clusters have the same temperature as their environment. In the atmosphere, homogeneous nucleation occurs mainly with the participation of two or more chemical compounds.

The kinetic method of the nucleation theory assumes that clusters increase or decrease in size with the addition or removal of one molecule for nonassociated vapours. Therefore, if $N_i(t)$ is the nonequilibrium number concentration of i-molecule clusters, the following equations describe the variation of the cluster concentration:

$$\frac{dN_i}{dt} = \beta_{i-1}s_{i-1}N_{i-1}(t) - \alpha_i s_i N_i(t) - \beta_i s_i N_i(t) + \alpha_{i+1}s_{i+1}N_{i+1}(t) \tag{2.10}$$

where β_i is the forward (condensation) molecular flux per unit time and area at which an i-cluster gains a molecule and α_i is the backward (evaporation) flux at which the cluster loses a molecule. (Equation 2.10 is a specific case of the GDE for single-molecule coagulation/fragmentation in the absence of other processes (see Equation 2.7) and constant single-molecule concentration.)

It is assumed that the cluster concentration is in steady state, and therefore all fluxes equal a stable flux J:

$$J_{i+1/2} = J, \quad \text{for all } i \tag{2.11}$$

The expression for the nucleation flux can be derived as (Debenedetti, 1996; Lazaridis, 2011):

$$J = N_1 \left(\sum_{i=1}^{\infty} \frac{1}{\beta_i s_i f_i} \right)^{-1} \tag{2.12}$$

The evaporation rate α_i is difficult to determine theoretically and therefore the kinetic problem for the evaluation of the nucleation flux becomes a problem of thermodynamics for the evaluation of the equilibrium droplet distribution. For this calculation, it is necessary to examine the energy required for cluster formation.

The radius of a cluster radius containing i^* molecules is given by:

$$r^* = \frac{2\sigma\upsilon'}{(-\Delta\mu)} \tag{2.13}$$

where υ' is the molecular volume in the liquid phase, σ the surface tension and $\Delta\mu$ the difference in the chemical potentials between the liquid phase (cluster) and the gaseous phase. Under the assumption that the liquid is incompressible and the vapour an ideal gas, the chemical potential difference may be expressed via the saturation ratio $-\Delta\mu = k_B T \ln S$.

The free energy of formation of an i-cluster is calculated using the capillarity approximation. In this approximation, the cluster free energy is determined by treating the cluster as an incompressible macroscopic spherical droplet with macroscopic (bulk and surface) properties. The droplet free minimum free energy, which is required for the formation of a cluster with i^* molecules, can be expressed as:

$$\Delta G^* = \frac{16\pi}{3} \left[\frac{\upsilon' \sigma^{3/2}}{(-\Delta\mu)} \right]^2 \tag{2.14}$$

It is further assumed that there exists an equilibrium distribution of clusters. Therefore, the cluster distribution can be expressed as:

$$n_i = n_1 \exp\left(-\frac{\Delta G_i}{k_B T} \right) \tag{2.15}$$

where the free energy of an i-cluster is ΔG_i, k_B is the Boltzmann constant and T is the temperature. The proportionality constant is taken to be the total gas (molecular) number density of the bulk metastable state n_1.

The nucleation rate according to the classical theory can be expressed as:

$$J = J_o \exp\left(-\frac{\Delta G^*}{k_B T} \right) \tag{2.16}$$

where J_o is a kinetic prefactor and ΔG^* is the free energy of formation of the critical droplet. The expression of the kinetic prefactor for unary nucleation may be written as (Lazaridis and Drossinos, 1997):

$$J_o = A\beta Z \rho_\upsilon \tag{2.17}$$

where A is the surface area of the droplet, β the average growth rate, Z the Zeldovich nonequilibrium factor and ρ_υ the number density of the condensable vapour. The growth rate, also known as the impingement rate per unit area, is calculated from the kinetic theory of gases (with a unity accommodation coefficient).

The Zeldovich nonequilibrium factor arises from the number fluctuations in the critical cluster:

$$Z = \frac{\upsilon' \sigma^{1/2}}{2\pi (r^*)^2 (k_B T)^{1/2}} \tag{2.18}$$

The critical droplet free energy expressed in terms of the critical radius is:

$$\Delta G^* = \frac{4\pi}{3} r^{*2} \sigma \tag{2.19}$$

For spherical clusters, the number i^* of molecules in the critical cluster becomes:

$$i^* = \frac{32\pi}{3} \left[\frac{(\upsilon')^{2/3} \sigma}{(-\Delta\mu)} \right]^3 \tag{2.20}$$

where $\Delta\mu$, which is less than 1, is the difference between the chemical potentials in the stable and metastable states (liquid and gaseous phase).

The final expression that gives the nucleation rate per unit volume according to the CNT becomes:

$$J_{CNT} = J_o^{CNT} \exp\left(-\frac{\Delta G^*}{k_B T}\right) = \upsilon' n_{vap}^2 \left(\frac{2\sigma}{\pi m_{mol}}\right)^{1/2} \exp\left[-\frac{16\pi}{3} \frac{\sigma^3 \upsilon'^2}{(k_B T)^3 (\ln S)^2}\right] \tag{2.21}$$

where the number density in the vapour phase is used: $n_{vap} = p_{vap}/(k_B T)$.

Note that the CNT expression is occasionally divided by the saturation ratio (Seinfeld and Pandis, 2006), a correction that arises from a different treatment of nucleation kinetics.

Detailed descriptions of the CNT and of other approaches to the study of nucleation are given by Debenedetti (1996) and Drossinos and Housiadas (2006).

2.3.2 Multicomponent Nucleation

'Multicomponent nucleation' refers to the nucleation process where mixtures of gaseous species are involved even under unsaturated conditions. This is contrary to unary nucleation, where supersaturation is required. Multicomponent nucleation has many applications in atmospheric conditions where several gaseous condensable species exist at low concentrations.

In a recent work, Kevrekidis *et al.* (1999) used the method introduced by Langer (1969) to derive the nucleation rate in binary systems. Their result is formally identical to Stauffer's (1976) result for binary nucleation, but much easier to evaluate. Accordingly, the binary nucleation rate for nonassociated vapours is (see also Drossinos and Housiadas, 2006):

$$J_{nuc} = -\frac{n_{vap} \exp(-\Delta G^*/k_B T)}{2(D_{12}^2 - D_{11} D_{22})^{1/2}} \left[D_{11} R_{11} + D_{22} R_{22} - \sqrt{(D_{11} R_{11} - D_{22} R_{22})^2 + 4 D_{12}^2 R_{11} R_{22}} \right] \tag{2.22}$$

where $n_{vap} = n_{vap,1} + n_{vap,2}$ is the total number density of condensable vapours and R_{ij} is the droplet growth tensor. The variable D_{ij} is the matrix of second-order derivatives of the droplet free energy with respect to the number of molecules n_i of each species evaluated at the saddle point. As in unary nucleation, the growth matrix is expressed as the product of the droplet surface area times the impingement rate of a molecule of species i, $R_{ij} = \delta_{ij}\beta_i 4\pi(r^{*2})^2$, where δ_{ij} is the Kronecker symbol (Drossinos and Housiadas, 2006).

2.3.3 Heterogeneous Nucleation

In the majority of cases, suspended particles exist in the atmosphere and phase transitions (nucleation) from the gaseous to the liquid phase occur on their surfaces. Suppose that a liquid embryo is formed on a particle surface. Heterogeneous nucleation upon insoluble particles takes place at lower saturation ratios than homogeneous nucleation. The minimum work required is given by Lazaridis, Kulmala and Laaksonen (1991):

$$\Delta G_{het} = \sigma_{gl} F_{gl} + (\sigma_{gs} - \sigma_{ls}) F_{gs} + (P - P')V' + [\mu'(T, P') - \mu(T, P)]n \tag{2.23}$$

where σ refers to the interfacial tensions between different phases, the indexes g, s and l to the phase being considered (g refers to the gaseous, s to the solid and l to the liquid phase) and n to the number of molecules in the cluster.

According to the classical theory, the free energy of formation of the critical cluster on a flat surface is modified as follows (Lazaridis, Kulmala and Laaksonen, 1991; Drossinos and Housiadas, 2006):

$$\Delta G^*_{het} = \Delta G^*_{hom} f(\cos \vartheta) \tag{2.24}$$

where the angle ϑ, which can vary from 0 to 180°, is the contact angle between the nucleating cluster and the solid substrate. For water nucleation, the solid is considered hydrophobic or hydrophilic according to whether the contact angle is greater or less than 90°.

As in the case of homogeneous nucleation, the heterogeneous nucleation rate J_{het} can be expressed as a product of a kinetic prefactor J^0_{het} times an Arrhenius factor:

$$J_{het} = J^o_{het} \exp\left(-\frac{\Delta G^*_{het}}{k_B T}\right) \tag{2.25}$$

The kinetic prefactor can be expressed as:

$$J^o_{het} = R_{growth} NZ \tag{2.26}$$

where R_{growth} is the cluster growth rate, Z the Zeldovich nonequilibrium factor and N the product of the total number of molecules adsorbed per unit seed particle surface area (N_{ads}) and the available surface area (A_{ads}) for adsorption per seed particle (Lazaridis, Kulmala and Laaksonen, 1991). A useful approximation is to consider A_{ads} the surface area that yields the maximum nucleation rate (πd_p^2), where d_p is the seed particle diameter, an approximation that allows multiple clusters per seed particle. Therefore, the heterogeneous nucleation rate can be written as:

$$J_{het} = \pi d_p^2 R_{growth} N_{ads} Z \exp\left(-\frac{\Delta G^*_{het}}{k_B T}\right) \tag{2.27}$$

The adsorbed molecules (N_{ads}) have a concentration equal to $N_{ads} = \beta_{ads}\tau$, where β_{ads} is the collision rate of vapour molecules per unit surface area and τ the average time an adsorbed vapour molecule remains on the particle surface. The latter may be written as $\tau = \tau_o \exp\left(\frac{L}{R_g T}\right)$, where L is the heat of adsorption and R_g the gas constant. The characteristic time scale τ_o is the inverse vibrational frequency of two harmonically bound molecules $\left(\tau_o = 2\pi m_\mu^{1/2}\left(\frac{d^2U}{dr^2}\right)^{-1/2}\right)$, where U is the intermolecular interaction potential and m_μ is the reduced mass.

The particle activation probability $P_d(t)$ is the ratio of the concentration of activated particles $N_d(t)$ to the initial (inactivated particle) concentration $N_{d,in}$:

$$P_d(t) = \frac{N_d(t)}{N_{d,in}} = 1 - \exp(-kt) \tag{2.28}$$

where k is the activation rate per preexisting particle. If particle activation occurs via heterogeneous nucleation, the activation rate constant is taken to be the heterogeneous nucleation rate. Particles are considered to be activated once their activation probability is greater than or equal to 0.5. A detailed presentation of classical heterogeneous nucleation and its application to particle activation in condensation particle counters can be found in Giechaskiel *et al.* (2011).

2.3.4 Atmospheric Nucleation

New particle formation in the atmosphere has been observed in the vicinity of polluted sources and in clean, remote regions (Kulmala *et al.*, 2004). Nucleation bursts (homogeneous nucleation) may be responsible for the occurrence of new particle formation in clean environments where the background aerosol concentration is low (Seinfeld and Pandis, 2006).

A main characteristic of a nucleation event is the increase in the aerosol number concentration at the ambient atmosphere. Figure 2.2 shows a comparison of the number size distribution versus time for two days (with and without a nucleation event) as measured at the Acrotiri research station (Chania, Greece). It may be noted that during the nucleation burst, the total number of airborne atmospheric particles increased significantly and rapidly. More precisely, the particle concentration increased on average by 46%, with a maximum value of 72 427 cm^{-3} (92% increase) after 6 hours. This nucleation event was observed on 10 April 2010 (Lazaridis *et al.*, 2008).

Another characteristic case of a nucleation event is presented in Figure 2.3. On 7 July 2009, a nucleation event was registered starting at about 10.00 am and lasting until 17.30 pm (local time). It was calculated that the maximum number concentrations of particles with aerodynamic diameter lower than 51.3 nm and lower than 101.1 nm corresponded to 37 and 71% of the total particle number concentration, respectively. Moreover, during the event the number concentration was 793 ± 281 particles cm^{-3}, whereas for over 3 hours the concentration was higher than the threshold of 1000 particles cm^{-3}, with the maximum value exceeding 1200 particles cm^{-3}.

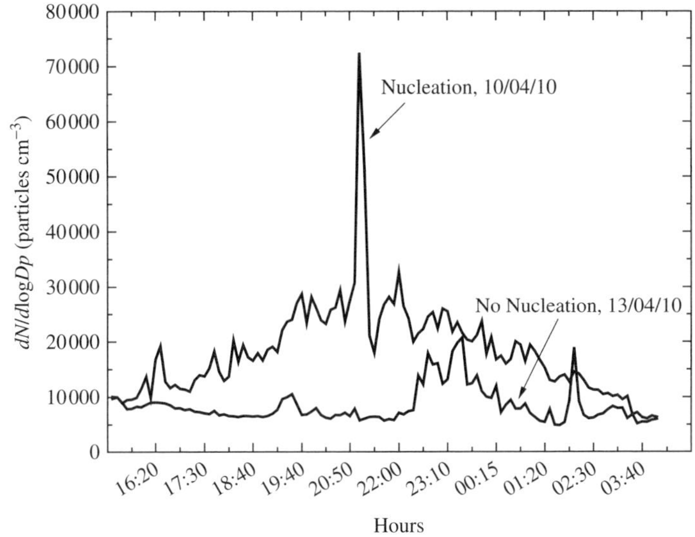

Figure 2.2 *Comparison of a nucleation event and a non-event day observed at the Akrotiri research station, Crete, Greece.*

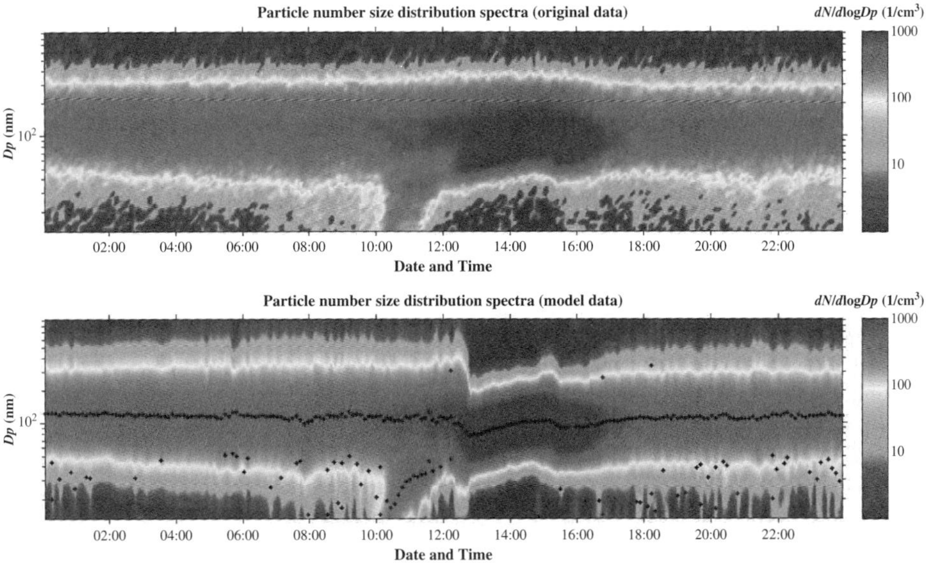

Figure 2.3 *New particle formation event observed during the summer of 2009 (7 July) at the Akrotiri research station (observations and modelled data), Crete, Greece. See plate section for colour version.*

2.4 Growth by Condensation

The particle size in the atmosphere changes primarily through water vapour condensation and evaporation processes. The rate of particle size increase depends on the relative humidity, the particle size and the relative sizes of particles compared to the gas mean free path.[1] Other supersaturated vapour species in the atmosphere may participate in particle growth.

Condensation or evaporation is driven by the pressure difference between the ambient vapour pressure and the vapour pressure at the surface of the particle. The direction of net vapour flux (evaporation or condensation) depends on their relative magnitudes. At the beginning of the particle growth process the nucleated particle diameter is smaller than the gas mean free path. Under these conditions, the particle growth rate depends on the rate of the random molecular impacts between the particles and the water vapour molecules. The rate of particle diameter increase is given by Lazaridis (2011):

$$\frac{d(d_p)}{dt} = \frac{2Ma_c(p_\infty - p_d)}{\rho_p N_a \sqrt{2\pi mkT}} \quad \text{for} \quad d_p < \lambda \tag{2.29}$$

where p_∞ is the partial water vapour pressure around the particle, but at a large distance from its surface, p_d is the partial water vapour at the particle surface, a_c is the condensation coefficient that specifies the percentage of molecules which adhere at the particle surface after impingement, M is the molecular liquid weight, ρ_p is the liquid density and N_a is the Avogadro number.

When the particle size becomes considerably larger than the gas mean free path, the rate of particle growth is related to the rate of diffusion of molecules to the particle surface. The rate of increase of the particle diameter (d_p) can be written as:

$$\frac{d(d_p)}{dt} = \frac{4MD_v}{R\rho_p d_p}\left(\frac{p_\infty}{T_\infty} - \frac{p_d}{T_d}\right)\phi \quad \text{for} \quad d_p > \lambda \tag{2.30}$$

where R is the gas constant and φ is the Fuchs correction factor. Equation 2.30 is based on molecular diffusion theory at the droplet surface. The diffusion equation, as well as the concept of pressure difference, is not valid for distances less than one gas mean free path from the droplet's surface. A number of approximations based on the so-called 'flux-matching techniques' have been introduced, among them the Fuchs correction factor, also known as the Knudsen correction factor. The Fuchs correction factor is given by:

$$\varphi = \frac{2\lambda + d_p}{d_p + 5.33(\lambda^2/d_p) + 3.42\lambda} \tag{2.31}$$

A quick increase in the droplet size also leads to a temperature increase, due to latent heat release during condensation.

The atmosphere mainly consists of soluble particles. We will look at a common atmospheric soluble particle, sodium chloride, in order to discuss how its size changes. When salt dissolves in water, the boiling point of the solution increases and its freezing point decreases, due to the decrease in the water vapour pressure. The affinity of the dissolved

[1] The mean free gaseous path λ is defined as the mean distance of a molecule transport between two sequential impacts. Under standard ambient temperature and pressure conditions, the air mean free path is approximately 66 nm.

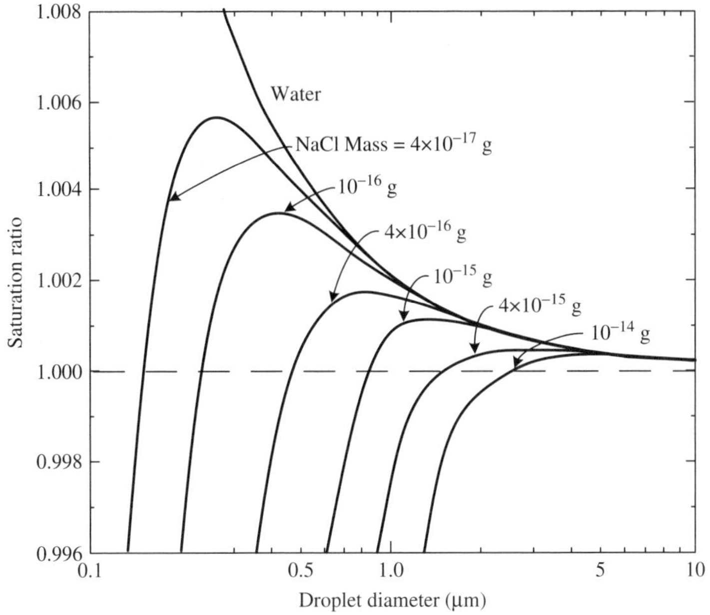

Figure 2.4 *Saturation ratio versus droplet size for pure water and for water droplets containing dissolved sodium chloride at temperature 293 K (20°C). The region above each curve is the condensation region and the region below it the evaporation region.*

salt for the water molecules allows the formation of stable droplets in saturated or undersaturated environments.

Two competitive mechanisms occur during the changes in size of a soluble particle. The salt concentration inside the droplet increases with a decrease in its size. Concurrently, the Kelvin effect should be taken into account; this leads to an increase in the vapour pressure at the droplet surface with a decrease in its size. The relation between the Kelvin ratio K_R and the size of a droplet that contains dissolved chemical components is written as:

$$K_R = \frac{p_d}{p_s} = \left(1 + \frac{6imM_W}{M_s\rho\pi d_p^3}\right)\exp\left(\frac{4\sigma M_w}{\rho RTd_p}\right) \tag{2.32}$$

where m is the mass of the dissolved chemical constituent of mass M_s, M_w the molecular weight of the solvent (usually water), ρ the density of the solvent and i the number of ions into which a salt molecule dissociates. Figure 2.4 shows that the presence of dissolved salt dramatically changes the curve of pure water (solid line).

2.5 Coagulation and Agglomeration

Aerosols in the atmosphere can collide as a result of their Brownian motion or of hydrodynamic, electrical or gravitational forces. These collisions result in either particle coagulation, where the colliding primary particles fuse together and lose their identity, or agglomeration, where they retain their identity and shape. Both mechanisms are crucial to the development of the particle size distribution in the atmosphere (Friedlander, 2000; Drossinos and Housiadas, 2006). The effect of particle collisions in the atmosphere is well described by the Smoluchowski equation, which is usually expressed in terms of particle volume coordinates (Williams and Loyalka, 1991). The Smoluchowski coagulation equation may be written in terms of continuous or discrete distributions; for example, see Equation 2.5 (Williams and Loyalka, 1991; Seinfeld and Pandis, 2006). Particle coagulation leads to a decrease in the number of particles and an increase in their size (diameter).

2.5.1 Brownian Coagulation

Here we will look at Brownian motion resulting from the relative movement of the particles, also called thermal coagulation. A simple case will be examined for spherical particles of diameter larger than 0.1 μm. We assume that particles coagulate after each collision and that the initial change in particle size is small. Furthermore, we focus on a single particle (which does not undergo Brownian motion) and consider the diffusive flux of the other particles to its surface.

The particle number *N(t)* at time *t* is given by:

$$N(t) = \frac{N_o}{1 + N_o K_o t} \tag{2.33}$$

where N_o is the initial particle number and K_o is the coagulation coefficient, which is given by the relation $K_o = 4\pi d_p D$ (m^3/s) (d_p is the particle diameter and D is the diffusion coefficient). Equation 2.33 can be easily obtained from Equation 2.6 by noting that $N(t) = \sum_{k=1}^{\infty} n_k(t)$ and that in Equation 2.6 the central particle is also in Brownian motion (in which case the diffusion coefficient in the expression for K_o describes the relative motions of two particles $D = D_{12} = 2D_1$ where D_1 is the single-particle diffusion coefficient).

The coagulation rate, as obtained by differentiating Equation 2.33, is proportional to N^2, which denotes high rates for increased particle concentrations (Lazaridis, 2011). Equation 2.33 is not applicable for concentration variations at distances of approximately one gas mean free path from the particle's surface; therefore, its region of validity is restricted to particles larger than 0.1 μm. The importance of noncontinuum effects increases as the diameter decreases; they become significant for diameters smaller than 0.4 μm. A correction for the coagulation coefficient was proposed by Fuchs:

$$K = K_o \beta \tag{2.34}$$

At standard conditions (temperature of 298.15 K and pressure of 100 kPa), numerical values for the functions β, K_o and K are given in Table 2.1.

2.5.2 Agglomeration

Particles that retain their identity upon collision agglomerate to form complex structures. Such aggregates are generated in emissions from, for example, combustion in diesel engines, coal combustion in power plants and commercial production of fine particles. Aggregate morphology and hydrodynamic properties influence their size distribution, their precipitation and deposition behaviour and their agglomeration.

Table 2.1 *Coagulation coefficients at standard conditions.*

Diameter (μm)	Correction coefficient, β	K_o (m^3/s)	K (m^3/s)
0.004	0.037	$168 \cdot 10^{-16}$	$6.2 \cdot 10^{-16}$
0.01	0.14	$68 \cdot 10^{-16}$	$9.5 \cdot 10^{-16}$
0.04	0.58	$19 \cdot 10^{-16}$	$10.7 \cdot 10^{-16}$
0.1	0.82	$8.7 \cdot 10^{-16}$	$7.2 \cdot 10^{-16}$
0.4	0.95	$4.2 \cdot 10^{-16}$	$4.0 \cdot 10^{-16}$
1.0	0.97	$3.4 \cdot 10^{-16}$	$3.4 \cdot 10^{-16}$
4	0.99	$3.1 \cdot 10^{-16}$	$3.1 \cdot 10^{-16}$
10	0.99	$3.0 \cdot 10^{-16}$	$3.0 \cdot 10^{-16}$

Adapted from Hinds (1999).

In general, the primary particles that compose an aggregate are polydisperse. However, for simplicity, and in the absence of appropriate theoretical models, we will analyse only aggregates composed of identical spherical monomers. Forrest and Witten (1979), by analysing the agglomeration of ultrafine smoke particle, were the first to suggest that the resulting agglomerates form fractal-like (quasifractal) structures obeying the following scaling law (over a finite size range):

$$N = k_f \left(\frac{R_g}{R_1} \right)^{d_f} \tag{2.35}$$

where N is the number of primary particles that form the aggregate, d_f the fractal (or Hausdorff) dimension, k_f the fractal prefactor, R_g the radius of gyration and R_1 the radius of the primary particles. The fractal dimension provides a quantitative measure of the degree to which a structure fills physical space: a compact three-dimensional object has $d_f = 3$, whereas a line has $d_f = 1$. The fractal prefactor, whose importance was first appreciated by Wu and Friedlander (1993), is an indicator of the aggregate's local structure. For equal-mass identical monomers, the radius of gyration, a geometric measure of the mass distribution about the aggregate centre of mass, becomes the root-mean-square distance of the monomers from the cluster centre of mass:

$$R_g^2 = \frac{1}{N} \sum_{i=1}^{N} (r_i - R_{CM})^2 + a^2 \tag{2.36}$$

where r_i is the position of the ith monomer and the cluster centre of mass is $R_{CM} = 1/N \sum_{i=1}^{N} r_i$. The additional term a^2 is frequently added to ensure that the scaling law remains valid even as the number of monomers tends to unity. It can be chosen to be either the monomer radius or, preferably, the monomer radius of gyration $a = R_1 \sqrt{3/5}$.

The fractal dimension depends on the agglomeration mechanism: computer simulations have been used extensively to study the fractal dimension of aggregates formed via particle–cluster or cluster–cluster aggregation in two and three dimensions, the agglomeration mechanism being diffusive, reaction-limited or ballistic. For example, three-dimensional simulations of diffusion-limited cluster–cluster growth predict $d_f \sim 1.80$ and particle–cluster growth gives $d_f \sim 2.50$ (see for example Table 1.6 in Colbeck, 1998).

The most studied aggregates are those generated via diffusion-limited cluster-cluster aggregation (DLCA) or reaction-limited cluster-cluster aggregation (RLCA). For both groups, the agglomeration mechanism is diffusion, the difference arising from the cluster–monomer sticking probability, which is unity for DLCA clusters and approximately 0.001 for RLCA clusters. The most frequently used scaling law parameters are ($d_f = 1.8$, $k_f = 1.3$) for DLCA aggregates and (2.05, 0.94) for RLCA clusters.

It should be noted that the fractal-like scaling law is valid in a statistical sense, in that the aggregate properties become self-similar after an ensemble average over many aggregates. In that sense, aggregates resemble random isotropic fractals exhibiting inherent randomness that becomes self-similar only in a statistical sense. Moreover, since the aggregates have a finite extent, they are never exactly self-similar. They exhibit self-similarity over a finite range of scales; for that reason, they are referred to as 'fractal-like aggregates'. These observations become more precise when we consider the orientationally averaged monomer–monomer correlation function $g(r)$. Its physical interpretation is that it gives the probability (per unit volume) of finding a monomer at distance r from another, arbitrarily chosen monomer. An analytic expression for $g(r)$ is highly desirable, as structural and dynamical aggregate properties depend on it. A frequently used expression is:

$$g(r) = \frac{A}{R_1^{d_f}} r^{d_f - 3} e^{-\left(\frac{r}{\xi}\right)^{\gamma}} \tag{2.37}$$

where A is a constant (to be determined from the normalisation condition that the spatial integral of $g(r)$ over the system volume equal the number of monomers). The algebraic decay arises from the scaling behaviour and the stretched exponential – the cut-off function – models finite-size effects. The correlation length ξ is a measure of the fractal's diffuse interface and γ is the stretching exponent. The most accurate expression for $g(r)$ to date was proposed by Lattuada, Wu and Morbidelli (2003a), who appreciated the importance of small-scale structure (absent in Equation 2.37, which becomes more accurate as the number of monomers increases). The stretching exponent has been determined to be 2.02 (Heinson, Sorensen and Chakrabarti, 2012) or 2.20 (Lattuada, Wu and Morbidelli, 2003b) for DLCA clusters and 2.16 for RLCA clusters (Lattuada, Wu and Morbidelli, 2003b). The limit $\gamma \to \infty$ corresponds to a sharp aggregate interface, where the scaling law, Equation 2.35, is satisfied exactly up to the sharp interface.

The mobility of N-monomer fractal-like aggregates is usually expressed in terms of the mobility radius R_m, defined via the Stokes friction coefficient:

$$f_N = \frac{1}{B_N} = \frac{k_B T}{D_N} = \frac{6\pi\mu R_m}{C_c\left(R_m\right)} \tag{2.38}$$

where B_N is the aggregate mechanical mobility, D_N the Stokes–Einstein diffusion coefficient, k_B the Boltzmann constant, μ the fluid viscosity, C_c the Cunningham slip correction factor (see Equation 2.44) and T the absolute temperature. In the continuum momentum and mass-transfer regimes (unity slip correction factor), the mobility radius equals the hydrodynamic radius R_h. Herein, we shall concentrate on empirical expressions for the mobility radius in the continuum regime.

Recently, Melas *et al.* (2013), using a methodology proposed by Isella and Drossinos (2011), determined the hydrodynamic radius of orientationally averaged aggregates via the calculation of molecule–aggregate collision rates. The latter were obtained from the solution of a diffusion equation (a Laplace equation instead of the creeping-flow Stokes equations) with appropriate boundary conditions. An analysis of their results suggested the following empirical relationship:

$$\frac{R_h}{R_1} = 0.248\left(2 - N^{-\frac{1}{3}}\right)\frac{R_g}{R_1} + 0.69N^{0.415} \tag{2.39}$$

The particularity, and usefulness, of Equation 2.39 is that neither the fractal dimension nor the fractal prefactor are separately required to estimate the hydrodynamic radius. This contrasts with most empirical fits available in the literature (see, for example, the review of the mobility of fractal aggregates in Sorensen, 2011), where the hydrodynamic radius is expressed in terms of cluster statistical properties like d_f and k_f. According to Equation 2.39, the hydrodynamic radius may be calculated for a single cluster if the monomer positions are known (from simulations or experimental measurements), since the independent variables do not depend on ensemble-averaged properties.

Predictions of Equation 2.39 were compared to predictions of an expression proposed by Kirkwood and Riseman (1948) in their pioneering analysis of the translational diffusion coefficient of flexible macromolecules. The Kirkwood–Riesman expression for the hydrodynamic radius is a geometric expression as well, in that it depends on monomer positions only; it is:

$$\frac{NR_1}{R_h} = 1 + \frac{1}{N}\sum_{j=1}^{N}\sum_{m=1,m\neq j}^{N}\frac{R_1}{|r_j - r_m|} \tag{2.40}$$

Predictions of Equation 2.40 were found to be slightly lower than numerical simulations and empirical-fit predictions (Equation 2.39) for small aggregates, approaching the same limit as the number of monomers increases.

Given an analytical expression for the hydrodynamic radius of fractal-like aggregates, be it Equation 2.39 or 2.40, the Smoluchowski collision kernel between fractal-like aggregates in the continuum regime can be determined. The Smoluchowski kernel for agglomerating spheres (*i,j*) resulting from their thermal motion (Brownian agglomeration) is:

$$K_{ij} = 4\pi(r_{pi} + r_{pj})(D_i + D_j) \tag{2.41}$$

where the first factor is the sum of the collision radii r_p (for spherical particles equal to their geometric radii) and the second factor is the sum of the corresponding Stokes–Einstein diffusion coefficients D_i. For fractal-like aggregates, the collision radii may be taken to be the radii of gyration and calculated from the scaling law (Equation 2.35), and the diffusion coefficients may be expressed in terms of the hydrodynamic radii (according to Equation 2.38) and, for example, Equation 2.39, to obtain:

$$K_{ij} = \frac{2k_B T R_1}{3\mu k_f^{1/d_f}}\left(N_i^{1/d_f} + N_j^{1/d_f}\right)\left(\frac{1}{R_{hi}} + \frac{1}{R_{hj}}\right) \tag{2.42}$$

We have assumed that the two agglomerating aggregates have identical morphology, namely identical fractal dimensions and prefactors. More complicated expressions have been proposed for the modifications of the Smoluchowski kernel in the transition regime for colliding fractal-like structures; see, for example, Thajudeen, Gopalakrishnan and Hogan (2012).

2.6 Deposition Mechanisms

Deposition mechanisms are responsible for the migration of aerosol particles suspended in a fluid and their consequent transfer and deposition to a surface or transfer across boundaries within the flow (Drossinos and Housiadas, 2006). Deposition is an external process in the GDE. Extensive presentations of single-particle deposition have appeared in the literature (Friedlander, 2000; Drossinos and Housiadas, 2006). A study of single-particle deposition in a fluid is presented here without reference to aspects related to convective transport and deposition.

2.6.1 Stokes Law

A particle moving in a fluid feels a drag force, referred to as Stokes drag, that acts in the direction opposite to its own. The flow around a moving aerosol particle is usually in the Stokes regime ($Re_p < 1$). The particle Reynolds number (Re_p) can be expressed as $Re_p = \frac{d_p V}{v_g}$, where V is the mean velocity of the particle relative to the fluid and v_g is the kinematic viscosity (Drossinos and Housiadas, 2006).

The net force exerted by the fluid on the moving particle may be derived from the solution of the Navier–Stokes equations, which describe fluid motion; the exerted force is obtained by integrating the normal and tangential forces over the particle surface. In addition, for small particles, momentum transfer (from the fluid to the particle) does not occur in the continuum momentum transfer regime, leading to a decrease of the Stokes drag force. The total resisting force acting on a small spherical particle of velocity V in the fluid is expressed as (Hinds, 1999):

$$F_D = \frac{3\pi\eta V d_p}{C_c} \tag{2.43}$$

where C_c is a correction factor, referred to as the Cunningham slip correction factor, given by:

$$C_c = 1 + \frac{\lambda}{d_p}\left[2.34 + 1.05\exp\left(-0.39\frac{d_p}{\lambda}\right)\right] \tag{2.44}$$

where λ is the mean free path.

2.6.2 Gravitational Settling

The gravitational settling velocity for a spherical particle in still air is obtained from Stokes drag force to give:

$$V_s = \frac{\rho_p d_p^2 g C_c}{18\eta} \qquad \text{Re} < 0.1 \tag{2.45}$$

where g is the acceleration due to gravity and η is the fluid viscosity. The terminal settling velocity in Equation 2.45 increases rapidly with particle size. The slip correction factor has been incorporated to render this expression valid for any particle size.

A quantity that is often used in aerosol science is the mechanical mobility B (in $\text{m N}^{-1}\,\text{s}^{-1}$), which is the ratio of the terminal velocity of a particle to the steady force producing that velocity (Hinds, 1999):

$$B = \frac{V_s}{F_D} = \frac{C_c}{3\pi\eta d} \tag{2.46}$$

A correction factor is applied to Stokes law for nonspherical particles, called the dynamic shape factor χ. Accordingly, Equation 2.45 becomes:

$$V_s = \frac{\rho_p d_e^2 g C_c}{18\eta\chi} \tag{2.47}$$

where the equivalent volume diameter d_e replaces the particle diameter. The volume equivalent diameter d_e is the diameter of a sphere having the volume of the particle. In most cases, the dynamic shape factor χ has values greater than one, which means that nonspherical particles settle with smaller settling velocity than their equivalent spherical particles. χ usually varies between 1 (sphere) and 2. Dynamic shape factor values for various particle shapes can be found in Hinds (1999).

A characteristic particle diameter that is used extensively in aerosol science and technology is the equivalent aerodynamic diameter, which expresses particle size in a homogeneous manner. It is defined as the diameter of a sphere of density 1 g/cm^3 and the same settling velocity as the particle under study. The aerodynamic diameter d_{ae} is useful because it can be correlated to the residence time of particles in the atmosphere and their deposition in the human RT. It is related to the volume equivalent diameter via:

$$d_{ae} = d_e \sqrt{\frac{\rho}{\chi\rho_o} \times \frac{C_c(d_e)}{C_c(d_{ae})}} \tag{2.48}$$

where ρ is the particle density, whereas $\rho_o = 1$ g/m^3.

Particles of diameter smaller than 0.5 μm are better described by the thermodynamic equivalent diameter, also known as the mobility equivalent diameter, see Equation 2.38, which is the diameter of a spherical particle that has the same diffusion coefficient as the particle under study. The equivalent thermodynamic diameter can be calculated with the help of parameters that control the diffusion of particles through:

$$d_{th} = \frac{k_B T C_c(d_{th})}{3\pi\mu D} \tag{2.49}$$

where T is the absolute temperature (310.15 K), k_B is the Boltzmann constant (0.014×10^{-23} J/grad) and μ is the dynamic viscosity of air (1.90×10^{-4} Poise).

Furthermore, the thermodynamic diameter is connected to the equivalent aerodynamic diameter of the particle via:

$$d_{th} = d_{ae} \sqrt{\frac{\chi\rho_0}{\rho} \times \frac{C_c(d_{ae})}{C_c(d_{th})}} \tag{2.50}$$

2.6.3 Deposition by Diffusion

Particle diffusion results from its Brownian motion, which is the random motion of the particle in the fluid as a result of its continuous bombardment by gaseous molecules. Diffusion of particles is the net transport under the influence of a concentration gradient. The particles move from regions of high concentration to regions of low concentration. Fick's first law of diffusion describes Brownian motion; it can be written as:

$$J = -D\nabla n \tag{2.51}$$

where J is the number flux vector (particles/m^2 s), n the number particle concentration (particles/m^3) and D a diffusion coefficient (m^2/s) (Drossinos and Housiadas, 2006).

The particle diffusion coefficient can be expressed as (Hinds, 1999):

$$D = \frac{k_B T C_c}{3\pi\eta d} \tag{2.52}$$

where $k_B = 1.381 \times 10^{-23}\,\mathrm{J\,K^{-1}}$ is the Boltzmann constant. Equation 2.52 is called the Stokes-Einstein equation. Another way of expressing the diffusion coefficient in terms of particle mobility is

$$D = k_B T B \tag{2.53}$$

2.6.4 Deposition by Impaction

Impaction occurs when particles do not follow the fluid streamlines, due to their inertia, and consequently collide on a surface. This happens in a curvilinear motion inside the human RT or inside a size classifier (e.g. cascade impactors). The parameter that describes the deviation of a particle trajectory from the fluid streamlines is the dimensionless Stokes number (*Stk*), defined as the characteristic velocity times the relaxation time (τ_v) divided by a characteristic length of the flow (L). A detailed description of the Stokes number is presented in Drossinos and Housiadas (2006).

The Stokes number can be expressed as:

$$Stk = \frac{\rho d_e^2 C_c V}{18\eta\chi L} \tag{2.54}$$

For a cascade impactor, the characteristic length is defined with respect to the nozzle radius. When $Stk \ll 1$, particles follow the fluid streamlines perfectly, whereas when $Stk \gg 1$, particles move along straight lines, deviating from the fluid streamlines if they curve.

2.6.5 Phoretic Effects

The motion of aerosol particles depends on the external forces that act upon them. The most commonly encountered external forces that influence a particle's mobility, and thus lead to particle transport, are related to *thermophoresis* and *diffusiophoresis*.

A thermal gradient in a fluid induces a thermophoretic force on aerosol particles, since gaseous molecules exert different impulses to particles on the colder and warmer sides. The thermal gradient results in a net force that moves the particles from the high- to the low-temperature region. The thermophoretically induced particle flux is expressed as:

$$J_c = n\upsilon_{th} \tag{2.55}$$

where n is the particle concentration and υ_{th} the thermophoretic velocity. The thermophoretic velocity can be expressed as:

$$\upsilon_{th} = -K_{th}\nu\frac{\nabla T}{T} \tag{2.56}$$

where K_{th} is a dimensionless parameter, called the thermophoretic coefficient, that depends on properties of both the gas and the particle (Drossinos and Housiadas, 2006).

The dynamics of aerosols, and more specifically their deposition as they are transported in a flowing fluid, is of great importance in technological applications such as aerosol filtration and instrumentation. A thorough discussion of these processes is examined in the scientific literature (Hinds, 1999; Friedlander, 2000; Drossinos and Housiadas, 2006). A brief discussion of aerosol deposition in the atmosphere and in the human RT is presented in this chapter.

2.6.6 Atmospheric Aerosol Deposition

Aerosol deposition in the atmosphere is a complex process; consequently, it is common practise to parameterise it using the concept of deposition velocity. Aerosols are removed from the atmosphere through the mechanisms of dry and wet deposition. The deposition velocity is defined as the ratio of the deposition flux of the specified pollutant (Seinfeld and Pandis, 2006) to the pollutant concentration (hence, the deposition velocity is nothing other than an appropriately defined mass-transfer coefficient). Two general approaches can be used to determine the dry deposition velocity. The first method uses available experimental data for different aerosol species. The second method is based on a calculation of the transfer of materials from the atmosphere to the earth's surface through different resistance mechanisms; such resistance mechanisms are the aerodynamic resistance, the surface resistance and the transfer resistance.

Particle deposition varies with particle size and can be expressed as:

$$v_d^i = \frac{1}{r_a + r_d^i + r_a r_d^i v_g^i} + v_g^i \tag{2.57}$$

where the term on the left-hand side (LHS) is the deposition velocity (m/s) of particles in size bin i and the variable r on the RHS is an appropriately chosen resistance; the subscript a identifies the aerodynamic resistance (s/m), the subscript d the deposition layer resistance (s/m) of particles in size bin i. The variable v with subscript g is the gravitational settling velocity (m/s) of particles in size bin i.

A review of field measurements aimed at determining the deposition velocity of various species is presented in Seinfeld and Pandis (2006). However, measurements of dry deposition have not yet provided a complete understanding of the dry deposition process, since the problem is quite complex, involving many factors that cannot be accounted for in the various field studies. These uncertainties include meteorological conditions (and specifically temporal and spatial characteristics of atmospheric turbulence), surface characteristics and aerosol properties.

Wet scavenging is also an efficient mechanism for the removal of aerosols from the atmosphere. Below-cloud scavenging rate has an approximate value of 3×10^{-5} s^{-1} and in-cloud scavenging is about 10 times larger. The wet deposition velocity can be expressed as the product of an average scavenging rate (Λ) and the vertical height h, where a uniform distribution of pollutant is assumed between the earth surface and height h. A detailed discussion of wet deposition characteristics is presented in Finlayson Pitts and Pitts (2000).

2.6.7 Deposition in the Human Respiratory Tract

Besides a definition of a detailed morphometric model for the respiratory system, understanding of particle deposition in the human RT requires information on the particle size distribution characteristics. The most widely used inhalation model, in which the main aspects of the respiratory system are introduced, is the ICRP model (International Commission on Radiological Protection, 1994).

Particle deposition in the RT depends on the size and physicochemical properties of the particles and the physiology of the person. Inertial impaction, settling and Brownian diffusion are the most important deposition mechanisms in the RT (Housiadas and Lazaridis, 2010; Lazaridis, 2011) (see also Figure 2.5). In addition, interception and electrostatic deposition are important in a number of cases (Hinds, 1999). Total particle deposition efficiency is calculated as a superposition of independent deposition efficiencies arising from different deposition mechanisms.

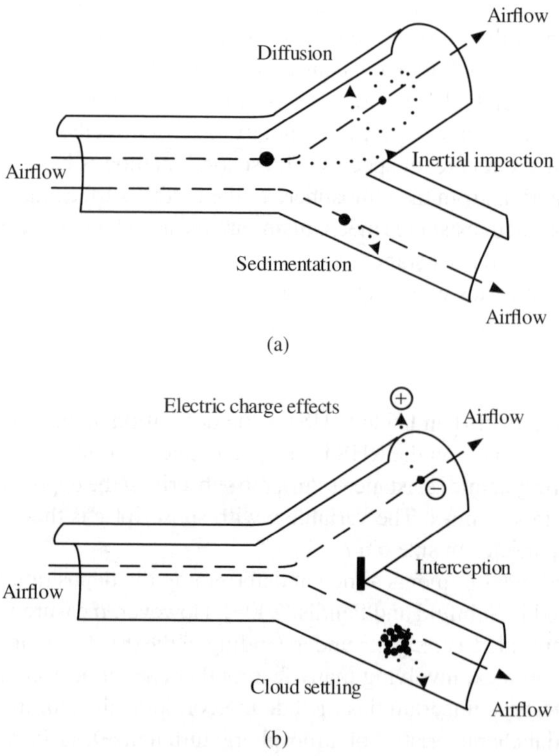

Figure 2.5 Schematic description of deposition mechanisms ((a) diffusion, sedimentation and impaction; (b) interception and electrostatics) of particles in the RT. (Adapted from Ruzer and Harley, 2005.)

The dose of particles in each part of the RT is calculated using the following expression:

$$H = n_0 c_A B \left(\alpha \sum_i n_{fine,i} + (1-\alpha) \sum_i n_{coarse,i} \right) \tag{2.58}$$

where H is the dose rate (μg/h), n_0 the inhalability ratio, c_A the aerosol concentration in air, B the ventilation rate, α the fine mode mass fraction, $n_{fine,\,i}$ the fine-particle retention in lung region i and $n_{coarse,\,i}$ the coarse-particle retention in lung region i (International Commission on Radiological Protection, 1994).

An important factor that determines the intake efficiency at which ambient aerosols enter the mouth and nose is the inhalability (I). The entrance of particles inside the RT depends on the flow characteristics close to the mouth and nose and on the particle size. 'Inhalability' is defined as the ratio of the particle mass concentration inspired through the nose or the mouth (M_o) for a specific size to the ambient mass concentration ($M_{o,amb}$) (Hinds, 1999; Lazaridis, 2011):

$$I = \frac{M_o}{M_{o,amb}} \tag{2.59}$$

Regional deposition profiles in the lung are usually given in terms of deposition fractions (DF) in specific functional compartments: the nasopharyngeal region (the extrathoracic compartment, ET), the tracheobronchial region (TB) and the alveolar region (AL) (Housiadas and Lazaridis, 2010). For an average person performing an average activity (average parameter values), the deposition of particles can be described by the following empirical equations (Hinds, 1999):

For the *extrathoracic region*:

$$DF_{ET} = I \left(\frac{1}{1+\exp\left(6.84 + 1.183 \ln d_p\right)} + \frac{1}{1+\exp(0.924 - 1.885 \ln d_p)} \right) \tag{2.60}$$

where I is the inhalability at air velocity 0 m s^{-1} and d_p is the particle diameter (μm).
For the *tracheobronchial region*:

$$DF_{TB} = \left(\frac{0.00352}{d_p} \right) [\exp(-0.234(\ln d_p + 3.40)^2) + 63.9 \exp(-0.819(\ln d_p - 1.61)^2)] \tag{2.61}$$

For the *alveolar region*:

$$DF_{AL} = \left(\frac{0.0155}{d_p} \right) [\exp(-0.416(\ln d_p + 2.84)^2) + 19.11 \exp(-0.482(\ln d_p - 1.362)^2)] \tag{2.62}$$

The total deposition fraction can be expressed as the sum of the fractions:

$$DF = I \left(0.0587 + \frac{0.911}{1+\exp\left(4.77 + 1,485 \ln d_p\right)} + \frac{0.943}{1+\exp(0.508 - 2.58 \ln d_p)} \right) \tag{2.63}$$

Equations 2.60–2.63 are applicable for particles of diameters between 0.001 and 100 μm, and the results agree with the predictions of the ICRP model, with a deviation of ±0.03%.

2.7 Resuspension

Particle resuspension resulting from turbulent fluid flow is important in the interaction of the atmosphere with various surfaces and in numerous industrial processes. Current models of particle resuspension are based either on force- or energy-balance models. In force-balance models, resuspension occurs when aerodynamic lift forces become greater than the adhesive forces. Force-balance models underestimate the resuspension rate as they neglect energy transfer from the turbulent fluid to a bound particle (Ziskind, Fichman and Gutfinger, 1995; Lazaridis, Drossinos and Georgopoulos, 1998). In energy-balance approaches, the deposited particle is considered to be bound to the surface by an anharmonic potential (Reeks, Reed and Hall, 1988). In the presence of turbulent flow, the fluid transfers energy to the particle through turbulent eddies, and when the particle has accumulated enough energy to escape from the potential well, it resuspends. This section examines the energy-balance approach in the study of monolayer and multilayer resuspension, based on the works of Reeks, Reed and Hall (1988), Lazaridis and Drossinos (1998) and Lazaridis, Drossinos and Georgopoulos (1998).

2.7.1 Monolayer Resuspension

The resuspension rate J, the rate at which a particle can cross the barrier of the adhesive potential well, can be expressed in terms of the height of the potential barrier Q and the average potential energy of particles ($<U>$) in the potential well (Lazaridis *et al.*, 1998):

$$J = \frac{\omega_0}{2\pi} \exp\left(-\frac{Q}{2\langle U\rangle}\right) \tag{2.64}$$

where ω_o is a linear function of the natural frequency of vibration of the bound particle.

The fraction of particles f_R that remain on the surface can be calculated from a first-order kinetic equation (Lazaridis *et al.*, 2008):

$$f_R(t) = \exp(-Jt) \tag{2.65}$$

which shows that f_R decays exponentially with the time exposure to flow.

Since both the surface and the particle surfaces are rough, surface roughness results in a distribution of adhesive forces that can be incorporated through a renormalisation of the particle radius. Defining the normalised particle radius as $r' \equiv r_{eff}/r_p$, the probability distribution of the effective particle radius becomes a probability distribution for r'. It has been argued (Reeks, Reed and Hall, 1988) that the probability density function of the adhesive forces may be approximated by a lognormal distribution:

$$\phi(r')dr' = \frac{1}{(2\pi)^{1/2}\ln\sigma} \exp\left\{-\frac{1}{2(\ln\sigma)^2}[\ln(r'/\overline{r}')]^2\right\} d(\ln r') \tag{2.66}$$

where $\overline{r}'$ is the geometric mean of r' and σ is the standard deviation of the lognormal distribution.

Using the fraction of particle remaining f_R, the fractional resuspension rate $\Lambda(t)$ may be derived as (Lazaridis, Drossinos and Georgopoulos, 1998):

$$\Lambda(t) = -\dot{f}_R(t) = \int_0^\infty J(r') \exp[-J(r')t]\phi(r')dr' \tag{2.67}$$

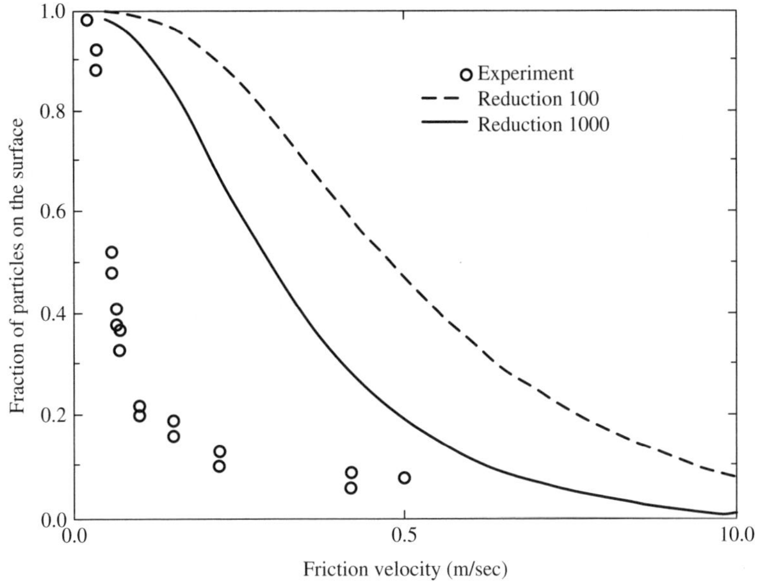

Figure 2.6 *Comparison of model predictions and experimental data for 10 μm alumina spheres on a stainless-steel surface, using the mean lift force of Leighton and Acrivos, Equation 2.68 ($\bar{r}' = 100$ or 1000, $\sigma = 10.4$, t = 1 s). Adapted with permission from Lazaridis, Drossinos and Georgopoulos (1998). Copyright © 1998, Elsevier Ltd.*

Different expressions can be used for the mean lift force responsible for particle resuspension. Among other expressions for the mean lift force, that derived by Leighton and Acrivos (1985) for a sphere resting on a flat surface in wall-bounded shear is:

$$\langle F_{LA} \rangle = 9.22 \rho_f v_f^2 (r_p^+)^4 \tag{2.68}$$

The energy-balance resuspension model has been applied to calculate the resuspension rate of alumina particles from stainless-steel surfaces. The energy spectrum of the turbulent flow – the one-dimensional Fourier transform of the velocity autocorrelation tensor – used in the calculations was taken from Hall (1994). A comparison with experimental data is shown in Figure 2.6; it can be seen that the model under-predicts the resuspension rate. Additional research is required to determine the effect of surface roughness and drag force on the resuspension of single particles, in order to more accurately model experimental data.

2.7.2 Multilayer Resuspension

Particle resuspension from a multilayer deposit induced by turbulent flow is also an important process; the resuspension rate can be derived by solving a set of coupled, first-order kinetic equations (Lazaridis and Drossinos, 1998).

Currently, two multilayer resuspension models exist: one proposed by Lazaridis and Drossinos (LD; Lazaridis and Drossinos, 1998) and one proposed by Friess and Yadigaroglou (FY; Friess and Yadigaroglu, 2001). Both models assume an idealised description of the deposit. The multilayer deposit is considered to consist of n layers, each of which has $N_i(t)$ noninteracting particles (the index i denotes the particle layer; $i = 1$ refers to the first, top, layer). The particles are assumed to be identical, nondeformable spherical particles. Only interlayer interactions are considered; intralayer interactions are neglected. Particles are stacked one on top of another. Initially, all lattice sites are occupied such that the number of particles per layer is constant and equal to N_0. The difference between the two models is best appreciated by taking the single-particle resuspension rate J independent of the layer number. Accordingly, the two models describe the rate of change of N_i by:

$$\frac{dN_i}{dt} = -\left[1 - \frac{N_{i-1}(t)}{N_{i-1}^0}\right] JN_i, \quad \text{for} \quad i \geq 2, \text{(LD)} \tag{2.69}$$

$$\frac{dN_i}{dt} = -\left[1 - \frac{N_{i-1}(t)}{N_i(t)}\right] JN_i, \quad \text{for} \quad i \geq 2, \text{(FY)} \tag{2.70}$$

Both models express the rate of change of the number of particles in the ith layer as the product of the number of particles in that layer, $N_i(t)$, times the resuspension rate per particle J times the fraction of particles exposed (to the turbulent flow). The models differ in how the fraction of exposed particles is calculated. LD argue that the rate of change of $N_i(t)$ is determined solely from the fraction of particles resuspended (missing or holes) from the $(i-1)$th layer; this is, whenever a particle is removed from the layer above, the particle below immediately resuspends (the superscript 0 refers to the initial number of particles in a given layer). Hence, the model is valid for large resuspension rates or, equivalently, gives the maximum multilayer resuspension rate. FY, on the other hand, argue that the resuspension rate depends on the relative difference between the number of particles in the $(i-1)$th layer and the ith layer. As such, the FY model is more general and provides the appropriate description of generic resuspension from a multilayer deposit.

An interesting, and distinguishing, feature of the LD model is that the single-particle resuspension rate is allowed to vary with particle position in the deposit: it is taken to be layer-dependent. The system of coupled, first-order kinetic equations that describes the rate of change of the number of particles in the ith layer can thus be expressed as:

$$\frac{dN_i}{dt} = -\left[1 - \frac{N_{i-1}(t)}{N_{i-1}^0}\right] J_i N_i \quad \text{for} \quad i \geq 2 \tag{2.71}$$

where the superscript 0 denotes the initial number of particles in a given layer.

The kinetic equation for the first layer can be written as:

$$\frac{dN_1}{dt} = -J_1 N_1 \tag{2.72}$$

The solution of this set of equations gives the resuspension rate at different layers. It is shown by Lazaridis and Drossinos (1998) that for a geometrical arrangement of deposited particles with a coordination number of 2 (particles stacked on top of each other), particles from the top layers resuspend at lower friction velocities than particles adjacent to the surface.

Simulation results for a two-layer deposit of SnO_2 particles of radius 10 μm are shown in Figure 2.7 as function of friction velocity. Top-layer particles detach faster than particles in

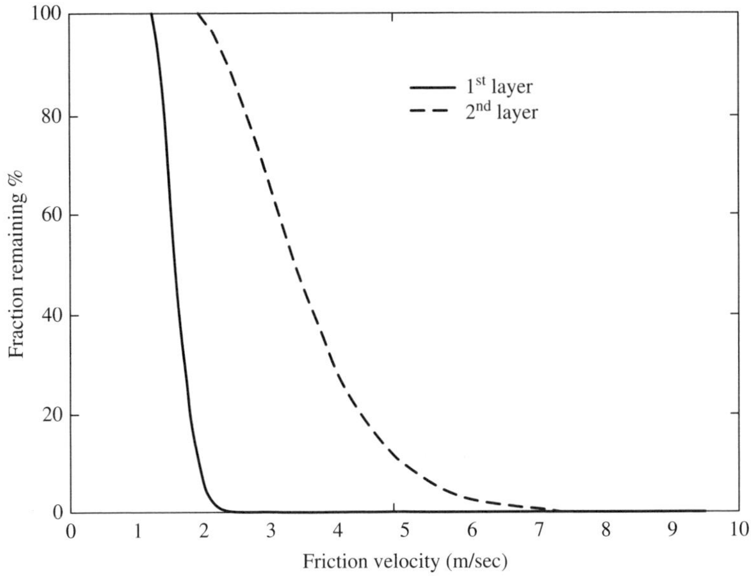

Figure 2.7 *Percentage of SnO_2 particles that remain on a stainless-steel surface versus friction velocity for a two-layer deposit ($r_p = 10\,\mu m$, t = 1 s). Adapted with permission from Lazaridis and Drossinos (1998). Copyright © 1998, Taylor and Francis.*

layers below (and, in particular, in the second layer). Results from Lazaridis and Drossinos (1998) suggest that the fraction of particles remaining in a given layer decays algebraically with time due to resuspension induced by the turbulent flow.

References

Becker, R. and Döring, W. (1935) Kinetische behandlung der keimbildung übersättigten dampfen. *Annalen der Physik (Leipzig)*, **24**, 719.

Colbeck, I. (1998) Introduction to aerosol science, in *Physical and Chemical Properties of Aerosols* (ed. I. Colbeck), Blackie Academic & Professional and Chapman & Hall, London.

Colbeck, I. and Lazaridis, M. (2010) Aerosols and environmental pollution. *Naturwissenschaften*, **97**, 117–131.

Debenedetti, P.G. (1996) *Metastable Liquids: Concepts and Principles*, Princeton University Press, Princeton, NJ.

Drossinos, Y. and Housiadas, C. (2006). Aerosol flows. In *Multiphase Flow Handbook*, (ed. C.T. Crowne), CRC Press, Boca Raton, FL, pp. 6-1–6-58.

Finlayson Pitts, B.J. and Pitts, J.N. (2000) *Chemistry of the Upper and Lower Atmosphere*, Academic Press, San Diego, CA.

Forrest, S.R. and Witten, T.A. (1979) Long-range correlations in smoke-particle aggregation. *Journal of Physics A: Mathematical and General*, **12**, L109–L117.

Friedlander, S.K. (2000) *Smoke, Dust, and Haze: Fundamentals of Aerosol Dynamics*, 2nd edn, Oxford University Press, New York.

Friess, H. and Yadigaroglu, G. (2001) A generic model for the resuspension of multilayer aerosol deposits by turbulent flow. *Nuclear Science and Engineering*, **138**, 161–176.

Fuchs, N.A. (1964) *The Mechanics of Aerosols*, Dover Publication, New York.

Giechaskiel, B., Wang, X., Gilliland, D. and Drossinos, Y. (2011) The effect of particle chemical composition on the activation probability in *n*-butanol condensation particle counters. *Journal of Aerosol Science*, **42**, 20–37.

Hall, D. (1994) Nuclear Electric Report RPTG/P(93) 11.

Heinson, W.R., Sorensen, C.M. and Chakrabarti, A. (2012) A three parameter description of the structure of diffusion limited cluster fractal aggregates. *Journal of Colloid and Interface Science*, **375**, 65–69.

Hinds, W.C. (1999) *Aerosol Technology*, Wiley-Inter Science, New York.

Housiadas, C. and Lazaridis, M. (2010) Inhalation dosimetry modeling, in *The Book Human Exposure to Pollutants via Dermal Absorption and Inhalation* (eds M. Lazaridis and I. Colbeck), Springer.

International Commission on Radiological Protection (1994) *Human Respiratory Tract Model for Radiological Protection*, Annals of the ICRP, Vol. **24** (4), ICRP Publication 66, Pergamon, Oxford, p. 482 p.

Isella, L. and Drossinos, Y. (2011) On the friction coefficient of straight-chain aggregates. *Journal of Colloid and Interface Science*, **356**, 505–512.

Kevrekidis, P.G., Lazaridis, M., Drossinos, Y. and Georgopoulos, P.G. (1999) A unified kinetic approach to binary nucleation. *Journal of Chemical Physics*, **111**, 8010–8012.

Kirkwood, J.G. and Riseman, J. (1948) The intrinsic viscosities and diffusion constants of flexible macromolecules in solution. *Journal of Chemical Physics*, **16**, 565–573.

Kulmala, M., Vehkamaki, H., Petajä, T. *et al.* (2004) Formation and growth rates of ultrafine atmospheric particles: a review of observations. *Journal of Aerosol Science*, **35**, 143–176.

Langer, J.S. (1969) Statistical theory of the decay of metastable states. *Annals of Physics*, **54**, 258–275.

Lattuada, M., Wu, H. and Morbidelli, M. (2003a) Hydrodynamic radius of fractal clusters. *Journal of Colloid and Interface Science*, **368**, 96–105.

Lattuada, M., Wu, H. and Morbidelli, M. (2003b) A simple model for the structure of fractal aggregates. *Journal of Colloid and Interface Science*, **368**, 106–120.

Lazaridis, M. (2011) *First Principles of Meteorology and Air Pollution*, Springer, Dordrecht.

Lazaridis, M. and Drossinos, Y. (1997) Energy fluctuations in steady-state binary nucleation. *Journal of Physics A: Mathematical and General*, **30**, 3847.

Lazaridis, M. and Drossinos, Y. (1998) Multilayer resuspension of small particles by turbulent flow. *Aerosol Science and Technology*, **28**, 548–560.

Lazaridis, M., Drossinos, Y. and Georgopoulos, P.G. (1998) Turbulent resuspension of small non-deformable particles. *Journal of Colloid and Interface Science*, **204**, 24–32.

Lazaridis, M., Dzumbova, L., Kopanakis, I. *et al.* (2008) PM_{10} and $PM_{2.5}$ levels in the Eastern Mediterranean (Akrotiri research station, Crete, Greece). *Water, Air, and Soil Pollution*, **189**, 85–101.

Lazaridis, M., Kulmala, M. and Laaksonen, A. (1991) Binary heterogeneous nucleation of a water-sulphuric acid system: the effect of hydrate interaction. *Journal of Aerosol Science*, **22**, 823–830.

Leighton, D. and Acrivos, A. (1985) The lift on a small sphere touching a plane in the presence of a simple shear flow. *Zeitschrift fur Angewandte Mathematik und Physik*, **36**, 174–178.

Melas, A.D., Isella, L., Konstandopoulos, A.G. and Drossinos, Y. (2013) Morphology and mobility of synthetic colloidal aggregates, *Journal of Colloid and Interface Science* (in press).

Reeks, M.W., Reed, J. and Hall, D. (1988) On the resuspension of small particles by a turbulent flow. *Journal of Physics D: Applied Physics*, **21**, 574.

Ruzer, L.S. and Harley, N.H. (2005) *Aerosol Handbook: Measurement, Dosimetry, and Health effects* (eds.), CRC Press, Boca Raton, FL.

Seinfeld, J.H. and Pandis, S.N. (2006) *Atmospheric Chemistry and Physics: From Air Pollution to Climate Change*, 2nd edn, John Wiley & Sons, Inc., Hoboken, NJ.

Smoluchowski, M. (1916) Drei vortrage uber diffusion, Brownsche molekularbewegung und koagulation von kolloidteilchen. *Physik Zeits*, **17**, 557.

Sorensen, C.M. (2011) The mobility of fractal aggregates: a review. *Aerosol Science and Technology*, **45**, 755–769.

Stauffer, D. (1976) Kinetic theory of two-component ("hetero-molecular") nucleation and condensation. *Journal of Aerosol Science*, **7**(4), 319–333.

Thajudeen, T., Gopalakrishnan, R. and Hogan, C.J. Jr., (2012) The collision rate of nonspherical particles and aggregates for all diffusive Knudsen numbers. *Aerosol Science and Technology*, **46**, 1174–1186.

Williams, M.M.R. and Loyalka, S.K. (1991) *Aerosol Science: Theory and Practice: With Special Applications to the Nuclear Industry*, Pergamon, Oxford.

Wu, M.K. and Friedlander, S.K. (1993) Note on the power law equation for fractal-like aerosol agglomerates. *Journal of Aerosol Science*, **159**, 246–248.

Zeldovich, Y.B. (1942) On the theory of new phase formation: cavitation. *Journal of Experimental and Theoretical Physics*, **12**, 525. (Translated in Ostriker, J.P., Barenblatt, G.I. and Sunayev, R.A. (eds) (1992) *Selected works of Yakov Borisovich Zeldovich*, Chemical Physics and Hydrodynamics, Vol. **1**, Princeton University Press, Princeton, p. 120.)

Ziskind, G., Fichman, M. and Gutfinger, C. (1995) Resuspension of particulates from surfaces to turbulent flows-review and analysis. *Journal of Aerosol Science*, **26**, 613.

3

Recommendations for Aerosol Sampling

Alfred Wiedensohler[1], Wolfram Birmili[1], Jean-Philippe Putaud[2], and John Ogren[3]
[1]Leibniz Institute for Tropospheric Research, Germany
[2]European Commission, Joint Research Centre, Italy
[3]NOAA ESRL GMD, USA

3.1 Introduction

Atmospheric aerosols are measured in many parts of the globe in order to evaluate their various effects on climate, visibility, and human health. 'Particulate matter' (PM) can also be used to describe the atmospheric particulate phase. Atmospheric particles can occur at various diameters from 1 nm to 100 μm. Within this range, the atmospheric particle size distribution is usually considered a continuous function. The most important subdivisions concern fine particles (<1 μm) and coarse particles (>1 μm). Fine particles originate mainly from gas-to-particle conversion and combustion, while coarse particles are typically the result of abrasion or resuspension. The fine particle range can be further divided into the accumulation mode (0.1–1.0 μm), Aitken mode (20–100 nm), and nucleation mode (1–20 nm) ranges. A widely used convention is 'ultrafine particles' for the size range smaller than 100 nm. Particle mass concentrations are usually determined for certain aerodynamic size ranges (PM_{10}, $PM_{2.5}$, PM_1) that simulate different depths of particle penetration into the human airways. The chemical composition of aerosol particles can strongly depend on their origin and thus on their size. In the atmosphere, particles may exhibit considerable spatial and temporal variations. This can cause problems, for example when defining representative sampling cut points for observational networks.

Aerosol Science: Technology and Applications, First Edition. Edited by Ian Colbeck and Mihalis Lazaridis.

Aerosol particles have different atmospheric residence times depending on their deposition rates, which are, again, size-dependent. Particles at the extreme ends of the particle size distribution suffer from coagulation losses (ultrafine particles) and from sedimentation and impaction losses (coarse particles). Accumulation-mode particles are the most long-lived with respect to atmospheric deposition, and they are also the most inert fraction during particle sampling. Due to their high number concentration and efficient interaction with visible solar radiation, accumulation-mode particles play the most relevant role in aerosol radiative effects, which include light scattering, light absorption, and direct radiative forcing. Only in regions that are particularly influenced by sea spray or mineral dust sources can coarse particles dominate the total particle mass concentration and light scattering. For human health, fine and ultrafine aerosol particles are considered the most relevant fraction. These particles can penetrate deep into the lung and initiate respiratory and cardiovascular disease.

3.2 Guidelines for Standardized Aerosol Sampling

3.2.1 General Recommendations

Practical aerosol sampling should avoid any artifacts caused by changing weather or ambient conditions and/or changes within the buildings or container laboratories from which the particles are sampled. Recommendations for standardizing aerosol measurements regardless of the type of observation site and local climate have been issued by, for example, WMO-GAW (World Meteorological Organization – Global Atmosphere Watch), EMEP (European Monitoring and Evaluation Program), and European research infrastructure programs (EUSAAR – European Supersites for Atmospheric Aerosol Research and ACTRIS – Aerosols, Clouds, and Trace gases Research InfraStructure Network). While individual recommendations may be very detailed with respect to, for example, aerosol inlet configuration and conditioning steps under special ambient conditions, the larger standardization framework is usually considered general enough to cater for the requirements of the wide range of existing scientific and monitoring approaches. In practice, aerosol measurements may be divided among short-term measurements for process studies and continuous long-term observations used to obtain seasonal and decadal trends in regional climate or human health. Generating aerosol data of defined and comparable quality should, however, be a priority for both kinds of activity.

An ideal aerosol sampling system allows an undisturbed sample flow from the environment right to the instrumentation, while removing some unwanted air ingredients, such as hydrometeors or excessive moisture. An ideal sampling system:

- excludes precipitation and fog droplets from the sampled aerosol,
- provides a representative ambient aerosol sample under as little diffusional and inertial losses as possible,
- provides aerosol particles at a low relative humidity (RH) (<40%), and
- minimizes the evaporation of volatile particulate species.

The most common set-ups combine an outdoor aerosol inlet, smooth transport pipes, an aerosol conditioner to dry the sampling flow, and a final flow splitter to distribute the aerosol among the various instruments and samplers. Aerosol instrumentation should generally be housed in a room that provides a clean laboratory environment and temperatures between 15 and 30 °C. Optimum indoor temperatures range between 20 and 25 °C.

3.2.2 Standardization of Aerosol Inlets

3.2.2.1 Size Cut-Offs

The cut-off size of the aerosol inlet and the height above ground are usually guided by the purpose of the measurement network. The most widely used options are currently PM_{10}, $PM_{2.5}$, and PM_1, implying upper aerodynamic cut-off diameters at 10, 2.5, and 1 μm under ambient conditions. These inlets are based on particle separation by either an impactor or a cyclone.

Observational networks such as WMO-GAW recommended an upper cut point of 10 μm at ambient conditions (WMO, 2003). The rationale is that particles larger than 10 μm tend to be of local origin and are thus not representative for the regional-scale aerosol and its impact on climate effects. Total suspended matter (TSM) inlets, in contrast, turn out to be sensitive toward wind speed and cannot provide representative samples of larger particles. To obtain additional sizing information, aerodynamic size cuts at 2.5 μm (ambient conditions) and 1 μm (dry conditions) are recommended by WMO-GAW in order to distinguish between fine and coarse particles. The recommendations of the WMO-GAW Report 153 (WMO, 2003) have also been adopted by EMEP and the European Infrastructure Projects EUSAAR and ACTRIS.

For health-related aerosol sampling, PM_{10} and $PM_{2.5}$ are the legal criteria used to sample particulate mass concentrations in government monitoring networks (EC, 1999), although the number of ultrafine particles or black carbon concentrations has been recommended as an additional health-relevant metric. For a justification of the sampling criteria of health-related aerosol measurement and future perspectives, see Vincent (2005).

3.2.2.2 Whole-Air Inlet for Extreme Ambient Conditions

Alternative inlet designs might be considered for measurements in an extreme climate. Sampling sites that experience frequent clouds or freezing may prefer to use a heated whole-air inlet to capture cloud and fog droplets within the sample. This inlet concerns sites that are located in the polar regions or on high alpine mountains. Figure 3.1 illustrates the concept of such a heated whole-air inlet based on the design of the inlet of the Jungfraujoch station in Switzerland, as described in Weingartner, Nyeki, and Baltensperger (1999). Heating prevents clogging of the inlet with ice. Inside the inlet, cloud and fog droplets are evaporated, so that all aerosol particles, whether activated or not, will be included in the measurement. For such whole-air inlets, it is desirable to scrutinize the relationship between the ambient wind velocity and variations in the size-cut characteristics.

3.2.2.3 Tubing and Flow Splitters

Inside the measurement station, the aerosol flow is usually distributed among several instruments. For aerosol particles, care should be taken with the choice of tubing and the

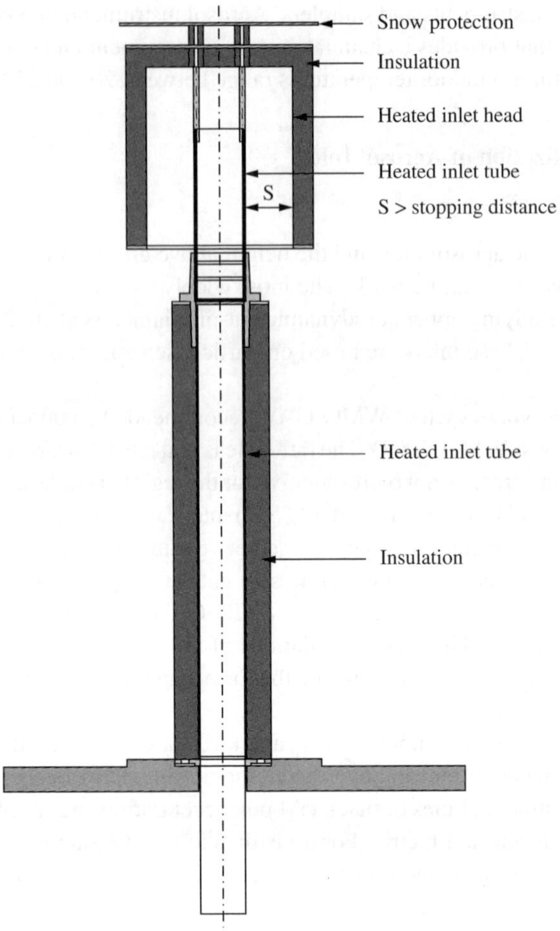

Figure 3.1 *Sketch for a heated whole-air inlet.*

design of flow-distribution devices. Pipes conducting aerosol should be manufactured from metal, preferably stainless steel. It is vital for the sampling of particles that the pipes be of conductive material and electrically grounded. Otherwise, static charges may remove significant portions of the aerosol to be sampled. Short pieces of tubing might be replaced by conductive silicone tubing, which is elastic and conducting at the same time. A perfect inlet installation also avoids sources of turbulence (bends, connectors) as far as possible (because turbulence enhances particle losses due to diffusion) and keeps the sampling lines as short as possible.

Figure 3.2 illustrates a custom-designed isokinetic flow splitter, in which the sample flow velocity is nearly the flow velocity of the main flow. A key feature of the splitter is that a sample is removed from the core of the main aerosol flow rather than from streamlines near the wall of the main pipe. This principle ensures a representative sampling, especially of coarse and nanoparticles.

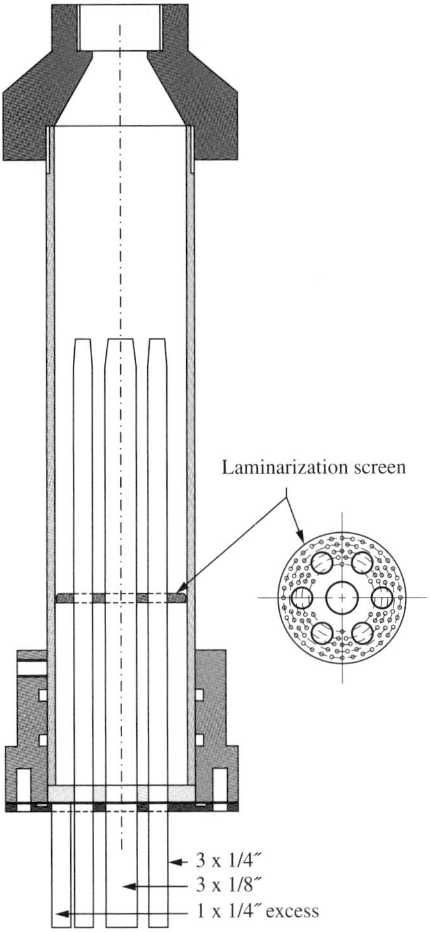

Figure 3.2 *Sketch for an isokinetic flow splitter.*

3.2.3 Humidity Control

One particular problem during summertime or in warm and humid atmospheres is the high dew-point temperature of the air. If this dew point exceeds 20–25 °C, as in tropical or subtropical regions, water may condense in the aerosol sampling lines or hydrophilic particles may grow significantly in size and change their properties. The philosophy of long-term observational networks (WMO-GAW, EMEP, EUSAAR, ACTRIS) is therefore to limit the RH in aerosol instrumentation to a maximum of 40% when determining physical and optical properties. Below 40% RH, the hygroscopic growth is limited and usually causes a less than 10% change in particle diameter compared to that under dry conditions.

3.2.3.1 Ambient Dew-Point Temperature

The choice of measures for drying a sample aerosol at an observation site depends on the difference between the maximum occurring ambient dew-point temperature and the temperature under which the measurement is usually performed. *In situ* physical and optical aerosol measurements are preferably made in an air-conditioned laboratory at around 23 ± 2 °C. Assuming an indoor temperature of 21 °C, the following decisions might be made:

Ambient dew-point temperature below 7 °C: Drying is not necessary, because the RH in the sampling lines at room temperature will always be lower than 40%. A temperature difference of 14 K between the indoor and outdoor environments will decrease RH sufficiently. If the ambient temperature and RH are 7 °C and 100%, for example, with a dew point of 7 °C, then the RH of the aerosol at a room temperature of 21 °C will be 40%.

Ambient dew-point temperature between 7 and 21 °C: In this range, the aerosol flow must be dried at least upstream of each instrument, because RH can exceed 40% in the sampling lines at room temperature. If the ambient temperature and RH are 25 °C and 73%, for example, with a dew point of 20 °C, then the RH of the aerosol at room temperature will be 93%, with no condensation occurring in the sampling pipes.

Ambient dew-point temperature above 21 °C: In this case, RH will always exceed 100% in the sampling lines at a room temperature of 21 °C. This worst-case will lead to condensation of water in the sampling lines and might even damage instrumentation. The main aerosol inlet flow must therefore be dried before entering the room. If the ambient temperature and RH are 30 °C and 65%, for example, at a dew point temperature of 23 °C, then the RH will theoretically amount to 108% at room temperature, so that condensation will occur.

3.2.3.2 Drying Technology

We present four methods for conditioning the aerosol sample flow to an RH below 40%. Each has its advantages and disadvantages. The choice might depend on the requirements of the measurement program and/or the technical facilities at the observation site.

Membrane Dryers. Membrane dryers are elastic tubes based on water vapor-permeable polytetrafluoroethylene (PTFE). Commercially available membranes include Nafion® and Gore-Tex® tubes. Nafion is a sulfonated tetrafluoroethylene that works as a permeable membrane in which water vapor molecules are transported. Gore-Tex is a porous fluoropolymer membrane with roughly one billion pores per square centimeter. The small pore size enables water vapor molecules to pass through. Liquid water, however, cannot penetrate because of the hydrophobic nature of the membrane. In both types, the transport of water vapor is driven by the gradient of its partial pressure across the tube membrane. The air flow to be dried inside of the tube will reach an equilibrium RH if the residence time is sufficiently long. This final RH of the air flow is limited by the dew-point temperature of the surrounding purge gas. In order to dry an air flow sufficiently, the dew-point temperature of the purge gas should be below −20 °C. In practice, only Nafion driers are commercially available for the drying of air flows. These dryers consist of either a single Nafion tube or a bundle of up to 100 individual tubes. While single-tube types can be recommended for the drying of aerosol flows of approximately 1 l/min, bundle dryers should be used with care. The total flow rate used in a bundle dryer can be higher but the flow rate is limited by the pressure drop. In addition, particle losses due to diffusion need to be taken into account.

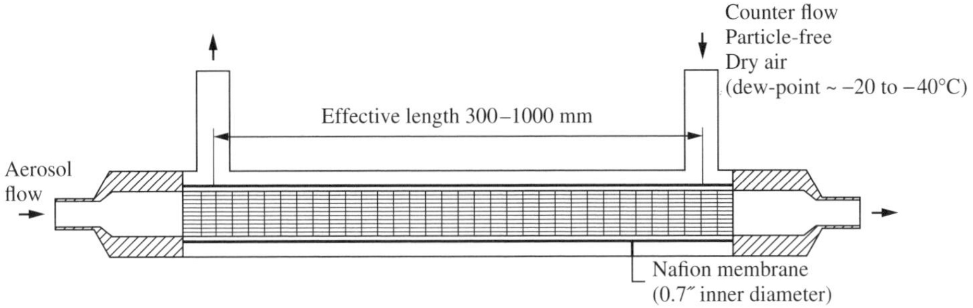

Figure 3.3 *Sketch for an aerosol dryer with low particle losses, based on a Nafion membrane with a 0.7 in. (17.78 mm) inner diameter.*

By design, the bundle Nafion dryer acts as a diffusion battery, so losses of particles smaller than 0.05 μm are significant. Moreover, the entrance of the bundle dryer acts as an impactor for coarse particles larger than 1 μm. The bundle Nafion dryers can thus be recommended only for the particle size range 0.05–1.0 μm.

Custom-designed Nafion dryers can be considered as an alternative. Figure 3.3 illustrates the construction of a single-tube Nafion dryer with minimum pressure drop and particle losses. Here, a Nafion tube with an inner diameter of 12 mm is used. The pressure drop and particle losses are very low in this device.

The advantage of a Nafion dryer is that is does not need regular maintenance. The membrane only requires exchange when the drying efficiency drops. The disadvantage is that purge air with a dew-point temperature below −20 °C must be made available at the measurement site. This requires additional costs for an external compressed air system, including adsorption dryer and filters.

Diffusion Dryers. Diffusion dryers are widely used and available in commercial as well as custom-built designs. The principle is that a stainless-steel screen forms an aerosol tube and is surrounded by silica gel spheres. When the aerosol passes through the metal screen tube, water vapor is adsorbed by the silica. The advantage of an aerosol diffusion dryer is its inexpensive construction and use. However, the silica needs to be replaced and/or regenerated once it is saturated with water. The aerosol flow rate through the diffusion dryer may typically be on the order of a few liters per minute. The higher the flow rate, the more often the silica needs to be exchanged. The main disadvantage of the diffusion dryer is thus its labor-intensive use, especially in environments with high ambient dew-point temperatures. Another disadvantage is enhanced losses by diffusion. The equivalent length required to calculate the losses by diffusion is much longer than the actual length (Wiedensohler *et al.*, 2012).

An automatic diffusion dryer for long-term operation in humid environments is shown in Figure 3.4. The entire dryer is housed in a cabinet and can be placed on the roof of the measurement laboratory or container. The aerosol enters the cabinet and is fed to one of two parallel stainless-steel columns, with an inner diameter of 70 mm and a total length of 800 mm. Each of these columns has one or more stainless-steel mesh tube surrounded by silica gel. The first column is used for aerosol drying while the second is regenerated at ambient pressure by dry air with a low dew-point temperature, preferably below −20 °C.

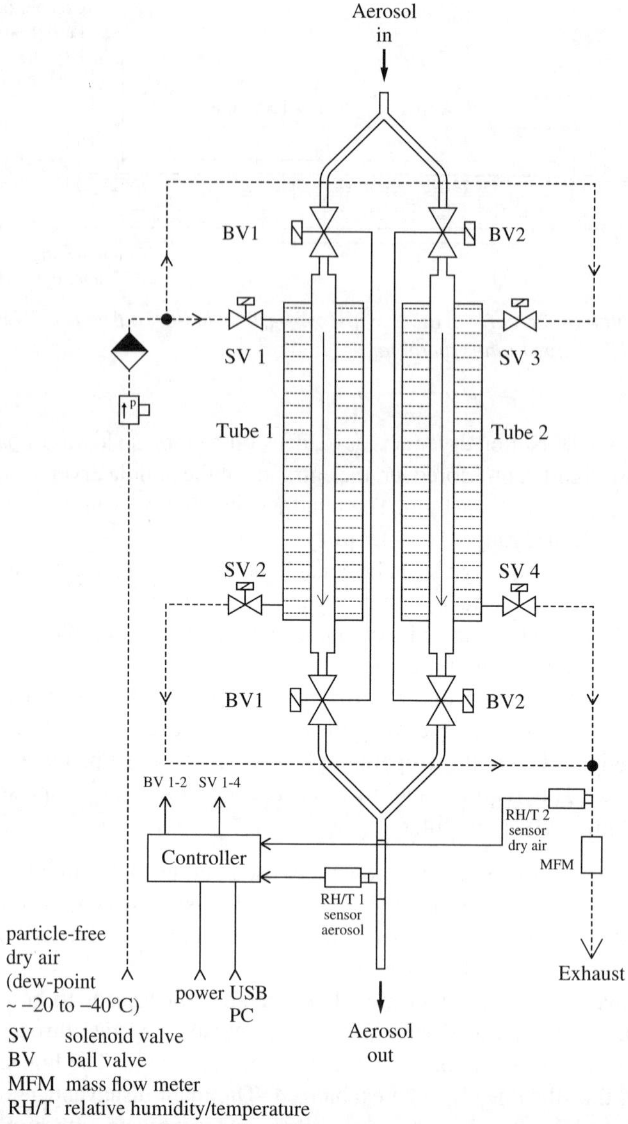

Figure 3.4 *Sketch for a regenerative silica-based diffusion dryer. Reproduced from Tuch* et al. *(2009) with permission of Copernicus Publications.*

During regeneration, dry air is flushed through the column, separated from the atmosphere. The temperature and RH of the dried aerosol and the currently regenerating column are continuously measured. When the RH of the dried aerosol increases above a certain threshold, the columns are switched. For this purpose, each column can be shut at the top and bottom by motor-actuated ball valves. The silica gel of the first column is then generated while the second is used to dry the aerosol.

Drying by Dilution. This method requires the continuous provision of particle-free dry air. It also introduces several uncertainties into the system. First, the mixing between sample and zero air usually requires turbulence, which causes some associated particle diffusion losses. This issue might be of secondary importance if the measurement program concentrates on accumulation-mode particles; that is, if particles at the upper and lower ends of the particle size distribution can be ignored. Second, the dilution ratio must be monitored with great accuracy in order to prevent propagation of the associated error into all measured ambient concentrations. However, if the ambient aerosol concentration is already low, such as in a tropical rainforest, dilution might be a counterproductive measure.

Drying by Heating. Heating of an aerosol sample leads to a reduction in RH in the sampling line. It is insufficient to heat only the aerosol flow while keeping the instrument at room temperature, because the dew-point temperature will remain unchanged. In the case that both the aerosol flow and the instrument can be heated to reduce RH, a modest heating might be applied in order to reach a temperature that is at maximum 10 K above ambient temperature, but not higher than 40 °C. One must realize that heating of the aerosol flow may irreversibly change the sample in that some of the most volatile particulate species will be evaporated. For this reason, we discourage heating of the aerosol sampling flow. Only a heated whole-air inlet can be considered to evaporate cloud and fog droplets or to present accumulation of ice at the inlet.

3.3 Concrete Sampling Configurations

In this section, we compare exemplary aerosol inlet system configurations. The transport efficiencies of these systems are calculated by equations and parameterizations for particle motion. The equations presented hereafter and their derivations can be found in the comprehensive works by Hinds (1999) and Baron and Willeke (2001), among other sources.

3.3.1 General Aspects of Particle Motion

The main challenge in the transportation of an aerosol to collectors and aerosol measuring instrumentation is to avoid particle losses. Particle-loss mechanisms are size-dependent and are generally caused by particle diffusion, impaction, and sedimentation. Generally, losses due to particle diffusion are critical for ultrafine particles smaller than 0.1 µm. In contrast, particle losses due to sedimentation are related to horizontal and sloping pipes, while losses due to impaction are related to bends. The configuration of the whole sampling configuration and the regime of the main air flow are strongly dependent on the purpose of the observational network.

The regime of an air flow in a pipe, laminar versus turbulent, is characterized by its Reynolds number (Re). The flow in a pipe is laminar up to an Re of approximately 2000. Above this value, the flow becomes gradually more and more turbulent. The Re of the flow can be determined by:

$$\frac{\mathrm{Re}_{flow} = \rho_G \times u_{flow} \times D_{pipe}}{\eta_G} \tag{3.1}$$

where ρ_G is the gas density, u_{flow} the flow velocity, D_{pipe} the inner diameter of the pipe, and η_G the gas viscosity.

The inertia of a particle in a flow is characterized by its Stokes number (Stk):

$$Stk = \frac{\tau \times u_{flow}}{D_{pipe}}$$

where:

$$\tau = \frac{\rho_P \times D_P^2 \times C_C}{18\eta_G}$$

and τ is the relaxation time of the particle, u_{flow} the flow velocity, D_{pipe} the inner diameter of the pipe, ρ_P the particle density, D_p the particle diameter, C_C the Cunningham correction factor, and η_G the gas viscosity.

3.3.2 Laminar Flow Sampling Configuration

Generally, a laminar aerosol sampling is recommended in order to minimize particle losses due to diffusion and inertia over a wide size range, especially for nucleation and coarse-mode particles. Furthermore, it allows the pressure drop from the inlet to the instruments to be kept in the range of a few hectopascals. Minimum losses due to particle diffusion in a laminar flow can be achieved by keeping the length of the pipe as short as possible and the flow rate as high as possible. Losses of supermicrometer particles can be minimized by avoiding bends or horizontally oriented sampling pipes. In order to design a laminar sampling configuration, the size-dependent particle penetration can be calculated using Equations 3.2 and 3.3 (Hinds, 1999):

$$P = 1 - 5.5\mu^{2/3} + 3.77\mu \tag{3.2}$$

For μ <0.007:

$$P = 0.819 \times \exp(-11.5\mu) + 0.0975 \times \exp(-70.1\mu) + 0.0325\, exp\,(-179\mu) \tag{3.3}$$

For μ >0.007:

$$\mu = \frac{D \times L_{pipe}}{Q} \tag{3.4}$$

where D is the particle diffusion coefficient, L_{pipe} the length of the pipe, and Q the volume flow rate.

In cases where bends cannot be avoided in the sampling pipe, the size-dependent particle penetration can be calculated by:

$$P = 1 - Stk \times \frac{\theta^\circ}{180^\circ}\pi \tag{3.5}$$

where θ is the angle of the bend.

Size-dependent losses due to sedimentation of supermicrometer particles in horizontal or sloping pipes can be calculated by:

$$P = 1 - \frac{2}{\pi}[2\kappa\sqrt{1-\kappa^{\frac{2}{3}}} - \kappa^{\frac{1}{3}}\sqrt{1-\kappa^{\frac{2}{3}}} + arc\sin\left(\kappa^{\frac{1}{3}}\right)] \tag{3.6}$$

with:

$$
\begin{aligned}
\kappa &= \epsilon \times \sin(\theta) \\
\epsilon &= \frac{3}{4} Z \\
Z &= \frac{L_{pipe}}{D_{pipe}} \times \frac{u_S}{u_{flow}}
\end{aligned}
\tag{3.7}
$$

where L_{pipe} is the length of the pipe, D_{pipe} the inner diameter of the pipe, u_S the sedimentation velocity, u_{flow} the mean flow velocity, and θ the angle of the pipe against the horizontal plain.

3.3.2.1 *Examples of Sampling Configurations with a Laminar Flow*

A low-flow laminar sampling configuration might be preferred if only a limited total sampling flow is required. Physical and optical aerosol instruments, for instance, are operated at low flow rates, and a total flow rate of 1 m^3/h might be sufficient in many situations. In this case, a commercially available low-flow PM_{10} aerosol inlet could be employed, with the option of an additional $PM_{2.5}$ or PM_1 cyclone behind the inlet head. A standard sampling configuration might include measurements of particle number size distribution and of light-scattering and absorption coefficients.

A special case is the construction used for tall towers. Tall-tower samplings require the construction of lengthy sampling pipes, as seen in the Zotino Tall Tower Observation Facility (ZOTTO). Here, the aerosol sample is taken through a 300 m long pipe into the laboratory on the ground with modest particle losses (Birmili *et al.*, 2007). A laminar flow with a rate of 40 l/min provides a penetration efficiency of 20% for 10 nm particles. Particle losses due to sedimentation are naturally minimized by the vertical configuration of the pipe. One of the greatest difficulties with the use of such lengthy pipes may be the creation of an airtight sampling pipe from many individual pieces.

3.3.3 Turbulent Flow Sampling Configuration

High-flow turbulent aerosol sampling configurations may be used at monitoring sites that have a primary focus on the particles responsible for radiative climate forcing. In order to design a turbulent sampling configuration, the size-dependent particle penetration can be calculated using the following equations. The size-dependent particle losses due to diffusion can be estimated to:

$$\delta = \frac{28.5 D_{pipe} \times D^{1/4}}{Re_{flow}^{\frac{7}{8}} \left(\frac{\eta_G}{\rho_G}\right)^{1/4}}$$

where D is the diffusion coefficient, D_{pipe} the inner diameter of the pipe, η_G the gas viscosity, ρ_G the gas density, and Re_{flow} the Re of the flow. The particle size-dependent deposition velocity u_{dep} to the wall is then given by:

$$u_{dep} = \frac{D}{\delta}$$

The particle size-dependent penetration can be calculated to:

$$P = exp\left(\frac{-4 \times u_{dep} \times L_{pipe}}{D_{pipe} \times \overline{u}_{flow}}\right)$$

where u_{flow} is the mean flow velocity and L_{pipe} the length of the pipe. The particle penetration through a bend depends on *Stk* and the curvature of the bend. The size-dependent particle penetration can be approximated by:

$$P = exp\left(-2.823 \times Stk\frac{\theta^{\circ}}{180^{\circ}}\pi\right)$$

The penetration in a horizontally oriented pipe due to sedimentation is described by:

$$P = exp\left(-4Z \times \cos\left(\frac{\theta^{\circ}}{180^{\circ}}\pi\right)\right)$$

3.3.3.1 Example of a Sampling Configuration with a Turbulent Flow

A sampling configuration with a turbulent flow is used at NOAA's long-term aerosol monitoring stations and is designed to provide up to 120 l/min of conditioned aerosol flow from a shared inlet to analyzers and sample-collection devices. This design is optimized to provide quantitative measurements of particles in the size range 0.02–2.0 μm aerodynamic diameter, with an additional goal of achieving 50% collection of particles up to 10 μm. The design can support multiple analyzers and filter samplers that require flow rates up to 30 l/min each in order to be operated in parallel.

The tradeoffs required to achieve these design goals include turbulent flow conditions in the sample lines, subisokinetic conditions at transitions to smaller-diameter sample lines, and nonisoaxial conditions in which the flow is split into four separate lines. Despite these tradeoffs, calculations of particle losses due to turbulent diffusion, impaction, and sedimentation show that the design criterion for size-dependent sampling efficiency is met in many implementations of the system, as shown in Figure 3.5.

The design of the inlet can be briefly summarized as follows:

- A 20 cm diameter polyvinyl chloride (PVC) sampling stack is supported by a triangular meteorological tower, generally extending 10 m above adjacent structures.
- An inverted stainless-steel pot is used as a rain hat.
- A 5 cm diameter heated stainless-steel tube extracts 150 l/min aerosol sample flow with a Re = 4500 from the center of the 1000 l/min main stack flow with Re = 7500.
- The heater is controlled by a downstream RH sensor to maintain the RH at no more than 40%, with a thermostat disabling the heater if the air temperature reaches 40 °C.
- Air leaving the heated tube is split into four analytical sample lines (1.9 cm diameter, 30 l/min each, Re = 2700) and one bypass line (30 l/min).
- The sample lines are at an angle of 3.75° from the axis of the heated sample tube.
- The 1.9 cm diameter sample lines are made of stainless steel and/or conductive silicone tubing of various lengths, depending on the particular station.

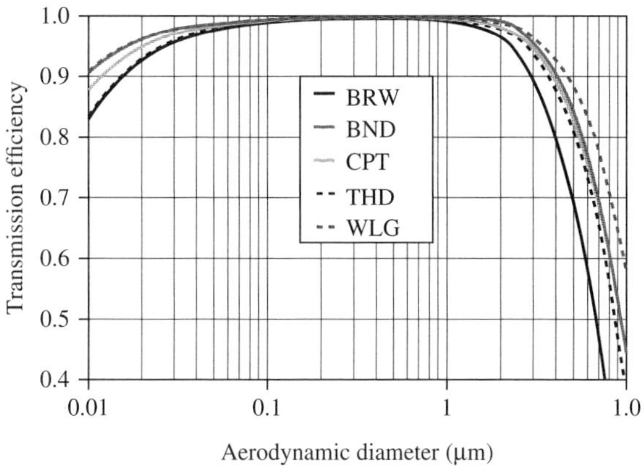

Figure 3.5 *Calculated sampling efficiencies for some existing aerosol inlet systems: Barrow, USA (BRW), Bondville, USA (BND), Cape Point, South Africa (CPT), Trinidad Head, USA (THD), and Mount Waliguan, China (WLG). Details are given at http://www.esrl.noaa.gov/gmd/aero/instrumentation/instrum.html.*

The advantages and disadvantages of this turbulent sampling configuration are:

- high aerosol flow rate,
- short residence time in the sampling system,
- reduced losses due to sedimentation in horizontal pipes,
- increased losses of ultrafine particles due to enhanced diffusion,
- increased losses of coarse particles due to enhanced impaction,
- limited ability to actively dry the aerosol flow.

3.4 Artifact-Free Sampling for Organic Carbon Analysis

Aerosol particles may change their size and chemical composition if the surrounding gas phases change. This is especially obvious for reactive species and for semivolatile aerosol constituents; that is, species that coexist in the condensed (liquid or solid) and gas phase in significant amounts, such as H_2O, NH_4NO_3, and a number of semivolatile organic compounds (SVOCs). The concentrations of semivolatiles species in the condensed and/or gas phase are controlled by equilibrium constants. This principle is used in aerosol technology to dry aerosol particles, for example, as described previously. It also leads to undesired modifications of the chemical composition of the particles called 'artifacts'. Positive and negative sampling artifacts result in gains and losses of PM, respectively. Chemical sampling artifacts result in the modification of the structures of reactive molecules. Sampling artifacts are especially critical when particles are collected for a longer time period on filters or impactor foils before being analyzed.

Positive sampling artifacts result from the trapping by the collection media or the PM already collected of molecules that were in the gas phase in the atmosphere. Sulfur dioxide, nitric acid, ammonia, and a vast range of organic molecules have been shown to be trapped significantly by various filter types (and subsequently to affect the determination of particulate species). Two methods can be used to prevent or at least reduce positive sampling artifacts:

- Concentration of particles in a reduced aerosol flow rate by use of a virtual impactor. The drawback of this method is that smaller particles whose trajectories are not affected by aerodynamic effects will not be sampled. There is also a risk to loose particles on the virtual impactor nozzle.
- Denuding of the air sample from gaseous species that could significantly bias the determination of particulate species. The limitation of this technique is that the residence time of the sample in the denuder must be long enough to allow diffusion of the targeted gas molecules to the denuder walls; this is not always possible for high-volume flow samplers. The use of denuders can also increase the magnitude of negative sampling artifacts, which should also be addressed. The use of impregnated denuders has been well documented for nitric acid and ammonia, for example, and is recommended by EMEP (www.emep.int). For SVOC, the use of a multichannel carbon monolith denuder has been shown to reduce positive sampling artifacts to less than 20% at many sites across Europe.

Negative artifacts result from the evaporation of PM from the condensed phase to the gas phase due to changes in pressure, temperature, RH, and gas-phase concentrations in the aerosol sample in relation to the atmosphere. Negative artifacts are to a large extent avoided when aerosol particles are analyzed online, without a collection step (e.g. in single particle analyzers), or collected in liquid droplets, as in the Steam Jet Aerosol Collector (SJAC; Khlystov, Wyers, and Slanina 1995) and the Particle Into Liquid Sampler (PILS; Weber *et al.*, 2001), which also makes it possible to perform an online particle chemical analysis. When filters or impactors are used, negative artifacts can be determined by measuring the concentrations of species of interest in the gas phase downstream of the filter or impactor. This is usually done by analyzing backup filters made of a material that is able to trap these species. Difficulties related to this approach include:

- The need to assess the amount of material collected by the backup filter originally coming from the atmospheric gas phase (denuder breakthrough).
- The risk of contamination of these backup filters, which are by essence very prone to absorbing molecules from the gas phase.

The use of backup filters in conjunction with denuders to determine artifact-free NH_4NO_3 is recommended in the EMEP manual. For SVOC, the issue is much more complex, since suitable material for backup filters is not commercially available and the breakthrough of denuders is significant and variable. Recent experiments performed in Europe show that negative artifacts for particulate organic matter are smaller than 10% and that their amplitude is not significantly magnified by the use of a denuder.

Chemical artifacts can occur when the PM collected is exposed, for example on a filter during sampling to an air stream with trace gas concentrations different from those in the atmosphere where the particles where originally suspended. Examples include acidic aerosol, which can be neutralized by NH_3, and poly-aromatic hydrocarbons (PAHs), which can be oxidized or nitrified when O_3- or NO_2-richer air is drawn through the filter. Also, chemical artifacts can be minimized by scrubbing undesired gases out of the air stream using suitable denuders.

It should be noted that sampling artifacts affect not only the chemical composition of the atmospheric PM but possibly also particle physical and optical properties, including the mass concentration, especially in locations and/or time periods where semivolatile species are major constituents of the aerosol.

Acknowledgements

We thank Andrea Haudek (TROPOS) for preparing the technical sketches in Figures 3.1–3.4.

References

Baron, P.A. and Willeke, K. (eds) (2001) *Aerosol Measurement – Principles, Techniques, and Applications*, 2nd edn, John Wiley & Sons, Inc.

Birmili, W., Stopfkuchen, K., Hermann, M. *et al.* (2007) Particle penetration through a 300 m inlet pipe for sampling atmospheric aerosols from a tall meteorological tower. *Aerosol Science and Technology*, **41**, 811–817.

EC (1999) Council directive 1999/30/EC of 22 April 1999 relating to limit values for sulfur dioxide, nitrogen dioxide and oxides of nitrogen, particulate matter and lead in ambient air. *Official Journal of the European Communities*, **L 163**, 41–60.

Hinds, W.C. (1999) *Aerosol Technology: Properties, Behavior, and Measurement of Airborne Particles*, 2nd edn, John Wiley & Sons, Inc., New York.

Khlystov, A., Wyers, G.P. and Slanina, J. (1995) The steam-jet aerosol collector. *Atmospheric Environment*, **29**, 2229–2234.

Tuch, T.M., Haudek, A., Müller, T. *et al.* (2009) Design and performance of an automatic regenerating adsorption aerosol dryer for continuous operation at monitoring sites. *Atmospheric Measurement Techniques*, **2**, 417–422.

Vincent, J.H. (2005) Health-related aerosol measurement: a review of existing sampling criteria and proposals for new ones. *Journal of Environmental Monitoring*, **7**, 1037–1053.

Weber, R.J., Orsini, D., Daun, Y. *et al.* (2001) A particle-into-liquid collector for rapid measurement of aerosol bulk chemical composition. *Aerosol Science and Technology*, **35**, 718–727.

Weingartner, E., Nyeki, S. and Baltensperger, U. (1999) Seasonal and diurnal variation of aerosol size distributions ($10 < D < 750$ nm) at a high-alpine site (Jungfraujoch 3580 m asl). *Journal of Geophysical Research-Atmospheres*, **104**(D21), 26809–26820.

Wiedensohler, A., Birmili, W., Nowak, A. *et al.* (2012) Mobility particle size spectrometers: harmonization of technical standards and data structure to facilitate high quality long-term observations of atmospheric particle number size distributions. *Atmospheric Measurement Techniques*, **5**, 657–685.

WMO (2003). WMO/GAW Aerosol Measurement Procedures Guidelines and Recommendations. WMO TD No. 1178 – GAW Report No. 153, World Meteorological Organization, Geneva.

4
Aerosol Instrumentation

Da-Ren Chen[1] *and David Y. H. Pui*[2]
[1]*Department of Mechanical and Nuclear Engineering, Virginia Commonwealth University, USA*
[2]*Mechanical Engineering Department, University of Minnesota, USA*

4.1 Introduction

Aerosol particles may consist of microscopic bits of materials from molecular clusters only a few nanometers in diameter to particles of a near macroscopic size of a fraction of a millimeter's diameter. The nominal size range of interest in aerosol research and applications is from 2 nm to 100 µm. The lower size limit, 2 nm, corresponds to the smallest particles that can be detected by an aerosol instrument such as the ultrafine condensation particle counter (CPC) (described in Section 4.4.1). However, nanoparticle and ultrafine aerosol measurements have recently attracted much attention through the national initiative on nanotechnology (http://nano.gov/). Ultrafine aerosols are formed spontaneously in the atmosphere (i.e. a nucleation event) and are emitted from transportation systems, power plants, and manufacturing processes. Engineered nanoparticles, formed from gas-phase synthesis or via the liquid-phase route, are important building blocks for nanoscale materials and devices. Nanoparticles often possess special physical (i.e. mechanical, electrical, optical, and magnetic), chemical, or biological properties compared with micrometer-sized particles of the same composition. To understand particle formation processes, it is necessary to characterize particles to less than 1 nm. The largest particles are those that will fall out of carrier gases too quickly to be considered an aerosol, such as rain drops. Particle size determination is important because many aerosol properties are particle-size-dependent. For example, in health studies particle deposition in different regions of the human respiratory tract is a function of particle size.

Aerosol Science: Technology and Applications, First Edition. Edited by Ian Colbeck and Mihalis Lazaridis.

Aerosol instruments measure different aerosol properties. These include aerosol number concentration, mass concentration, surface area concentration, size distribution, opacity, and chemical composition. Particle size distribution is usually presented as particle concentration in terms of number, surface, or volume (mass) weightings as a function of particle size. Although many instruments provide data as a function of the particle 'size,' this so-called 'size' is actually derived from one of many particle properties, including the optical, aerodynamic, mechanical, or force-field mobility of particles. It is thus important to pay attention when comparing particle sizes measured according to different principles. Data corrections are often needed in order to account for the differences between them.

Two categories of aerosol instrument are available for aerosol characterization. One is aerosol sampling instruments. These collect particles for microscopic, gravimetric, or chemical analysis. The instruments in this category often give a time-average concentration measurement. The sampling flow rate and time requirements typically depend on the analytic techniques that are to be applied on collected particle samples. Because of their time-averaging nature, the possibility of chemical interaction between particles–particles and particles–substrates may become an issue.

The other category is direct-reading instruments, in which aerosol sampling and analysis are carried out and the property of interest obtained instantaneously. Because of the availability of modern electronics, direct-reading instruments are capable of real-time measurement. The fast response of direct-reading instruments makes it possible to follow rapid changes in both particle size and concentration or to obtain good counting statistics, because they allow repeated measurements to be performed in a short space of time. These instruments usually rely on indirect sensing techniques and therefore much effort must be made to ensure precise instrument calibration. The accuracy of these instruments is dependent upon the relationship between the sensor signal and the aerosol property. Although this relationship can often be derived from first principles, it is very common to establish an empirical relationship using 'well-calibrated or standard' aerosols. The potential issue of this calibration approach is that the measured aerosol may have a different signal–property relationship to the 'calibrated or standard' aerosol, which can result in inaccuracy in the measurement.

In this chapter, we will first present the general strategy applied in aerosol instrumentation and then discuss particle sampling inlet and transport as they relate to aerosol measurements. In the later sections, selected aerosol instruments and measurement techniques commonly applied in aerosol research and applications will be described. Due to the page limit, we will not be able to cover all principles and instruments used; those who wish more detail opn the subject of aerosol instrumentation are referred to the reference book edited by Kulkarni, Baron, and Willeke (2011). A series of monographs published by ACGIH Air Sampling Committee would be a good resource for those who wish to learn the practical aspects of specific aerosol instruments (e.g. ACGIH 2008a, 2008b, 2009).

4.2 General Strategy

The common strategy for aerosol characterization is depicted in Figure 4.1. Aerosol particles are sampled from the ambient environment or from specific sources into an instrument via a sampling inlet. Size-selective sampling inlets are often used either to meet the specific

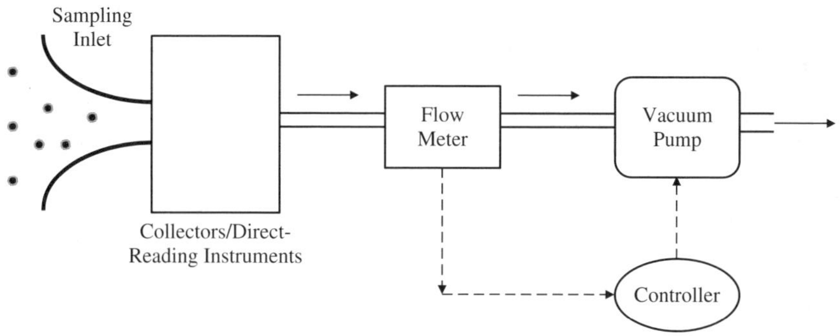

Figure 4.1 *Schematic diagram of the characterization strategy for aerosol instrumentation.*

requirements of aerosol applications and standards, such as $PM_{2.5}$ sampling, or to reduce potential interference from particles of larger sizes. Once sampled into the instrument, particles are either deposited on/in a collector (i.e. substrate or filter) in an aerosol sampling instrument or directed into the sensing volume of a direct-reading instrument. A vacuum pump is used to enable the sampling task and a flow meter monitors the sampling flow rate in volume or mass. A feedback control loop is often implemented to accurately control the sampling flow rate. Since the physical properties of aerosols are often presented in the unit of concentration (i.e. number/mass concentration), the total sampling volume in the sampling period (calculated by multiplying the measured volumetric flow rate with the sampling time) is a key parameter for aerosol instruments.

4.3 Aerosol Sampling Inlets and Transport

The aspiration efficiency, defined as the percentage of particles penetrating through a sampling inlet opening, is often used to characterize the performance of sampling inlets. The aspiration efficiency of a sampling inlet depends on the particle size, inlet design, and ambient environments. Size-selective sampling inlets are often used in aerosol instruments because of the sampling requirements, for example, $PM_{2.5}$ sampling for particles less than 2.5 μm, or to minimize the potential interference from larger particles. Particles larger than a certain size are often removed for the light-scattering instruments. The presence of large particles scatters many photons and prevents the instruments from accurately detecting particles of much smaller sizes. These size-selective inlets can be accomplished by using separation techniques based on particle inertia. Examples of such inlets are single-stage impactors and cyclones (as shown in Figure 4.2). In a single-stage impactor, the velocity of particles is accelerated via a convergent nozzle facing a solid substrate below it. When particle-laden flow approaches the substrate, the flow is deflected and makes its turn. However, particles with high inertia (i.e. large particles) cannot follow the flow turn and are deviated from the flow streamlines and impacted on the substrate. Particles with low inertia (i.e. small particles) flow with the carrier gas and escape from the impaction. As a result, particles with sizes larger than a certain critical value will be collected in a single-stage impactor. Cyclones separate particles via centrifugal forces. The centrifugal force acting

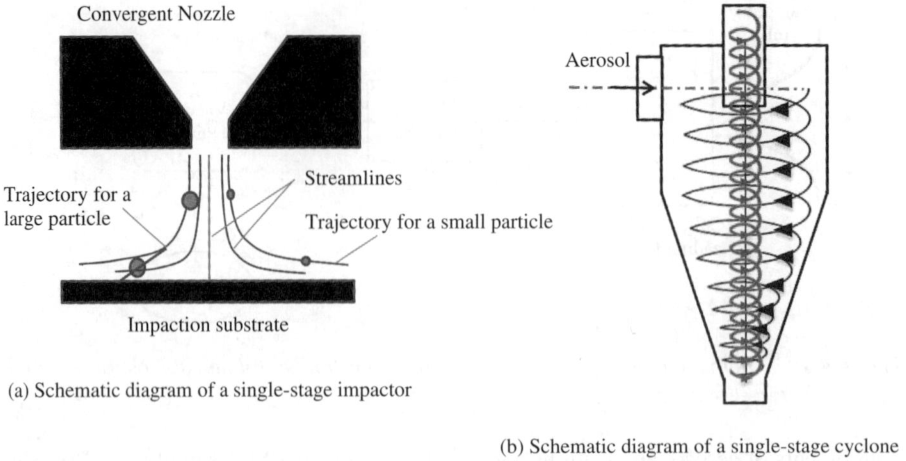

Figure 4.2 *Schematic diagrams of (a) a single-stage impactor and (b) a single-stage cyclone as size-selective inlets.*

on the particles is created by tangentially injecting particle carrier flow into the cyclone. The higher the tangential velocity of carrier flow, the higher the centrifugal force in the cyclone. Thus, particles with high inertia are collected on the cyclone wall, while those with low inertia penetrate through the device. For sampling of aerosol from the ambient, it is important to minimize the effect of wind on the sampling efficiency of aerosol inlets. An omnidirectional sampling inlet has thus been developed to accomplish this objective (Liu and Pui, 1981). In addition to the aspiration efficiency of the aerosol sampling inlet, the particle loss in the transport line from the inlet to either collectors or sensing volumes of aerosol instruments is also important in aerosol characterization. The penetration of particles through a transport line depends on the particle size, flow rate, and tubing types.

The overall sampling efficiency of an aerosol instrument is the combined result of the aspiration efficiency of the sampling inlet and the penetration efficiency of particles through the transport lines. In order to retrieve an accurate size distribution, the sampling efficiency of aerosol instruments must be taken into consideration in the data reduction schemes. Those who are interested in the subject of aerosol sampling should refer to Vincent (2007).

4.4 Integral Moment Measurement

Measurement of the integral moment of a total particle size distribution is accomplished by either aerosol sampling instruments or direct-reading instruments. Through aerosol sampling instruments, particles are collected on a substrate surface and then characterized by electron microscopes (i.e. scanning electron microscope/transmission electron microscope, SEM/TEM). Given a known sampling volume and sampling efficiency, the total number concentration of particles can be derived. In order to measure the total mass concentration, particles are typically collected by filters; by measuring the filter mass before and after sampling, the mass concentration of particles can be obtained.

Direct-reading, integral concentration detectors can be used to measure one integral parameter of an entire size distribution of particles, such as the total number, the surface and mass concentration, or the total light-scattering or extinction coefficients. Several widely used instruments are described in this section, including the CPC for number concentration determination, the quartz-crystal microbalance (QCM) and the tapered-element oscillating microbalance (TEOM), used for mass concentration measurement, and other techniques based on light scattering.

4.4.1 Total Number Concentration Measurement: Condensation Particle Counter (CPC)

The CPC, or condensation nucleus counter (CNC), is widely used to measure particles in the diameter range from approximately 0.005 to 1.0 μm. This instrument operates by passing the aerosol stream through a vapor-supersaturated region produced by direct-contact cooling, which causes vapor condensation on the particles. The particles are then grown to a size where they can be detected optically by light scattering. CPCs include a continuous-flow, direct-contact type (Bricard *et al.*, 1976; Sinclair and Yue, 1982) and a type that mixes a hot vapor stream and a cool aerosol stream to achieve a supersaturation condition (Kousaka *et al.*, 1982).

Figure 4.3 shows a schematic diagram of a continuous-flow CPC (Agarwal and Sem, 1980). Butyl alcohol is used as the working fluid in this instrument. An air stream is saturated with working fluid vapor in a saturator kept at 35 °C. The subsequent cooling of this vapor-saturated air stream in a thermoelectrically cooled condenser tube kept at 10 °C produces the required supersaturation for vapor condensation on the particles. The particles are grown to the super-micrometer range after passing through the condenser tube. They are

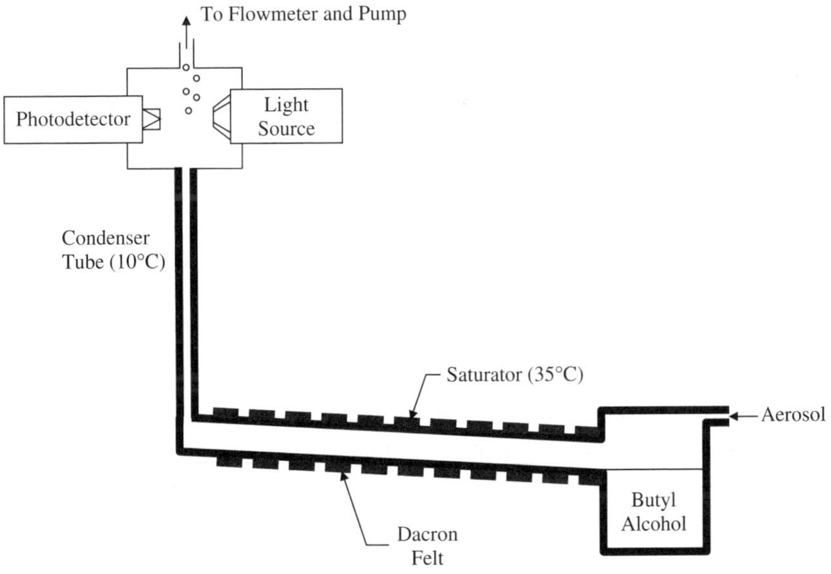

Figure 4.3 *Schematic diagram of a continuous-flow condensation particle counter (CPC).*

then detected optically by light scattering. Particles with a number concentration below a certain level are counted individually by CPCs in the single-particle-counting mode. Some CPCs include the so-called 'photometric mode', based on total light scattering, to count particles with a number concentration above a certain level. The number concentration range measured by CPCs runs from 0.01 #/cm^3 up to 10^7 #/cm^3 for individual particle counting. With the photometric mode, the maximum number concentration detectable by CPCs goes up to 10^8 #/cm^3.

Calibration studies of CPC have shown that below a particle size of 0.005 μm, the response of the instrument begins to drop off with reducing particle size (Zhang and Liu, 1990; McDermott, Ockovic, and Stolzenburg, 1991; Stolzenburg and McMurry, 1991; Keston, Reineking, and Porstendorfer, 1991; Noone and Hansson, 1990). The drop-off is primarily caused by a lack of 100% activation due to inhomogeneous vapor concentration distribution in the condenser (Engilmez and Davies, 1984). By introducing a clean sheath air around the aerosol stream in the CPC, Stolzenburg and McMurry (1991) were able to increase the counting efficiency of the instrument to over 70% at a particle size of 0.003 μm, resulting in the commercial ultrafine condensation particle counter (UCPC).

Water-based CPCs are the latest development in this type of instrument (Hering and Stolzenburg, 2005; Hering *et al.*, 2005). In order to condense water vapor on particles, these CPCs first cool the sampled aerosol and then deliver the aerosol stream into a moisture-saturated, heated 'condenser'. In the heated condenser, the water vapor diffuses from the water-saturated porous wall to the core region of the tube; it condenses on the particles when the cooled aerosol stream is introduced in the tube. Particle event detection is the same as that used in butanol-based CPCs. Special attention must be paid to the application of such CPCs. Their calibration reveals that their lower detection limit can be as good as that of butanol-based CPCs when they detect particles with slightly hydrophilic materials. For highly hydrophobic particles, the lower detection limit can be as large as 20–30 nm.

4.4.2 Total Mass Concentration Measurement: Quartz-Crystal Microbalance (QCM) and Tapered-Element Oscillating Microbalance (TEOM)

Sensors that measure the mass concentration of sampled particles in near real-time have been developed. In the QCM, a quartz crystal is used as the sensing element. By depositing particles on a quartz crystal, its natural vibrating frequency can be affected. The shift in the vibrating frequency is then used as a measure of the deposited particle mass, which is proportional to the frequency shift. Particle deposition can be achieved either by electrostatic precipitation (Lundgren, Carter, and Daley, 1976; Sem and Tsurubayashi, 1975; Sem, Tsurubayashi, and Homma, 1977) or by inertial impaction (Chuan, 1976). The sensitivity of the QCM is approximately 10^9 Hz/g, corresponding to a frequency shift of 1 Hz for a 10 MHz AT-cut quartz crystal.

The operating principle of TEOM is the same as that of QCMs (Patashnick and Rupprechl, 1980). Instead of relying on the natural frequency of quartz crystal, the vibration of the hollow tapered element is initiated and maintained by an electronic feedback system. The oscillation of the tapered element is monitored by a light-emitting diode and a phototransistor aligned perpendicularly to the oscillation plan of the tapered element. Figure 4.4 shows a typical arrangement for a TEOM. The sampled aerosol stream is collected by a filter installed on the top of the tapered element. The collected particle mass is then inferred from the frequency difference before and after each sampling. Unlike

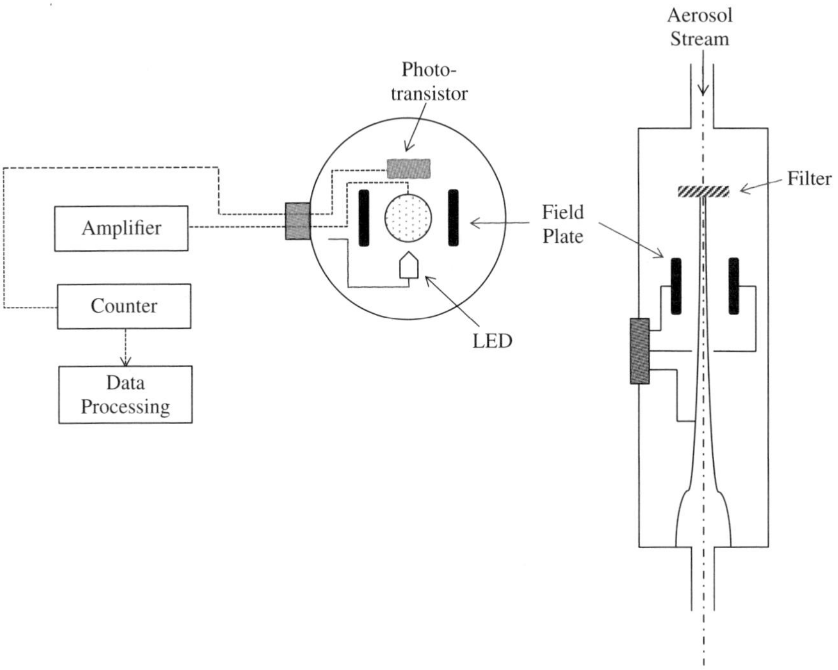

Figure 4.4 *Schematic diagram of a tapered-element oscillating microbalance (TEOM).*

in QCMs, the collected particle mass is not directly proportional to the frequency shift; instead, it is proportional to the difference of the inverse square of frequencies before and after sampling. The TEOM design extends the measurement of particle mass concentration to the grams-per-cubic-meter range.

4.4.3 Light-Scattering Photometers and Nephelometers

For atmospheric studies, the total light-scattering coefficient of airborne particles is important as it is related to the atmospheric visibility or visual range. Measurement of the total light-scattering coefficient is usually made with a photometer or integrating nephelometer. For aerosols that differ only in concentration and have the same size distribution, the integral light-scattering measurement can be converted to mass concentration (Waggoner and Charlson, 1976).

Figure 4.5 shows the schematic diagram of an integrating nephelometer. The particles are illuminated in a sensing volume and scattered light from the particles reaches the photoreceptor at a wide, off-axis angle. Although the instrument was originally used to measure visual range, it has found application in the study of the urban and rural atmospheric aerosol. In some cases, the scattering has been shown to be well correlated with the atmospheric mass concentration (Waggoner and Charlson, 1976; Butcher and Charlson, 1972). Note that one should pay more attention when applying the nephelometer in an environment with sooty particles, since the scattering will be attenuated by light absorption. In this case, the apparent concentration will be lower than expected.

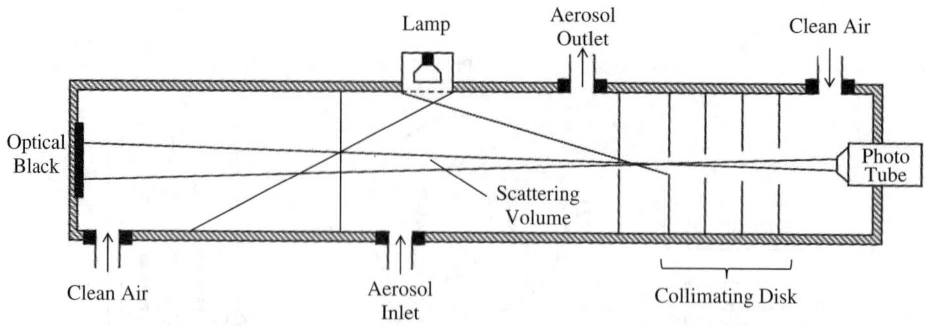

Figure 4.5 *Schematic diagram of an integrating nephelometer.*

Photometers have been commercially produced that employ a laser or incandescent light source and optics similar to dark-field microscopy. Unlike with nephelometers, a focused beam of light converges on the aerosol cloud; only light scattered in a well-defined range of angles is collected on the photodetector. The readout is in mass or number concentration, but the calibration may change with the composition and size distribution of the particles. Based on the solutions to Maxwell's equations, forward-scattering photometers are less sensitive to a change in the refractive index of particles than are photometers at other sensing angles, such as 30°, 45°, and 90°.

4.5 Particle Surface Area Measurement

Recent studies on nanoparticle toxicity have found that the biological response of human and animal lung cells after nanoparticle exposure has a strong correlation to the surface area of the nanoparticles (Oberdorster, 1996; Donaldson, Li, and MacNee, 1998). Different techniques have been explored to measure the total surface areas of aerosol particles. One method, the 'epiphaniometer', tags particles with radioactive material and measure the radioactivity downstream of the particle-tagging chamber using an α-detector (Pandis *et al.*, 1991). The response correlates well to the total surface area of the particles. A radioactive source of ^{227}Ac is used, which provides a constant supply of ^{219}Rn gas, which in turn decays to ^{211}Pb. ^{211}Pb is what actually tags the particles. Unfortunately, the stringent regulation that is increasingly being applied to the use and control of radioactive substances prevents the popular use of devices of this type. Further, this technique does not measure the actual particle surface area. Instead, the so-called 'active surface area' or 'Fuchs surface area' is characterized (Pandis *et al.*, 1991).

Another technique to obtain the total particle surface area, proposed by Woo *et al.*, (2001), involves the use of multiple integral moment instruments. The feasibility study of this technique used a CPC, for total particle number concentration measurement (i.e. the zeroeth moment of the particle size distribution), an EAD (electrical aerosol detector), for total particle length measurement (i.e. the first moment), and a nephelometer, for total particle mass concentration measurement (i.e. the third moment of the particle distribution). The total particle surface area (i.e. the second moment of the particle size distribution) was then derived from the measured data, with the assumption of lognormal particle size distribution.

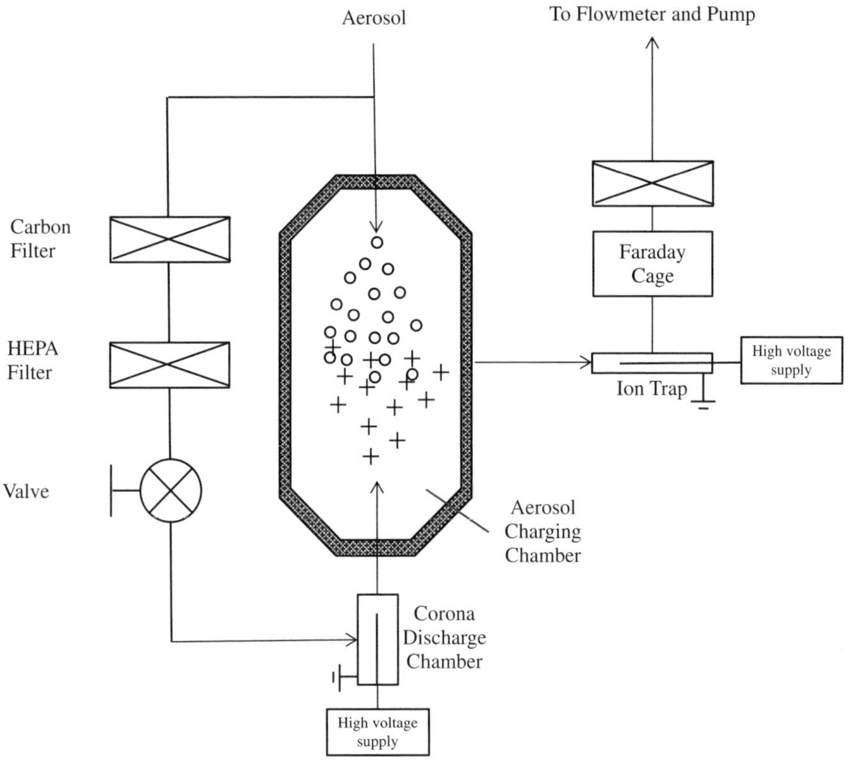

Figure 4.6 *Schematic diagram of a commercially available, nanometer surface area monitor (NSAM).*

A new development in nanoparticle surface area measurement, the nanometer surface area monitor (NSAM) (shown in Figure 4.6), is based on the particle surface area being inhaled and deposited in the human lung. The NSAM configuration is based on that of the EAD (Shin *et al.*, 2007). The sampled aerosol flow is first split into two streams: one, laden with particles, flows into the aerosol charging chamber, while the other has its particles and organic vapors removed by filters and is used as an ion carrier, passing through the corona discharge chamber. The particles are electrically charged in the charging chamber when the two streams collide as they flow from opposite directions. The electrical charges acquired by the particles are then detected by a sensitive aerosol electrometer of the Faraday cage type after passing through the ion trap, which is now used as a particle precipitator. By tweaking the particle loss in the ion trap (i.e. by varying the voltage set on the ion trap), the response of the EAD becomes a linear function of the particle size. Applying the published lung deposition curves for the alveolar (A) and tracheobronchial (TB) regions of the human lung, the surface area concentration of lung-deposited particles as measured by the NSAM is also a linear function of particle size, ranging from 10 to 600 nm (Fissan *et al.*, 2007). A factor is empirically determined which correlates the NSAM readout and the calculated surface area concentrations of particles deposited in the A and TB regions.

4.6 Size-Distribution Measurement

In order to characterize particle size distributions, particles are required to be collected on the SEM/TEM stub in aerosol samplers via either impaction or electrostatic precipitation. Microscopic techniques (i.e. SEM/TEM) can then be applied to characterize the size and number of particles collected. Any of the direct-reading, integral concentration-measuring techniques described in this chapter can be used in combination with an appropriate particle size-classification device to measure the size distribution of aerosols. In this section, we present the techniques used in size-distribution measurement based on either particle–light interaction or the difference in particle properties, namely, particle inertia, electrical mobility, and diffusivity.

4.6.1 Techniques based on Particle–Light Interaction

4.6.1.1 Optical Particle Counter (OPC)

Optical particle counters (OPCs) are widely applied for size-distribution measurement in both the indoor and the outdoor environments. Figure 4.7 shows the operating principle of the OPC. Single, individual particles are carried by an air stream through an illuminating volume in the counters. The interaction of a single particle with the light beam scatters photons in all angular directions. The scattered light in a certain range of angles is then optically collected and directed to a photodetector, which generates a voltage pulse whenever a particle passes through the illuminating volume. The height of each pulse is recorded and converted into the particle size based on the calibration curve for a given counter. The number of pulses in different pulse-height bins is counted to yield a pulse-height histogram, which is then converted to a histogram for particle size distribution. Commercial counters

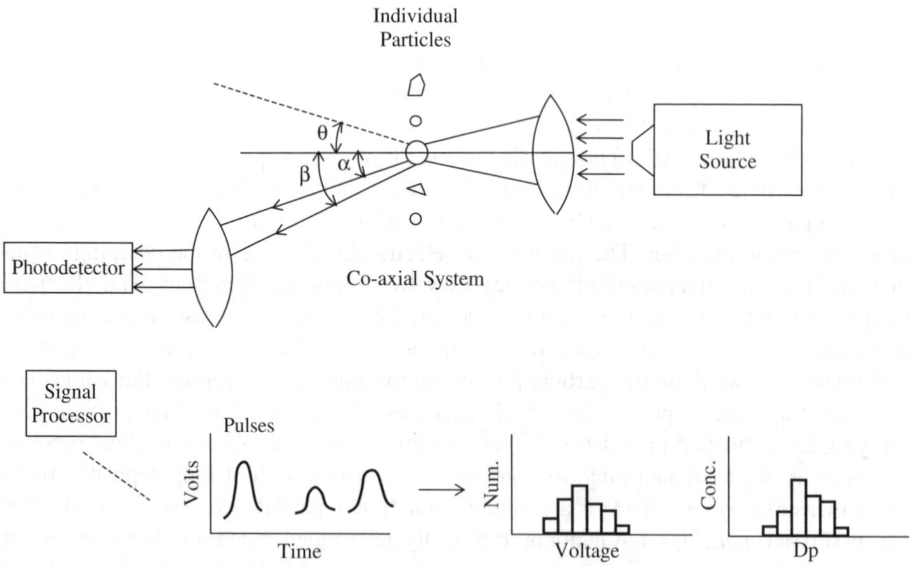

Figure 4.7 Operating principle of an optical particle counter (OPC).

using an incandescent light source have been developed for particle size-distribution measurement in the range 0.3 to approximately 10 μm. The modern counters mostly apply laser illumination in order to achieve lower detection limits, down to 0.1 μm.

The functional relationship between the pulse height of an OPC and the particle size depends on both the instrument design (i.e. optical design, illumination source, and electronic gain) and the particle properties (i.e. particle size and shape, refractive index, and orientation of nonspherical particles with the incident beam).

For an OPC with an axisymmetric scattering geometry ($\theta = 0°$) and near-forward, narrow-angle ($\alpha < \beta < 30°$) detection, the simple geometry provides strong signals but with high background noises. It is also susceptible to strong multivalued response; that is, to different particle sizes giving the same pulse height. The design is relatively insensitive to variations in real and imaginary parts of the refractive index. For a counter with wide-angle detection, the response is more sensitive to the parts, but much less prone to multivalued response, particularly for white-light illumination (Liu, Szymanski, and Pui, 1986).

In addition to the absolute voltage-size response of an OPC, other instrument characteristics, such as resolution, count coincidence, response to irregular particles, and inlet efficiency, are also important in the selection of OPCs for specific applications. Experimental studies are generally required to obtain these characteristics. For example, Liu, Szymanski, and Pui (1986) reported on the performance evaluation of several commercially available white-light counters using monodisperse spherical particles. Liu, Berglund, and Agarwal (1974), Wen and Kasper (1986), and Liu and Szymanski (1987) evaluated the counting efficiencies of several commercial OPCs. Later works mostly involve the evaluation of laser OPCs (Kim and Boatman, 1990; van der Meulen and van Elzakker, 1986; Chen, Cheng, and Yeh, 1984; Szymanski and Liu, 1986).

Recent developments in OPCs include increasing the count coincidence level while keeping the sampling flow rate high, measuring particle refraction indices (Szymanski and Schindler, 1998), and determining the particle shape using the scattering intensity from multi-angle light scattering (Dick, McMurry, and Bottiger, 1994; Sachweh *et al.*, 1995, 1998).

4.6.2 Techniques based on Particle Inertia

4.6.2.1 Particle Relaxation-Size Analyzers

In addition to direct measurement, light scattering can also be used in combination with other sizing principles to obtain the particle size distribution (Mazumder and Kirsch, 1977). One example is the use of an accelerating nozzle in combination with light-scattering measurement in an aerodynamic particle sizer (APS), described by Agarwal *et al.*, (1982). Figure 4.8 shows the schematic diagram of an APS. In this instrument, a single particle is accelerated through a small convergent nozzle, with the velocity obtained depending on the particle size. Due to the particle inertia effect, the larger the particle size, the lower the particle velocity. The velocity at the nozzle exit is then measured by detecting the time required to pass through two laser beams set a fixed distance apart. The difference between the measured particle velocity and the nozzle-exit velocity of carrier gas is then used to infer the particle size. This principle enables the 'aerodynamic size' of particles in the size range 0.5–30.0 μm to be measured (Chen, Cheng, and Yeh, 1985; Baron, 1986).

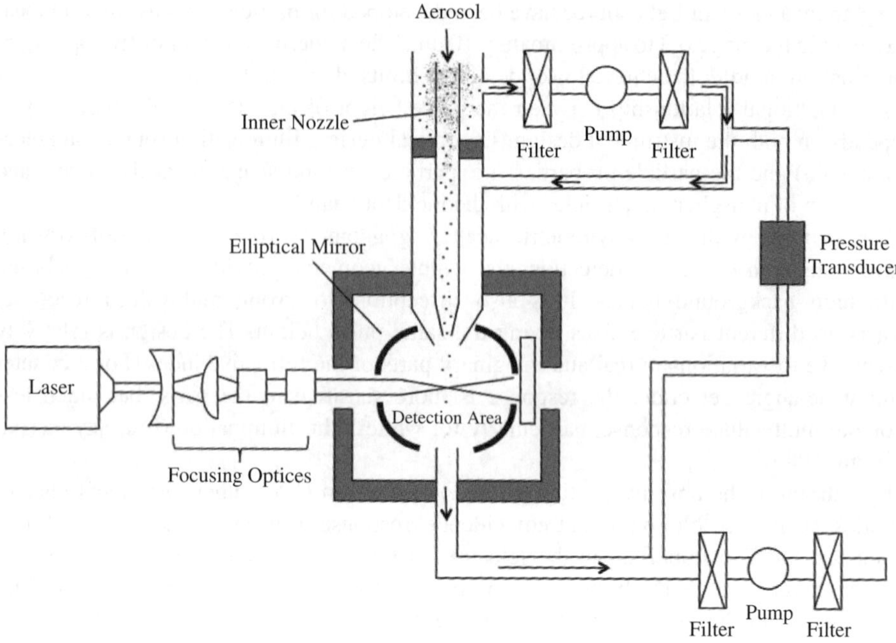

Figure 4.8 *Schematic diagram of an aerodynamic particle sizer (APS).*

Recent APS development incorporates an ultraviolet (UV) pulse laser to detect the viability of bioaerosol. The operating principle of APS is also applied in aerosol mass spectrometry for *in situ* particle sizing and composition measurement (Sylvia and Prather, 1998; Salt, Noble, and Prather, 1996; Silva and Prather, 1997).

4.6.2.2 Cascade Impactors

Impactors collect particles based on their inertia. As previously described, particles with a diameter above a certain value (i.e. the particle cut-off size) are collected on the solid substrate (i.e. impaction substrate) in a single-stage impactor. The particle cut-off size of an impactor can be reduced by accelerating particles to a higher velocity via the use of nozzles with smaller diameters. A cascade impactor normally consists of a series of single-stage impactors, with the largest nozzle at the top stage and the smallest nozzle at the last stage (shown in Figure 4.9). The particles passing through the last stage are often collected by a filter. With this arrangement, particles with diameters larger than the cut-off size of the top stage (i.e. Stage 1) will be collected on the top impaction substrate; particles with diameters between the cut-off size of Stage 1 and that of Stage 2 will be collected on the impaction substrate of Stage 2; and so on. Particles collected on the *n*th stage will have sizes between the cut-off size of the $(n-1)$th stage and that of the *n*th stage. After a period of sampling, the mass of particles collected on all the stages can be individually measured via a sensitive microbalance. From the sampling flow rate and time, the mass concentration at each impaction stage can be obtained. An example of such a cascade impactor is the MOUDI^{TM} (Micro-Orifice Uniform Deposition Impactor; Marple, Rubow, and Behm, 1991). MOUDIs

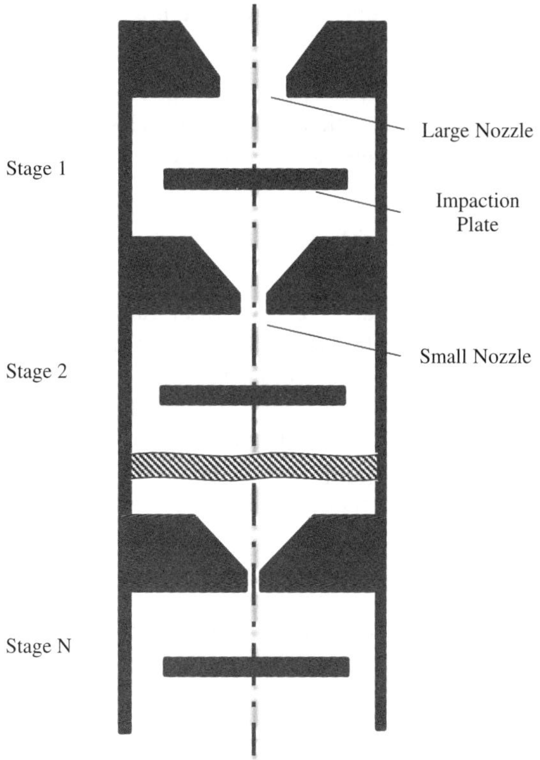

Figure 4.9 *Schematic diagram of a cascade impactor.*

of the first generation have 10 stages, with particle cut-off sizes ranging from 10 μm for the top stage to 56 nm for the last stage. MOUDI operation requires manual weighing of the particle mass collected on individual impaction stages (by obtaining the mass difference in the impaction substrate before and after the sampling).

Through combination with a real-time aerosol detector, novel cascade impactors can be developed to improve the instrument response speed or to increase the dynamic range of concentration measurement. One method is to combine the QCM technique with the cascade impaction technique for rapid mass-distribution measurement (Chuan, 1976; Wallace and Chuan, 1977). Two drawbacks of this technique are that the crystals need frequent cleaning and that the concentration range must be limited in order to avoid overloading of the crystal stages.

The issues encountered in a QCM cascade impactor can be partially remedied by the development of an electrical low-pressure impactor (ELPI) (Keskinen, Pietarinen, and Lehtimaki, 1992). The ELPI make use of a diffusion charger to electrically charge the sampled aerosol. It can then be impacted on to the collection stages and the charges carried by impacted aerosol at each stage can be measured by current-detecting electrometers. The impactor provides the cut-off-size information and the current provides the number-concentration information. The parallel current measurement of ELPI enables its fast response and reduced maintenance compared with QCM cascade impactors.

4.6.3 Techniques based on Particle Electrical Mobility

The high electric mobility of submicron particles in an electric field makes it efficient to separate and classify electrically charged particles. If the electrical mobility of a particle is a monotonic function of its size then either particle size classification or size-distribution measurement can be conducted on the basis of the particle electrical mobility.

4.6.3.1 Electrical Aerosol Analyzers (EAAs)

A widely used size-distribution-measurement instrument that uses the electrical mobility technique is the electrical aerosol analyzer (EAA), originally developed by Whitby and Clark (1966) and further improved by Liu and Pui (1975). Figure 4.10 shows its operating principle. The particles to be characterized are first electrically charged in the unipolar aerosol charger and then enter the mobility classifier, which precipitate particles with high electrical mobility while passing those with low mobility. The 'cut-off' mobility of the particles is determined by the voltage applied on the EAA inner column. By stepping this voltage and measuring the corresponding current carried by the charged particles with an aerosol electrometer, a voltage–current curve is generated, which can be further analyzed to obtain the desired particle-size distribution.

The EAA has low resolution on particle sizing and has been replaced with the high-resolution scanning mobility particle sizer (SMPS), described in the next section.

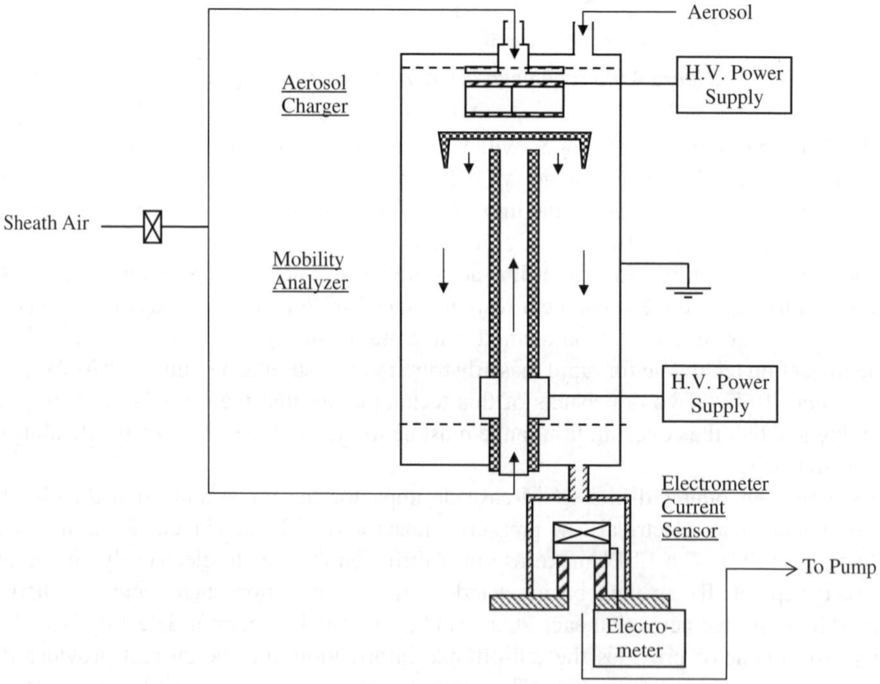

Figure 4.10 *Schematic diagram of an electrical aerosol analyzer (EAA).*

4.6.3.2 Differential Mobility Analyzers (DMAs) and Fast-Mobility Particle Sizers

Further development in the electrical mobility measurement technique includes the replacement of the integral type of mobility analyzer (i.e. EAA) with a differential type. The differential mobility analyzer (DMA), described by Liu and Pui (1974) and Knutson and Whitby (1975), enables the high-resolution particle sizing of the mobility classification technique. The sizing resolution of a DMA is further improved by the use of a bipolar charger rather than a unipolar one. Knutson (1976) was the first to propose the bipolar charging differential mobility analysis method for size-distribution measurement. While the DMA technique is capable of high-resolution sizing, the use of an aerosol electrometer limits the low particle concentration measurement through the signal-to-noise ratio of the pre-amp used in the electrometer. Accurate particle-size-distribution measurement with the DMA becomes possible with the inclusion of a CPC (or UCPC). A commercial differential mobility particle sizer (DMPS) is given schematically in Figure 4.11. A DMPS consists of a bipolar aerosol charger, a DMA, and a CPC. A platform with a microcomputer and all necessary accessories is used to control the instrument in order to step the DMA voltage, automatically acquiring the CPC and applying the data-reduction scheme to the raw data collected.

A significant advance on the DMA technique was developed by Wang and Flagan (1990). They made use of a scanning electric field, in place of changing the electric field in discrete steps as in the traditional DMPS operation, to considerably speed up the cycle time of the mobility analyzer. SMPS is the commercial instrument that incorporates this scanning technique. Further, battery-operated, portable versions of this type of analyzer have been developed and are now available commercially.

Making use of high sizing classification and size-distribution measurement of the DMA technique, tandem DMA set-ups (i.e. arranging two DMAs in series and inserting a particle conditioner in between them) have been applied to investigate the properties of submicrometer particles. The first DMA is used to classify particles of desired sizes and

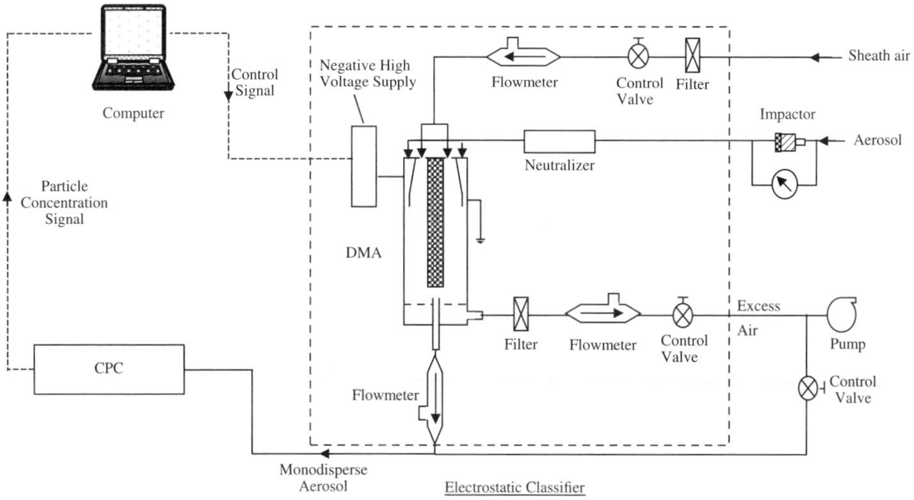

Figure 4.11 *Configuration of a differential/scanning mobility particle sizer.*

the second detects small changes in the selected particle size caused by the conditioner. Liu *et al.*, (1978) used a humidifier to study the size change of sulfuric acid aerosol under various humidities. Rader and McMurry (1986) made further refinements in the technique and applied it in precise droplet growth and evaporation studies.

Nanoparticles have recently received significant attention because of their special electrical, optical, and/or magnetic properties. Nanoparticles are of increasing interest to the industrial hygiene community as some engineered nanoparticles are suspected carcinogens. The DMA technique is a valuable tool for investigating nanometer aerosol, because of its capacity for high-resolution measurement and classification of nanoparticles. Although the aforementioned DMA works well in the size range 20–500 nm, it becomes increasingly difficult to perform accurate measurement and classification for nanoparticles with diameters smaller than 10 nm, due to the deposition loss and diffusion broadening of the nanometer aerosol in the DMA (Kousaka *et al.*, 1986; Fissan *et al.*, 1996). In order to improve performance, a new DMA column has been developed that is optimized for the nanoparticle size range, 2–150 nm (Chen *et al.*, 1998). A different route of DMA operation at high sheath flow rate has been explored by Fernández de la Mora and his research group, in order to increase sizing accuracy (Fernández de la Mora and de Juan, 1998; Martínez-Lozano and Fernández de la Mora, 2006). A sizing resolution less than 0.94% has been demonstrated for particles with a diameter close to 1 nm.

Fast-mobility particle sizers have recently been developed in the area of submicrometer particle characterization. The demand for fast mobility sizers is attributed to the need and desire to characterize the dynamics of combustion aerosols, for example from engine emissions. Nanoparticles are often emitted from the combustors in power generation plants, diesel engines used in heavy duty trucks, jet engines, and incinerators. Unlike the SMPS, fast-mobility particle sizers such as the DMS (differential mobility spectrometer; Reavell, 2002) and the EEPS (engine exhaust particle sizer; Mirme, 1994) make use of parallel processing to the DMA in order to obtain a fast instrument response time. The basic configuration of a DMS is given in Figure 4.12. Charged particles are introduced into the

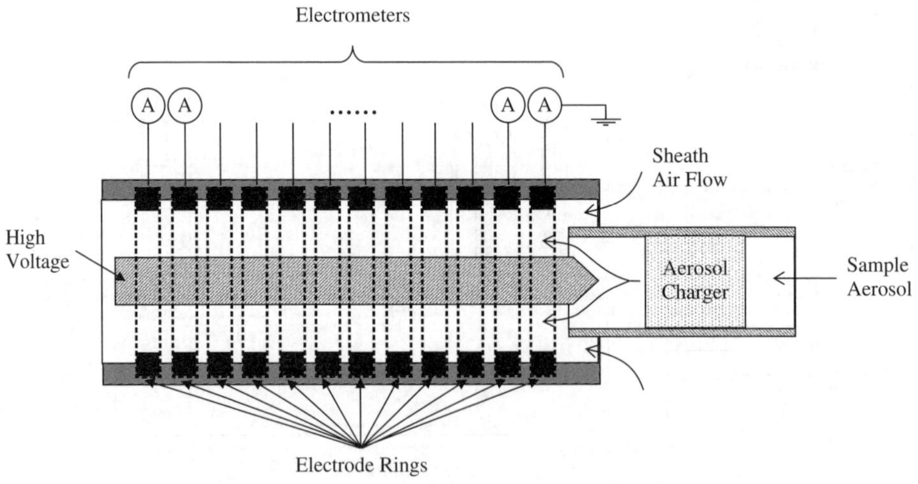

Figure 4.12 *Schematic diagram of a fast-response differential mobility spectrometer (DMS).*

instrument from the top of the inner column. They drift outward to the outer cylinder with the application of a fixed voltage of the same polarity as that of the charged particles on the inner column. Multiple electrodes, individually connected to a sensitive electrometer, are installed along the longitudinal length of the outer cylinder to detect the electrical current of deposited particles on the electrodes. A unipolar aerosol charger is used in DMS, whereas two unipolar chargers (one for positive ions and tone for negative ions) are used in EEPS, in order to prevent particles from overcharging.

4.6.3.3 Aerosol Particle Mass (APM) Analyzer and Couette Centrifugal Particle Mass Analyzer (Couette CPMA)

Another recent development of the DMA technique is implemented in the aerosol particle mass (APM) analyzer (Ehara, Hagwood, and Coakley, 1996) and its improved version, the Couette centrifugal particle mass analyzer (Couette CPMA) (Olfert and Collings, 2005; Olfert *et al.*, 2006). These instruments measure the charge-to-mass ratio of sampled particles. The masses of characterized particles can be derived if the electrical charges on individual particles are known. The basic configuration of such devices is similar to that of cylindrical DMAs, but instead of the stationary cylinders of a conventional DMA, the APM has two rotating coaxial DMA cylinders (at varying angular velocities). By mechanical rotation and voltage application, both electrostatic and centrifugal force fields are established in the annular spacing between two DMA cylinders. With the balance of electrostatic and centrifugal forces acting on the particles, only particles with a certain charge-to-mass ratio will pass through the APM.

The combination of a DMA and an APM (i.e. DMA–APM technique) makes it possible to measure the particle density (McMurry *et al.*, 2002; Park *et al.*, 2003a) and fractal dimension of aggregates (Maricq and Xu, 2004; Park, Kittelson, and McMurray, 2003b).

4.6.4 Techniques based on Particle Diffusion

4.6.4.1 Diffusion Batteries

Because the rate of particle diffusion to a solid surface is a function of particle size, particle penetration through a diffusion collector can be used for size-distribution measurement of submicrometer particles. The technique has been used for many years for particles below 0.1 μm (Sinclair, 1986). A simple diffusion battery consists of a capillary-tube bundle through which the aerosol is passed. A CPC is then used to measure the aerosol concentration upstream and downstream of the bundle. If the particles to be characterized are electrically charged, aerosol electrometers can also be applied in the penetration measurement. The particle size can be derived from the particle penetration measurement. By arranging a number of these collectors in series, the size distribution of the aerosol can be measured (Cheng, Keating, and Kanapilly, 1980; Yeh, Cheng, and Orman, 1982).

The diffusion battery has also been used as a particle separator for size-selective particle sampling (Figure 4.13). George and Hinchliffe (1972) and Sinclair, George, and Knutson (1977) made use of screen diffusion batteries to measure submicron radioactive aerosols.

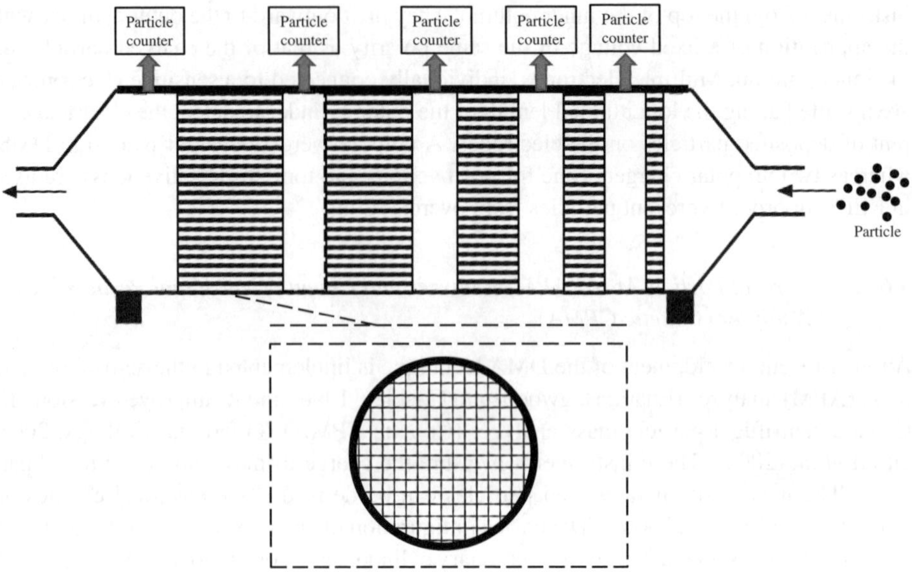

Figure 4.13 *Schematic diagram of an automatic diffusion battery.*

4.7 Chemical Composition Measurement

Chemical composition characterization of particles has mostly been performed to date by collecting particles (either on the collector surface or in filters) for a period of sampling time and offline analysis of the collected samples using analytic chemical instruments. Because of the concerns over particle aging and potential artifacts resulting from the accumulation of particles on filters during the collection time, instruments capable of continuous real-time analysis of the chemical compositions of particles – ideally as a function of particle size – have recently been devloped.

Near-real-time aerosol chemical composition instruments, operating with a measuring cycle of 10–60 minutes, have been developed to characterize the chemical content of an ensemble of particles in a size range defined by the sampling inlet of the instruments. For example, a particle-into-liquid sampler (PILS) utilizes automated ion chromatography to quantify the average major anion or cation particle content (Weber *et al.*, 2001). Some instruments collect ensembles of particles, thermally decompose them, and apply gas-phase chemiluminescence or absorption spectroscopy for the semicontinuous measurement of sulfate and nitrate (Stolzenburg and Hering, 2000). Other examples are instruments that analyze organic and total carbon content based on particle collection, thermal pyrolysis, and oxidation followed by carbon dioxide detection (Bae *et al.*, 2004; Turpin, Cary, and Huntzicker, 1990). Due to the time-averaging nature of these near-real-time instruments, none of them produces significant information on the variation in particle chemical composition with particle size or a comprehensive picture of total particle composition.

Over the past decade, several research groups have made major strides in adapting mass-spectrometer techniques to meet this challenge. The work related to the mass-spectrometry application of aerosol has been the subject of a number of review articles (Coe and Allan,

2006; Johnston, 2006; Noble and Prather, 2000; Suess and Prather, 1999; Sullivan and Prather, 2005). One type of system uses a laser to both vaporize and ionize individual particles sampled into a mass spectrometer's source region. This class of instrument focuses on single-particle measurements. An example is the aerosol time-of-flight mass spectrometer (ATOFMS), a detailed description of which can be found in Gard *et al.*, (1998).

Although these instruments provide a great deal of chemical information on an individual particle basis, there are several unresolved issues. Because of the high particle loss during aerosol transport and the extremely low efficiency in targeting particles for vaporization and ionization, the representativeness of the collected data is in question. Another issue is the lower size-detection limit of sampled particles. All of these instruments use the light-scattering technique to size sampled particles (inferring the particle size either from the scattered light signal or from the time it takes a particle to travel between two laser beams). The sizing techniques limit the lower size to 0.2 μm if scattered light is used to infer the particle size. The commercial ATOFMS has a lower size limit of 0.5 μm, because of the use of the particle time-of-flight technique for particle sizing. The last issue is portability, because of the importance of laser alignment to instrument performance. Realignment is often required after moving or transporting the instruments. These issues have led to the development of a second class of instrument.

Aerosol mass spectrometers (AMSs) make use of thermal vaporization of collected particles, followed by various ionization techniques (Allen and Gould, 1981; Sinha *et al.*, 1982; Hearn and Smith, 2004). Separation of the vaporization and ionization steps enables quantitative detection of particle chemical composition. Four versions of the AMS (the quadrupole aerosol mass spectrometer (Q-AMS), time-of-flight aerosol mass spectrometer (ToF-AMS), compact time-of-flight aerosol mass spectrometer (C-ToF-AMS), and high-resolution time-of-flight aerosol mass spectrometer (HR-ToF-AMS)) are currently commercially available.

The basic configuration of a Q-AMS is shown in Figure 4.14. A detailed description of the instrument can be found in the pioneering work of Jayne *et al.*, (2002). It consists of

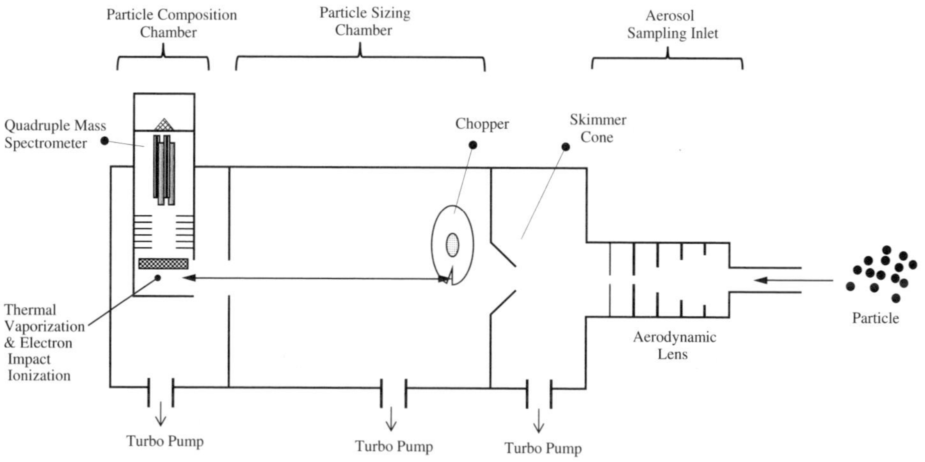

Figure 4.14 *Schematic diagram of a commercially available quadrupole aerosol mass spectrometer (Q-AMS).*

three main sections: an aerosol sampling inlet, a particle sizing chamber, and a particle composition chamber. Each section is separated by critical apertures and is differentially pumped. A focused particle beam is formed in the sampling inlet by passing the particles through a series of aerodynamic lenses (Liu *et al.*, 1995a,b). Once formed, the beam enters the sizing chamber (a flying tube maintained at $\sim 10^{-5}$ torr by a turbo pump) through a skimmer cone. The size-dependent particle velocities resulting from the gas expansion provide a means of obtaining the aerodynamic diameter of the particles by measuring their time-of-flight. The particle velocity is measured by modulating a rotating wheel chopper in the chamber and knowing the flight distance. Detection of particle composition is then carried out in the particle detection chamber. The particle beam is directed into a resistively heated closed-end tube. Upon collision, the volatile and semivolatile constituents in/on the particles vaporize. The vapor molecules are then charged by an electron-impact ionizer mounted at the entrance of a quadrupole mass spectrometer. With the quadrupole tuned to a representative mass, bursts of ions are measured. Quantification is made possible by the universal and reproducible nature of electron impact as an ionization technique, the similar (and quantifiable) efficiency of all nonrefractory species, and the absence of matrix effects (Jimenez *et al.*, 2003).

A later improvement of the AMS replaces the quadrupole mass analyzer with a compact time-of-flight mass spectrometer (ToF-AMS or C-ToF-AMS) (Drewnick *et al.*, 2005). Although the ToF-AMS has better than unit mass resolution, this resolution is not generally sufficient to separate ions of different elemental composition at a nominal m/z. In an HR-ToF-AMS, a custom-designed Tofwerk AG (H-ToF Platform, Thun, Switzerland), high-resolution, orthogonal time-of-flight mass spectrometer replaces the quadrupole mass filter used in Q-AMS (DeCarlo *et al.*, 2006).

4.8 Conclusion

Airborne particles in the size range from 1 nm to 100 μm are of interest in aerosol research and applications. Both aerosol sampling and direct-reading instruments can be applied to characterize the physical and chemical properties of particles in the size range of interest. Aerosol sampling instruments involve particle collection in the samplers followed by offline characterization via EM techniques or analytic chemical instruments. The particle collection in these instruments is typically accomplished by inertial impaction, electrostatic precipitation, or Brownian diffusion techniques. None of the single direct-reading-measurement techniques is capable of covering the entire particle size range of interest. The typical rule of thumb is that most direct-reading instruments cover a 2-decade optimized size range.

Techniques with different measuring principles are thus required to characterize the physical and chemical properties of particles in the entire size range. In general, the measuring principles based on the particle inertia are commonly applied to characterize particles in the super-micrometer size range. Examples of instruments that work well in this size range are cascade impactors and APS. Light-scattering techniques are often used to measure particles with sizes in the super- and submicrometer ranges. OPCs are one example of utilizing light scattering techniques. For particles with diameters less than 1 μm, especially those less than

500 nm, the measuring principle based on particle electrical mobility is often implemented. SMPSs are an example in the submicrometer and nanometer size ranges.

For mass-concentration measurement, particles may be captured on filters and then gravimetric weighed using a microbalance or the TEOM technique. They may also be deposited on a surface for measurement by QCMs. Long sampling times are usually required for filter technique, and frequent cleaning of the surface for QCMs. A faster approach is to characterize the mass concentrations of particles by measuring the light-scattering intensity of the aerosol and then inferring its mass concentration through calibration. The accuracy of the light-scattering technique significantly improves if the measured aerosols have nearly the same size and composition distributions but differ in concentration. For number-concentration measurement, CPCs may be used for particle diameters less than 2 μm. With modern electronics, the maximum number concentration of particles that can be individually counted is 10^7 #/cm^3. OPCs may be used to size and count particles with diameters larger than 0.1 μm. The lower detection limit of OPCs depends on the types of light source (i.e. white-light or lasers) used in the counters and the refractive indices of the particles. With lasers of a short wavelength, some LPCs (laser particle counters) can detect metal particles as small as 25 nm diameter. For particles less than 0.6 μm diameter, NSAMs are capable of measuring the surface area concentration of particles deposited in the A and TB regions of the human lung. Note that the NSAM does not mimic the deposition of particles in the A and TB regions; the correlation between the NSAM's readout and the calculated surface concentration of lung-deposited particles is completely empirical.

For particle larger than 0.1 μm, OPCs or particle-relaxation-size analyzers may be used to measure size distribution. The latter measure aerodynamic particle sizes. The lower detection limits of white-light and laser particle counters are typically at 0.3 and 0.1 μm, respectively. The lower detection limit of a commercially available APS is 0.5 μm. For aerosols in the submicron and nanometer size ranges, DMPS (and EAA) or a low-resolution diffusion battery may be used to characterize the particle size distributions. In order to trace the dynamics of submicron particles in the sub-second time scale, fast-mobility particle sizers may be utilized. Cascade impactors, separating particles based on their inertia, may be used to measure particle size distributions. In combination with QCMs or an electrometer, they can directly measure the particle size distributions with limited numbers of size bins. The impactors are useful for particles in the upper submicron and super-micron size ranges, although the lower detection size limit of such instruments has been extended to 30 nm.

Offline techniques (i.e. collection of particles and later analysis using analytic chemical instruments) have been practiced in the chemical composition characterization of particles. Significant development in particle composition measurement have made it possible to perform such characterizations automatically and continuously, and to directly characterize the size-dependent particle composition distributions. PILS is one of the semicontinuous instruments used for this measurement task. Instruments for the direct characterization of size-dependent particle composition distributions have also been made commercially available. The AMS characterizes the compositions of ensembles of particles. The ATOFMS is designed to characterize particles with sizes larger than 0.5 μm on a single-particle basis. Unfortunately, the ATOFMS is no longer commercially available.

Particle instruments have also been developed to meet specific applications. For examples, portable CPCs have been developed for respirator fit testing. The addition of UV laser detection in an APS enables the detection of bioaerosol. The development of water-based CPC is very welcome in semiconductor processing and cleanroom applications. Recent advances have been made in the fields of real-time chemical-composition-measuring instruments and portable size analyzers for field measurements.

References

ACGIH (2008a) Nanoparticles and Ultrafine Aerosol Measurements. Air Sampling Instruments Committee, Publication ASI3.

ACGIH (2008b) Aerosol Sampler Calibration. Air Sampling Instruments Committee, Publication ASI2.

ACGIH (2009) Performance Standards and Criteria for Air Sampling and Monitoring Instruments. Air Sampling Instruments Committee, Publication ASI14.

Agarwal, J.K., Remiaz, R.J., Quant, F.J. *et al.* (1982) Real-time aerodynamic particle size analyzer. *Journal of Aerosol Science*, **13**, 222.

Agarwal, J.K. and Sem, G.J. (1980) Continuous flow, single-particle-counting condensation nucleus counter. *Journal of Aerosol Science*, **11**, 343.

Allen, J., Gould, R.K. (1981): Mass spectrometric analyzer for individual aerosol particles *Review of Scientific Instruments*, **52**: 804–809.

Bae, M.S., Schauer, J.J., DeMinter, J.T. *et al.* (2004) Hourly and daily patterns of particle-phase organic and elemental carbon concentrations in the urban atmosphere. *Journal of the Air and Waste Management Association*, **54**, 823–833.

Baron, P.A. (1986) Calibration and use of the aerodynamic particle sizer (APS-3300). *Aerosol Science and Technology*, **5**, 55.

Bricard, J., Delattre, P., Madelaine, G., *et al.* (1976): Detection of ultra-fine particles by means of a continuous flux condensation nuclei counter. In: *Fine Particles*, (ed. B.Y.H. Liu), Academic Press, New York, pp. 566–580.

Butcher, S.S. and Charlson, R.J. (1972) *An Introduction to Air Chemistry*, Academic Press, New York.

Chen, B.T., Cheng, Y.S. and Yeh, H.C. (1984) Experimental responses of two optical particle counters. *Journal of Aerosol Science*, **15**, 457.

Chen, B.T., Cheng, Y.S. and Yeh, H.C. (1985) Performance of a TSI aerodynamic particle sizer. *Aerosol Science and Technology*, **4**, 89.

Chen, D.R., Pui, D.Y.H., Hummes, D. *et al.* (1998) Design and evaluation of a nanometer aerosol differential mobility analyzer (Nano-DMA). *Journal of Aerosol Science*, **29**, 497.

Cheng, Y.S., Keating, J.A. and Kanapilly, G.M. (1980) Theory and calibration of a screen-type diffusion battery. *Journal of Aerosol Science*, **11**, 549.

Chuan, R.L. (1976) Rapid measurement of particulate size distribution in the atmosphere, in: *Fine Particles*, (ed. B.Y.H. Liu), Academic Press, New York, pp. 763–775.

Coe, H. and Allan, J.D. (2006) Mass spectrometric methods for aerosol composition measurements, in *Analytical Techniques for Atmospheric Measurement* Chapter 6, Blackwell Publishing.

DeCarlo, P.F., Kimmel, J.R., Trimborn, A. *et al.* (2006) Field-deployable, high-resolution, time-of-flight aerosol mass spectrometer. *Analytical Chemistry*. doi: 10.1021/ac061249n.

Dick, W.D., McMurry, P.H. and Bottiger, J.R. (1994) Size- and composition-dependent response of DAWN – a multiangle single-particle optical detector. *Aerosol Science and Technology*, **20**, 345.

Donaldson, K., Li, X.Y. and MacNee, W. (1998) Ultrafine (nanometer) particle mediated lung injury. *Journal of Aerosol Science*, **29**(5–6), 553–560.

Drewnick, F., Hings, S.S., DeCarlo, P. *et al.* (2005) A new time-of-flight aerosol mass spectrometer (TOF-AMS) – Instrument description and first field deployment. *Aerosol Science and Technology*, **39**, 637–658.

Ehara, K., Hagwood, C. and Coakley, K.J. (1996) Novel method to classify aerosol particles according to their mass-to-charge ratio – Aerosol particle mass analyzer. *Journal of Aerosol Science*, **27**, 217–234.

Engilmez, N. and Davies, C.N. (1984) An investigation into the loss of particles by diffusion in the Nolan-Pollak condensation nucleus. *Journal of Aerosol Science*, **15**, 177–181.

Fernández de la Mora, J. and de Juan, L. (1998) High resolution size analysis of nanoparticles and ions: running a Vienna DMA of near optimal length at Reynolds numbers up to 5000. *Journal of Aerosol Science*, **29**, 617–626.

Fissan, H., Hummes, D., Stratman, F. *et al.* (1996) Experimental comparison of four differential mobility analyzers for nanometer aerosol measurements. *Aerosol Science and Technology*, **24**, 1.

Fissan, H., Neumann, S., Trampe, A. *et al.* (2007) Rationale and principle of an instrument measuring lung deposited nanoparticle surface area. *Journal of Nanoparticle Research*, **9**, 53–59.

Gard, E.E., Kleeman, M.J., Gross, D.S. *et al.* (1998) Direct observation of heterogeneous chemistry in the atmosphere. *Science*, **279**, 1184–1187.

George, A.C. and Hinchliffe, L. (1972) Measurement of uncombined radon daughter in uranium mining. *Health Physics*, **23**(6), 791–803.

Hearn, J.D. and Smith, G.D. (2004) A chemical ionization mass spectrometry method for the online analysis of organic aerosols. *Analytical Chemistry*, **76**, 2820–2826.

Hering, S. and Stolzenburg, M.R. (2005) A method for particle size amplification by water condensation in a laminar, thermally diffusive flow. *Aerosol Science and Technology*, **39**, 428–436.

Hering, S., Stolzenburg, M.R., Quant, F.R. *et al.* (2005) A laminar-flow, water-based condensation particle counter (WCPC). *Aerosol Science and Technology*, **39**, 659–672.

Jayne, J.T., Leard, D.C., Zhang, X. *et al.* (2002) Development of an aerosol mass spectrometer for size and composition analysis of submicron particles. *Aerosol Science and Technology*, **33**, 49–70.

Jimenez, J.L., Jayne, J.T., Shi, Q. *et al.* (2003) Ambient aerosol sampling with an aerosol mass spectrometer. *Journal of Geophysical Research*, **108**(D7), 8425. doi: 8410.1029/2001JD001213.

Johnston, M.V. (2006) Sampling and analysis of individual particles by aerosol mass spectrometry. *Journal of Mass Spectrometry*, **35**, 585–595.

Keskinen, J., Pietarinen, K. and Lehtimaki, M. (1992) Electrical low pressure impactor. *Journal of Aerosol Science*, **23**, 353.

Keston, J., Reineking, A. and Porstendorfer, I. (1991) Calibration of a TSI model 3025 ultrafine condensation nucleus counter. *Aerosol Science and Technology*, **15**, 107.

Kim, Y.J. and Boatman, J.F. (1990) Size calibration corrections for the active scattering aerosol spectrometer probe (ASASP-100X). *Aerosol Science and Technology*, **12**, 665.

Knutson, E.O. (1976) Extended electric mobility method for measuring aerosol particle size and concentration, in *Fine Particles*, (ed. B.Y.H. Liu), New York: Academic Press, pp. 739–762.

Knutson, E.O. and Whitby, K.T. (1975) Aerosol classification by electrical mobility: apparatus, theory and applications. *Journal of Aerosol Science*, **6**, 443.

Kousaka, Y., Niida, T., Okyuama, K. *et al.* (1982) Development of a mixing type condensation nucleus counter. *Journal of Aerosol Science*, **13**, 231.

Kousaka, Y., Okuyama, K., Adachi, M. *et al.* (1986) Effect of Brownian diffusion on electrical classification of ultrafine aerosol particles in differential mobility analyzer. *Journal of Chemical Engineering of Japan*, **19**, 401.

Kulkarni, P., Baron, P.A. and Willeke, K. (2011) *Aerosol Measurement – Principles, Techniques and Applications*, 3rd edn, John Wiley & Sons, Inc, Hoboken, NJ.

Liu, B.Y.H., Berglund, R.N. and Agarwal, J.K. (1974) Experimental studies of optical particle counters. *Atmospheric Environment*, **8**, 717.

Liu, B.Y.H. and Pui, D.Y.H. (1974) A submicron aerosol standard and the primary, absolute calibration of the condensation nucleus counter. *Journal of Colloid and Interface Science*, **47**, 155.

Liu, B.Y.H. and Pui, D.Y.H. (1975) On the performance of the electrical aerosol analyzer. *Journal of Aerosol Science*, **6**, 249.

Liu, B.Y.H. and Pui, D.Y.H. (1981) Aerosol sampling inlets and inhalable particles. *Atmospheric Environment*, **15**, 589–600.

Liu, B.Y.H., Pui, D.Y.H., Whitby, K.T. *et al.* (1978) The aerosol mobility chromatograph: a new detector for sulfuric acid aerosols. *Atmospheric Environment*, **12**, 99.

Liu, B.Y.H.; Szymanski, W.W. (1987) Counting efficiency, lower detection limit and noise levels of optical particle counters. Proceedings of the 33rd Annual Technical Meeting, Institute of Environmental Sciences, San Jose, CA.

Liu, B.Y.H., Szymanski, W.W. and Pui, D.Y.H. (1986) Response of laser optical particle counter to transparent and light absorbing particles. *ASHRAE Transactions*, **92**, 1.

Liu, P., Ziemann, P.J., Kittelson, D.B. *et al.* (1995a) Generating particle beams of controlled dimensions and divergence: I. Theory of particle motion in aerodynamic lenses and nozzle expansions. *Aerosol Science and Technology*, **22**, 293–313.

Liu, P., Ziemann, P.J., Kittelson, D.B. *et al.* (1995b) Generating particle beams of controlled dimensions and divergence: II. Experimental evaluation of particle motion in aerodynamic lenses and nozzle expansions. *Aerosol Science and Technology*, **22**, 314–324.

Lundgren, D.A., Carter, L.D. and Daley, P.S. (1976) Aerosol mass measurement using piezoelectric crystal sensors, in *Fine Particles* (ed B.Y.H. Liu), Academic Press, New York.

Maricq, M.M. and Xu, N. (2004) The effective density and fractal dimension of soot particles from premixed flames and motor vehicle exhaust. *Journal of Aerosol Science*, **35**, 1251–1274.

Marple, V.A., Rubow, K.L. and Behm, S.M. (1991) A Micro-Orifice Uniform Deposit Impactor (MOUDI): description, calibration, and use. *Aerosol Science and Technology*, **14**(4), 434–446.

Martínez-Lozano, P. and Fernández de la Mora, J. (2006) Resolution improvements of a nano-DMA operating transonically. *Journal of Aerosol Science*, **37**, 500–512.

Mazumder, M.K. and Kirsch, K.J. (1977) Single particle aerodynamic relaxation time analyzer. *The Review of Scientific Instruments*, **48**, 622.

McDermott, W.T., Ockovic, R.C. and Stolzenburg, M.R. (1991) Counting efficiency of an improved 30A condensation nucleus counter. *Aerosol Science and Technology*, **14**, 278.

McMurry, P.H., Wang, X., Park, K. *et al.* (2002) The relationship between mass and mobility for atmospheric particles: a new technique for measuring particle density. *Aerosol Science and Technology*, **36**, 227–238.

van der Meulen, A. and van Elzakker, B.G. (1986) Size resolution of laser optical particle counters. *Aerosol Science and Technology*, **5**, 313.

Mirme, A. (1994) Electric aerosol spectrometry. PhD thesis. University of Tartu.

Noble, C.A. and Prather, K.A. (2000) Real-time single particle mass spectrometer: a historical review of a quarter century of the chemical analysis of aerosol. *Mass Spectrometry Reviews*, **19**, 248–274.

Noone, K.J. and Hansson, H.C. (1990) Calibration of the TSI 3760 condensation nucleus counter for nonstandard operating conditions. *Aerosol Science and Technology*, **13**, 478.

Oberdorster, G. (1996) Significance of particle parameters in the evaluation of exposure-dose-response relationships of inhaled particles. *Particulate Science and Technology*, **14**(2), 135–151.

Olfert, J.S. and Collings, N. (2005) New method for particle mass classification – the Couette centrifugal particle mass analyzer. *Journal of Aerosol Science*, **36**, 1338–1352.

Olfert, J.S., Reavell, K.S., Rushton, M.G. *et al.* (2006) The experimental transfer function of the Couette centrifugal particle mass analyzer. *Journal of Aerosol Science*, **37**, 1840–1852.

Pandis, S.N., Baltensperger, U., Wolfenbarger, J.K. *et al.* (1991) Inversion of aerosol data from the epiphaniometer. *Journal of Aerosol Science*, **22**, 417–428.

Park, K., Cao, F., Kittelson, D.B. *et al.* (2003a) Relationship between particle mass and mobility for diesel exhaust particles. *Environment Science and Technology*, **37**, 577–583.

Park, K., Kittelson, D.B. and McMurray, P.H. (2003b) A closure study of aerosol mass concentration measurement: comparison of values obtained with filters and by direct measurement of mass distributions. *Atmospheric Environment*, **37**, 1223–1230.

Patashnick, H. and Rupprechl, G. (1980) A new real-time aerosol mass monitoring instrument: the TEOM, in *Proceedings: Advances in Particle Sampling and Measurement* EPA-600/9-80-004 (ed W.B. Smilh) U.S. Environmental Protection Agency, p. 264.

Rader, D.J. and McMurry, P.H. (1986) Application of the tandem differential mobility analyzer to studies of droplet growth or evaporation. *Journal of Aerosol Science*, **17**, 771.

Reavell, K. (2002) Determination of real-time particulate size spectra and emission parameters with a differential mobility spectrometer. International ETH Conference on Nanoparticle Measurement.

Sachweh, B.A., Dick, W.D. and McMurry, P.H. (1995) Distinguishing between spherical and nonspherical particles by measuring the variability in azimuthal light scattering. *Aerosol Science and Technology*, **23**, 373.

Sachweh, B., Umhauer, H., Ebert, F. *et al.* (1998) In situ optical particle counter with improved coincidence error correction for number concentrations up to 107 particle cm-3. *Journal of Aerosol Science*, **29**, 1075.

Salt, K., Noble, C. and Prather, K.A. (1996) Aerodynamic particle sizing versus light scattering intensity measurement as methods for real-time particle sizing coupled with time-of-flight mass spectrometry. *Analytical Chemistry*, **68**, 230.

Sem, G.J. and Tsurubayashi, K. (1975) A new mass sensor for respirable dust measurement. *American Industrial Hygiene Association Journal*, **36**, 791.

Sem, G.J., Tsurubayashi, K. and Homma, K. (1977) Performance of the piezoelectric microbalance respirable aerosol sensor. *American Industrial Hygiene Association Journal*, **38**, 580.

Shin, W.G., Pui, D.Y.H., Fissan, H. *et al.* (2007) Calibration and numerical simulation of the nanoparticle surface area monitor (TSI Model 3550 NSAM). *Journal of Nanoparticle Research*, **9**, 61–69.

Silva, P.J. and Prather, K.A. (1997) On-line characterization of individual particles from automobile emissions. *Environment Science and Technology*, **31**, 3074.

Sinclair, D. (1986) Measurement of nanometer aerosols. *Aerosol Science and Technology*, **5**, 187.

Sinclair, D., George, A.C., Knutson, E.O. (1977) Application of diffusion batteries to measurement of submicron radioactive aerosols. 1977 ANS Winter Meeting on Airborne Radioactivity, San Francisco, CA, December, 1977.

Sinclair, D. and Yue, P.C. (1982) Continuous flow condensation nucleus counter, II. *Aerosol Science and Technology*, **1**, 217.

Sinha, M.P., Giffin, C.E., Norris, D.D. *et al.* (1982) Particle analysis by mass spectrometry. *Journal of Colloid and Interface Science*, **87**, 140–153.

Stolzenburg, M., Hering, S.V. (2000) Method for the automated measurement of fine particle nitrate in the atmosphere, *Environment Science and Technology*, **34** (5): 907–914.

Stolzenburg, M.R. and McMurry, P.H. (1991) An ultrafine aerosol condensation nucleus counter. *Aerosol Science and Technology*, **14**, 48.

Suess, D.T. and Prather, K.A. (1999) Mass spectrometry of aerosols. *Chemical Reviews*, **99**, 3007.

Sullivan, R.C. and Prather, K.A. (2005) Recent advances in our understanding of atmospheric chemistry and climate made possible by on-line aerosol analysis instrumentation. *Analytical Chemistry*, **77**, 3861–3886.

Sylvia, H.W. and Prather, K.A. (1998) Time-of-flight mass spectrometry methods for real time analysis of individual aerosol particles. *Trend in Analytical Chemistry*, **17**, 346.

Szymanski, W.W. and Liu, B.Y.H. (1986) On the sizing accuracy of laser optical particle counters. *Particle and Particle Systems Characterization*, **3**, 1.

Szymanski, W.W.; Schindler, C. (1998) Response characteristics of a new dual optics particle spectrometer. Proceedings 5th International Congress on Optical Particle Sizing, Minneapolis, Mniiesota, August 10–14, 1998, pp. 219–220.

Turpin, B.J., Cary, R.A. and Huntzicker, J.J. (1990) An in-situ time-resolved analyzer for aerosol organic and elemental carbon. *Aerosol Science and Technology*, **12**, 161–171.

Vincent, J.H. (2007) *Aerosol Sampling: Standards, Instrumentation and Applications*, John Wiley & Sons, Ltd, Chichester.

Waggoner, A.P. and Charlson, R.J. (1976) Measurement of aerosol optical parameters, in *Fine Particles* (ed. B.Y.H. Liu), Academic Press, New York.

Wallace, D. and Chuan, R. (1977) *A cascade Impaction Instrument Using Quartz Crystal Microbalance Sensing Elements for 'Real-Time' Particle Size Distribution Studies*, Special Publication 464, US National Bureau of Standards, Washington, DC, p. 199.

Wang, S.C. and Flagan, R.C. (1990) Scanning electrical mobility spectrometer. *Aerosol Science and Technology*, **13**, 230.

Weber, R.J., Orsini, D., Daun, Y.N. *et al.* (2001) A particle-into-liquid collector for rapid measurement of aerosol bulk chemical composition. *Aerosol Science and Technology*, **35**, 718–727.

Wen, H.Y. and Kasper, G. (1986) Counting efficiencies of six commercial particle counters. *Journal of Aerosol Science*, **17**, 947–961.

Whitby, K.T. and Clark, W.E. (1966) Electrical aerosol particle counting and size distribution measuring system for the 0.015 to 1 μ size range. *Tellus*, **18**, 573.

Woo, K., Chen, D.R., Pui, D.Y.H. *et al.* (2001) Use of continuous measurements of integral aerosol parameters to estimate particle surface area. *Aerosol Science and Technology*, **34**, 57–65.

Yeh, H.C., Cheng, Y.S. and Orman, M.M. (1982) Evaluation of various types of wire screens as diffusion battery cells. *Journal of Colloid and Interface Science*, **86**, 12.

Zhang, Z.Q. and Liu, B.Y.H. (1990) Dependence of the performance of TSI 3020 condensation nucleus counter on pressure, flow rate and temperature. *Aerosol Science and Technology*, **13**, 493.

5

Filtration Mechanisms

Sarah Dunnett
Department of Aeronautical and Automotive Engineering, Loughborough University, UK

5.1 Introduction

Filtration may be defined as a process of separating particles from a gas (usually air) flow by means of a porous medium. The needs for such a process are many; for example, in many working environments people are exposed to harmful particles, and so protection must be provided. This might be in the form of a large-scale industrial filter or of personal protective equipment such as respirator filters. Some industries, including parts of the pharmaceutical and electronic sectors, rely on clean air of exceptional quality, which is provided through filtration. This is also true of the medical sector, where the control of infection is crucial. Filtration can also improve the quality of air in homes, workplaces, schools and transport.

There are two principal types of filtration: surface filtration and depth filtration. In surface filtration, particles are collected on a permeable surface; as the deposit builds, it forms a cake on the filter, which increases the resistance to flow. Depth filtration removes particles both from the surface and throughout the depth of the medium. One of the most important types of depth filter is one made up of fibres. Such filters consist of pads of fibres in an open three-dimensional network. Porosity is high, generally from 70 to 99%. An example of such a filter is shown in Figure 5.1. Because of their high porosity, fibrous filters generally offer a low resistance to flow. Gas flows through the open structure of the filter and particles suspended in it are captured on the fibres. Due to the distances between the fibres, any captured particle is unlikely to come in contact with more than one fibre, and hence the modelling of filter performance is often undertaken using single-fibre theory.

Aerosol Science: Technology and Applications, First Edition. Edited by Ian Colbeck and Mihalis Lazaridis.

Figure 5.1 *Scanning electron microscope image of a stainless-steel filter. Reproduced with permission from Heim et al. (2005). Copyright © 1995. Mount Laurel, NJ. Reprinted with permission.*

In addition to fibrous filters there are other particulate filters, such as granular filters, fabric filters and membrane filters. Granular filters are composed of packed beds of particles and are used extensively in the water and sewage industries. These are covered in detail in Tien and Ramarao (2007). Fabric filters are made from textile fibres, which are processed into a relatively compacted form. Most of the dust does not penetrate into the material but is captured on the surface. Membrane filters are made of perforated material or highly compacted fibrous material, resulting in a lower porosity than that in fibrous filters. They tend to act principally by surface filtration.

Fibrous filters are the most important type of filter for aerosol sampling, due to their ability to obtain high separation efficiencies with a relatively low pressure drop. Use is sometimes made of electrostatic interactions between particles and fibres, which enhance filter performance. Filters that are made of permanently charged fibres have become known as 'electrets'. In this work, a summary is made of the main factors affecting the performance of fibrous filters and significant work in the area is reviewed.

There are two main parameters used to assess the performance of a filter: collection efficiency, E, and penetration, P. The first is employed when filters are used for the recovery of particulates or when particulate is collected for analysis. It refers to the fraction of particles entering the filter that are retained:

$$E = \frac{N_{in} - N_{out}}{N_{in}} \tag{5.1}$$

where N_{in} and N_{out} are the number of particles entering and leaving the filter, respectively.

Penetration is important when air quality is being considered. It refers to the fraction of particles entering the filter that exit again:

$$P = \frac{N_{out}}{N_{in}} = 1 - E \tag{5.2}$$

The penetration of monodisperse aerosol through homogeneous filters of depth X can be given by $P = e^{-\gamma X}$ (Hinds, 1999), where γ is the layer efficiency. This layer efficiency is related to the single-fibre efficiency, η, which is the ratio of the number of actual particles removed by the fibre to the number that would be removed by a 100% efficient fibre, by (Hinds, 1999):

$$\eta = \frac{\gamma \pi d_f}{4\alpha} \tag{5.3}$$

where α is the filter packing fraction, which is the fraction of the perceived volume of the filter that is actually occupied by fibres, and d_f is the fibre diameter. Hence the efficiency of a fibrous filter of thickness X is related to the single-fibre efficiency by:

$$E = 1 - P = 1 - \exp\left[-\frac{4\alpha\eta X}{\pi d_f}\right] \tag{5.4}$$

The difficulty in applying Equation 5.4 is how to determine the value of the single-fibre efficiency, η. This has been the subject of many previous studies and is discussed in Sections 5.2.2–5.2.6. This efficiency is dependent upon the mechanism by which the particles deposit upon the fibre.

5.2 Deposition Mechanisms

The objective of filtration is to remove harmful aerosols from a gas, so the effectiveness of the mechanisms of aerosol deposition to the surfaces contained in a filter is fundamental to its use. Table 5.1 gives a fairly complete set of possible mechanisms, including the conditions under which they are effective and the filter types in which they are employed. The alternatives to the filters included in Table 5.1, whose use for particle removal is generally limited to large-scale operations, are centrifuges, used in vacuum cleaners, and electrostatic precipitators, used in large industrial plants. A comprehensive theoretical analysis of most of the mechanisms is given in Williams and Loyalka (1991). The last four mechanisms listed in the table are not generally practical for filtration. In the case of thermophoresis, the creation of a large temperature gradient is required, as aerosol particles will move to lower temperatures due to more energetic collisions with molecules on their high-temperature sides, leading to deposition. This is not practical in filters. Diffusiophoresis and vapour condensation and particle growth also require a temperature gradient to the filter surface in order to produce condensation. In the case of bubbling through liquids, splitting up of a gas flow into small bubbles passing through a liquid would have the advantage of simple removal of aerosol particles deposited by various mechanisms into bubble walls during their passage. A problem could arise, however, from aerosol created by the bursting of bubbles on their emergence from the liquid.

Due to these limitations, the main mechanisms by which filters remove aerosol particles from the fluid stream are the first five listed in the table: (i) diffusion, (ii) interception, (iii) inertial impaction, (iv) gravitational settling and (v) electrostatic attraction. This review will concentrate on these mechanisms and their application in fibrous filters.

The first four are mechanical capture mechanisms that work without the influence of attractive forces between the particles and the fibres. Diffusional deposition is important for small particles whose Brownian motion causes them to deviate from the flow streamlines and come into contact with the filter. Interception and inertial impaction are important

Table 5.1 Deposition mechanisms and filter use.

Mechanism or process	Conditions for which mechanism is effective	Filter type
Diffusion	Small or very small particles	Fibrous filters Ceramic filters at high T
Interception	Intermediate-size particles	Fibrous filters
Inertial impaction	Large-size particles, or smaller ones in high-velocity accelerated flows	Fibrous filters Centrifuges
Gravitational settling	Very large, or heavy, particles	Not effective in practice
Electrostatic attraction	Determined by charge; most effective for small or intermediate-size particles	Electret filters Electrostatic precipitators
Thermophoresis	Temperature gradient to surface needed	Not used in practice
Diffusiophoresis	Vapour condensation on surface needed	Not used in practice
Vapour condensation and particle growth	Particle size increased for easier removal	Used for particle observation but not in filtration
Bubbling through liquids	Depends on secondary mechanism to reach bubble walls	Can be effective with small bubbles

for larger particles. Interception occurs when particles follow the streamlines of the flow and pass close to a filter fibre. Inertial impaction occurs when particles deviate from the flow streamlines due to their inertia and impact upon a fibre. Gravitational settling is only significant for those particles whose settling velocity is larger than the convective velocity of the flow through the filter and hence is limited to very large or heavy particles. In filters constructed from charged fibres there is an additional mechanism of deposition: electrostatic attraction. These five mechanisms are shown schematically in Figure 5.2.

In order to understand the performance of filters for the different particle-deposition mechanisms, it is important to obtain a good understanding of the flow through them. Section 5.2.1 reviews the flow models developed to describe the flow field and their comparison with experimental measurements. Following sections consider the different deposition mechanisms in turn.

5.2.1 Flow Models

In theory it is possible to calculate the rate at which particles deposit directly on the surfaces of the fibres by solving the appropriate equations of motion. However, as the flow within the filter is complex, this is not possible without simplifications. The fluid flowing through a filter can be treated as continuous so long as the obstacles in its path are large compared with the mean free path of the fluid molecules. This condition is fulfilled if the Knudsen number, Kn, is negligible:

$$Kn = \frac{\lambda}{d_f} \tag{5.5}$$

where λ is the mean free path of the fluid molecules.

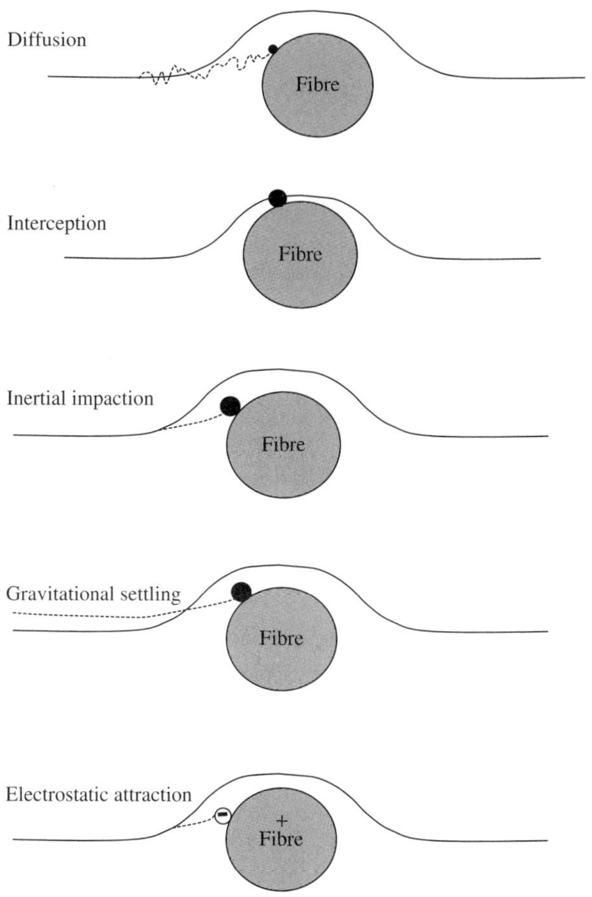

Figure 5.2 *Five deposition mechanisms.*

In the majority of filters, $Kn \ll 1$, and hence the fluid can be assumed to be continuous. For example, under normal conditions in air, λ is 0.065 μm, and hence the condition is satisfied for fibrous filters as long as the fibres are not submicron size.

Another important parameter when studying the fluid flow through filters is the Reynolds number, Re:

$$\mathrm{Re} = \frac{d_f \rho U_0}{\mu} \tag{5.6}$$

where ρ and μ are the density and viscosity of the fluid and U_0 is its mean velocity. This parameter is the ratio of the inertia and viscous forces of the fluid and in filtration theory it is generally small. For example, for air passing through a fibrous filter with fibres of diameter 10 μm at a velocity of 0.1 m/s, the value of Re is approximately 0.065. This shows that the flow is dominated by the fluid's viscosity. Hence in filtration theory, an assumption of negligible Re is usually made. The equations of motion describing the fluid flow are then given by:

$$\nabla p = \mu \nabla^2 \underline{U}$$
$$\nabla . \underline{U} = 0 \tag{5.7}$$

where p is the fluid pressure. In order to solve these equations, the majority of the analytical and numerical studies in filtration theory have been based on cell models for the fluid flow (Brown, 1993). In these models, the 3D filter structure in which fibres are randomly distributed is approximated by a 2D regularly staggered array of fibres, as shown in Figure 5.3. In two dimensions, Equation 5.7 can be written in terms of the stream function, ψ, as:

$$\nabla^4 \psi = 0 \tag{5.8}$$

The hexagonal cell shown in Figure 5.3 is approximated by a circular cell. Solutions to Equation 5.8 of the form:

$$\psi = \left(\frac{A_1}{r} + A_2 r + A_3 r \ln r + A_4 r^3 \right) \sin \theta \tag{5.9}$$

in the cell were obtained by Kuwabara (1959) and Happel (1959), where the A_i's are constants dependent upon α, the difference between the two models lies in the boundary condition applied on the cell boundary. Experiments performed by Kirsch and Fuchs (1967) found that these models gave an accurate description of the flow field near a fibre, with Kuwabara's model being the most accurate. This cell model approach has

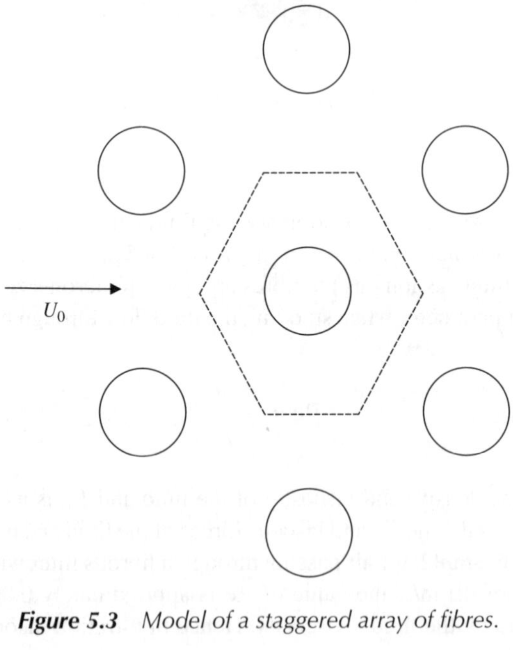

Figure 5.3 *Model of a staggered array of fibres.*

been adopted extensively in filtration theory; see, for example, Asgharian and Cheng (2002), Kirsh (2005), Raynor (2008), Dunnett and Clement (2009) and Chernyakov, Kirsh and Kirsh (2011). However, due to the assumptions made, such approaches do have their limitations; for example, the full filter space cannot be covered by circular cells and hence the description of the flow away from the fibres is poor. Therefore, other models have been developed; see for example Kirsh (2006) and Hellou, Martinez and El Yazidi (2004), who included higher-order terms in the solution to Equation 5.8 than those taken in Equation 5.9. In recent years, with the increasing power of computers, computational fluid dynamics (CFD) models have become more widespread. This technique allows, in principal, any filter structure to be modelled, as a mesh can be set up throughout the filter and the equations of fluid motion approximated at discrete mesh points. Generally, however, due to the computational demands of the technique, simple lattice structures are studied. For example, when Li and Marshall (2007) modelled the flow through a 2D array of cylindrical fibres, the use of CFD allowed flows with Re up to O(1) to be considered. Wang and Pui (2009) used CFD to model a staggered array of elliptical fibres. In a study by Qian, Zhang and Huang (2009), the flow field of a filter with a staggered array of fibres and varying solid volume fraction was modelled using CFD. In work by Wang *et al.* (2006), a virtual 3D fibrous filter structure was generated based on information obtained from analysis of microscopic images of filter media and the flow field obtained inside the filter using CFD. Experimental results for the filter efficiency were also obtained and were found to lie between those obtained by the CFD model and those using the Kuwabara cell model. In Maze *et al.* (2007), a similar approach was adopted in order to study the performance of filters composed of nanofibres operating at reduced pressures. The disturbance to the flow field caused by the fibres was neglected. In a study by Jaganathan, Tafreshi and Pourdeyhimi (2008), CFD simulations were performed on the microstructure of a real fibrous media. Digital volumetric imagining was used to obtain a 3D image of the filter. In a recent study by Hosseini and Tafreshi (2010), 2D random fibrous geometries were generated numerically and their efficiencies determined using CFD. These were compared with the 3D results obtained in the studies of Wang *et al.* (2006). Although computing power has increased considerably over recent years, that required by 3D simulations is still excessive, which may well prohibit extensive studies.

An efficient alternative approach introduced in order to determine the flow in filters by Filippova and Hanel (1997) is the lattice-Boltzmann method, which is a discrete computational method based upon an approximation of the Boltzmann equation. The technique is shown to simulate in an effective way the low-Re flow around complex geometries. It has been adopted by Przekop, Moskal and Gradon (2003) to describe the structure of deposited particulate matter in filters and by Przekop and Gradon (2008) to consider the filtration of nanoparticles.

Once the flow through a filter is understood, the behaviour of the particles within that flow can be studied. As mentioned earlier, there are five main particle-deposition mechanisms that are important for filtration. These will now be considered in turn for a clean filter. The different mechanisms are generally considered separately for ease of study, but in reality some of the mechanisms will act simultaneously. However, depending upon the situation, it is likely that one of the mechanisms will be dominant. Single-fibre efficiency, η, has been studied in detail for all of the deposition mechanisms. The simplest way of estimating η when more than one mechanism is important is to assume that the contributions from the different mechanisms are additive (Ramarao, Tien and Mohan, 1994):

$$\eta = \eta_D + \eta_R + \eta_I + \eta_G + \eta_{EL} \tag{5.10}$$

where η_D, η_R, η_I, η_G and η_{EL} are the single-fibre efficiencies due to diffusion, interception, inertial impaction, gravity and electrostatic deposition, respectively. An alternative and more accurate approach is to assume that the aerosol penetration, $1 - \eta$, is the product of the penetration due to the different mechanisms:

$$\eta = 1 - (1 - \eta_D)(1 - \eta_R)(1 - \eta_I)(1 - \eta_G)(1 - \eta_{EL}) \tag{5.11}$$

It is possible when considering the problem numerically to include all the mechanisms in the model and hence determine η directly (Ramarao, Tien and Mohan, 1994; Oh *et al.*, 2002). Results obtained this way have shown that for intermediate-sized particles there is a coupling between Brownian diffusion and inertial impaction (Ramarao, Tien and Mohan, 1994).

5.2.2 Diffusional Deposition

As stated in Table 5.1, this mechanism is important for small particles, which due to Brownian motion do not follow the streamlines of the flow and may diffuse from the flow to the fibres and deposit. The intensity of the Brownian motion increases for smaller particles and as a consequence so does the efficiency of removal. In this case, the nondimensional particle concentration, n, is determined by the nondimensional Fokker–Planck equation (Fuchs, 1964):

$$\frac{\partial n}{\partial t} + \underline{U}.\nabla n = \frac{1}{Pe}\nabla^2 n \tag{5.12}$$

where *Pe* is the Peclet number, which is a measure of the relative magnitude of the diffusional motion of the particles and of the convective motion of the air past the fibre:

$$Pe = \frac{U_0 d_f}{D} \tag{5.13}$$

where *D* is the coefficient of diffusion of the particles.

In this case, the single-fibre efficiency is given by:

$$\eta_D = \frac{2(1+R)\int_0^{\pi}\left(\frac{\partial n}{\partial r}\right)_{r=r_1} d\theta}{Pe} \tag{5.14}$$

where *R* is the interception parameter, $R = d_p/d_f$ and $r_1 = 1 + R$.

An exact solution of Equation 5.12 has not been found but by assuming the Kuwabara flow field given by Equation 5.9 and taking the diffusion layer around the fibre to be small, Lee and Liu (1982) found η_D to be given by:

$$\eta_D = A\left[\frac{1-\alpha}{Ku}\right]^{1/3} Pe^{-2/3} \tag{5.15}$$

where Ku is the hydrodynamic factor of Kuwabara flow:

$$Ku = -0.5 \ln \alpha - 0.75 + \alpha - 0.25\alpha^2 \tag{5.16}$$

and A is a constant, which was 2.6 in Lee and Lui's study and has been found to be 2.9 in others (Brown, 1993). This expression has been verified experimentally. The semiempirical Equation 5.15 applies to a continuum flow regime when Kn, given by Equation 5.5, is negligible. When the fibre diameter, d_f, is of the same magnitude as the mean free path of the fluid molecules, λ, the effect of the slip flow must be taken into account when estimating η_D. In this case, an expression for η_D is given by Payet *et al.* (1992) (Equation 5.17), which agrees well with experimental results for particles of size 80–400 nm and fibre diameter 1 μm:

$$\begin{aligned}
\eta_D &= 1.6\left[\frac{1-\alpha}{Ku}\right]^{1/3} Pe^{-2/3} C_1 C_2 \\
C_1 &= 1 + 0.388 Kn \left(\frac{(1-\alpha)\,Pe}{Ku}\right)^{1/3} \\
C_2 &= \frac{1}{1 + 1.6\left(\frac{(1-\alpha)}{Ku}\right)^{1/3} Pe^{-2/3} C_1}
\end{aligned} \tag{5.17}$$

There have been a number of recent studies on the performance of filters with submicron particles, including one by Qian, Zhang and Huang (2009), who numerically determined the filter efficiency for different operating conditions and found Brownian diffusion played a significant role for particles with particle diameter, d_p, <0.2 μm. The diffusional deposition of polydisperse aerosol has been considered analytically by Kwon, Kim and Lee (2002), who obtained expressions for the filters efficiency. In the CFD study by Hosseini and Tafreshi (2010), η_D was determined for a 2D model with randomly distributed fibres and the situations in which slip flow was and was not important were considered. As non-circular fibres may perform better than circular ones, due to their increased surface area, a numerical simulation was undertaken by Wang and Pui (2009) to investigate the single-fibre efficiency of elliptical fibres. They found that long and slim fibres performed better for particles dominated by the diffusion effect. A numerical study has also been undertaken by Kirsh (2007, 2011); the 2007 paper modelled the deposition of submicron particles in a model filter with fibres covered with porous shells and the 2011 paper looked at a 3D filter.

In recent years, the filtration of nanoparticles (d_p <100 nm) has become an important issue, as they are becoming produced more widely, for example in combustion emissions. As their size decreases, particles will begin to behave more like molecules and rebound from fibres may become significant. The preceding single-fibre theory has assumed that particles adhere to a fibre on impact and the lower bounds of particle size for which this theory holds are thus important. Works by Heim *et al.* (2005), Japuntich *et al.* (2007), Kim, Harrington and Pui (2007) and Wang, Chen and Pui (2007) have found that the single-fibre efficiency theory predicts filter efficiency well down to particles as small as 2.5 nm. The effect upon the filtration of nanoparticles of polydisperse filters was considered by Podgorski (2009), Podgorski, Maisser and Wladyslaw (2011) and Yamada, Takafumi and Yoshio (2011), who found that the classical theory significantly underestimates penetration in this case. A study was made of the effect of particle shape upon efficiency for particles in the size range 50–300 nm by Boskovic *et al.* (2008), who found it to be important at low filter-face velocity but less so as the velocity increased.

As the size of the particle increases relative to the thickness of the diffusion layer around the fibre, the effects of particle interception with the fibre will increase. It has been found (Stechkina and Fuchs, 1966; Dunnett and Clement, 2009) that this can be characterised by the parameter $s = R/\delta$, where δ represents the nondimensional thickness of the diffusion layer around the fibre. Diffusion dominates for $s < 1$ and interception for $s > 1$.

5.2.3 Deposition by Interception

Deposition by interception occurs as a result of the finite size of the particles. As stated in Table 5.1, it holds for intermediate-sized particles, which have negligible inertia, gravitational settling and Brownian motion and hence follow the flow streamlines. Deposition occurs when a streamline, called the 'limiting streamline', comes within one particle radius of the fibre. If y_C is the distance from the centreline at which the limiting streamline enters the solution domain then the single-fibre efficiency due to interception can be obtained from:

$$\eta_R = \frac{2y_C}{d_f} \tag{5.18}$$

Using the Kuwabara flow field, Kuwabara (1959) and Lee and Liu (1982) obtained an expression for η_R of the form:

$$\eta_R = \frac{1+R}{2Ku}\left[2\ln(1+R) - 1 + \alpha + \left(\frac{1}{1+R}\right)^2\left(1 - \frac{\alpha}{2}\right) - \frac{\alpha}{2}(1+R)^2\right] \tag{5.19}$$

where Ku is the hydrodynamic factor of Kuwabara flow, defined in Equation 5.16.

The Kuwabara flow field used to determine Equation 5.19 assumes that the fibres in the filter are staggered such that they are separated from each other by the same horizontal and vertical distances. Liu and Wang (1997) modified Equation 5.19 to include the effects of fibre arrangement. The performance of elliptical fibres for particles in the interception regime was investigated by Raynor (2008), who found η_R to depend upon the fibre aspect ratio and the orientation of the fibre axis to the flow, as well as on R and α. The capture of fibrous aerosol by interception was investigated by Asgharian and Cheng (2002), who found interception to be the sole loss mechanism for $St_m < 0.2$, where the modified Stokes number, St_m, is given by $St_m = \frac{St}{2Ku}$. The Stokes number, St, is the ratio of the particle stopping distance to the fibre diameter and is given by:

$$St = \frac{d_p^2 \rho_p U_0 C}{18\mu d_f} \tag{5.20}$$

where C is the Cunningham correction factor, which is necessary because of aerodynamic slip at the particle surface:

$$C = 1 + \frac{2\lambda}{d_p}\left[1.246 + 0.42 \times e^{-0.44 d_p/\lambda}\right] \tag{5.21}$$

For larger values of St_m, it has been found that interception cannot be treated separately from impaction.

5.2.4 Deposition due to Inertial Impaction

For particles for which Brownian diffusion is negligible, the general nondimensional equation of motion for a spherical aerosol particle is given by Equation 5.22:

$$St\frac{d\underline{U}_p}{dt} = \underline{U} - \underline{U}_p - N_g\underline{i}_3 \tag{5.22}$$

where $\underline{U}_p$ is the particle velocity, *St* and N_g are nondimensional parameters related to the importance of inertia and gravity upon particle transport, respectively, and i_3 is the unit vector in the direction in which gravity is acting.

The gravitational parameter N_g is the ratio of the particle settling velocity, V_s, to the mean filter velocity, U_0:

$$N_g = \frac{V_s}{U_0} = \frac{d_p^2 \rho_p C g}{18\mu U_0} \tag{5.23}$$

where g is the acceleration due to gravity.

If *St* and N_g are both negligible then Equation 5.22 gives $\underline{U} = \underline{U}_p$; that is, the particles follow the fluid streamlines and hence interception is the mechanism of deposition (see Section 5.2.3). The ratio of *St* and N_g is the Froude number, Fr:

$$Fr = \frac{St}{N_g} = \frac{U_0^2}{g d_f} \tag{5.24}$$

If Fr≫1 then gravity effects are negligible relative to inertia and Equation 5.22 reduces to:

$$St\frac{d\underline{U}_p}{dt} = \underline{U} - \underline{U}_p \tag{5.25}$$

This equation does not have a general solution, due to the complexity of $\underline{U}$, and hence no general analytical expression for the single-fibre efficiency resulting from inertial impaction, η_I, has been determined. Expressions have however been obtained for the limiting cases of small and large values of *St* (Brown, 1993):

$$\eta_I = \frac{J.St}{(2Ku)^2} \quad \text{for small } St \tag{5.26}$$

$$\eta_I = 1 - \frac{k_1}{St} \quad \text{for large } St \tag{5.27}$$

where $J = (29.6 - 28\alpha^{0.62})R^2 - 27.5R^{2.8}$ for $0.01 < R < 0.4$ and $0.0035 < \alpha < 0.111$ and k_1 is a constant that depends on the flow field; it is 0.805 for the Kuwabara flow field, with $\alpha = 0.05$.

For moderate values of *St*, empirical formulae have been obtained with fitted coefficients (Brown, 1993):

$$\eta_I = \frac{St^3 \times a^3}{St^3 a^3 + 0.77\left(1 + \frac{K_3}{\sqrt{Re}} + \frac{K_4}{Re}\right) St^2 a^2 + 0.58}, \qquad a = 1 + K_1\alpha + K_2\alpha^2 \tag{5.28}$$

Parameters K_1–K_4 have been fitted by least squares to experimental data and given as (Nguyen and Beeckmans, 1975):

$$K_1 = 37, K_2 = 91, K_3 = 12, K_4 = 60 \quad \text{for a model filter}$$

$$K_1 = 4, K_2 = 2250, K_3 = 4, K_4 = 65 \quad \text{for a real filter}$$

Wang and Pui (2009) obtained numerical results for large St and compared them with Equation 5.27, finding agreement to be good for $St > 3$. More recently, an analytical expression for η_I has been derived for the situation of $R \ll 1$, $St\alpha^{1/2} \ll 1$ by Zhu, Lin and Cheung (2000):

$$\eta_I = \frac{2(1-\alpha)\sqrt{\alpha}}{Ku} St.R + \frac{(1-\alpha)\alpha}{Ku} St^2 \tag{5.29}$$

Wang and Pui (2009) considered numerically the efficiency of filtration by elliptical fibres for the combined effects of interception and inertial deposition and found that blunt and near-circular fibres performed best for particles in this regime.

5.2.5 Gravitational Deposition

The effect of gravity on filtration efficiency depends on the direction of the airflow and can work to either increase or decrease deposition. In general, the settling velocity of a particle under the influence of gravity, V_s, will tend to increase deposition, whereas the mean convective velocity of the fluid through the filter, U_0, will tend to carry the particles past the fibre. The relative size of these two velocities, N_g, is generally equal to the single-fibre efficiency due to gravitational deposition, η_G (Brown, 1993):

$$\eta_G = N_g \tag{5.30}$$

In general, this mechanism is small compared with the others, except when particles are large or heavy and U_0 is low. When U_0 is greater than about 0.1 m/s, impaction is generally more important than settling.

Kirsh (2005) found that it is important to take into account the effects of gravity when determining the radius of the most penetrating particle while considering the diffusional deposition of heavy submicron particles.

5.2.6 Electrostatic Deposition

The deposition mechanisms considered in Sections 5.2.2–5.2.5 do not involve any attractive forces between the particles and the fibres. One technique applied to augment these mechanisms is the use of electrostatic forces. This is particularly useful for intermediate-sized particles, where the main mechanical deposition mechanism is interception. The filtration efficiency has a minimum in this region, due to the weakness of the particle Brownian motion and the particle inertia. With the addition of electrostatic forces, a filter can achieve a certain efficiency at a lower packing fraction, and hence a lower resistance to the air flow in the filter. Filters that are composed of permanently charged fibres have become known as 'electret filters' and are commonly used for air-cleaning applications that require high efficiency and a low pressure drop (Romay, Liu and Chae, 1998).

When a filter's fibres are charged, in addition to the deposition mechanisms discussed in Sections 5.2.2–5.2.5, particle deposition occurs because of polarisation forces. The electric field of the fibre induces a dipole in the particle, the strength of which depends upon the particle volume and the dielectric constant of its constituent material. As the electric field decreases with distance from the fibre, the attractive force on the negative part of the induced dipole exceeds the repulsive force on the positive part, causing the particle to be attracted by the fibre. For oppositely charged particles, this mechanism is increased by coulomb forces. The efficiency of the deposition mechanism depends on the ratio of the drift velocity, V_d, which moves the particle to the fibre, to the convective velocity of the flow, which tends to move the particle past the fibre. Hence all electrostatic filters are more efficient at low filtration velocities. The drift velocity of a particle is a product of the electrostatic force acting on it and of its mechanical mobility and is given by:

$$\underline{V}_d = \frac{qC}{3\pi \mu d_p} \underline{E} \tag{5.31}$$

where q is the net charge on the particle and E is the electric field due to the charge on the fibre. In early work on electrostatic deposition, the assumption was usually made that the fibres were uniformly charged, but details of the techniques used to charge filters suggest that this is not the case. More complicated configurations of charge have thus been considered, such as bipolarly charged fibres (Podgorski and Balazy, 2008). Reviews of work on the performance of clean filters are given by Brown (1993) and Wang (2001).

A dimensionless parameter, N_{sub}, which is the ratio of the drift velocity at the fibre surface to the freestream velocity, U_0, is used to describe the capture efficiency. The subscript sub is related to the charge on the fibre and the particles. In the following expressions for N_{sub}, the Cunningham correction factor is assumed to be unity.

5.2.6.1 Uniformly Charged Fibres

Assuming that a filter fibre has a uniform charge, Q, per unit length, then:

$$N_{Qq} = \frac{Qq}{3\pi^2 \epsilon_0 \mu d_p d_f U_0} \qquad \text{for charged particles with charge } q$$

$$N_{Q0} = \frac{Q^2 d_p^2}{3\pi^2 \epsilon_0 \mu d_f^3 U_0} \left(\frac{D_p - 1}{D_p + 2} \right) \qquad \text{for neutral particles} \tag{5.32}$$

where ϵ_0 is the permittivity of free space, $10^{-9}/36\pi$ Farads per metre (F/m), and D_p is the dielectric constant of the particle material (Brown, 1993).

The single-fibre efficiency due to electrostatic deposition for these two cases has been shown to be:

$$\eta_{Qq} = \pi N_{Qq} \qquad \text{assuming that only coulomb forces act between the particle and fibre}$$

$$\eta_{Q0} = \pi N_{Q0} \quad \text{for small } \eta_{Q0}$$

$$\eta_{Q0} = \left(\frac{3\pi N_{Q0}}{2}\right)^{1/3} \quad \text{for large } \eta_{Q0} \tag{5.33}$$

An empirical expression for η_{Q0} has been obtained for $0.03 < N_{Q0} < 0.91$ using Kuwabara's flow field with $\alpha = 0.03$ (Brown, 1993):

$$\eta_{Q0} = 0.84 N_{Q0}^{0.75} \tag{5.34}$$

5.2.6.2 Fibres with Nonuniform Charge

In real filters, the charge on a fibre is generally nonuniform. Although the calculation of electric fields due to the complex real-world configurations of charge are complicated, some simplified situations have been modelled. For example, if a fibre carries a line dipole charge with surface charge distribution $\sigma\cos\theta$ then:

$$N_{\sigma q} = \frac{\sigma q}{3\pi\epsilon_0 \mu (1 + D_f) d_p U_0} \quad \text{for charged particles with charge } q$$

$$N_{\sigma 0} = \frac{2\sigma^2 d_p^2}{3\epsilon_0 \mu d_f (1 + D_f)^2 U_0}\left(\frac{D_p - 1}{D_p + 2}\right) \quad \text{for neutral particles} \tag{5.35}$$

where D_f is the dielectric constant of the fibre material. The corresponding expressions for the single-fibre efficiency are given by (Brown, 1993):

$$\eta_{\sigma q} = 0.59 Ku^{-0.17} N_{\sigma q}^{0.83} \quad \text{for } 0.1 < N_{\sigma q} < 10$$

$$\eta_{\sigma 0} = 0.54 Ku^{-0.6} N_{\sigma 0}^{0.4} \quad \text{for } 1 < N_{\sigma 0} < 100 \tag{5.36}$$

Experimental results by Romay, Liu and Chae (1998) obtained on three commercially available fibrous electret filters for single-fibre efficiencies were fitted to power-law functions of $N_{\sigma q}$; it was found that for charged particles the exponent of $N_{\sigma q}$ was between 0.69 and 0.9 and for uncharged particles the exponent of $N_{\sigma 0}$ varied from 0.38 to 0.48, agreeing well with the predictions of Brown (1993).

The expression for neutral particles was later extended to values of $N_{\sigma 0}$ less than unity for Kuwabara flow (see Kim *et al.*, 2005):

$$\eta_{\sigma 0} = 1.48 N_{\sigma 0}^{0.93} \quad \text{for } 10^{-4} < N_{\sigma 0} < 10^{-2}$$

$$\eta_{\sigma 0} = 0.51 Ku^{-0.35} N_{\sigma 0}^{0.73} \quad \text{for } 10^{-2} < N_{\sigma 0} < 1 \tag{5.37}$$

More recently, Podgorski and Balazy (2008) simulated numerically the case for neutral particles and fitted their data to obtain two expressions for $\eta_{\sigma 0}$:

$$\eta_{\sigma 0} = \frac{0.771 N_{\sigma 0}}{1 + 0.973(KuN_{\sigma 0})^{0.785}} \qquad \text{for } 10^{-3} < N_{\sigma 0} < 10^{3}$$

$$\eta_{\sigma 0} = \frac{0.528 Ku^{0.26} N_{\sigma 0}^{0.74}}{1 + 0.366(KuN_{\sigma 0})^{0.63}} \qquad \text{for } 5 \times 10^{-3} < N_{\sigma 0} < 10^{3} \tag{5.38}$$

The first equation is believed to be more accurate for nanoparticles filtered in weakly charged electret media and the second equation for an intermediate strength of electrostatic interactions and particles larger than 0.3 μm.

5.2.6.3 Neutral Fibre, Charged Particles

In this case, the dimensionless parameter describing particle capture is given by:

$$N_{0q} = \left(\frac{D_f - 1}{D_f + 1}\right) \frac{q^2}{12\pi^2 \mu U_0 \epsilon_0 d_p d_f^2} \tag{5.39}$$

and the single -fibre efficiency by:

$$\eta_{0q} = \frac{2}{Ku^{1/2}} N_{0q}^{0.5} \tag{5.40}$$

An experimental study by Huang *et al.* (2006) investigated the capture of four different kinds of aerosol particle under different kinds of charging condition.

5.2.6.4 External Electric Fields

Single-fibre efficiency can be enhanced by placing the fibre in an external electric field, $\underline{E}$. In this case, the fibre is polarised and any particles passing it will be affected by both the external electric field and the polarisation field. Expressions for N and η for both charged and neutral particles are given in Brown (1993) and reproduced here:

$$N_{pq} = \frac{Eq}{3\pi \mu d_p U}$$

$$N_{p0} = \frac{2(D_p - 1)(D_f - 1) d_p^2 \epsilon_0 E^2}{3(D_p + 2)(D_f + 1) d_f \mu U} \tag{5.41}$$

where E is the size of the electric field.

$$\eta_{pq} = N_{pq} \frac{\left(\frac{D_f - 1}{D_f + 1}\right) + 1}{N_{pq} + 1}$$

$$\eta_{p0} = \frac{N_{p0}}{2} \tag{5.42}$$

An experimental investigation of the use of external electric fields with textile fibre filters for the capture of micrometre-size particles was undertaken by Thorpe and Brown (2003).

5.2.6.5 Combinations of Electrostatic and Mechanical Effects

The efficiency of an electret filter with rectangular fibres was studied experimentally by Kanaoka *et al.* (1987) for cases in which electrostatic and Brownian diffusion mechanisms were important. Empirical expressions were obtained for the single-fibre efficiency for charged and uncharged particles:

$$\eta = 1.07Pe^{-2/3} + 0.06N_{Q0}^{0.4} \qquad \text{uncharged particles}$$

$$\eta = 1.07Pe^{-2/3} + 0.06N_{Q0}^{0.4} + 0.067N_{Qq}^{0.75} - 0.017(N_{Q0}N_{Qq})^{0.5} \quad \text{charged particles} \tag{5.43}$$

There is some uncertainty about the coefficients in these equations (Brown, 1993) due to the assumptions made. However, the exponents are consistent with the theory.

Numerical simulations, including mechanical and electrostatic effects, have been undertaken by Oh *et al.* (2002), Cao, Cheung and Yan (2004) and Cheung, Cao and Yan (2005). In the work by Oh *et al.* (2002), Kuwabara's flow field was assumed and the fibres were uniformly charged, while Cao, Cheung and Yan (2004) and Cheung, Cao and Yan (2005) considered rectangular bipolarly charged fibres.

5.3 Factors Affecting Efficiency

5.3.1 Particle Rebound

The expressions given for single-fibre efficiency in this chapter have assumed that particles adhere to fibres on contact, but in some situations particles will rebound from the fibre surface and hence remain uncaptured. Taking into account rebound, single-fibre efficiency, η_{Re}, is a product of the collision efficiency, η (the efficiency assuming all particles that impact adhere), and an adhesion efficiency or probability, h:

$$\eta_{Re} = \eta.h \tag{5.44}$$

where η is the single-fibre efficiency discussed in Sections 5.2.2–5.2.6; that is, the assumption is made that $h \approx 1$. This assumption is only valid for particles between about 10 nm and a few micrometres in diameter, where diffusion and interception are the main deposition mechanisms (Wang and Kasper, 1991). For larger particles, rebound following impaction decreases efficiency, and for the smaller nanosized particles, bounce occurs as a result of the thermal impact velocity. This process of rebound is different from re-entrainment, where deposited particles are removed from the fibre due to drag forces acting on them or as a result of bombardment by airborne particles.

In the case of rebound, the adhesion efficiency is dependent upon the particle impact velocity, V_i. A particle will rebound from the surface if V_i is greater than a critical value given by:

$$V_{cr} = \left[\frac{2\Phi\left(1 - e^2\right)}{me^2}\right]^{0.5} \tag{5.45}$$

Equation 5.45 is valid when the adhesion energy of the approaching particle is the same as that of the rebounding particle. If the adhesion energy of the rebounding particle is much greater than that of approach, the $1 - e^2$ in the numerator is replaced by e^2 (Brown, 1993).

In Equation 5.45, Φ is the energy of adhesion, m is the particle mass and e is the coefficient of restitution. Due to the number of unknown quantities present in the description of h, such as e and Φ, most expressions in the literature are of an empirical nature. The one most commonly adopted comes from Ptak and Jaroszcyk (1990) and is valid for $0.4 < Re < 5.75$:

$$h = \frac{190}{(18St^2/R)^{0.68} + 190} \tag{5.46}$$

In an experimental study on a single fibre for $0.4 < St < 4.82$ (Rembor, Maus and Umhauer, 1999), it was found that Equation 5.46 overestimated h by a factor of up to 6. Later experimental work (Kasper *et al.*, 2009), for single fibres of size 8, 20 and 30 μm and $0.4 < St < 9$ obtained:

$$h_{8\mu m} = \frac{St^{-3}}{St^{-3} + 2.1St^{-2}Re^{0.503}}$$

$$h_{20/30\mu m} = \frac{St^{-3}}{St^{-3} + 0.0365Re^{2.46} + 1.91} \tag{5.47}$$

The efficiency of a nonwoven metal filter was considered by Klouda *et al.* (2011) for $U_0 = 10$ m/s and compared with expressions obtained using Equation 5.46. For the smaller particles, $d_p < 0.7$ μm, agreement was reasonable, but for $d_p > 0.7$ μm, agreement deteriorated.

For small particles, V_i is characterised by its thermal velocity and the mean thermal impact velocity is given by:

$$V_i = \left[\frac{48k_BT}{\pi^2\rho_p d_p^3}\right]^{0.5} \tag{5.48}$$

where k_B is Boltzmann's constant and T is temperature (Wang and Kasper, 1991). The theoretical model developed by Wang and Kasper suggests that the thermal impact velocity of a particle will exceed the critical sticking velocity in the size range between 1 and 10 nm, depending on elastic and surface adhesion parameters. Experimental evidence for the effect of thermal rebound upon filtration is however scarce. For example, Kim, Harrington and Pui (2007) measured the penetration of particles of size 3–20 nm through a wide range of filter media and found no significant evidence of rebound. Heim *et al.* (2005) examined the filtration of charged and uncharged particles in the range 2.5–20.0 nm on grounded metal fibres and meshes and also detected no thermal rebound effects. However, Kim *et al.* (2006) measured the filtration efficiency through glass fibrous filters for nanoparticles down to 1 nm and detected rebound effects for particles below 2 nm. An experimental study was performed at temperatures up to 500 K by Shin *et al.* (2008) for particles in the range 3–20 nm and no thermal rebound was detected.

Recently, Mouret *et al.* (2011) extended the theoretical model of Wang and Kasper and concluded that thermal rebound effects should only be observable in the subnanometre particle range.

5.3.2 Particle Loading

As particles are collected in a filter, they alter its structure and affect both the pressure drop and the collection efficiency. Hence an understanding of the mechanisms of deposit and the effects upon filter performance is important. However, this is a very complex process and is still an area of active research.

If a fibre is taken from a loaded filter and examined, it will be seen that dendrites are formed on its surface. Dendrites are essentially branchlike structures made up of deposited particles (see Figure 5.4). The formation of dendrites on a fibre is dependent upon the dominant deposition mechanism. In the case of diffusional deposition, dendrites are formed over the entire fibre surface, and for interception they are captured on the leading surface of the fibre. In the case of inertial impaction, particles will be mainly deposited close to the front stagnation point of the fibre. Electrostatic forces tend to cause deposition at any point on the fibre surface, similar to diffusional deposition (Oh *et al.*, 2002). Electric forces also encourage the collapse of dendrites as they tend to attract both captured and airborne particles. External electric fields cause particles to form long dendrites with relatively few branches (Brown, 1993). Considering only mechanical deposition mechanisms, Figure 5.5 shows the

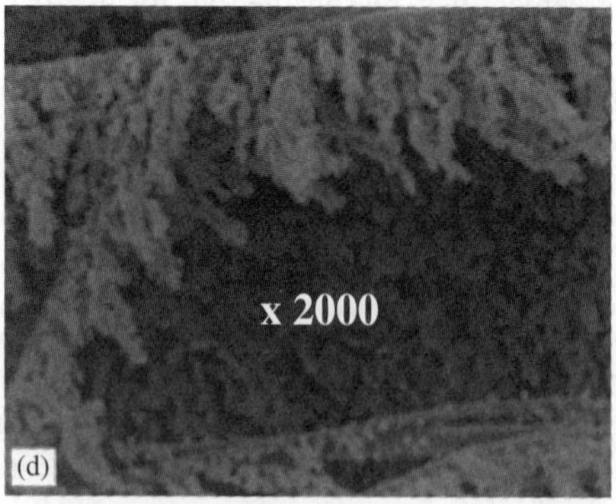

Figure 5.4 Photograph of dendrite formation. Reproduced with permission from Thomas et al. (1999). Copyright © 1999, Elsevier Ltd.

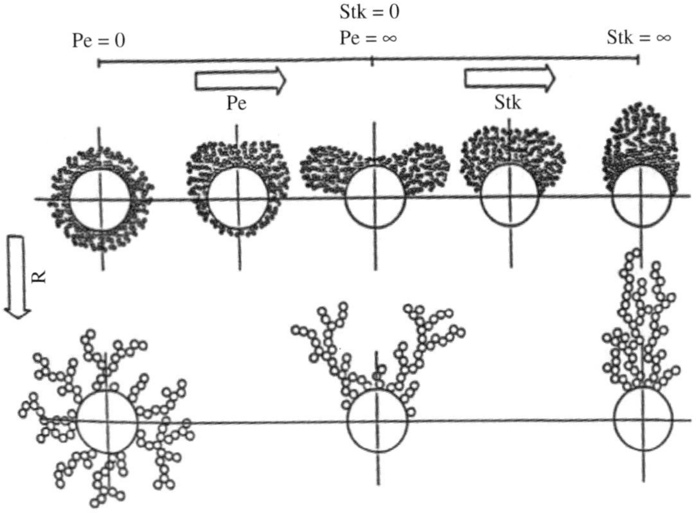

Figure 5.5 *Schematic relationship between deposit structure and filtration conditions in terms of the nondimensional filtration parameters Peclet number, Pe, Stokes number, St, and interception number, R. Reproduced from Kasper, Schollmeier and Meyer (2010), with permission from Elsevier.*

relationship between deposit structure and filtration operating conditions. As can be seen in the figure, for small *Pe* and small R, when diffusional deposition is the dominant mechanism, the particles will be evenly distributed around the fibre with a relatively open pore structure. As interception becomes more important, the basic distribution of the deposit will remain the same but the structure will be more open. For large *St*, when inertial impaction is dominant, the deposition will purely be on the front of the fibre, with the majority close to the stagnation point. As stated by Kasper, Schollmeier and Meyer (2010), this figure does not reflect the entire reality, especially when particle bounce is significant.

Numerical simulations of dendrite formation on fibres have been performed for many years; for a review of much of this work, see Brown (1993). Such simulations concern the early stages of deposition for the various mechanisms and are computationally very expensive. More recent work in this area has been done by Li and Marshall (2007), who included rebound effects for particles in the inertial impaction regime. A dimensionless adhesion parameter was found to have a dominant effect on the predicted particle deposition rate. The lattice-Boltzmann approach was adopted by Przekop, Moskal and Gradon (2003) to describe the structure of deposits on a cylindrical collector for particles in the diffusional regime. One of the reasons that the study of particle loading is computationally expensive is the need to recalculate the flow field when deposition occurs; Karadimos and Ocone (2003) investigated this and found that the single-fibre efficiency is overestimated when recalculation is not performed. CFD simulations by Lehmann and Kasper (2002) demonstrate the importance of particle rebound upon deposit structure in the inertial regime and show that simulations must be three-dimensional in order to be accurate. Lehmann and Kasper (2002) also used the simulations to determine the packing density of the deposits formed.

The deposition of particles on an electret fibre was simulated by Kanaoka, Hiragi and Tanthapanichakoon (2001). Despite the increasing power of computers, numerical models of particle loading are still limited in what they can achieve.

Experimental work on the performance of loaded filters has shown that the rise in the pressure drop across the filter as load increases can be described in two main steps: during the first stage, the evolution is slow, but in the second, the increase becomes markedly rapid and linear (Thomas *et al.*, 1999), as shown in Figure 5.6. Thomas *et al.* (1999) developed a model to describe the filter performance during loading, which was found to agree well with experimental data (Thomas *et al.*, 2001). In an experimental study by Song, Park and Lee (2006), an empirical expression was obtained for the pressure drop as a function of particle diameter and Cunningham correction factor. Using the experimental results of Walsh and Stenhouse (1997) on the performance of an electret filter under dust loading conditions for charged and uncharged particles (Sae-lim, Tanthapanichakoon and Kanaoka, 2006), a semiempirical expression was obtained for η as a sum of the electrical and mechanical contributions.

Recently, detailed experimental work was undertaken to gain a greater understanding of the performance of single fibres (Kasper *et al.*, 2009; Kasper, Schollmeier and Meyer, 2010). The single-fibre efficiency was found to be described by a power law of the type:

$$\frac{\eta(M)}{\eta(0)} = 1 + bM^c \tag{5.49}$$

where b and c are empirical fit coefficients and M is the accumulated mass per unit filter volume. The experimental process undertaken by Kasper's group generated much useful structural information about the deposit, including qualitative information about deposit morphology as a function of particle size, flow velocity and fibre diameter for particles in which inertia, interception and bounce are important. It was found that the transition from compact deposits to the more dendritic structures shown in Figure 5.5 is driven not by interception but by particle bounce.

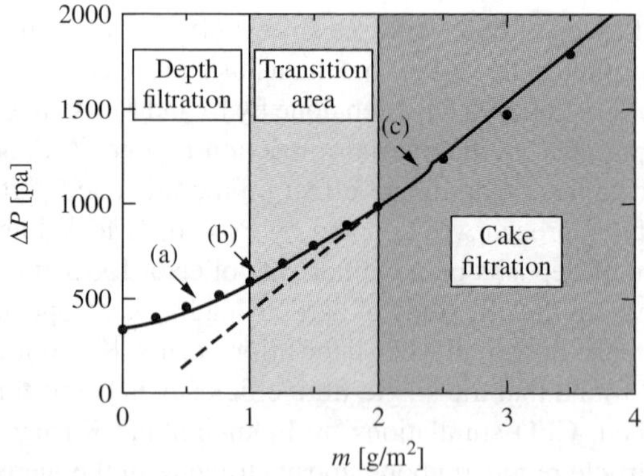

Figure 5.6 *Evolution of pressure drop, P, with mass loading, m. Reproduced with permission from Thomas et al. (1999). Copyright © 1999, Elsevier Ltd.*

5.4 Filter Randomness

Practically all the theoretical results given here have been obtained for a single fibre, or a regular array of fibres, arranged perpendicular to the gas flow. A realistic fibrous filter, such as that in Figure 5.1, shows a random array of fibres not exactly at right angles to the flow. There has been relatively little research on the effect of this randomness, but a pioneering study by Schweers and Löffler (1994) showed that it is essential to take account of it in calculating the efficiency of a filter. Their numerical model considered transmission through a random array of cells containing fibres specified by random geometry (angle to the flow) and local packing fraction. The problem has also been approached by a direct statistical model treating transmission through a series of cells of random variable efficiencies (Clement and Dunnett, 2000), which shows that randomness always reduces the overall filtration efficiency compared to that obtained from mean values of the variables that specify the cells. Much more experimental information on the randomness of filters, such as that for packing density distributions investigated by Lehmann, Hardy and Kasper (2003), needs to be obtained in order to further this research.

5.5 Applications

Fibrous filters are efficient devices for capturing particles and hence have many applications. This section touches upon just a few of these, and references will be provided that give more detail.

People are exposed to harmful particulates in the workplace and at home. A wide range of control systems have been designed to reduce exposure, from small personal devices to large-scale industrial installations. Small-scale respirator filters are commonly used to protect individuals against harmful biological and nonbiological aerosol. Work on the performance of respirators against nonbiological particles has been done by Wake and Brown (1988), Ortiz, Soderholm and Valdez (1988), Brown, Wake and Gray (1988) and Eshbaugh *et al.* (2009). The recent increase in the production of engineered nanoparticles in workplaces has led to studies on the performance of respirators for these small particles, many of which are reviewed in Mostofi *et al.* (2010). Another area of increasing importance is the filtration of bioaerosols: particles of biological origin, including viruses, bacteria, fungi and pollen. Interest has increased in this area since the severe acute respiratory syndrome (SARS) outbreak in 2003 and the global H1N1 viral infection in 2009 and with the increasing threat of bioterrorism. A review of some of the work on the performance of respirators in filtering bioaerosols can be found in Rengasamy, Zhuang and BerryAnn (2004) and Xu *et al.* (2011).

For people who suffer from allergic respiratory disease, air filtration is often recommended as part of a group of measures designed to reduce exposure to indoor allergens. In such situations, air filtration can be provided by whole-house filtration via the home-heating, ventilation or air-conditioning system, by portable room air cleaners, or through a combination of these methods. A review of studies on the various types of filtration is given in Sublett (2011).

High-efficiency particulate air (HEPA)-type filters, which remove airborne particles sized 0.3 mm and above with 99.97% efficiency, are used in many hospital environments (such as operating theatres) to reduce infection rates (Dharan and Pittet, 2002). As HEPA filtration is relatively inexpensive and studies have shown that it can significantly reduce airborne levels and/or infection rates for several aerolised pathogens, Curtis (2008) states that it should probably be used in all hospital rooms. The use of portable HEPA filters has also been found to significantly reduce airborne levels of some viruses (Curtis, 2008).

Diesel exhaust is a major contributor to ultrafine particles in places with high traffic intensity. Studies suggest that commuters travelling in their vehicles may experience a high exposure to ultrafine particles (Zhu *et al.*, 2007). In order to reduce this exposure, many automobiles are now equipped with a cabin air filter. Some work on the efficiency of these filters is given in Rudell *et al.* (1999) and Qi *et al.* (2008). Another increasingly popular mode of transport that carries the risk of exposing passengers to particulates is air travel. Confined space, limited ventilation, prolonged exposure times and recirculating air are risk factors for the transmission of respiratory tract infections. Use of recirculated air means that filtration is required (Leder and Newman, 2005). The majority of large, modern, commercial aircraft use HEPA filters, which have been found to be effective at removing airborne contamination from the recirculated air in such vehicles (Bull, 2008).

5.6 Conclusions

In this work, an attempt has been made to summarise the main factors affecting the performance of fibrous filters and to discuss their application. The main significant findings from past studies have been mentioned. Although filters and the process of filtration have been studied for many years, especially since the 1960s and the development of computer systems and experimental facilities (Spurny, 1997), this is still an active area of research. For example, the current interest in nanotechnology, despite offering great technological promise, also carries potential health risks and raises the question of how effective filters are at protecting people. The need to protect against global threats such as an influenza epidemic or bioterrorism requires further understanding of the filtration of microorganisms. In addition to new paths of research opening as a result of changing human needs, there are still fundamental areas of filtration performance that are not fully understood. For example, although there are formulae, validated by experimental work, that describe the performance of a clean filter well, our understanding of the performance of filters that contain deposit is less complete and still requires further research.

In conclusion, although we have advanced considerably in our understanding of fibrous filters, there are still many questions to be answered before this understanding is complete.

Nomenclature

C	Cunningham correction factor
D	Diffusion coefficient
D_f	Dielectric constant of the fibre material
D_p	Dielectric constant of the particle material
d_f	Filter fibre diameter
d_p	Particle diameter

E	Electric field
E	Collection efficiency
e	Coefficient of restitution
Fr	Froude number
h	Adhesion probability or efficiency
Kn	Knudsen number
Ku	Kuwabara factor ($= -0.5\ln(\alpha) - 0.75 + \alpha - 0.25\alpha^2$)
k_B	Boltzmann constant
M	Mass of deposited material per unit volume of filter material
m	Particle mass
N_g	Gravitational parameter
N_{in}	Number of particles entering the filter
N_{out}	Number of particles leaving the filter
N_{Qq}	Dimensionless parameter describing capture of charged particles by a charged fibre
N_{Q0}	Dimensionless parameter describing capture of neutral particles by a charged fibre
$N_{\sigma q}$	Dimensionless parameter describing capture of charged particles by a line dipole charged fibre
$N_{\sigma 0}$	Dimensionless parameter describing capture of neutral particles by a line dipole charged fibre
N_{pq}	Dimensionless parameter describing capture of charged particles by a polarised fibre
N_{p0}	Dimensionless parameter describing capture of neutral particles by a polarised fibre
N_{0q}	Dimensionless parameter describing capture of charged particles by a neutral fibre
N	Nondimensional particle concentration
P	Penetration
p	Fluid pressure
Pe	Peclet number
q	Charge held by a particle
Q	Charge per unit length of fibre
R	Interception parameter, ratio of particle and fibre diameters
Re	Reynolds number of the flow
r	Polar coordinate
s	Deposit mechanism parameter ($= R/\delta$)
St	Stokes number
St_m	Modified Stokes number ($= St/(2Ku)$)
T	Temperature
U_0	Freestream velocity

U	Fluid velocity
U_p	Particle velocity
V_d	Particle drift velocity
V_i	Particle impact velocity
V_S	Particle settling velocity
x,y	Cartesian components
α	Filter packing fraction
δ	Nondimensional thickness of the diffusion layer
ϵ_0	Permittivity of free space
Φ	Adhesion energy
γ	Layer efficiency
η	Single-fibre efficiency
η_D	Single-fibre efficiency by diffusional deposition
η_G	Single-fibre efficiency by gravitational deposition
η_I	Single-fibre efficiency by inertial deposition
η_R	Single-fibre efficiency by interception
η_{EL}	General expression for single-fibre efficiency by electrostatic deposition
η_{Re}	General expression for single-fibre efficiency, taking into account particle rebound
η_{Qq}	Single-fibre efficiency by charged fibre with charged particles
η_{Q0}	Single-fibre efficiency by charged fibre with neutral particles
$\eta_{\sigma q}$	Single-fibre efficiency by line dipole fibre with charged particles
$\eta_{\sigma 0}$	Single-fibre efficiency by line dipole fibre with neutral particles
η_{pq}	Single-fibre efficiency by polarised fibre with charged particles
η_{p0}	Single-fibre efficiency by polarised fibre with neutral particles
η_{0q}	Single-fibre efficiency by neutral fibre with charged particles
λ	Mean free path of fluid molecules
μ	Fluid viscosity
θ	Polar coordinate
ρ	Fluid density
ρ_p	Particle density
σ	Surface charge density of fibre
ψ	Stream function

References

Asgharian, B. and Cheng, Y.S. (2002) The filtration of fibrous aerosol. *Aerosol Science and Technology*, **36**, 10–17.

Boskovic, L., Agranovski, I.E., Altman, I.S. and Braddock, R.D. (2008) Filter efficiency as a function of nanoparticle velocity and shape. *Journal of Aerosol Science*, **39**, 635–644.

Brown, R.C. (1993) *Air Filtration. An Integrated Approach to the Theory and Application of Fibrous Filters*, Pergamon Press, Oxford.

Brown, R.C., Wake, D., Gray, R. *et al.* (1988) Effect of industrial aerosols on the performance of electrically charged filter material. *Annals of Occupational Hygiene*, **32**, 271–294.

Bull, K. (2008) Cabin air filtration: helping to protect occupants from infectious diseases. *Travel Medicine and Infectious Disease*, **6**, 142–144.

Cao, Y.H., Cheung, C.S. and Yan, Z.D. (2004) Numerical study of an electrets filter composed of an array of staggered parallel rectangular split-type fibres. *Aerosol Science and Technology*, **38**, 603–618.

Chernyakov, A.L., Kirsh, A.A. and Kirsh, V.A. (2011) Efficiency of inertial deposition of aerosol particles in fibrous filters with regard to particle rebounds from fibers. *Colloid Journal*, **73**(3), 389–393.

Cheung, C.S., Cao, Y.H. and Yan, Z.D. (2005) Numerical model for particle deposition and loading in electrets filter with rectangular split-type fibers. *Computational Mechanics*, **35**, 449–458.

Clement, C.F. and Dunnett, S.J. (2000) The use of random variables in fibrous filtration theory, European Aerosol Conference 2000. *Journal of Aerosol Science*, **31**, S200–201.

Curtis, L.T. (2008) Prevention of hospital-acquired infections: review of non-pharmacological interventions. *Journal of Hospital Infection*, **69**, 204–219.

Dharan, S. and Pittet, D. (2002) Environmental controls in operating theatres. *Journal of Hospital Infection*, **51**, 79–84.

Dunnett, S.J. and Clement, C.F. (2009) A numerical model of fibrous filters containing deposit. *Engineering Analysis with Boundary Elements*, **33**(5), 601–610.

Eshbaugh, J.P., Gardner, P.D., Richardson, A.W. and Hofacre, K.C. (2009) N95 and P100 respirator filter efficiency under high constant and cyclic flow. *Journal of Occupational and Environmental Hygiene*, **6**(1), 52–61.

Filippova, O. and Hanel, D. (1997) Lattice-Boltzmann simulation of gas-particle flow in filters. *Computers and Fluids*, **26**(7), 697–712.

Fuchs, N.A. (1964) *The Mechanics of Aerosols*, Dover Publication, New York.

Happel, J. (1959) Viscous flow relative to arrays of cylinders. *AIChE Journal*, **5**(2), 174–177.

Heim, M., Mullins, B.J., Wild, M. *et al.* (2005) Filtration efficiency of aerosol particles below 20 nanometers. *Aerosol Science and Technology*, **39**, 782–789.

Hellou, M., Martinez, J. and El Yazidi, M. (2004) Stokes flow through microstructural model of fibrous media. *Mechanics Research Communications*, **31**, 97–103.

Hinds, W.C. (1999) *Aerosol Technology: Properties, Behaviour, and Measurement of Airborne Particles*, John Wiley & Sons, Inc., New York.

Hosseini, S.A. and Tafreshi, H.V. (2010) Modeling particle filtration in disordered 2-D domains: a comparison with cell models. *Separation and Purification Technology*, **74**, 160–169.

Huang, B., Yao, Q., Li, S.Q. *et al.* (2006) Experimental investigation on the particle capture by a single fibre using microscopic image technique. *Powder Technology*, **163**, 125–133.

Jaganathan, S., Tafreshi, H.V. and Pourdeyhimi, B. (2008) A realistic approach for modelling permeability of fibrous media: 3-D imaging coupled with CFD simulation. *Chemical Engineering and Technology*, **63**, 244–252.

Japuntich, D.A., Franklin, L.M., Pui, D.Y. *et al.* (2007) A comparison of two nano–sized particle air filtration tests in the diameter range of 10 to 400 nanometres. *Journal of Nanoparticle Research*, **9**, 93–107.

Kanaoka, C., Emi, H., Hiragi, S. and Myojo, T. (1986) Morphology of particulate agglomerates on a cylindrical fiber and collection efficiency of a dustloaded fiber. *Aerosols – Formation and Reactivity*. Proceedings 2nd International Aerosol Conference, Berlin 1986, Pergamon Journals Ltd, Oxford, pp. 674–677.

Kanaoka, C., Emi, H., Otani, Y. and Iiyama, T. (1987) Effect of charging state of particles on electret filtration. *Aerosol Science and Technology*, **7**(1), 1–13.

Kanaoka, C., Hiragi, S. and Tanthapanichakoon, W. (2001) Stochastic simulation of the agglomerative deposition process of aerosol particles on an electret fiber. *Powder Technology*, **118**, 97–106.

Karadimos, A. and Ocone, R. (2003) The effect of the flow field recalculation on fibrous filter loading: a numerical simulation. *Powder Technology*, **137**, 109–119.

Kasper, G., Schollmeier, S. and Meyer, J. (2010) Structure and density of deposits formed on filter fibers by inertial particle deposition and bounce. *Journal of Aerosol Science*, **41**, 1167–1182.

Kasper, G., Schollmeier, S., Meyer, J. and Hoferer, J. (2009) The collection efficiency of a particle-loaded single filter fiber. *Journal of Aerosol Science*, **40**, 993–1009.

Kim, C.S., Bao, L., Okuyama, K. *et al.* (2006) Filtration efficiency of a fibrous filter for nanoparticles. *Journal of Nanoparticle Research*, **8**, 215–221.

Kim, S.C., Harrington, M.S. and Pui, D.Y.H. (2007) Experimental study of nanoparticles penetration through commercial filter media. *Journal of Nanoparticle Research*, **9**, 117–125.

Kim, J.C., Otani, Y., Noto, D. *et al.* (2005) Initial collection performance of resin wool filters and estimation of charge density. *Aerosol Science and Technology*, **39**, 501–508.

Kirsch, A.A. and Fuchs, N.A. (1967) The fluid flow in a system of parallel cylinders perpendicular to the flow direction at small Reynolds number. *Journal of the Physical Society of Japan*, **22**(5), 1251–1255.

Kirsh, V.A. (2005) Diffusional deposition of heavy submicron aerosol particles on fibrous filters. *Colloid Journal*, **67**(3), 313–317.

Kirsh, V.A. (2006) Stokes flow in periodic systems of parallel cylinders with porous permeable shells. *Colloid Journal*, **68**(2), 173–181.

Kirsh, V.A. (2007) Deposition of aerosol nanoparticles in filters composed of fibres with porous shells. *Colloid Journal*, **69**(5), 655–660.

Kirsh, V.A. (2011) Diffusional deposition of nanoparticles in 3D model fibre filter. *Russian Journal of Physical Chemistry A*, **85**(11), 2089–2093.

Klouda, G.A., Fletcher, R.A., Gillen, J.G. and Verkouteren, R. (2011) Aerosol collection efficiency of a graded metal-fibre filter at high airflow velocity (10 m/s). *Aerosol Science and Technology*, **45**(3), 336–342.

Kuwabara, S. (1959) The forces experienced by randomly distributed parallel circular cylinders or spheres in a viscous flow at small Reynolds numbers. *Journal of the Physical Society of Japan*, **14**(4), 527–532.

Kwon, S.B., Kim, H.T. and Lee, K.W. (2002) Analytical solutions to diffusional deposition of polydisperse aerosols in fibrous filters. *Aerosol Science and Technology*, **36**(6), 742–747.

Leder, K. and Newman, D. (2005) Respiratory infections during air travel. *Internal Medicine Journal*, **35**, 50–55.

Lee, K.W. and Liu, B.Y.H. (1982) Theoretical study of aerosol filtration by fibrous filters. *Aerosol Science and Technology*, **1**, 147–161.

Lehmann, M.J., Hardy, E.H. and Kasper, G. (2003) Local packing density distribution within a fibrous filter – determination by MRI. *Abstracts of the European Aerosol Conference*, 513–514.

Lehmann, M.J. and Kasper, G. (2002) CFD simulations of single fibre loading, in (eds M.J. Lehmann and G. Kasper), *Particle Loading and Kinetics of Filtration in Fibrous Filters*, pp. 91–98, Institutfur Mechanische Verfahrenstechnik und Mechanik Universitat Karlsruhe, ISBN: 3-9805-220-2-4.

Li, S.Q. and Marshall, J.S. (2007) Discrete element simulation of micro-particle deposition on a cylindrical fiber in an array. *Journal of Aerosol Science*, **38**, 1031–1046.

Liu, G.Z. and Wang, P.K. (1997) Pressure drop and interception efficiency of multifibre filters. *Aerosol Science and Technology*, **26**(4), 313–325.

Maze, B., Tafreshi, H.V., Wang, Q. and Pourdeyhimi, B. (2007) A simulation of unsteady-state filtration via nanofiber media at reduced operating pressures. *Journal of Aerosol Science*, **38**, 550–571.

Mostofi, R., Wang, B., Haghighat, F. *et al.* (2010) Performance of mechanical filters and respirators for capturing nanoparticles – limitations and future direction. *Industrial Health*, **48**, 296–304.

Mouret, G., Chazelet, S., Thomas, D. and Berner, D. (2011) Discussion about the thermal rebound of nanoparticles. *Separation and Purification Technology*, **78**, 125–131.

Nguyen, X. and Beeckmans, J.M. (1975) Single fibre capture efficiencies of aerosol particles in real and model filters in the inertial-interceptive domain. *Journal of Aerosol Science*, **6**, 205–212.

Oh, Y.W., Jeon, K.J., Jung, A.I. and Jung, Y.W. (2002) A simulation study on the collection of submicron particles in a unipolar charged fibre. *Aerosol Science and Technology*, **36**, 573–582.

Ortiz, L.W., Soderholm, S.C. and Valdez, F.O. (1988) Penetration of respirator filters by an asbestos aerosol. *American Industrial Hygiene Association Journal*, **49**(9), 451–460.

Payet, S., Bouland, D., Madelaine, G. and Renoux, A. (1992) Penetration and pressure drop of a HEPA filter during loading with submicron liquid particles. *Journal of Aerosol Science*, **23**(7), 723–735.

Podgorski, A. (2009) Estimation of the upper limit of aerosol nanoparticles penetration through inhomogeneous fibrous filters. *Journal of Nanoparticle Research*, **11**, 197–207.

Podgorski, A. and Balazy, A. (2008) Novel formulae for deposition efficiency of electrically neutral, submicron aerosol particles in bipolarly charged fibrous filters derived using Brownian dynamics approach. *Aerosol Science and Technology*, **42**(2), 123–133.

Podgorski, A., Maisser, A. and Wladyslaw, S.W. (2011) Penetration of monodisperse, singly charged nanoparticles through polydisperse filters. *Aerosol Science and Technology*, **45**(2), 215–233.

Przekop, R. and Gradon, L. (2008) Deposition and filtration of nanoparticles in the composites of nano and microsized fibers. *Aerosol Science and Technology*, **42**, 483–493.

Przekop, R., Moskal, A. and Gradon, L. (2003) Lattice-Boltzmann approach for description of the structure of deposited particulate matter in fibrous filters. *Journal of Aerosol Science*, **34**, 133–147.

Ptak, T. and Jaroszcyk, T. (1990) Theoretical-experimental aerosol filtration model for fibrous filters at intermediate Reynolds numbers. Proceedings 5th World Filtration Congress, Nice.

Qi, C., Stanley, N., Pui, D.Y.H. and Kuehn, T.H. (2008) Laboratory and on-road evaluations of cabin air filters using number and surface area concentration monitors. *Environmental Science and Technology*, **42**, 4128–4132.

Qian, F., Zhang, J. and Huang, Z. (2009) Effects of the operating conditions and geometry parameter on the filtration performance of a fibrous filter. *Chemical Engineering and Technology*, **32**(5), 789–797.

Ramarao, B.V., Tien, C. and Mohan, S. (1994) Calculation of single fibre efficiencies for interception and impaction with superposed Brownian motion. *Journal of Aerosol Science*, **25**(2), 295–313.

Raynor, P.C. (2008) Single-fiber interception efficiency for elliptical fibers. *Aerosol Science and Technology*, **42**(5), 347–368.

Rembor, H.J., Maus, R. and Umhauer, H. (1999) Measurements of single fibre efficiencies at critical values of the Stokes number. *Particle and Particle Systems Characterization*, **16**, 54–59.

Rengasamy, A., Zhuang, Z.P. and BerryAnn, M.S. (2004) Respiratory protection against bioaerosols: literature review and research needs. *American Journal of Infection Control*, **32**(6), 345–354.

Romay, F.J., Liu, B.Y.H. and Chae, S.J. (1998) Experimental study of electrostatic capture mechanisms in commercial electret filters. *Aerosol Science and Technology*, **28**(3), 224–234.

Rudell, B., Wass, U., Hörstedt, P. *et al.* (1999) Efficiency of automotive cabin air filters to reduce acute health effects of diesel exhaust in human subjects. *Occupational and Environmental Medicine*, **56**, 222–231.

Sae-lim, W., Tanthapanichakoon, W. and Kanaoka, C. (2006) Correlation for the efficiency enhancement factor of a single electret fibre. *Journal of Aerosol Science*, **37**, 228–240.

Schweers, E. and Löffler, F. (1994) Realistic modeling of the behaviour of fibrous filters through consideration of filter structure. *Powder Technology*, **80**, 191–206.

Shin, W.G., Mulholland, G.W., Kim, S.C. and Pui, D.Y.H. (2008) Experimental study of filtration efficiency of nanoparticles below 20 nm at elevated temperatures. *Journal of Aerosol Science*, **39**, 488–499.

Song, C.B., Park, H.S. and Lee, K.W. (2006) Experimental study of filter clogging with monodisperse PSL particles. *Powder Technology*, **163**, 152–159.

Spurny, K.R. (1997) *Advances in Aerosol Filtration*, Lewis Publisher, London.

Stechkina, I.B. and Fuchs, N.A. (1966) Studies in fibrous aerosol filters – I. Calculation of diffusional deposition of aerosols in fibrous filters. *Annals of Occupational Hygiene*, **9**, 59–64.

Sublett, J.L. (2011) Effectiveness of air filters and air cleaners in allergic respiratory diseases: a review of the recent literature. *Current Allergy and Asthma Reports*, **11**, 395–402.

Thomas, D., Contal, P., Renaudin, V. *et al.* (1999) Modelling pressure drop in HEPA filters during dynamic filtration. *Journal of Aerosol Science*, **30**, 235–246.

Thomas, D., Penicot, P., Contal, P. *et al.* (2001) Clogging of fibrous filters by solid aerosol particles experimental and modelling study. *Chemical Engineering Science*, **56**, 3549–3561.

Thorpe, A. and Brown, R.C. (2003) Performance of electrically augmented fibrous filters, measured with monodisperse aerosols. *Aerosol Science and Technology*, **37**(3), 231–245.

Tien, C.V. and Ramarao, B.V. (2007) *Granular Filtration of Aerosols and Hydrosols*, Elsevier, New York.

Wake, D. and Brown, R.C. (1988) Measurements of the filtration efficiency of nuisance dust respirators against respirable and non-respirable aerosols. *Annals of Occupational Hygiene*, **32**(3), 295–315.

Walsh, D.C. and Stenhouse, J.I.T. (1997) The effect of particle size, charge and composition on the loading characteristics of an electrically active fibrous filter material. *Journal of Aerosol Science*, **28**, 307–321.

Wang, C.S. (2001) Electrostatic forces in fibrous filters – a review. *Powder Technology*, **118**, 166–170.

Wang, J., Chen, D.R. and Pui, D.Y.H. (2007) Modelling of filtration efficiency of nanoparticles in standard filter media. *Journal of Nanoparticle Research*, **9**, 109–115.

Wang, H.C. and Kasper, G. (1991) Filtration efficiency of nanometre-size aerosol particles. *Journal of Aerosol Science*, **22**(1), 31–41.

Wang, Q., Maze, B., Tafreshi, H.V. and Pourdeyhimi, B. (2006) A case study of simulating submicron aerosol filtration via lightweight spun-bonded filter media. *Chemical Engineering Science*, **61**, 4871–4883.

Wang, J. and Pui, D.Y.H. (2009) Filtration of aerosol particles by elliptical fibers: a numerical study. *Journal of Nanoparticle Research*, **11**, 185–196.

Williams, M.M.R. and Loyalka, S.K. (1991) Particle deposition and resuspension. *Aerosol Science: Theory and Practice*, Chapter 7, Pergamon, Oxford, pp. 326–374.

Xu, Z., Wu, Y., Shen, F. *et al.* (2011) Bioaerosol science, technology, and engineering: past, present, and future. *Aerosol Science and Technology*, **45**(11), 1337–1349.

Yamada, S., Takafumi, S. and Yoshio, O. (2011) Influence of filter inhomogeneity on air filtration of nanoparticles. *Aerosol and Air Quality Research*, **11**, 155–160.

Zhu, Y., Eiguren-Fernandez, A., Hinds, W.C. and Miguel, A.H. (2007) In-cabin commuter exposure to ultrafine particles on Los Angeles freeways. *Environmental Science and Technology*, **41**, 2138–2145.

Zhu, C., Lin, C.H. and Cheung, C.S. (2000) Inertial impaction-dominated fibrous filtration with rectangular or cylindrical fibers. *Powder Technology*, **112**, 149–162.

6

Remote Sensing of Atmospheric Aerosols

Sagnik Dey[1] *and Sachchida Nand Tripathi*[2]
[1]*Centre for Atmospheric Sciences, IIT Delhi, India*
[2]*Department of Civil Engineering, IIT Kanpur, India*

6.1 Introduction

Atmospheric aerosols are liquid or solid particles suspended in the atmosphere, with radii varying from a few nanometers to tens of micrometers. These particles can be directly emitted by anthropogenic (e.g. fossil-fuel burning) and natural (e.g. dust, maritime aerosol, volcanic ash) sources or can form from precursor gases (e.g. secondary organic aerosol, sulfates, etc.). Interest in the study of atmospheric aerosols can be dated back to the 1920s (Brav, 1929). Since then, significant improvement has occurred in our understanding of the role of aerosols in air quality (Lelieveld *et al.*, 2002; Chow and Watson, 2007) and the earth's energy budget (Haywood and Boucher, 2000; Yu *et al.*, 2006). Geographically localized sources/sinks and transformations within their short lifetimes (e.g. the mixing of various aerosol species, the hygroscopic growth of particles) lead to extreme spatial and temporal heterogeneity in aerosol optical and microphysical properties.

Aerosols influence the earth's radiative budget through three different mechanisms: first, they directly scatter and absorb solar and terrestrial radiation (Charlson and Pilat, 1969; Charlson *et al.*, 1992; Chylek and Wong, 1995; Haywood and Boucher, 2000; Ramanathan *et al.*, 2001; Boucher and Pham, 2002; Bellouin *et al.*, 2005); second, they act as cloud condensation nuclei (CCN) and thereby affect cloud albedo and lifetime (Twomey, 1977; Albrecht, 1989; Lohmann and Feichter, 2005; Andreae and Rosenfeld, 2008); and third, their direct absorption of solar radiation can alter the thermal structure of the atmosphere

Aerosol Science: Technology and Applications, First Edition. Edited by Ian Colbeck and Mihalis Lazaridis.

and hence cloud formation (Hansen, Sato, and Ruedy, 1997; Ackerman *et al.*, 2000; Koren *et al.*, 2004, 2008; Johnson, Shine, and Forster, 2004). Perturbation of net (downward minus upward) radiation by aerosols through all the three mechanisms is quantified by aerosol radiative forcing. Global mean aerosol direct and indirect (only cloud albedo effect) radiative forcing is virtually certain to be negative, with a median value of $-1.3\,\mathrm{W\,m^{-2}}$ and of -2.2 to $-0.5\,\mathrm{W\,m^{-2}}$, respectively, at 90% confidence range (IPCC, 2007). However, large uncertainty continues to exist in quantifying aerosol radiative forcing because of the poor representation of aerosols in climate models, due to inadequate data on aerosol spectral optical properties at large spatiotemporal scale and the complex interaction between aerosols, clouds, and meteorology (Stevens and Feingold, 2009).

Aerosol concentration in the atmospheric column is quantified by aerosol optical depth (AOD), which is a measure of the columnar extinction (scattering and absorption) of solar radiation by aerosols. Precise measurements of AOD, single scattering albedo (SSA) (a measure of the scattering and absorbing nature of the particles), and phase function (a measure of angular distribution of scattered radiation) are required at a global scale to reduce the uncertainty in aerosol radiative forcing. Advancements in remote sensing techniques in the last 3 decades have provided much-needed data from the local to the global scale, which have been extensively used by scientists, policy makers, and the general public. Over the years, remote sensing applications of aerosols have evolved through experience. As more direct *in situ* observations have become available for the evaluation of remote-sensing-based retrievals, algorithms have been modified to reduce the retrieval error. The sensors, which were originally launched for other purposes, have also been utilized to derive information about aerosol properties. New techniques (e.g. polarimetric measurements, multiangle view) have been proposed to address the unresolved issues in aerosol remote sensing. The purpose of this chapter is to provide a comprehensive overview of these state-of-the-art retrieval techniques, their applicability for climate studies, and future requirements in this field.

6.2 Surface-Based Remote Sensing

The most accurate information about aerosol characteristics can be obtained through ground-based remote sensing, as it does not have to deal with surface reflectance issues. Such techniques have expanded in the last 2 decades, providing valuable data for the validation of satellite-based products and models.

6.2.1 Passive Remote Sensing

There are a number of international and national networks that maintain ground-based monitoring of aerosol properties using passive remote sensing techniques. AOD can be determined from the ground by radiometers that measure the direct spectral transmission of solar radiation through the atmosphere. The solar irradiance I at any given wavelength λ can be expressed as $I_\lambda = I_{0\lambda}\exp(-m\tau)$, where I_0 is the solar irradiance at the top of the atmosphere, m is the air mass, and τ is the total optical depth. τ may be composed of scattering by gaseous molecules (commonly known as Rayleigh scattering), extinction by aerosols, and absorption by trace gases. The contribution by components other than aerosols can be accurately estimated, and subtracting them from τ gives AOD.

The most popular and widespread ground-based network of aerosol remote sensing is the Aerosol Robotic Network (AERONET), which maintains well-calibrated sunphotometers and sky-radiance radiometers (Holben *et al.*, 1998, 2001). The radiometers measure direct sun radiance in eight spectral bands (between 340 and 1020 nm) across 10 seconds. A sequence of three such measurements is taken at 15-minute intervals in order to minimize the influence of clouds on the measured radiance. Sky-radiance measurements in the almucanter and principal plane are carried out at 440, 670, 870, and 1020 nm wavelengths in order to acquire a large range of scattering angles for the retrieval of size distribution, phase function, and SSA (Dubovik and King, 2000). AERONET-retrieved AOD is highly accurate, with an uncertainty $<\pm 0.01$ for wavelengths greater than 440 nm and $<\pm 0.02$ for shorter wavelengths. The aerosol optical properties are retrieved with greater accuracy for $AOD_{440} \geq 0.5$ with solar zenith angle $> 50°$ (Dubovik *et al.*, 2000). AERONET data have been used by numerous researchers to examine aerosol characteristics and composition in several regions of the world.

6.2.1.1 *Aerosol Optical Properties*

Major aerosol types are characterized by their microphysical (particle size and shape) and optical (ability to scatter and/or absorb radiation) properties. For example, both urban/industrial and biomass-burning aerosols are small in size and spherical in shape, but the biomass aerosols have much lower SSA (i.e. more absorbing) than urban/industrial aerosols. Natural aerosols (e.g. dust and maritime) are large in size, but dust particles have lower SSA than maritime particles and they are nonspherical in shape. Thus, if these properties can be measured by remote sensing techniques, dominant aerosol types can be identified. However, it must be noted that various individual particles often mix with one another, altering the optical properties of composite aerosols. The optical properties for key aerosol types have been summarized by Dubovik *et al.* (2002), utilizing the AERONET measurements. These results demonstrate the range of variability in aerosol optical properties across geographic regions, characterized by various anthropogenic and natural sources. Both biomass-burning and urban aerosols are characterized by a dominance of accumulation-mode particles and spectrally decreasing SSA. Desert dust aerosols are characterized by a dominance of coarse-mode particles and spectrally increasing SSA. However, absorption by dust particles varies depending on the hematite content (Koven and Fung, 2006; Mishra and Tripathi, 2008) and mixing with other species (e.g. Dey, Tripathi, and Mishra, 2008; Mishra *et al.*, 2012). On the other hand, maritime aerosols are dominantly coarse-mode particles and scattering in nature. Oceans are natural test beds for quantifying the seasonal changes in regional aerosol properties, due to the seasonal transport of aerosols from the continents. The need to establish a robust climatology of maritime particles has led to the development of a Maritime Aerosol Network under AERONET (Smirnov *et al.*, 2009). The globally averaged maritime AOD at 500 nm and the Angstrom exponent (AE, for the wavelength range 440–870 nm) derived from island-based AERONET measurements are 0.11 and 0.6, respectively. Mean annual AOD is observed to be low (<0.10) in the Pacific Ocean, but varies spatially and seasonally over the Atlantic and Indian Oceans (Figure 6.1). Cape Verde and Barbados site in the Atlantic

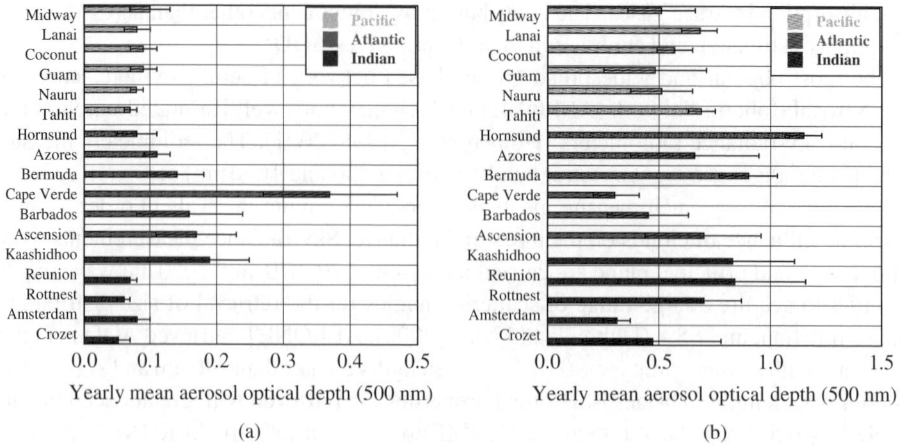

Figure 6.1 *Mean annual (a) AOD at 500 nm and (b) Angstrom parameter, at various island-based AERONET sites. Reproduced with permission from Smirnov* et al. *(2009). Copyright © 2009, John Wiley and Sons Ltd. See plate section for colour version.*

Ocean are strongly affected by Saharan dust transport, while the Kaashidhoo site in the Indian Ocean is affected by pollution transport from the Indian subcontinent.

AE derived from spectral AOD measurements by ground-based radiometers also provides a qualitative indication of the dominant size, with values greater than 2 indicating small particles associated with combustion products and values less than 1 indicating large particles such as dust or maritime aerosols (Schuster, Dubovik, and Holben, 2006). Temporal changes in the dominant aerosol types are inferred based on changes in AE (Dey *et al.*, 2004; Moorthy, Babu, and Satheesh, 2007; Kaskaoutis *et al.*, 2010). However, AE also changes with wavelength, and the second-order AE (quantifying the change of AE with wavelength) provides additional information about aerosol size distribution. For example, AE calculated from longer wavelength pairs ($\lambda = 670$ and 870 nm) is sensitive to the fine-mode fraction, while AE calculated from shorter wavelength pairs ($\lambda = 380$ and 440 nm) is sensitive to the fine-mode effective radius (Schuster, Dubovik, and Holben, 2006). The curvature in the spectral distribution of AODs depends strongly on atmospheric turbidity (Kedia and Ramachandran, 2009). At low AOD (~0.2), AE increases fourfold when the wavelength range is changed from shorter (400–500 nm) to longer (750–875 nm). The variability becomes high when mixing of various aerosol species occurs, altering the spectral AOD and absorption properties. An analysis of multiyear AERONET data at two urban sites in Asia (Kanpur and Beijing) revealed a similar spectral variation of SSA at a low (~0.15) fine-mode fraction (Figure 6.2). This indicates the dominance of desert dust particles at these urban sites, which have similar hematite contents despite having different dust source regions (Eck *et al.*, 2010). Both these sites have nearly equal SSAs at a high (~0.85) fine-mode fraction, suggesting a similar nature of urban aerosols. At an intermediate fine-mode fraction (~0.45), characteristic of the mixing of fine- and coarse-mode particles, the SSA at near-infrared wavelength is smaller in Beijing than in Kanpur. This implies that coarse-mode dust particles are more absorbing in Beijing than in Kanpur, which may result from a greater rate of black carbon (BC) coating of dust particles in Beijing. In another

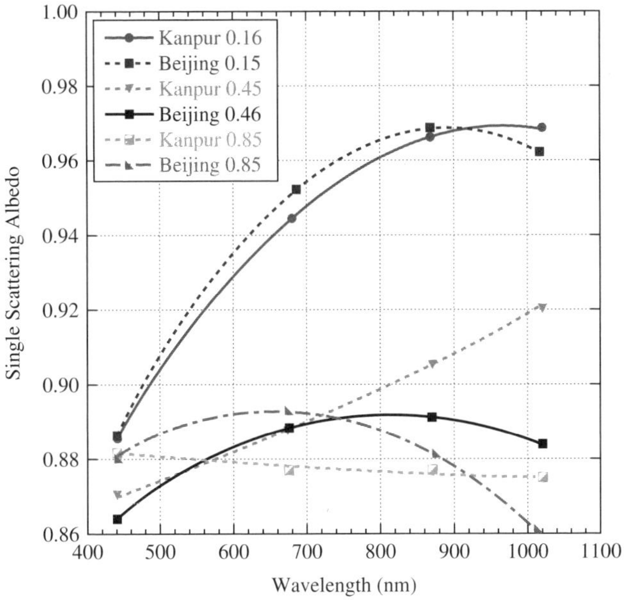

Figure 6.2 *Comparison of spectral SSAs at Kanpur (26.5°N, 80.2°E) and Beijing (39.9°N, 116.4°E) for low (~0.15), medium (~0.45), and high (~0.85) fine-mode fractions. Reproduced with permission from Eck* et al. *(2010). Copyright © 2010, American Geophysical Union. See plate section for colour version.*

AERONET-based study, Giles *et al.* (2012) observed higher absorption by dust particles mixed with smoke than by pure dust.

6.2.1.2 Aerosol Composition

The scarcity of aerosol composition data at the global scale hinders in the improvement of climate models and this has led scientists to explore the utilization of measured spectral aerosol optical properties (by sunphotometers or any other radiometers) to infer aerosol composition. In recent years, two different approaches have been adopted. In the first, a combination of aerosol parameters is used to infer aerosol composition. For example, Gobbi *et al.* (2007) have proposed a method for characterizing urban, biomass-burning, and dust aerosols using the spectral curvatures of AOD and the AE (an indicator of dominant aerosol size). Fine-mode-fraction and SSA measurements by sunphotometers are used to separate out dust, maritime, nonabsorbing, and absorbing anthropogenic aerosols (Lee *et al.*, 2010). Statistics show that North and South Asia, South America, and South Africa are dominated by absorbing aerosols emitted by anthropogenic activities and biomass burning (annual occurrences of 66.2, 89.2, 76.8, and 94.5%, respectively). The dust emission from the Saharan Desert dominates the air mass in North Africa (73.8% annual occurrence), nonabsorbing urban/industrial aerosols dominate North American sites, while Europe has mixed air mass. A similar approach was taken by Srivastava *et al.* (2012) to distinguish air masses dominated by various aerosol types (e.g. polluted dust, clean dust, polluted continental, high-absorbing, and low-absorbing) over the polluted Indo-Gangetic basin in Northern India during the pre-monsoon season (Mar-May).

In the second approach, composite aerosols are assumed to be a mixture of various individual species. The composite refractive indices are estimated theoretically by varying the volume fractions of the individual components constrained by AERONET-retrieved refractive indices. The volume fractions are then multiplied by the density of the species of interest and AERONET-retrieved total volume concentration to derive the columnar mass concentration. Schuster *et al.* (2005) implemented this idea by applying the Maxwell–Garnett mixing rule to an aerosol mixture containing BC, ammonium sulfate, and water and calculated the complex refractive indices. The simulations constrained by AERONET measurements have allowed an estimation of columnar BC concentrations at 46 AERONET sites across the world. The largest BC concentrations at biomass-burning sites and the lowest concentrations at the remote island sites (Figure 6.3) qualitatively justify the approach. Dey *et al.* (2006) have extended the methodology by including partly absorbing organic carbon and dust in the mixture in order to examine their impacts on the retrieval of BC at a site in the Indo-Gangetic Basin in northern India. Absorbing organic carbon concentrations have also been quantified using the same approach (Arola *et al.*, 2011). Koven and Fung (2006) have adopted the approach to infer the hematite content (which determines the absorptive nature of dust) of dust aerosols at AERONET sites dominated by desert dusts. They concluded that the calculated hematite content is highly sensitive to mixing assumptions. External mixing overestimates hematite content, while internal mixing underestimates it (Figure 6.4). A higher hematite content is observed in Saharan and East Asian dust than in Arabian dust. These results provide an alternative approach (in the absence of robust *in situ* data) to improve the quantification of dust absorption and its impact on direct radiative forcing.

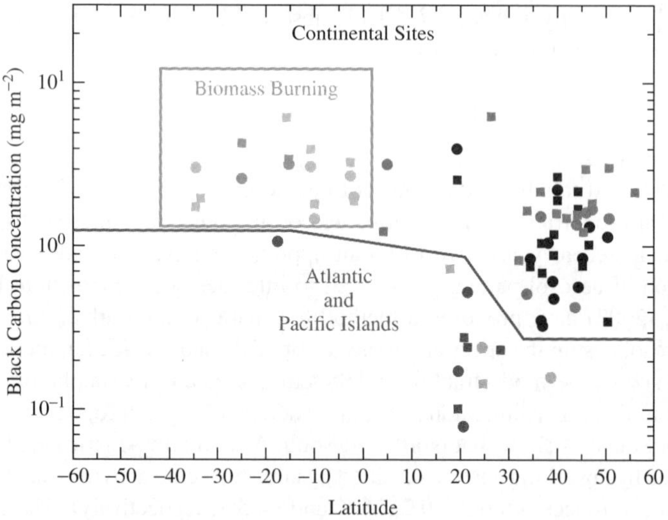

Figure 6.3 *Black carbon (BC) concentration at various AERONET sites in North America (red), Europe (gray), Asia (purple), South America (green), Africa (orange), Atlantic Islands (light blue), and Pacific Islands (dark blue) for the year 2000 (circles) and 2001 (squares). Reproduced with permission from Schuster* et al. *(2005). Copyright © 2005, American Geophysical Union. See plate section for colour version.*

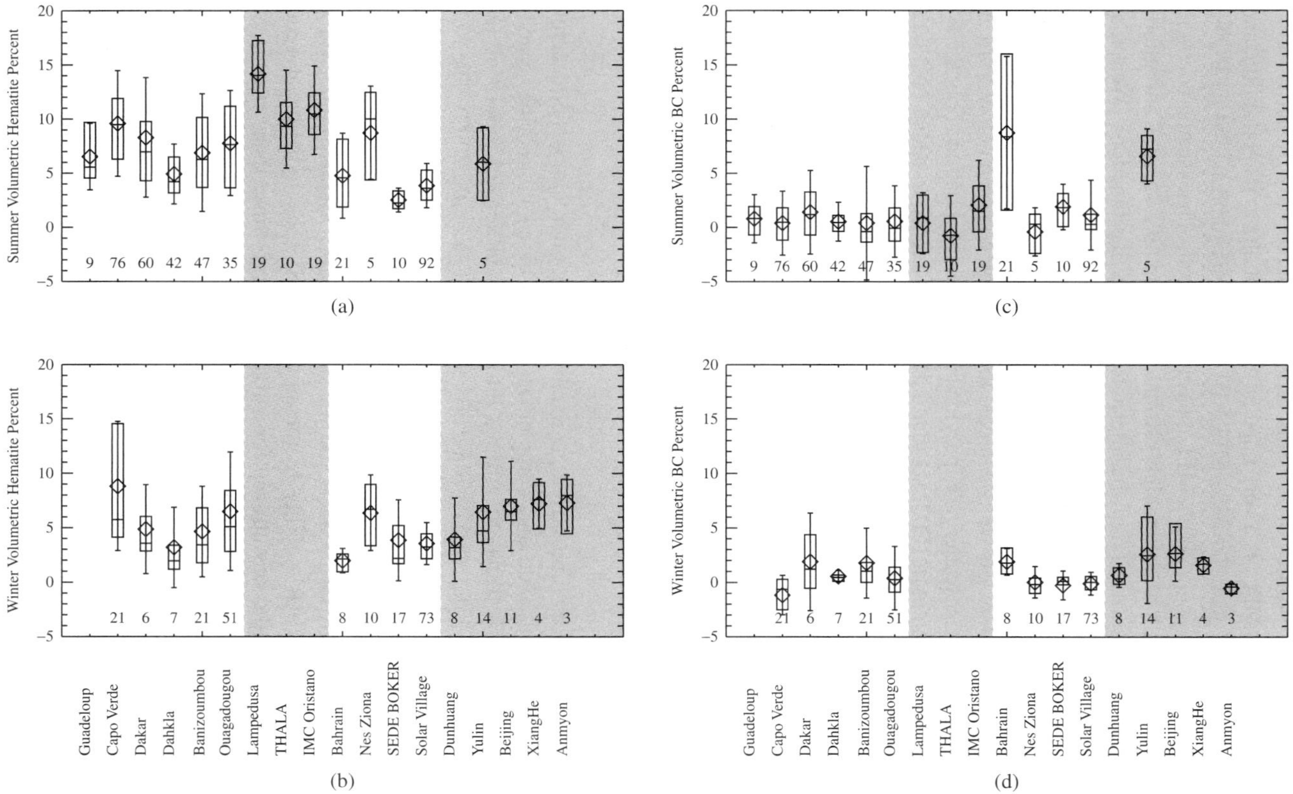

Figure 6.4 *Hematite content in aerosols at various AERONET stations during the summer–fall season (top panel) and the winter–spring season (bottom panel), using external (left panel) and internal mixing (right panel). Reproduced with permission from Koven and Fung (2006). Copyright © 2006, American Geophysical Union.*

6.2.2 Active Remote Sensing

Surface-based passive remote sensing can provide information about columnar aerosol properties, but the vertical profiles of aerosols can be measured only by active remote sensing. Micro-pulse lidar (MPL) is a very popular instrument due to its portability, and a network (the Micro-Pulse Lidar Network, MPLNET) has been established at AERONET sites over the years (Welton *et al.*, 2001). MPL measures the aerosol backscatter coefficient profile at 532 nm wavelength at a very high vertical resolution (30–75 m). Calibration of the MPL system involves correction of the afterpulse and overlap functions (Welton and Campbell, 2002). Ground-based lidars have been used to measure aerosol profiles in continuous and campaign modes (Welton *et al.*, 2000). In the Indian Ocean Experiment, MPL data were utilized to elucidate the temporal evolution of marine boundary layer heights during the winter season. Aerosols were observed to be mostly confined below 1 km altitude, while the extinction backscatter ratio provided additional information about the type of air mass sampled (Welton *et al.*, 2002). Lidar data were also used to detect dust layers and mixing with other spherical particles (Liu *et al.*, 2011a).

These datasets are also a valuable tool in evaluating spaceborne active remote sensing measurements (e.g. Campbell *et al.*, 2012; Misra *et al.*, 2012). Besides these studies on the vertical profiles of aerosols, lidar data are also used to constrain satellite-based estimates of the surface concentrations of aerosols for health-impact studies (Liu *et al.*, 2011b).

6.3 Satellite-Based Remote Sensing

Satellites provide the space–time coverage required to monitor atmospheric aerosol concentration at the global scale and complement ground-based monitoring. Satellite remote sensing applications began to retrieve information on aerosol properties (e.g. AOD or absorbing aerosol index (AAI)) some 30 years ago (Lee *et al.*, 2009). The Multi Spectral Scanner (MSS) onboard the Earth Resources Technology Satellite (ERTS-1) was the first sensor used to retrieve AOD (Griggs, 1975). However, the Advanced Very High Resolution Radiometer (AVHRR) onboard the TIROS-N satellite was the first to be used for operational aerosol products (Stowe *et al.*, 2002). Originally launched for weather observations, AVHRR's ability to measure reflected solar radiation at visible and mid-infrared wavelengths was utilized for aerosol retrieval. Since then, a number of passive sensors have been launched to measure aerosols over ocean and land (Figure 6.5). Early passive sensors were developed for other purposes (e.g. Total Ozone Mapping Spectrometer, TOMS was launched onboard the Nimbus 7 satellite in order to detect ozone, but its ultraviolet (UV) bands were used to measure absorbing aerosols), but their measurement capabilities have been utilized to retrieve aerosol properties, with varying success. Our understanding of the limitations of existing sensors in measuring aerosol properties has led to the development of more robust techniques. Recently, an active remote sensing technique was utilized to retrieve aerosol vertical distributions, adding a new dimension to aerosol remote sensing. This section summarizes the passive and active remote sensing applications in aerosol studies over the last 3 decades.

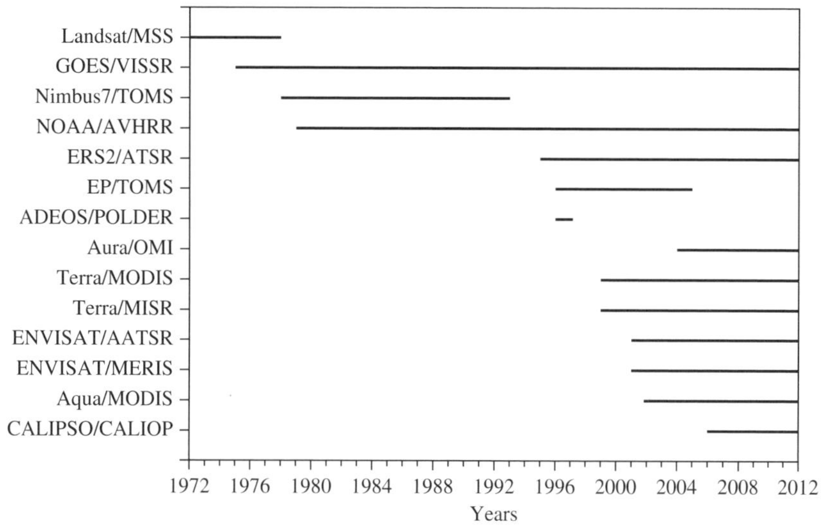

Figure 6.5 *Temporal coverage of various satellites/sensors utilized to monitor aerosols.*

6.3.1 Passive Remote Sensing

Satellite observations are carried out from different orbits, view angles, and solar zenith angles, which vary with time of the day and season. Satellite-based retrievals of aerosol properties by passive sensors account for the viewing geometry in the radiative transfer model. The retrieval algorithms follow several steps. The first is the cloud screening, where a pixel is identified as 'clear' or 'cloudy'. High reflectance of clouds in the pixel overshadows the 'aerosol' reflectance and hence aerosol cannot be detected in a cloudy pixel by passive remote sensing (Ackerman *et al.*, 1998). Various cloud-masking techniques have been developed, based on the sensor characteristics. The next step is to separate the surface reflectance from clear-sky atmospheric reflectance. The contribution of molecular scattering and absorption to atmospheric reflectance is estimated using a radiative transfer model. Look-up tables (LUTs) are prepared in order to expedite the retrieval, with TOA radiance theoretically calculated for a set of aerosol models, taking into account the sun and sensor geometry. Aerosol properties are then retrieved by comparing the measured TOA radiance with the theoretical radiance values from the LUTs. Retrieved aerosol products at high resolution are reported as level 2 data. All level 2 pixels within a larger domain (typically at 1.0° or 0.5° grids) are averaged and reported as level 3 data, which are distributed on 'daily' and 'monthly' timescales. Aerosol remote sensing is currently carried out using UV and visible–near-infrared (VIS-NIR) wavelengths.

6.3.1.1 Aerosol Remote Sensing Using Near-UV

Aerosol detection capability in the near-UV wavelengths first became apparent with the development of the AAI, as a byproduct of the TOMS version 7 ozone algorithm (McPeters *et al.*, 1996). The low and relatively invariant UV albedo of the earth's surface allows easy detection of aerosols over ocean and land (Herman and Celarier, 1997). TOMS provides daily coverage of most of the earth's surface at 50×50 km resolution at nadir. Aerosol measurements are carried out at the three longest wavelengths (340, 360, and 380 nm), at which gaseous absorption is weak and backscattered radiation is primarily influenced by Rayleigh scattering, surface reflection, and scattering from aerosols and clouds. AAI is defined as:

$$AAI = -100 \log_{10} \left[(I_{340}/I_{380})_{meas} - (I_{340}/I_{380})_{calc} \right]$$

where I_{meas} is the measured backscattered radiance at a given wavelength and I_{calc} is the radiance calculated theoretically at that wavelength using an atmospheric model, assuming a pure gaseous atmosphere. The difference between the measured and calculated radiances is attributed to aerosols (in clear sky). Nonabsorbing aerosols (e.g. sulfates and maritime particles) yield negative AAI values, while absorbing aerosols (e.g. dust and smoke) yield positive values.

AAI is sensitive to the height of the aerosol layer (Mahowald and Dufresne, 2004) and hence has been extensively utilized in the study of global distributions of UV-absorbing mineral dust aerosols. Using 13 years of TOMS AAI data, Prospero *et al.* (2002) have identified the major global dust sources based on the frequency of high AAI (>0.7). On a global scale, the concentration of absorbing aerosols is higher in July than in January (Figure 6.6), and large sources are confined to the northern hemisphere. The absorbing aerosols are mostly mineral dust, except in southern Africa and the adjacent ocean, where the high AAI values are attributed to smoke (Hao and Liu, 1994). The index has also been used to study regional distributions of dust (e.g. Dey *et al.*, 2004) and smoke (e.g. Gleason, Hsu, and Torres, 1998).

In addition to AAI, AOD is also retrieved using UV wavelengths (Torres *et al.*, 1998). The radiance measured by the sensor at the top of an aerosol-laden atmosphere consists of three terms: the radiance scattered upward by the aerosol layer, the fraction of Rayleigh-scattered and surface-reflected radiance below the aerosol layer, and the fraction of Rayleigh-scattered and surface-reflected radiance transmitted through the aerosol layer that is unaffected by aerosol absorption (Torres *et al.*, 2002). Near-UV remote sensing of aerosols is different from VIS/NIR remote sensing in two aspects: first, the contribution of Rayleigh scattering to total radiance cannot be neglected in the near-UV region and varies depending on the height of the absorbing and scattering aerosol layer; and second, the low surface reflectance at near-UV allows for the retrieval of aerosols using similar algorithms over both water and land, with less error. Since the sensitivity of TOA radiance to aerosol height decreases with decreasing aerosol absorption, near-UV remote sensing is most suitable for the detection of absorbing aerosols (Torres *et al.*, 1998).

The longest global record of AOD is provided by aerosol retrieval using near-UV remote sensing technique by TOMS sensor onboard Nimbus 7 (October 1978–April 1993), Meteor3 (August 1991–December 1994), and EP-TOMS (July 1996–December 2005). Figure 6.7 shows the inter-annual variability of latitudinal distribution of near-UV AOD that captures the broad regional features. High AOD at ~50°N corresponds to boreal forest fires, while tropical biomass burning and desert dusts contribute to large AOD in the tropics. The influence of two major volcanic eruptions (El Chichon in 1982 and Mt Pinatubo in 1991) on global AOD is also evident. This data set can be utilized for further regional-scale study (e.g. Massie, Torres, and Smith, 2004).

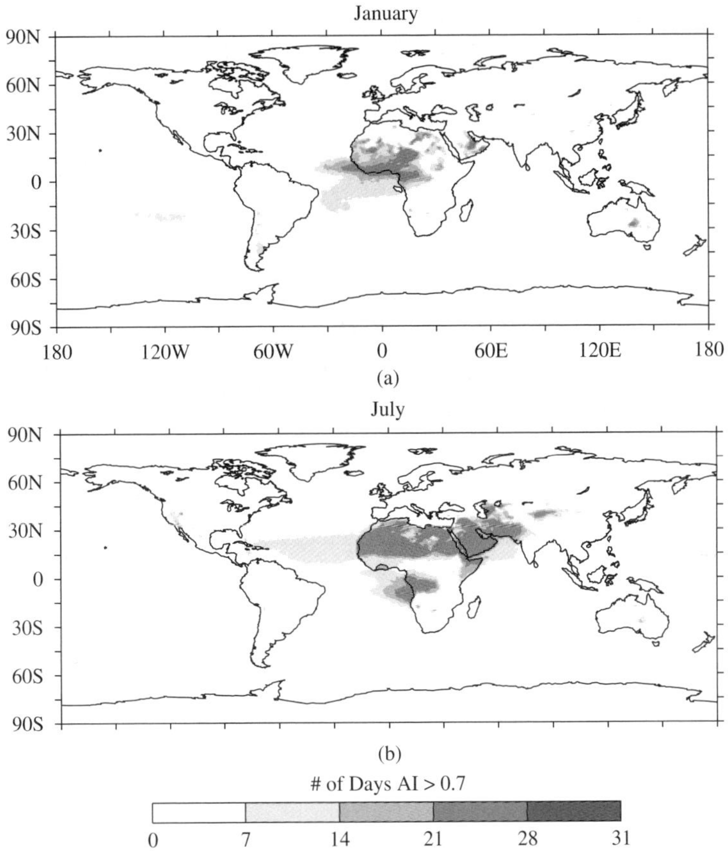

Figure 6.6 *Global distributions of dust and smoke: monthly frequency of occurrence of TOMS absorbing aerosol product over the period 1980–1992. (a) January. (b) July. The scale is the number of days per month on which the absorbing aerosol index (AAI) equals or exceeds 0.7. The large dark area in southern Africa in July is a result of biomass burning. In January, there is biomass burning in the region just north of the equator in Africa; part of the plume over the equatorial ocean is due to smoke. Essentially, all other distributions are a result of dust. Reproduced with permission from Prospero* et al. *(2002). Copyright © 2002, American Geophysical Union.*

6.3.1.2 Aerosol Remote Sensing Using VIS-NIR

Satellite-based passive remote sensing of aerosols using the VIS-NIR wavelengths has gained prominence over the years due to the straightforward use of retrieved mid-visible AOD in climatic applications. However, different retrieval techniques are employed over water and land, in order to account for the surface reflectance effect.

GACP Dataset. AVHRR instruments onboard NOAA weather satellites are another source of long-term records of aerosol properties. The radiances of channel 1 and 2 of AVHRR were used to retrieve AOD and AE over global oceans and archived as the Global Aerosol Climatology Project (GACP) dataset, covering the period August 1981 to

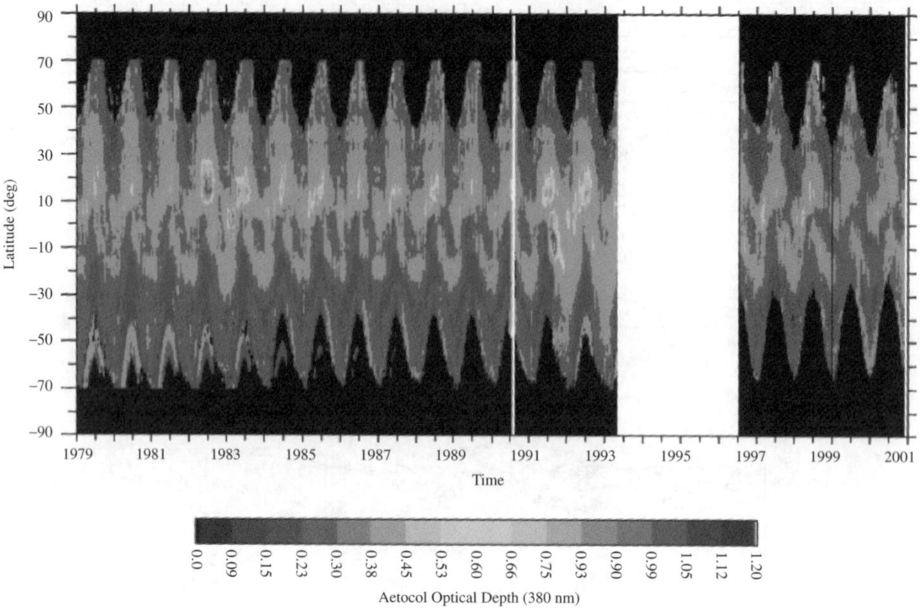

Figure 6.7 *Time series of the 380 nm aerosol optical depth during the period of operation of the TOMS instruments onboard the Nimbus 7 (1979–1992) and Earth Probe satellites (1996–2001). Weekly averages plotted over a 1 × 1° geographical grid. Reproduced from Torres* et al. *(2002) (with permission from American Meteorological Society). See plate section for colour version.*

December 2005 (Mishchenko *et al.*, 1999, 2007a, 2012; Geogdzhayev *et al.*, 2002, 2005). The GACP algorithm is based on a modified power law size distribution within the range 0.1–10.0 μm (Mishchenko *et al.*, 1999), where particles are assumed to be perfect spheres with complex refractive index $m = 1.5 + 0.003i$. This approach minimizes the long-term statistical errors of retrieved AOD and AE, but may have regional biases due to the mixing of aerosols from various sources.

The AOD retrieved by AVHRR shows similar global distribution (Figure 6.8) to AOD retrieved by Moderate Resolution Imaging Spectroradiometer (MODIS) and Multiangle Imaging SpectroRadiometer (MISR), which use multichannel retrieval algorithms. Dust outflows from the Saharan Desert to the Atlantic Ocean and large AODs over the Indian Ocean, near the coast and East Asia, are captured by all three sensors. However, the magnitude of AOD varies; AVHRR shows smaller AOD than the other two sensors in high-AOD regions affected by dust and transported continental aerosols. The different cloud screening and retrieval algorithms used by these sensors may also contribute to the discrepancy (Geogdzhayev *et al.*, 2004; Mishchenko *et al.*, 2007a). Nonetheless, the GACP dataset provides a valuable long-term record of AOD over global oceans. In a more recent work, Mishchenko *et al.* (2012) have shown that increasing the complex part of the refractive index in the AVHRR algorithm eliminates the previously identified long-term decreasing trend in global AOD (Mishchenko *et al.*, 2007b). This implies that for a real trend, the mean global aerosol absorption must be doubled as compared to the existing algorithm, which translates to a decrease of mean global SSA from ~0.95 to ~0.88.

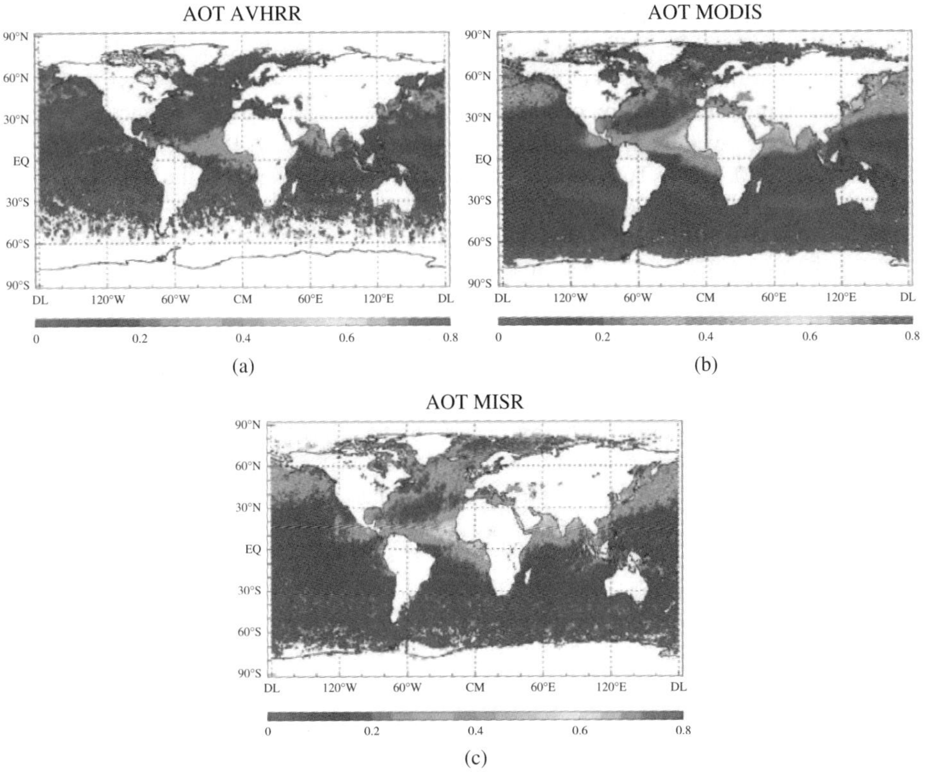

Figure 6.8 *Global distribution of aerosol optical depth for the period March to July 2000 from three passive sensors: AVHRR, MODIS, and MISR. Reproduced with permission from Geogdzhayev* et al. *(2002). Copyright © 2004, Elsevier Ltd. See plate section for colour version.*

MODIS and MISR Aerosol Products. A new era in satellite remote sensing of aerosols began with the launch of the polar-orbiting satellite EOS-Terra in the year 1999 (Figure 6.5). EOS-Terra carries five sensors, among them MODIS and MISR, which provided the first true global coverage of AOD. Later, MODIS was also launched onboard the Aqua satellite, in 2002, thereby increasing the temporal coverage. MODIS measures radiance at 36 wavelength bands (spanning from 0.415 to 14.5 μm) within a swath of ~2330 km at moderately high spatial resolution (0.25, 0.5, and 1.0 km, depending on the bands).

The MODIS 'dark-target' aerosol retrieval algorithm was developed to quantify clear-sky aerosol properties (AOD and fine-mode fraction) over land surfaces with a low VIS-NIR reflectance (Levy, Remer, and Dubovik, 2007). The algorithm uses two visible (0.47 and 0.65 μm) and one short-wave infrared (SWIR, 2.1 μm) bands, which are nearly transparent to gaseous absorption and show a consistent spectral relationship over the surface of land (Kaufman *et al.*, 1997). Aerosol is considered to be a mixture of fine- and coarse-mode particles, whose microphysical and optical properties are obtained based on a cluster analysis of AERONET data (Levy, Remer, and Dubovik, 2007). The algorithm considers three fine aerosol models that assume spherical particles with weak (SSA at 0.55 μm $\approx$ 0.95), moderate (SSA $\approx$ 0.91), and strong (SSA $\approx$ 0.86) absorption, and one coarse model that assumes nonspherical particles of SSA $\approx$ 0.95 (essentially representing dust). A LUT is created in which the TOA radiance values for different aerosol models plus Rayleigh scattering over a dark surface are theoretically estimated using radiative transfer model. Earlier (collection 4), the algorithm assumed negligible contribution of radiance by aerosols at a 2.1 μm channel, allowing the separation of surface reflectance from aerosol reflectance at the two visible channels, given a simple ratio of surface reflectance at VIS and SWIR channels (Remer *et al.*, 2005). The TOA radiance from LUT is compared with the measured radiance at 0.47 and 0.65 μm channels in order to retrieve AOD, which is then interpolated to calculate AOD at 0.55 μm and reported as level 2 data.

The present algorithm (collection 5) considers spectrally consistent atmospheric radiance for each aerosol model coupled with Rayleigh scattering (Levy, Remer, and Dubovik, 2007), which allows direct-retrieval AOD at 0.55 μm and fine-mode fraction. AE is determined from AOD measurements at 0.47 and 0.65 μm. The MODIS collection 5 aerosol product is reported as level 2 data at 10 km spatial resolution. Along with the aerosol products, the quality of the retrieval (denoted QA) is also reported, based on a series of tests in the form of a quality-control flag (QAC). QAC summarizes the relative confidence in the entirety of the aerosol product, ranging from 0 (low or no confidence) to 3 (high confidence). Larger QAC values are more useful in the scientific applications outlined in Section 6.3.3.

Global validation of the previous version of aerosol product (Remer *et al.*, 2005) revealed that >66% or one standard deviation of the over-land collocation points was contained within the expected-error (i.e. $\pm[0.05 + 0.15\mathrm{AOD}]$) envelope. In collection 5, 69% of the retrieved AODs lie within the expected-error envelope, with a closer match to the 1 : 1 line and a higher correlation relative to collection 4. However, categorization of the retrieved AOD using QAC in the global validation revealed that regression becomes more symmetric with the 1 : 1 line with an increase in QAC and that the percentage of data points lying within the expected error envelope increases from 50.34% for QC = 0 to 66.1, 67.75, and 72.6% for QAC = 1, 2, and 3, respectively (Levy *et al.*, 2010). The variation of the bias in MODIS-AOD in comparison to AERONET is shown in Figure 6.9 for all collocated QAC = 3 data points. Across a wide range of AODs, the mean bias (marked by black boxes) of MODIS-AOD is less than 0.01 and the 66% interval (i.e. $1-\sigma$, shown as red lines) and the expected error envelopes (shown as green lines) are almost identical. This further establishes the importance of quality flag for scientific applications.

MISR, launched aboard the NASA EOS-Terra satellite (along with MODIS), is unique in having a combination of a high spatial resolution, a wide range of along-track view angles, and a high accuracy of radiometric calibration and stability (Diner *et al.*, 1998). MISR measures upwelling short-wave radiance at 446, 558, 672, and 867 nm using nine cameras, spread out in the forward and aft directions along the flight path at 70.5°, 60.0°, 45.6°, 26.1°, and nadir. A 380 km-wide swath of the earth is successively viewed by each of the nine cameras within 7 minutes, allowing the measurement of a very large range of scattering angles, which enables the retrieval of aerosol microphysical properties (Kahn *et al.*, 2010). The multiangle view enhances the sensitivity to optically thin aerosol layers and allows the retrieval algorithm to differentiate surface and atmospheric contributions to TOA radiance.

MISR retrieves AOD and other microphysical properties (size and shape of the particles and SSA) at 17.6×17.6 km resolution as a level 2 product, by analyzing data from 16×16 pixels of 1.1 km resolution (Kahn *et al.*, 2009). The retrieval strategy assumes a laterally homogeneous distribution of aerosols within the 17.6×17.6 km region and so no retrieval

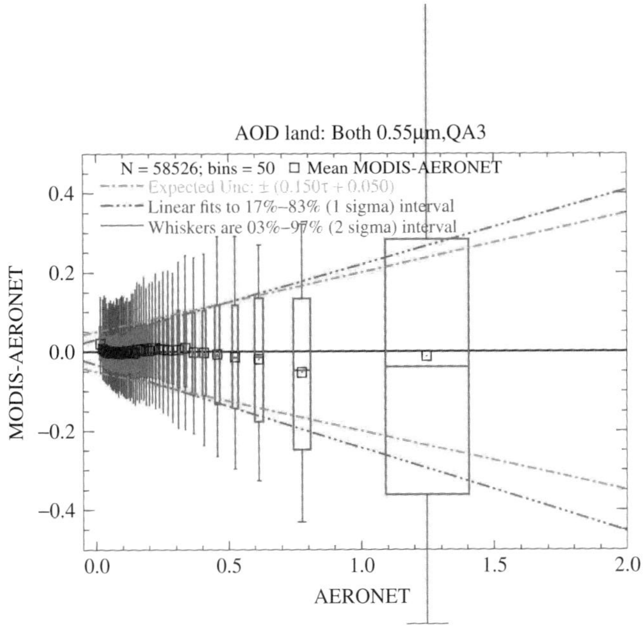

Figure 6.9 *Absolute error of AOD (MODIS-AERONET) vs. AERONET-AOD at 0.55 μm, for QAC = 3. Data sorted by AERONET AOD and grouped into 50 equal bins. Each box plot represents the statistics of the MODIS-AERONET differences in the bin. The means and standard deviations of the AERONET AOD are the centers and half-widths in the horizontal (red). The mean, medians, and 66% (1 − σ) interval of the MODIS-AERONET differences are the black squares, with the center and top–bottom red intervals in the vertical (also red). The blue whiskers are the 96% (2 − σ) intervals. The red dashed curves are linear best fits to 1 − σ intervals and the green dashed curves are the over-land expected errors for total AOD ± (0.05 + 0.15). Reproduced with permission from Levy* et al. *(2010). Copyright © 2010, Levy* et al. *See plate section for colour version.*

is performed over land when the surface topography is determined to be complex (e.g. in mountainous terrain). The latest (Version 22) algorithm assumes 74 types of modeled aerosol mixture in the LUT. The radiance of each of these mixtures is matched with the measured radiance and the particle properties of the mixture with the lowest chi-square value of all the successful mixtures passing the threshold test during radiance matching are reported as 'best estimates'. The aerosol product includes spectral AOD and SSA at all four MISR channels, the fraction of AOD due to 'fine' (particle radius < 0.35 μm), 'medium' (radius between 0.35 and 0.7 μm), and 'coarse' (radius > 0.7 μm) size, and the fraction of AOD due to 'spherical' and 'nonspherical' particles (Kahn *et al.*, 2009). This version improves upon the previous versions with the incorporation of a more realistic medium-mode desert dust (Kalashnikova *et al.*, 2005), spherical particles with mid-visible SSA in the range 0.8–0.9, and more multimodal size distributions, among other things. Many of these properties are unique to MISR.

The global validation of MISR AOD is summarized in Figure 6.10, where 3.7% of data are outliers and are not included in the statistics. Most of the outliers are due to sampling

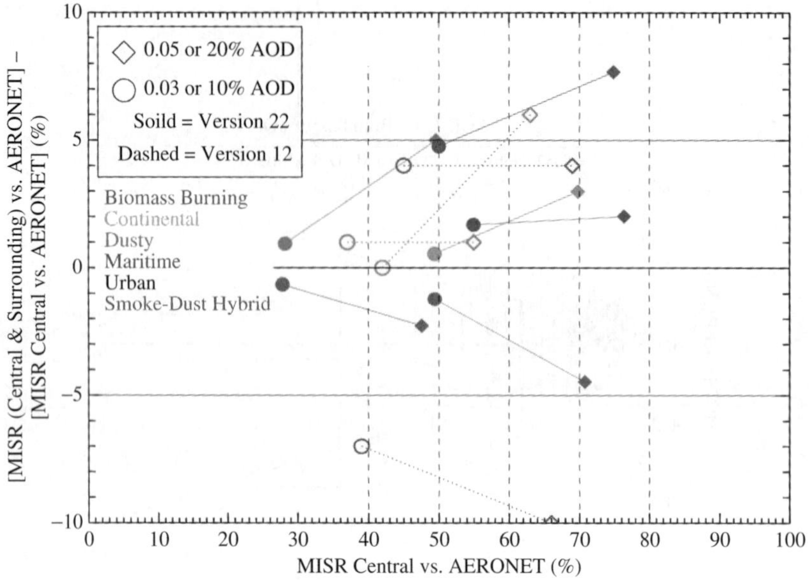

Figure 6.10 *MISR AOD bias relative to AERONET (in %) for all coincident samples stratified according to the dominant aerosol air-mass type. Filled diamonds and circles represent the class-averaged percentage meeting regular (0.05 or 20% × AOD) and more stringent (0.03 or 10% × AOD) criteria, respectively. Open symbols represent class average results for the earlier version of the product from Kahn* et al. *(2005). Figure adapted from Kahn* et al. *(2010) (with permission from AGU). See plate section for colour version.*

differences (Kahn *et al.*, 2010). The quality of the retrieval varies depending on the major aerosol type. For example, 70–75% of data fall within 0.05 or 20% × AOD of the validation data and about 50–55% meet the stricter criteria, except for the 'dust' and 'hybrid' categories. Version 22 values are 10, 7, and 6% higher than those for earlier versions for the 'biomass-burning', 'continental', and 'maritime' categories, respectively, suggesting improvements in the retrieval. The validation also suggests a requirement for a mixture of smoke and dust in the LUT. Validation with AERONET revealed that MISR overestimates AE, due to the absence of any channel greater than 867 nm and the limited use of medium-sized spherical particles in the LUT. Like MODIS, MISR also provides additional information (e.g. quality of retrieval, number of successful mixtures passing the chi-square test, retrieval success flag, etc.) that helps in understanding the quality of the retrieval and interpretation of the data for scientific applications.

Applicability and Intercomparison of Satellite-Based Aerosol Products. Due to the increased attention being given to the effect of aerosols on the climate, the number of satellite-based aerosol products has increased rapidly in the last decade. In addition to those mentioned in the previous sections, many other sensors (e.g. SeaWiFS, VIRS, AATSR, ATSR-2, POLDER) also provide aerosol products. However, the availability of multiple datasets for similar applications also increases the possibility of misinterpretion. Many recent studies have focused on the intercomparison of various satellite-based aerosol products and revealed a large discrepancy between them (e.g. Myhre *et al.*, 2004, 2005; Liu and Mishchenko, 2008; Li *et al.*, 2009; Mishchenko *et al.*, 2009). Discrepancy is not unexpected as the quality of the retrieval depends on numerous factors (including instrument calibration, cloud masking techniques, algorithms, and the treatment of surface reflectance). Retrieval algorithms have been modified over the years to improve the quality of the aerosol product but a lack of proper interpretation of data quality and its applicability still exists among the user community. Often AODs from multiple sensors, averaged over monthly or seasonal timescales, are intercompared without consideration of the spatial and temporal density of measurements relative to the gradients of true AOD (Levy *et al.*, 2009).

Kahn *et al.* (2009, 2011) have demonstrated the right approach for the interpretation of multi-sensor inter-comparison, using coincident samples. Key issues that should be considered are: (i) the collocation of measurements, (ii) the treatment of outliers, and (iii) the choice of an absolute or relative criterion, depending on the situation at hand. For example, despite being onboard the same satellite, Terra, MISR, and MODIS do not retrieve AODs at the exact same location. This can lead to sampling differences, particularly in regions with a large spatial variability in AOD (Kahn *et al.*, 2011). The three clusters of outliers in the MISR-MODIS comparison (Figure 6.6 of Kahn *et al.*, 2009) correspond to North-Central Africa, the eastern part of the Indo-Gangetic Basin, the Patagonia Desert in South America, and North-Central Australia. The discrepancy in North-Central Africa stems from the lack of a smoke–dust mixture in the MISR retrieval algorithm, while neither MODIS nor MISR considers the appropriate SSA when retrieving aerosols in the Indo-Gangetic Basin. The third cluster is attributed to the error in treatment of surface reflectance by the MODIS land algorithm. For a direct comparison, an absolute criterion is needed at very low AOD, where the sensitivity of the sensor is low. On the other hand, a relative criterion is useful for comparing AODs well above the limit of measurement sensitivity, in order to account for such issues as sampling differences and spatial variability. Often the combined use of MODIS and MISR provides better information (e.g. Kalashnikova and Kahn, 2008). These data are utilized in various applications, as summarized in Section 6.3.3.

6.3.2 Active Spaceborne Lidar

One of the major factors influencing the quality of retrieval of aerosol properties using passive remote sensing techniques is the impact of clouds. Undetected clouds within clear pixels enhance the TOA radiance. Even when a pixel is identified correctly as 'clear', aerosol retrieval in that pixel is influenced by enhanced reflectance from cloud edge (Wen, Marshak, and Cahalan, 2006). This artifact can impact the interpretation of aerosol–cloud interaction based on remote sensing data (Loeb and Schuster, 2008; Koren *et al.*, 2007). Aerosol and cloud can be simultaneously detected within a pixel using active remote sensing. Moreover, active remote sensing can provide information on aerosol vertical profiles in both day and night, unlike the daytime-only retrieval of columnar aerosol properties by passive sensors.

CALIOP was launched onboard the CALIPSO satellite as part of a constellation of satellites popularly known as the 'A-Train' (Stephens *et al.*, 2002) in the year 2006. It measures the vertical profiles of two orthogonal polarization components of attenuated backscatter at 532 nm (at 30 m vertical resolution and 1/3 km horizontal resolution below 8.2 km altitude, and at 60 m vertical and 1 km horizontal resolution between 8.2 and 20.2 km altitude) and of total backscatter at 1064 nm (at 60 m vertical resolution). Color ratio (ratio of 532 and 1064 nm backscatter) is utilized to distinguish between aerosol and clouds. Furthermore, utilizing the sensitivity of the volume depolarization ratio, VDR (the ratio of the layer integrated perpendicular to parallel components of 532 nm backscatter) to the irregularity of particle shape, nonspherical dusts can be separated out from other spherical particles (Liu *et al.*, 2008). Comparison with ground-based lidar reveals a poor match closer to the surface, primarily due to uncertainty in the assumption of the lidar ratio in the CALIPSO algorithm, combined with a low sensitivity as the signal gets attenuated closer to the surface by the presence of a heavy aerosol layer aloft (Misra *et al.*, 2012).

Nonetheless, CALIPSO measurements provide an opportunity to study aerosol vertical distributions on a global scale, which is critical in quantifying the aerosol-induced radiative effect. CALIPSO profiles have been used to examine aerosol transport (e.g. Liu *et al.*, 2008; Abdi Vishkaee *et al.*, 2012; Cabello *et al.*, 2012; Di Pierro, Jaegle, and Anderson, 2011), aerosol distribution (e.g. Chen *et al.*, 2012; Mishra and Shibata, 2012), aerosol radiative forcing (e.g. Chand *et al.*, 2009), and the validation of climate models (e.g. Koffi *et al.*, 2012).

6.3.3 Applications of Satellite-Based Aerosol Products

Perception of the importance of aerosols to the earth's climate has changed as a result of the capabilities of new-generation sensors in measuring aerosol properties on a global scale (Kaufman, Tanre, and Boucher, 2000). Satellite aerosol products are currently being applied for numerous scientific purposes (e.g. time-series analysis, aerosol characterization at various spatial and temporal scales, estimation of aerosol radiative forcing, aerosol–cloud interaction, validation of climate models, near-surface aerosol concentration for health studies). This section summarizes these applications and discusses caveats on their applicability.

6.3.3.1 Aerosol Characterization, Trends, and Radiative Forcing

Satellite-based aerosol products (columnar and vertical distributions) providing near-global coverage have helped reduce the uncertainty of aerosol direct radiative forcing, as is evident from IPCC AR3 (pre-Terra era) and AR4 (post-Terra era). While the GACP dataset provides fairly long-term (26 years) records of AOD over oceans, MODIS and MISR aerosol products provide near-global measurements of aerosol properties for the last 12 years (and continuing). These data have been utilized in the last decade to improve our understanding of the space–time distributions of aerosol properties. The transition from global dimming to brightening in recent years, as revealed by ground-based measurements of solar radiation (Wild *et al.*, 2005), matches quite well with the AOD trend obtained from the GACP dataset (Figure 6.11), where AOD shows a decreasing trend between 1991 and 2006. However, the global decreasing AOD trend may not be uniform, as revealed by studies conducted at regional scales. For example, Dey and Di Girolamo (2011) have observed an increasing trend in MISR-AOD over the Indian subcontinent for the period 2000–2010. The magnitude of the increase in AOD in seasonal hotspots is alarming (in the range 0.1–0.4 in the last decade; Figure 6.12) and has been attributed to a rapid increase in anthropogenic particles and to an additional contribution of dust in the rural and oceanic regions, characterized by a

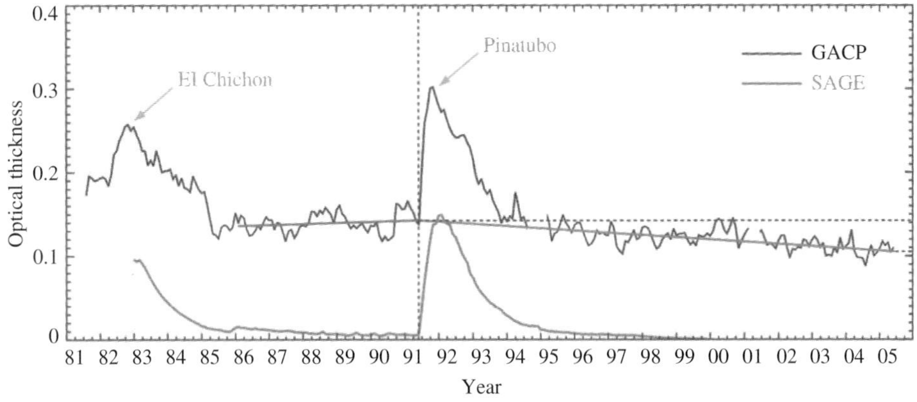

Figure 6.11 *GACP record of the globally averaged column AOD over the oceans and SAGE record of the globally averaged stratospheric AOD. Reproduced with permission from Mishchenko* et al. *(2007b). Copyright © 2007, American Association for Advancement of Science. See plate section for colour version.*

combination of particle size and shape from MISR aerosol product (Dey and Di Girolamo, 2010). The general conclusions are supported by other studies (e.g. Kishcha *et al.*, 2011) and by continued dimming in this region (Kumari *et al.*, 2007). Trend analysis using satellite data requires caution, because surface reflectance may change over time due to land-cover and land-use change, and hence bias the AOD retrieval. However, MISR's multiangle view provides a simultaneous retrieval of surface reflectance and AOD,

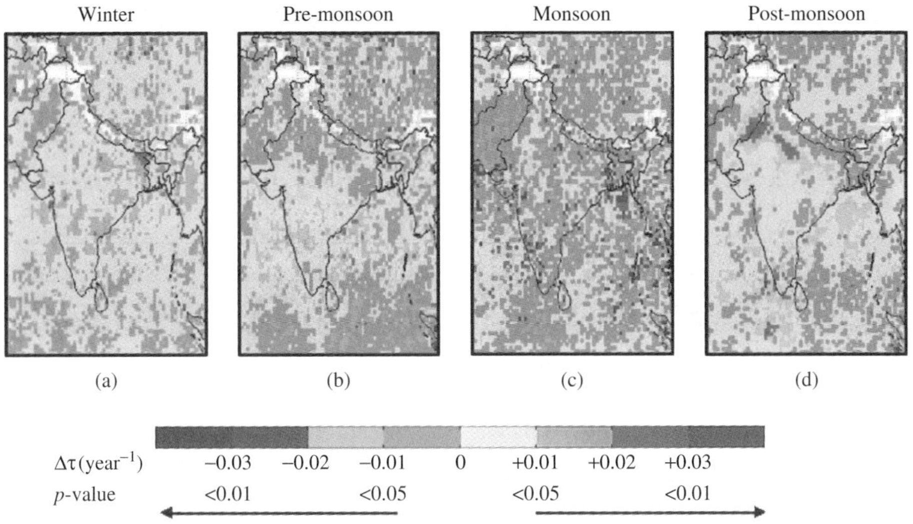

Figure 6.12 *Seasonal aerosol hotspots (significant rate of increase of AOD per year) over the Indian subcontinent for the period 2000–2010. Reproduced with permission from Dey and Di Girolamo (2011). Copyright © 2011, John Wiley and Sons Ltd. See plate section for colour version.*

limiting the chance of artificial bias in trend analysis. Satellite aerosol products have also been used for regional (e.g. Ramachandran and Cherian, 2008; Xia *et al.*, 2008; Dey and Di Girolamo, 2010; Marey *et al.*, 2011) and global (e.g. Remer *et al.*, 2008; Liu and Mishchenko, 2008) aerosol climatology.

The unique multiangle view of MISR further allows for the retrieval of aerosol plume and the tracking of its transport (Kahn *et al.*, 2007). First, mean wind is derived at the reflecting level at coarse resolution (70.4 km). Then matching is performed between the 26° forward and nadir views and independently between the 26° backward and nadir views, using the MISR-275 m spatial resolution red-band images over each 1.1 km horizontal region. Aerosol plume height is calculated from each fore and aft pair over the entire 70.4 km domain. The natural-color nadir view of the smoke plume emitted from wildfire in Oregon is shown in Figure 6.13a. Analysis has been carried out for the five patches, with very high (in the range 0.7–1.0) AOD (Figure 6.13b) being retrieved over the region, where particles are of medium size, as suggested by the AE values (Figure 6.13c). The plume height retrieved is 5 km (Figure 6.13d). Such a dataset is important to understanding the dispersion of pollutants, which is a key input for aerosol transport models (Kahn *et al.*, 2008; Val Martin *et al.*, 2010).

Aerosol radiative forcing estimates have improved as a result of the better characterization of aerosol properties from satellite remote sensing (Kaufman, 2006; Yu *et al.*, 2006, 2004). Bellouin *et al.* (2005) have quantified the anthropogenic fraction to total AOD using a fine-mode fraction constrained by AERONET and campaign measurements. They found the clear-sky aerosol direct radiative forcing at TOA to be $-1.9\ \mathrm{W\ m^{-2}}$, significantly different from that predicted by climate models (-0.5 to $-0.9\ \mathrm{W\ m^{-2}}$). Several other studies

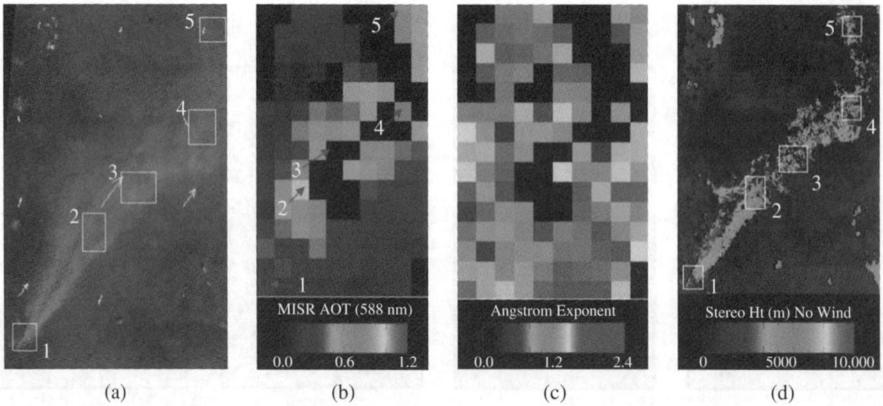

Figure 6.13 *Oregon fire, 4 September 2003, orbit 19 753, path 044, blocks 53–55, 19:00 UTC. (a) MISR nadir view of the fire plume, with five study site (patch) locations indicated as numbered white boxes, and MISR stereo-derived wind vectors superposed in yellow. (b) MISR mid-visible column AOD (version 17) retrieved at 17.6 km spatial resolution, with study site locations indicated by red arrows. There are no retrieval results for the black pixels, in most cases due to the high AOD and AOD variability of the plume core. (c) MISR-derived, column-average Angstrom exponent for the plume and surrounding area. (d) MISR stereo height product (version 13), without wind correction (labeled 'no wind' in the figure), for the same region. Reproduced with permission from Kahn* et al. *(2007). Copyright © 2007, John Wiley and Sons Ltd. See plate section for colour version.*

have also utilized the satellite aerosol products to reduce uncertainty in aerosol radiative forcing at regional (e.g. Elias and Roujean, 2008; Li *et al.*, 2010; Chung *et al.*, 2010) and global (e.g. Chung *et al.*, 2005; Chung, Ramanathan, and Decremer, 2012; Quaas and Boucher, 2005; Christopher *et al.*, 2006; Quaas *et al.*, 2008) scales. The utility of satellite-based aerosol products in constraining model-based estimates of aerosol direct radiative forcing has been summarized by Yu *et al.* (2006).

6.3.3.2 Aerosol–Cloud–Precipitation Interaction

Satellite-based aerosol products in tandem with cloud products are also used to examine aerosol–cloud interaction, which is the largest source of uncertainty in estimating anthropogenic climate forcing. Satellites complement the paucity of *in situ* observations to help understand this complex problem. The satellite-based approach relies on variations in cloud macrophysical (cloud fraction and cloud-top altitude) and microphysical (phase, effective radius, and water path) properties in response to changing aerosol characteristics. Detailed studies have been carried out to examine the impacts of aerosols on clouds over the Atlantic Ocean (Kaufman *et al.*, 2005; Koren *et al.*, 2005; Koren, Feingold, and Remer, 2010), Amazon (Koren *et al.*, 2008; Jones and Christopher, 2010), South Asia (Chylek *et al.*, 2006; Tripathi, Pattnaik, and Dey, 2007; Jones, Christopher, and Quaas, 2009; Dey *et al.*, 2011), and Africa (Loeb and Schuster, 2008; Chand *et al.*, 2009; Su *et al.*, 2010; Costantino and Breon, 2010) using satellite data. Satellite aerosol products have also been used to study aerosol-induced changes in precipitation (Sorooshian *et al.*, 2009, 2010; Duong, Sorooshian, and Feingold, 2011; Koren *et al.*, 2012). Many interesting conclusions, previously unknown, have been drawn from these studies. Notable ones include the invigoration of deep convective clouds in polluted environments and the intensification of rain, the transition from indirect to semidirect effects, and the dependence of precipitation susceptibility (i.e. change in precipitation in response to aerosol perturbation) on the cloud water path.

The aerosol–cloud relationships observed by satellites may also be attributed to passive remote sensing artifacts, codependence of aerosols and clouds on synoptic meteorology, and influence of aerosol on cloud retrieval and vice versa. Disentangling the meteorological effect on the observed cloud properties from the effect of aerosols is the most difficult task in the satellite-based approach (Lohmann, Koren, and Kuafman, 2006; Loeb and Schuster, 2008). Stevens and Feingold (2009) have reviewed the present approaches to address this critical problem and concluded that statistically meaningful relationships between aerosol, clouds, and precipitation are not always observed, due to the buffering effect of meteorology, making the problem even more difficult. They advocate treatment of individual cloud systems in order to better quantify the aerosol impact. The importance of regime-specific studies of aerosol–cloud interactions is also highlighted by Gryspeerdt and Stier (2012). Passive remote sensing artifacts involve the enhancement of AOD near clouds due to the swelling of aerosols in a humid environment, an increased number of aerosols from aerosol-generating processes, undetected sub-pixel clouds, instrument limitations, and 3D radiative interactions between clouds and surrounding clear areas (Varnai and Marshak, 2009; Redemann *et al.*, 2009; Marshak *et al.*, 2008; Koren *et al.*, 2007; Wen, Marshak, and Cahalan, 2006). Quantifying the relative contributions of these factors is very difficult. Tackett and Di Girolamo (2009) have analyzed CALIPSO data to quantify the relative contribution of 3D effects on cloud-induced enhancement in AOD near cloud fields over the Caribbean

Atlantic by utilizing nighttime CALIPSO data. Similar efforts are required in other regions, especially those dominated by small cumulus clouds, because cumulus clouds are typically smaller in size than the horizontal resolutions of level 2 pixels of commonly used passive sensors and may thus cause direct cloud contamination in aerosol retrieval (Myhre *et al.*, 2004).

6.3.3.3 Applications in Air Quality

The association between air quality and human health is well established in the literature. In recent years, the paucity of ground-based observations at a global scale has led to increased utilization of satellite-based aerosol products in the study of air quality (Gupta *et al.*, 2006; van Donkelaar *et al.*, 2010; Brauer *et al.*, 2012; Dey *et al.*, 2012; Cooper *et al.*, 2012). All these studies have converted satellite measurements of columnar AOD into surface fine particulate matter concentration (of particles smaller than 2.5 μm, $PM_{2.5}$) using different approaches. The relation between columnar AOD and $PM_{2.5}$ is influenced by many factors (aerosol layer thickness, scale height, cloud-free condition, composition, and meteorology) and hence any straightforward regression may be difficult to achieve. van Donkelaar *et al.* (2010) have utilized the GEOS-Chem model, constrained by satellite observations of columnar AOD and vertical distribution, to establish the global distribution of conversion factors. These factors account for influence on the AOD–$PM_{2.5}$ relationship and are used to convert columnar AOD to $PM_{2.5}$ (Figure 6.14). Global hotspots are easily identified in Figure 6.14 (e.g. the industrial areas in North America and Europe, the Indo-Gangetic Basin, the Saharan Desert, South American and Central African biomass-burning regions, West Asian dust sources, and Eastern China), where $PM_{2.5}$ concentration exceeds the largest threshold (35 μg m^{-3}) of the World Health Organization. A detailed comparison with *in situ* observations has revealed that these values are in fact underestimated. Dey *et al.* (2012), based on the analysis over the Indian subcontinent and applying the conversion factor of van Donkelaar *et al.* (2010) on MISR-AOD, concluded that the underestimation of

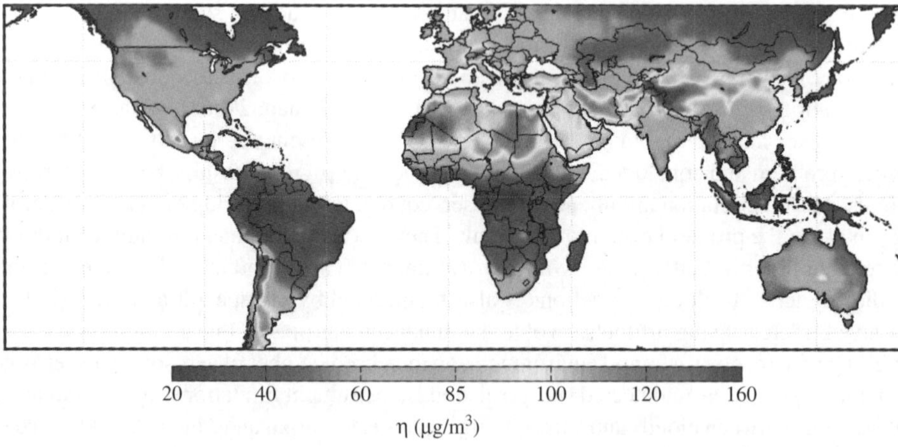

Figure 6.14 *Global view of satellite-derived $PM_{2.5}$ over land for the period 2001–2006. Reproduced from van Donkelaar* et al. *(2010). Copyright © 2010. This is an Open Access article. See plate section for colour version.*

MISR-$PM_{2.5}$ shows a linear correlation with coincident *in situ* data, which allowed the bias to be corrected. The bias correction increases the correlation between satellite-based and *in situ* $PM_{2.5}$ from 0.73 to 0.98 (Dey *et al.*, 2012). Such approaches provide the opportunity to examine the distribution of $PM_{2.5}$ at a larger spatial scale.

Hoff and Christopher (2009) have provided an excellent review of the present status of the satellite-based approaches to $PM_{2.5}$ estimation and of future requirements. The spatially heterogeneous relationship between AOD and $PM_{2.5}$ calls for more regional-scale analysis. The discrepancy between absolute values of AOD from various sensors is another source of uncertainty. Satellite-based estimates are biased toward clear-sky conditions, because aerosol retrieval is not possible through passive remote sensing under cloudy skies. Passive sensors onboard polar orbiting satellites provide discontinuous data (maximum once per day, because AOD is not retrieved from nighttime measurements), severely limiting their applicability in health-impact assessment. Geostationary satellites may provide continuous data at the desired temporal frequency, but limited spectral measurements might lead to larger uncertainty than the existing operational sensors. Currently, only the GOES satellite retrieves AOD at 4 km horizontal resolution at half-hourly interval, but its geographical coverage is restricted to the United States (Zhang *et al.*, 2011). Errors in AOD retrieval propagate into the accuracy of estimated $PM_{2.5}$, so an improved AOD is required, particularly in places dominated by aerosols of mixed type. The sampling gap in cloudy conditions is another concern that might be addressed by active remote sensing and the polarimetric technique of passive remote sensing (Kassianov *et al.*, 2011).

6.4 Summary and Future Requirements

In this chapter, we have summarized the quality of current ground-based and satellite aerosol products and their applicability for studying the climatic and health impacts of aerosols. The increasing demand for new data products with global coverage for such applications has pushed remote sensing techniques to a new level in the recent years. New-generation sensors have been designed, new techniques (e.g. multiangle and multispectral views) have been adopted, and retrieval algorithms have been modified based on prior experience and the requirements of the user community. These efforts have resulted in significant improvements in our understanding of aerosol characteristics and aerosol–climate feedbacks. Improved assessments of climate models have become possible with the availability of aerosol data at higher spatial and temporal scales. Satellites have closed the data gap in developing countries and helped identify previously unknown regional hotspots.

With such promise, demand for further improvements in data quality has increased. Future missions must be planned so that these requirements can be met. Aerosol retrieval is required at high temporal (more than once a day) and spatial (~100 m) resolutions in order to better quantify the artificial enhancement of AOD near clouds. Presently, the satellite-based aerosol product includes particle microphysical properties, along with the conventional product, AOD. However, these microphysical properties can be used qualitatively to identify major aerosol air mass. The quality of these retrieved products must be improved by updating the existing LUT with additional aerosol models. Properly planned filed campaigns may help in formulating appropriate aerosol models that can reduce the discrepancy in multisensor AOD. A unified aerosol product (combining the strengths of various sensors) would be ideal for scientific applications. Simultaneous retrieval of aerosol and cloud properties within a single pixel will further reduce the uncertainty in estimating aerosol–cloud interaction. Surface aerosol concentration (e.g. $PM_{2.5}$ and particulate matters of other size categories) should be included in routine aerosol products to facilitate health-impact assessments.

Acknowledgements

The authors are thankful to the various journals for allowing the use of published materials for this chapter. The efforts of the PIs and staffs of the AERONET sites used in this study are acknowledged. SNT acknowledges financial support from project 'Changing Water Cycle', jointly administered by MoES, India and NERC, UK. SD acknowledge research grants from DST, Govt. of India under contract SR/FTP/ES-191/2010 and DST/CCP/PR/11/2011 through research projects operational at IIT Delhi (IITD/IRD/RP02509 and IITD/IRD/RP2580).

References

Abdi Vishkaee, F., Flamant, C., Cuesta, J. *et al.* (2012) Dust transport over Iraq and northwest Iran associated with winter Shamal: a case study. *Journal of Geophysical Research*, **117**, D03201.

Ackerman, S., Strabala, K., Menzel, W. *et al.* (1998) Discriminating clear sky from clouds with MODIS. *Journal of Geophysical Research*, **103**(D24), 32141–32157.

Ackerman, A.S., Toon, O.B., Stevens, D.E. *et al.* (2000) Reduction of tropical cloudiness by soot. *Science*, **288**, 1042–1047.

Albrecht, B.A. (1989) Aerosols, cloud microphysics, and fractional cloudiness. *Science*, **245**, 1227–1230.

Andreae, M.O. and Rosenfeld, D. (2008) Aerosol-cloud-precipitation interactions. Part 1. The nature and sources of cloud-active aerosols. *Earth-Science Reviews*, **89**, 13–41.

Arola, A., Schuster, G., Myhre, G. *et al.* (2011) Inferring absorbing organic carbon content from AERONET data. *Atmospheric Chemistry and Physics*, **11**, 215–225.

Bellouin, B., Boucher, O., Haywood, J. and Reddy, M.S. (2005) Global estimates of aerosol direct radiative forcing from satellite measurements. *Nature*, **438**, 1138–1140.

Boucher, O. and Pham, M. (2002) History of sulfate aerosol radiative forcings. *Geophysical Research Letters*, **29**(9), 1308. doi: 10.1029/2001GL014048.

Brauer, M., Amman, M., Burnett, R.T. *et al.* (2012) Exposure assessment for estimation for the global burden of disease attributable to outdoor air pollution. *Environmental Science and Technology*, **46**(2), 652–660.

Brav, W.J. (1929) Aerosols – clouds, dust, fog and smoke. *Journal of Chemical Education*, **6**(9), 1497–1502.

Cabello, M., Orza, J.A.G., Barrero, M.A. *et al.* (2012) Spatial and temporal variation of the impact of an extreme Saharan dust event. *Journal of Geophysical Research*, **117**, D11204.

Campbell, J.R., Tackett, J.L., Reid, J.S. *et al.* (2012) Evaluating nighttime CALIOP 0.532 μm aerosol optical depth and extinction coefficient retrievals. *Atmospheric Measurement Techniques*, **5**, 2143–2160.

Chand, D., Wood, R., Anderson, T.L. *et al.* (2009) Satellite-derived direct radiative effect of aerosols dependent on cloud cover. *Nature Geoscience*, **2**, 181–184.

Charlson, R.J. and Pilat, M.J. (1969) Climate: the influence of aerosols. *Journal of Applied Meteorology*, **8**, 1001–1002.

Charlson, R.J., Schwartz, S.E., Hales, J.H. *et al.* (1992) Climate forcing by anthropogenic aerosols. *Science*, **255**, 423–430.

Chen, Y., Liu, Q., Geng, F. *et al.* (2012) Vertical distribution of optical and micro-physical properties of ambient aerosols during dry haze periods in Shanghai. *Atmospheric Environment*, **50**, 50–59.

Chow, J.C. and Watson, J.G. (2007) Review of measurement methods and compositions for ultrafine particles. *Aerosol and Air Quality Research*, **7**(2), 121–173.

Christopher, S.A., Zhang, J., Kuafman, Y.J. and Remer, L.A. (2006) Satellite-based assessment of top of atmosphere amthropogenic aerosol radiative forcing over cloud free oceans. *Geophysical Research Letters*, **33**, L15816.

Chung, C.E., Ramanathan, V., Carmichael, G. *et al.* (2010) Anthropogenic aerosol radiative forcing in Asia derived from regional models with atmospheric ad acrosol data assimilation. *Atmospheric Chemistry and Physics*, **10**, 6007–6024.

Chung, C.E., Ramanathan, V. and Decremer, D. (2012) Observationally constrained estimates of carbonaceous aerosol radiative forcing. *Proceedings of the National Academy of Sciences*, **109**(29), 11624–11629.

Chung, C.E., Ramanathan, V., Kim, D. and Podgorny, I.A. (2005) Global anthropogenic aerosol direct forcing derived from satellite and ground-based observations. *Journal of Geophysical Research*, **110**, D24207.

Chylek, P., Dubey, M.K., Lohmann, U. *et al.* (2006) Aerosol indirect effect over the Indian Ocean. *Geophysical Research Letters*, **33**, L06806.

Chylek, P. and Wong, J. (1995) Effect of absorbing aerosol on global radiation budget. *Geophysical Research Letters*, **22**, 929–931.

Cooper, M.J., Martin, R.V., van Donkelaar, A. *et al.* (2012) A satellite-based multi-pollutant index of global air quality. *Environmental Science and Technology*, **46**(12), 8523–8524.

Costantino, L. and Breon, F.M. (2010) Analysis of aerosol-cloud interaction from multi-sensory satellite observations. *Geophysical Research Letters*, **37**, L11801.

Dey, S. and Di Girolamo, L. (2010) A climatology of aerosol optical and microphysical properties over the Indian subcontinent from 9 years (2000–2008) of Multiangle Imaging SpectroRadiometer (MISR) data. *Journal of Geophysical Research*, **115**, D15204.

Dey, S. and Di Girolamo, L. (2011) A decade of change in aerosol properties over the Indian subcontinent. *Geophysical Research*, **38**, L14811.

Dey, S., Di Girolamo, L., van Donkelaar, A. *et al.* (2012) Variability of outdoor fine particulate ($PM_{2.5}$) concentration in the Indian subcontinent: a remote sensing approach. *Remote Sensing of Environment*, **127**, 153–161.

Dey, S., Di Girolamo, L., Zhao, G. *et al.* (2011) Satellite-observed relationships between aerosol and trade-wind cumulus cloud properties over the Indian Ocean. *Geophysical Research Letters*, **38**, L01804.

Dey, S., Tripathi, S.N. and Mishra, S.K. (2008) Probable mixing state of aerosols in the Indo-Gangetic Basub, Northern India. *Geophysical Research Letters*, **35**, L03808.

Dey, S., Tripathi, S.N., Singh, R.P. and Holben, B.N. (2004) Influence of the dust storms on the aerosol optical properties over the Indo-Gangetic Basin. *Journal of Geophysical Research*, **109**, D20122.

Dey, S., Tripathi, S.N., Singh, R.P. and Holben, B.N. (2006) Retrieval of black carbon and specific absorption over Kanpur city, Northern India during 2001–2003. *Atmospheric Environment*, **40**(3), 445–456.

Diner, D.J., Beckert, J.C., Reilly, T.H. *et al.* (1998) Multi-angle Imaging SpectroRadiometer (MISR) instrument description and experiment overview. *IEEE Transactions on Geoscience and Remote Sensing*, **36**, 1072–1087.

Di Pierro, M., Jaegle, L. and Anderson, T.L. (2011) Satellite observations of aerosol transport from East Asia to the Arctic: three case studies. *Atmospheric Chemistry and Physics*, **11**, 2225–2243.

van Donkelaar, A., Martin, R.V., Brauer, M. *et al.* (2010) Global estimates of ambient fine particulate matter concentrations from satellite-based aerosol optical depth: development and application. *Environmental Health Perspectives*, **118**(6), 847–855.

Dubovik, O., Holben, B.N., Eck, T.F. *et al.* (2002) Variability of absorption and optical properties of key aerosol types observed in worldwide locations. *Journal of the Atmospheric Sciences*, **59**, 590–608.

Dubovik, O. and King, M.D. (2000) A flexible inversion algorithm for retrieval of aerosol properties from sun and sky radiance measurements. *Journal of Geophysical Research*, **105**, 20673–20696.

Dubovik, O., Smirnov, A., Holben, B.N. *et al.* (2000) Accuracy assessments of aerosol optical properties retrieved from Aerosol Robotic Network (AERONET) sun and sky radiance measurements. *Journal of Geophysical Research: Atmospheres*, **105**(D8), 9791–9806.

Duong, H.T., Sorooshian, A. and Feingold, G. (2011) Investigating potential biases in observed and modeled metrics of aerosol-cloud-precipitation interactions. *Atmospheric Chemistry and Physics*, **11**, 4027–4037.

Eck, T.F., Holben, B.N., Sinyuk, A. *et al.* (2010) Climatological aspects of the optical properties of fine/coarse mode aerosol mixtures. *Journal of Geophysical Research*, **115**, D19205.

Elias, T. and Roujean, J.L. (2008) Estimation of the aerosol radiative forcing at ground level, over land, and in cloudless atmosphere, from METEOSAT-7 observation: method and case study. *Atmospheric Chemistry and Physics*, **8**, 625–636.

Geogdzhayev, I.V., Mishchenko, M.I., Liu, L. and Remer, L. (2004) Global two-channel AVHRR aerosol climatology: effects of stratospheric aerosols and preliminary comparisons with MODIS and MISR retrievals. *Journal of Quantitative Spectroscopy and Radiative Transfer*, **88**, 47–59.

Geogdzhayev, I.V., Mishchenko, M.I., Rossow, W.B. *et al.* (2002) Global two-channel AVHRR retrievals of aerosol properties over the ocean for the period of NOAA-9 observations and preliminary retrievals using NOAA-7 and NOAA-11 data. *Journal of the Atmospheric Sciences*, **59**, 262–278.

Geogdzhayev, I.V., Mishchenko, M.I., Terez, E.I. *et al.* (2005) Regional advanced very high resolution radiometer-derived climatology of aerosol optical thickness and size. *Journal of Geophysical Research*, **110**, D23205.

Giles, D.M., Holben, B.N., Eck, T.F. *et al.* (2012) An analysis of AERONET aerosol absorption properties and classifications representative of aerosol source regions. *Journal of Geophysical Research*, **117**, D17203.

Gleason, J.F., Hsu, N.C. and Torres, O. (1998) Biomass burning smoke measured using backscattered ultraviolet radiation: SCAR-B and Brazilian smoke interannual variability. *Journal of Geophysical Research*, **103**, 31969–31978.

Gobbi, G.P., Kaufman, Y.J., Koren, I. and Eck, T.F. (2007) Classification of aerosol properties derived from AERONET direct sun data. *Atmospheric Chemistry and Physics*, **7**, 453–458.

Griggs, M. (1975) Measurements of atmospheric aerosol optical thickness over water using ERTS-1 data. *Journal of the Air Pollution Control Association*, **25**, 622–626.

Gryspeerdt, E. and Stier, P. (2012) Regime-based aerosol-cloud interactions. *Geophysical Research Letters*, **39**(21).

Gupta, P., Christopher, S.A., Wang, J. *et al.* (2006) Satellite remote sensing of particulate matter and air quality over global cities. *Atmospheric Environment*, **40**, 5880–5892.

Hansen, J., Sato, M. and Ruedy, R. (1997) Radiative forcing and climate response. *Journal of Geophysical Research*, **102**, 6831–6864.

Hao, W.M. and Liu, M.H. (1994) Spatial and temporal distribution of tropical biomass burning. *Global Biogeochemical Cycles*, **8**, 495–503.

Haywood, J. and Boucher, O. (2000) Estimates of the direct and indirect radiative forcing due to tropospheric aerosols: a review. *Reviews of Geophysics*, **38**(4), 513–543.

Herman, J.R. and Celarier, E.A. (1997) Earth surface reflectivity climatology at 340–380 nm from TOMS data. *Journal of Geophysical Research*, **102**, 28003–28012.

Hoff, R.M. and Christopher, S. (2009) Remote sensing of particulate pollution from space: have we reached the promised land? *Journal of the Air and Waste Management Association*, **59**, 645–675.

Holben, B.N., Eck, T.F., Slutsker, I. *et al.* (1998) AERONET-A federated instrument network and data archive for aerosol characterization. *Remote Sensing of Environment*, **66**, 1–16.

Holben, B.N., Tanré, D., Smirnov, A. *et al.* (2001) An emerging ground-based aerosol climatology: aerosol optical depth from AERONET. *Journal of Geophysical Research*, **106**, 12067–12097.

IPCC (2007) Changes in atmospheric constituents and in radiative forcing, in *Climate Change 2007: The Physical Science Basis, Contribution of Working Group I to the Fourth Assessment Report of the Intergovernmental Panel on Climate Change* (eds S. Solomon, D. Qin, M. Manning *et al.*), Cambridge University Press.

Johnson, B.T., Shine, K.P. and Forster, P.M. (2004) The semi-direct aerosol effect: impact of absorbing aerosols on marine stratocumulus. *Quarterly Journal of the Royal Meteorological Society*, **130**, 1407–1422.

Jones, T.A. and Christopher, S.A. (2010) Statistical properties of aerosol-cloud-precipitation interactions in South America. *Atmospheric Chemistry and Physics*, **10**, 2287–2305.

Jones, T.A., Christopher, S. and Quaas, J. (2009) A six year satellite-based assessment of the regional variations in aerosol indirect effects. *Atmospheric Chemistry and Physics*, **9**, 4091–4114.

Kahn, R.A., Chen, Y., Nelson, D.L. *et al.* (2008) Wildfire smoke injection heights: two perspectives from space. *Geophysical Research Letters*, **35**, L04809.

Kahn, R.A., Gaitley, B.J., Garay, M.J. *et al.* (2010) Multiangle imaging spectroradiometer global aerosol product assessment by comparison with the aerosol robotic network. *Journal of Geophysical Research*, **115**, D23209.

Kahn, R.A., Gaitley, B., Martonchik, J. *et al.* (2005) MISR global aerosol optical depth validation based on two years of coincident AERONET observations. *Journal of Geophysical Research*, **110**, D10S04.

Kahn, R.A., Garay, M.J., Nelson, D.L. *et al.* (2011) Response to 'Towards unified satellite climatology of aerosol properties: 3. MODIS versus MISR versus AERONET'. *Journal of Quantitative Spectroscopy and Radiative Transfer*, **112**, 901–909.

Kahn, R.A., Li, W.H., Moroney, C. *et al.* (2007) Aerosol source plume physical characteristics from space-based multiangle imaging. *Journal of Geophysical Research*, **112**, D11205.

Kahn, R.A., Nelson, D.L., Garay, M. *et al.* (2009) MISR aerosol product attributes, and statistical comparisons with MODIS. *IEEE Transactions on Geoscience and Remote Sensing*, **47**(12), 4095–4114.

Kalashnikova, O.V. and Kahn, R.A. (2008) Mineral dust plume evolution over the Atlantic from combined MISR/MODIS aerosol retrievals. *Journal of Geophysical Research*, **113**, D24204. doi: 10.1029/2008JD010083.

Kalashnikova, O.V., Kahn, R., Sokolik, I.N. and Li, W.H. (2005) Ability of multiangle remote sensing observations to identify and distinguish mineral dust types: optical models and retrievals of optically thick plumes. *Journal of Geophysical Research: Atmospheres*, **110**(D18).

Kaskaoutis, D.G., Kalapureddy, M.C.R., Moorthy, K.K. *et al.* (2010) Heterogeneity in premonsoon aerosol types over the Arabian Sea deduced from ship-borne measurements of spectral AODs. *Atmospheric Chemistry and Physics*, **10**, 4893–4908.

Kassianov, E., Ovchinnikov, M., Berg, L.K. and Flynn, C. (2011) Aerosol retrievals under partly cloudy conditions: challenges and perspectives, in *Polarimetric Detection, Characterization and Remote Sensing* (eds M.I. Mishchenko, Y.S. Yatskiv, V.K. Rosenbush and G. Videen), Springer, Dordrecht, pp. 205–232.

Kaufman, Y.J. (2006) Satellite observations of natural and anthropogenic aerosol effects on clouds and climate. *Space Science Reviews*, **125**, 139–147.

Kaufman, Y.J., Koren, I., Remer, L.A. *et al.* (2005) The effect of smoke, dust and pollution aerosol on shallow cloud development over the Atlantic Ocean. *Proceedings of the National Academy of Sciences*, **102**(32), 11207–11212.

Kaufman, Y.J., Tanre, D. and Boucher, O. (2000) A satellite view of aerosols in the climate system. *Nature*, **419**, 215–223.

Kaufman, Y.J., Wald, A., Remer, L. *et al.* (1997) The MODIS 2.1 μm channel – correlation with visible reflectance for use in remote sensing of aerosol. *IEEE Transactions on Geoscience and Remote Sensing*, **35**(5), 1286–1298.

Kedia, S. and Ramachandran, S. (2009) Variability in aerosol optical and physical characteristics over the Bay of Bengal and the Arabian Sea deduced from Angstrom exponents. *Journal of Geophysical Research*, **114**, D14207.

Kishcha, P., Starobinets, B., Kalashnikova, O. and Alpert, P. (2011) Aerosol optical thickness trends and population growth in the Indian subcontinent. *International Journal of Remote Sensing*, **32**(24), 9137–9149.

Koffi, B., Schulz, M., Bréon, F.M., Griesfeller, J., Winker, D.M., Balkanski, Y., Bauer, S., Berntsen, T., Chin, M., Collins, W.D., Dentener, F., Diehl, T., Easter, R.C., Ghan, S.J., Ginoux, P.A., Gong, S., Horowitz, L.W., Iversen, T., Kirkevag, A., Koch, D.M., Krol, M., Myhre, G., Stier, P., and Takemura, T. (2012) Application of the CALIOP layer product to evaluate the vertical distribution of aerosols estimated by global models: part 1. AeroCom phase I results, *Journal of Geophysical Research*, **117**, D10201.

Koren, I., Altaratz, O., Remer, L.A. *et al.* (2012) Aerosol induced intensification of rain from the tropics to the mid-latitudes. *Nature Geoscience*, **5**, 118–122.

Koren, I., Feingold, G. and Remer, L.A. (2010) The invigoration of deep convective clouds over the Atlantic: aerosol effect, meteorology or retrieval artifact? *Atmospheric Chemistry and Physics*, **10**, 8855–8872.

Koren, I., Kaufman, Y.J., Remer, L.A. and Martins, J.V. (2004) Measurement of the effect of Amazon smoke on inhibition of cloud formation. *Science*, **303**, 1342–1345.

Koren, I., Kaufman, Y.J., Rosenfeld, D., Remer, L.A., and Rudich, Y. (2005) Aerosol invigoration and restructuring of Atlantic convective clouds, *Geophysical Research Letters*, **32**, L14828.

Koren, I., Martins, J.V., Remer, L.A. and Afargan, H. (2008) Smoke invigoration versus inhibition of clouds over the Amazon. *Science*, **321**(5891), 946–949.

Koren, I., Remer, L.A., Kaufman, Y.J. *et al.* (2007) On the twilight zone between clouds and aerosols. *Geophysical Research Letters*, **34**, L08805.

Koven, C.D. and Fung, I. (2006) Inferring dust composition from wavelength-dependent absorption in Aerosol Robotic Network (AERONET) data. *Journal of Geophysical Research: Atmospheres*, **111**(D14), D14205.

Kumari, B.P., Londhe, A.L., Daniel, S. and Jadhav, D.B. (2007) Observational evidence of solar dimming: offsetting surface warming over India. *Geophysical Research Letters*, **34**, L21810.

Lee, J., Kim, J., Song, C.H. *et al.* (2010) Characteristics of aerosol types from AERONET sunphotometer measurements. *Atmospheric Environment*, **44**, 3110–3117.

Lee, K.H., Li, Z., Kim, J., and Kokhanosky, A. (2009) Atmospheric aerosol monitoring from satellite observations: a history of three decades, in *Atmospheric and Biological Environmental Monitoring*, (eds Y.J. Kim, U. Platt, M.B. Gu, and H. Iwahashi), Springer, pp. 13–38.

Lelieveld, J., Berresheim, H., Borrmann, S. *et al.* (2002) Global air pollution crossroads over the Mediterranean. *Science*, **298**, 794–799.

Levy, R.C., Leptoukh, G.G., Kahn, R. *et al.* (2009) A critical look at the deriving monthly aerosol optical depth from satellite data. *IEEE Transactions on Geoscience and Remote Sensing*, **47**(8), 2942–2956.

Levy, R.C., Remer, L.A. and Dubovik, O. (2007) Global aerosol optical properties and application to moderate resolution imaging spectroradiometer aerosol retrieval over land. *Journal of Geophysical Research*, **112**, D13210.

Levy, R.C., Remer, L.A., Kleidman, R.G. *et al.* (2010) Global evaluation of the collection 5 MODIS dark-target aerosol products over land. *Atmospheric Chemistry and Physics*, **10**, 10399–10420.

Li, Z., Lee, K.H., Wang, Y. *et al.* (2010) First observation-based estimates of cloud-free aerosol radiative forcing across China. *Journal of Geophysical Research*, **115**, D00K18.

Li, Z., Zhao, X., Kahn, R. *et al.* (2009) Uncertainties in satellite remote sensing of aerosols and impact on monitoring its long-term trend: a review and perspective. *Annals of Geophysics*, **27**, 2755–2770.

Liu, D.Z., Liu, Z., Winker, D. and Trepte, C. (2008) A height resolved global view of dust aerosols from the first year of CALIPSO lidar measurements. *Journal of Geophysical Research*, **113**, D16214.

Liu, L. and Mishchenko, M.I. (2008) Toward unified satellite climatology of aerosol properties: direct comparisons of advanced level 2 aerosol products. *Journal of Quantitative Spectroscopy and Radiative Transfer*, **109**(14), 2376–2385.

Liu, Y., Wang, Z., Wang, J. *et al.* (2011a) The effect of aerosol vertical profiles on satellite-estimated surface particle sulfate concentrations. *Remote Sensing of Environment*, **115**, 508–513.

Liu, J., Zheng, Y., Li, Z. *et al.* (2011b) Transport, vertical structure and radiative properties of dust events in Southeast China determined from ground and space sensors. *Atmospheric Environment*, **45**, 6469–6480.

Loeb, N.G. and Schuster, G.L. (2008) An observational study of the relationship between cloud, aerosol and meteorology in broken low-level cloud conditions. *Journal of Geophysical Research: Atmospheres*, **113**(D14).

Lohmann, U. and Feichter, J. (2005) Global indirect aerosol effects: a review. *Atmospheric Chemistry and Physics*, **5**, 715–737.

Lohmann, U., Koren, I. and Kuafman, Y.J. (2006) Disentangling the role of microphysical and dynamical effects in determining cloud properties over the Atlantic. *Geophysical Research Letters*, **33**, L09802.

Mahowald, N.M. and Dufresne, J.L. (2004) Sensitivity of TOMS aerosol index to boundary layer height: implications for detection of mineral aerosol sources. *Geophysical Research Letters*, **31**, L03103.

Marey, H.S., Gille, J.C., El-Askary, H.M. *et al.* (2011) Aerosol climatology over Nile delta based on MODIS, MISR and OMI satellite data. *Atmospheric Chemistry and Physics*, **11**, 10637–10648.

Marshak, A., Wen, G., Coakley, J.A. Jr., *et al.* (2008) A simple model for the cloud adjacency effect and the apparent bluing of aerosols near clouds. *Journal of Geophysical Research*, **113**, D14S17.

Massie, S.T., Torres, O. and Smith, S.J. (2004) Total Ozone Mapping Spectrometer (TOMS) observations of increases in Asian aerosol in winter from 1979 to 2000. *Journal of Geophysical Research*, **109**, D18211.

McPeters, R.D., Hollandsworth, S.M., Flynn, L.E. *et al.* (1996) Long-term ozone trends derived from the 16-year combined Nimbus 7/Meteor 3 TOMS Version 7 record. *Geophysical Research Letters*, **25**, 3699–3702.

Mishchenko, M.I., Geogdzhayev, I.A., Cairns, B. *et al.* (2007a) Past, present, and future of global aerosol climatologies derived from satellite observations: a perspective. *Journal of Quantitative Spectroscopy and Radiative Transfer*, **106**, 325–347.

Mishchenko, M.I., Geogdzhayev, I.A., Rossow, W.B. *et al.* (2007b) Long-term satellite record reveals likely recent aerosol trend. *Science*, **315**(5818), 1543.

Mishchenko, M.I., Geogdzhayev, I.V., Cairns, B. *et al.* (1999) Aerosol retrievals over the ocean by use of channels 1 and 2 AVHRR data: sensitivity analysis and preliminary results. *Applied Optics*, **38**, 7325–7341.

Mishchenko, M.I., Geogdzhayev, I.V., Liu, L. *et al.* (2009) Toward unified satellite climatology of aerosol properties: what do fully compatible MODIS and MISR aerosol pixels tell us? *Journal of Quantitative Spectroscopy and Radiative Transfer*, **110**, 402–408.

Mishchenko M.I., Liu, L., Geogdzhayev, I.V., *et al.* (2012) Aerosol retrievals from channel-1 and -2 AVHRR radiances: long-term trends updated and revisited, *Journal of Quantitative Spectroscopy and Radiative Transfer*, **113**, 1974–1980, http://dx.doi.org/10.1016/j.jqsrt.2012.05.006 (accessed 13 July 2013).

Mishra, A.K. and Shibata, T. (2012) Climatological aspects of seasonal variation of aerosol vertical distribution over central Indo-Gangetic belt (IGB) inferred by the space-borne lidar CALIOP. *Atmospheric Environment*, **46**, 365–375.

Mishra, S.K. and Tripathi, S.N. (2008) Modeling optical properties of mineral dust over the Indian Desert. *Journal of Geophysical Research*, **113**, D23201.

Mishra, S.K., Tripathi, S.N., Aggarwal, S.G. and Arola, A. (2012) Optical properties of accumulation mode polluted mineral dust: effects of particle shape, hematite content and semi-external mixing with carbonaceous species. *Tellus*, **64B**, 18536.

Misra, A., Tripathi, S.N., Kaul, D.S. and Welton, E.J. (2012) Study of MPLNET-derived aerosol climatology over Kanpur, India, and validation of CALIPSO level 2 version 3 backscatter and extinction products. *Journal of Atmospheric and Oceanic Technology*, **29**, 1285–1294.

Moorthy, K.K., Babu, S.S. and Satheesh, S.K. (2007) Temporal heterogeneity in aerosol characteristics and the resulting radiative impact at a tropical coastal station – Part 1: microphysical and optical properties. *Annals of Geophysics*, **25**, 2293–2308.

Myhre, G., Stordal, F., Johnsrud, M. *et al.* (2004) Intercomparison of satellite retrieved aerosol optical depth over ocean. *Journal of the Atmospheric Sciences*, **61**, 499–513.

Myhre, G., Stordal, F., Johnsrud, M. *et al.* (2005) Intercomparison of satellite retrieved aerosol optical depth over ocean during the period September 1997 to December 2000. *Atmospheric Chemistry and Physics*, **5**, 1697–1719.

Prospero, J.M., Ginoux, P., Torres, O. *et al.* (2002) Environmental characterization of global sources of atmospheric soil dust identified with the Nimbus 7 Total Ozone Mapping Spectrometer (TOMS) absorbing aerosol product. *Reviews of Geophysics*, **40**(1). doi: 10.1029/2000RG000095.

Quaas, J. and Boucher, O. (2005) Constraining the first aerosol indirect radiative forcing in the LMDZ GCM using POLDER and MODIS satellite data. *Geophysical Research Letters*, **32**, L17814.

Quaas, J., Boucher, O., Bellouin, N. and Kinne, S. (2008) Satellite-based estimate of the direct and indirect aerosol climate forcing. *Journal of Geophysical Research*, **113**, D05204.

Ramachandran, S. and Cherian, R. (2008) Regional and seasonal variations in aerosol optical characteristics and their frequency distributions over India during 2001–2005. *Journal of Geophysical Research*, **111**, D20207.

Ramanathan, V., Crutzen, P.J., Kiehl, J.L. and Rosenfeld, D. (2001) Aerosols, climate, and the hydrological cycle. *Science*, **294**, 2119–2124.

Redemann, J., Zhang, Q., Russel, P.B. *et al.* (2009) Case studies of aerosol remote sensing in the vicinity of clouds. *Journal of Geophysical Research*, **114**, D06209.

Remer, L.A., Kaufman, Y.J., Tanre, D. *et al.* (2005) The MODIS aerosol algorithm, products and validation. *Journal of the Atmospheric Sciences*, **62**, 947–972.

Remer, L.A., Kleidman, R.G., Levy, R.C. *et al.* (2008) Global aerosol climatology from the MODIS satellite sensors. *Journal of Geophysical Research*, **113**, D14S07.

Schuster, G.L., Dubovik, O. and Holben, B.N. (2006) Angstrom exponent and bimodal size distributions. *Journal of Geophysical Research*, **111**, D07207.

Schuster, G.L., Dubovik, O., Holben, B.N. and Clothiaux, E.E. (2005) Inferring black carbon content and specific absorption from Aerosol Robotic Network (AERONET) aerosol retrievals. *Journal of Geophysical Research*, **110**, D10S17.

Smirnov, A., Holben, B.N., Slutsker, I., *et al.* (2009) Maritime aerosol network as a component of aerosol robotic network, *Journal of Geophysical Research*, **114**, D06204.

Sorooshian, A., Feingold, G., Lebsock, M.D. *et al.* (2009) On the precipitation susceptibility of clouds to aerosol perturbations. *Geophysical Research Letters*, **36**, L13803.

Sorooshian, A., Feingold, G., Lebsock, M.D. *et al.* (2010) Deconstructing the precipitation susceptibility construct: improving metholodology for aerosol-cloud precipitation studies. *Journal of Geophysical Research*, **115**, D17201.

Srivastava, A.K., Tripathi, S.N., Dey, S. *et al.* (2012) Inferring aerosol types over the Indo-Gangetic Basin from ground based sunphotometer measurements. *Atmospheric Research*, **109**, 64–75.

Stephens, G.L., Vane, D.G., Boain, R.J. *et al.* (2002) The CloudSat mission and the A-Train: a new dimension of space-based observations of clouds and precipitation. *Bulletin of the American Meteorological Society*, **83**, 1771–1790.

Stevens, B. and Feingold, G. (2009) Untangling aerosol effects on clouds and precipitation in a buffered system. *Nature*, **461**, 607–613.

Stowe, L.L., Jacobowitz, H., Ohring, G. *et al.* (2002) The Advanced Very High Resolution Pathfinder Atmosphere (PATMOS) climate dataset: initial analysis and evaluations. *Journal of Climate*, **15**, 1243–1260.

Su, W., Loeb, N.G., Xu, K.M. *et al.* (2010) An estimate of aerosol indirect effect from satellite measurements with concurrent meteorological analysis. *Journal of Geophysical Research*, **115**, D18219.

Tackett, J.L. and Di Girolamo, L. (2009) Enhanced aerosol backscatter adjacent to tropical trade wind clouds revealed by satellite-based lidar. *Geophysical Research Letters*, **36**, L14804.

Torres, O., Bhartia, P.K., Herman, J.R. *et al.* (1998) Derivation of aerosol properties from satellite measurements of back-scattered ultraviolet radiation: theoretical basis. *Journal of Geophysical Research*, **103**, 17,099–17,110.

Torres, O., Bhartia, P.K., Herman, J.R. *et al.* (2002) A long-term record of aerosol optical depth from TOMS observations and comparison to AERONET measurements. *Journal of the Atmospheric Sciences*, **59**, 398–413.

Tripathi, S.N., Pattnaik, A. and Dey, S. (2007) Aerosol indirect effect over the Indo-Gangetic plain. *Atmospheric Environment*, **41**(33), 7037–7047.

Twomey, S. (1977) The influence of pollution on the shortwave albedo of clouds. *Journal of the Atmospheric Sciences*, **34**(7), 1149–1152.

Val Martin, M., Logan, J.A., Kahn, R. *et al.* (2010) Smoke injection heights from fires in North America: analysis of five years of satellite observations. *Atmospheric Chemistry and Physics*, **10**, 1491–1510.

Varnai, T. and Marshak, A. (2009) MODIS observations of enhanced clear sky reflectance near clouds. *Geophysical Research Letters*, **36**, L06807.

Welton, E.J. and Campbell, J.R. (2002) Micropulse lidar signals: uncertainty analysis. *Journal of Atmospheric and Oceanic Technology*, **19**, 2089–2094.

Welton, E.J., Campbell, J.R., Spinhirne, J.D. and Scott, V.S. (2001) Global monitoring of clouds and aerosols using a network of micro-pulse lidar systems. *Proceedings of SPIE*, **4153**, 151–158.

Welton, E.J., Voss, K.J., Gordon, H.R. *et al.* (2000) Ground-based lidar measurements of aerosols during ACE-2: instrument description, results, and comparisons with other ground-based and airborne measurements. *Tellus*, **52B**, 636–651.

Welton, E.J., Voss, K.J., Quinn, P.K. *et al.* (2002) Measurements of aerosol vertical profiles and optical properties during INDOEX 1999 using micropulse lidars. *Journal of Geophysical Research*, **107**(D19), 8019.

Wen, G., Marshak, A. and Cahalan, R.F. (2006) Impact of 3D clouds on clear sky reflectance and aerosol retrieval in a biomass burning region of Brazil. *IEEE Geoscience and Remote Sensing Letters*, **3**, 169–172.

Wild, M., Gilgen, H., Roesch, A. *et al.* (2005) From dimming to brightening: decadal changes in solar radiation at Earth's surface. *Science*, **308**(5723), 847–850.

Xia, X.G., Wang, P., Wang, Y., *et al.* (2008) Aerosol optical depth over the Tibetan Plateau and its relation to aerosols over the Taklimakan desert, *Geophysical Research Letters*, **35**(16), L16804.

Yu, H., Dickinson, R.E., Chin, M., Kaufman, Y.J., Zhou, M., Zhou, L., Tian, Y., Dubovik, O., and Holben, B.N. (2004) Direct radiative effect of aerosols as determined from a combination of MODIS retrievals and GOCART simulations, *Journal of Geophysical Research*, **109**, D03206.

Yu, H., Kaufman, Y.J., Chin, M. *et al.* (2006) A review of measurement-based assessments of the aerosol direct radiative effect and forcing. *Atmospheric Chemistry and Physics*, **6**, 613–666.

Zhang, H., Lyapustin, A., Wang, Y. *et al.* (2011) A multi-angle aerosol optical depth retrieval algorithm for geostationary satellite over the United States. *Atmospheric Chemistry and Physics*, **11**, 11977–11991.

7

Atmospheric Particle Nucleation

Mikko Sipilä, Katrianne Lehtipalo, and Markku Kulmala
Department of Physics, University of Helsinki, Finland

7.1 General Relevance

Atmospheric aerosol particles impact the earth's radiation budget directly by scattering and absorbing solar radiation. Aerosol particles also act as condensation nuclei for cloud droplet formation. Therefore, aerosol properties can indirectly influence the earth's radiation budget by affecting cloud albedo (Twomey, 1974), extent, precipitation, and lifetime (Rosenfeld, 2000; Ramanathan *et al.*, 2001; Rosenfeld *et al.*, 2008). The atmospheric aerosol system is impacted by anthropogenic activities, and the total deviation in the earth's radiative balance (radiative forcing) caused by human impact from the preindustrial climate to the year 2005 has been estimated to range between −0.4 and −2.7 W m^{-2} (IPCC, 2007). The effect of changes in the aerosol system on radiative forcing is negative and therefore, at present, is partly counteracting the positive forcing of +2.64 W m^{-2} (in 2005) (IPCC, 2007) originating mainly from greenhouse gas emissions. A significant fraction of the total aerosol effect results from the cloud albedo effect – currently estimated to be between −0.3 and −1.8 W m^{-2} (in 2005) (IPCC, 2007) – forming the largest single source of uncertainty in our understanding of climate change. On the other hand, our level of understanding of many other effects, including the cloud lifetime effect, is so poor (e.g. Penner *et al.*, 2006) that no commonly accepted estimate exists. Understanding the physical and chemical processes affecting the earth's aerosol system and aerosol–cloud coupling is necessary to properly include aerosols in global climate models.

Nucleation is the principal step of new particle formation from precursor vapors (Seinfeld and Pandis, 2006; Kulmala, 2003) and, by definition, involves an energy barrier. During homogeneous nucleation, small clusters are formed from the supersaturated vapor(s) in the

Aerosol Science: Technology and Applications, First Edition. Edited by Ian Colbeck and Mihalis Lazaridis.

gas phase via collisions among the vapor molecules. Nucleation occurs when the particle or cluster passes the energy barrier and the molecular flux on to the particle surface exceeds the molecular flux from the surface, so that cluster continues its growth. A cluster that is just on the verge of becoming a particle (i.e. whose size corresponds to a maximum in the free energy diagram) is called a critical cluster. Nucleation can also occur on the surface of an existing particle, in a process called heterogeneous nucleation. The energy barrier for cluster formation in heterogeneous nucleation is typically lower than that for homogeneous nucleation (see e.g. Kulmala *et al.*, 2007a; Winkler *et al.*, 2008), and in the presence of any surface, heterogeneous nucleation is usually favored. Ion-induced nucleation is a special form of heterogeneous nucleation. In this case, the phase transition occurs on a molecular ion cluster. Nucleation can involve only a single gas-phase species, in which case it is called homomolecular nucleation, or two (binary nucleation), three (ternary nucleation), or even more precursor species.

Nucleation occurs frequently in the earth's atmosphere (Kulmala *et al.*, 2004). An example of such a 'nucleation event' measured at the SMEAR II station (Hari and Kulmala, 2005) in Hyytiälä, southern Finland (see also Mäkelä *et al.*, 1997), is depicted in Figure 7.1. The data in the figure were measured using a standard differential mobility particle sizer (DMPS) system (Aalto *et al.*, 2001). Here, it can be seen that the continuous population of so-called Aitken-mode (see e.g. Hinds, 1999) particles in the size range from some tens to roughly hundreds of nanometers is diluted in the morning. This dilution is mainly a result of the heating of the earth's surface layer by sun radiation,

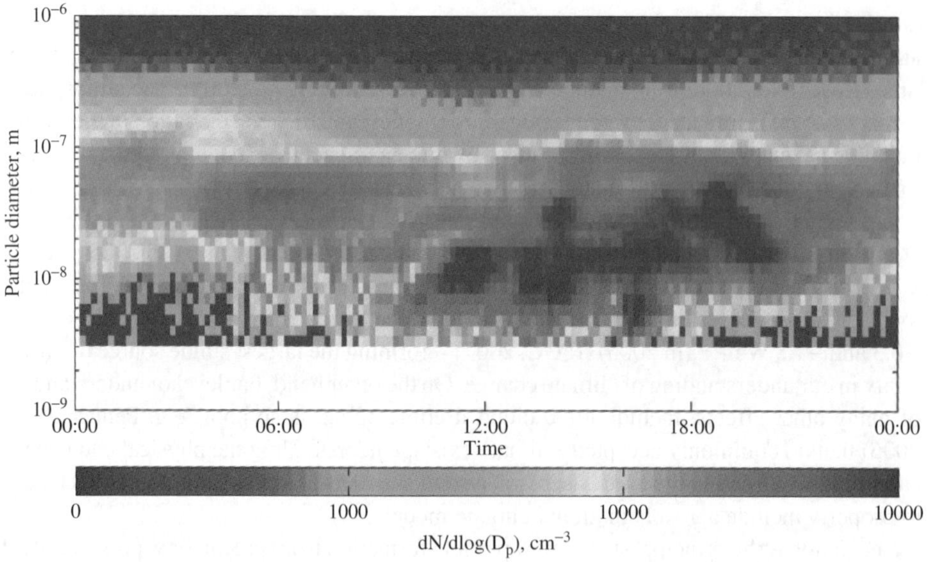

Figure 7.1 *Twenty-four-hour surface plot showing the development of the particle size distribution during a nucleation event day at the Hyytiälä SMEAR II forest field station, Finland. Shading depicts the number of particles in the (logarithmic) size channel. These data were obtained with an ordinary differential mobility particle sizer (DMPS) system, so no information on sub-3 nm concentrations can be obtained. See plate section for colour version.*

resulting in convective fluxes and mixing of the air in the boundary layer. After mixing, slightly before noon, newly formed particles appear at the smallest size channel (3 nm) of the detector. These particles grow with time via condensation and coagulation. Grown particles contribute to the Aitken-mode particle population again the next day. This type of daytime nucleation event occurs in Hyytiälä approximately 60–120 times per year (Kulmala *et al.*, 2010), with the frequency peaking in spring and autumn. It is notable that the nucleation process itself is not measured but that the existence of the phenomenon is concluded from the appearance of 3 nm particles. These 3 nm particles weight several thousand atomic mass units and most likely contain tens or hundreds of molecules. These particles are already stable and grow by coagulation and condensation of low volatile material, including, most likely, sulfuric acid and oxidized organic compounds. Nucleation in the atmosphere occurs at sizes well below 3 nm diameter and the process itself is hidden from the instrument measuring the evolution of the particle size distribution.

To be accurate, the appearance of new particles does not evidence the existence of a nucleation process. As previously mentioned, nucleation occurs when a cluster or particle passes the energy barrier maximum and starts to grow freely. If such a barrier does not exist – if the system is extremely highly supersaturated with respect to particle precursor vapors, for example – nucleation, by definition, will not occur. In such cases, beginning with single molecules, the rate of formation of clusters with any number of molecules is always faster than the rate of cluster decomposition by evaporation. On the other hand, the formation of new particles can also involve several local minima and maxima in the multidimensional free energy of formation space. In such cases, nucleation will happen several times during the growth of a single particle. Figure 7.2 schematically depicts the possible shapes of formation-free-energy distributions for a single-component (homomolecular) case. In a real atmosphere, homomolecular nucleation most likely does not occur, but the energy distribution is a multidimensional quantity. Despite the potential problems with the terminology, 'nucleation' is often used as a synonym for the formation of new particles. Particle size is usually defined, and typically ranges between 1 and 3 nm in diameter.

Even though new particle formation is a global phenomenon (Kulmala *et al.*, 2004), atmospheric nucleation mechanisms are still not understood in detail. In the following sections, we will discuss in more detail how atmospheric nucleation – or the formation of new growing molecular clusters – is studied and how recent progress in nucleation research has reformed our view of atmospheric nucleation mechanisms and the relevance of nucleation for the earth's atmospheric and climate system.

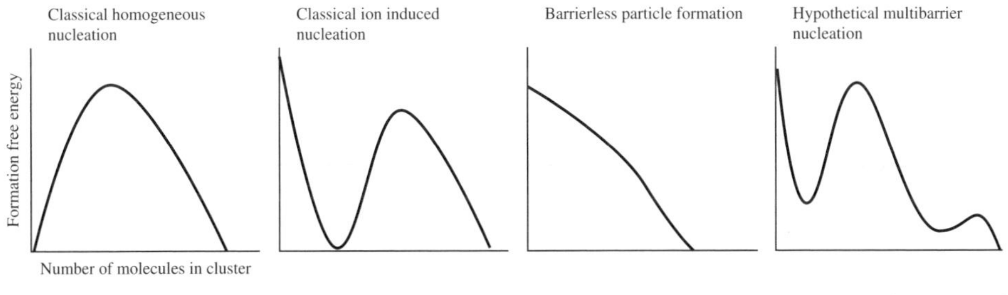

Figure 7.2 *Schematic examples of possible shapes of formation free energy as a function of the number of molecules in a cluster. See text for more details.*

7.2 Detection of Atmospheric Nanoparticles

A variety of different measurement methods are needed to obtain a complete picture of the atmospheric aerosol population. The most important properties of aerosol particles are their size distribution, concentration, and chemical composition. In order to understand particle formation, these and other parameters like formation rate and growth rate must be determined down to sizes close to 1 nm, which is the size at which particle formation takes place.

The size difference between molecules, clusters, and particles (about 1–3 nm) poses a challenge for both theory and experiments. Even the smallest aerosol particles are far too complex to be treated by means of quantum mechanics, but the theories for macroscopic particles start to fail at the molecular limit, and the thermodynamic properties of very small particles are poorly known or badly defined. Experimentalists can use mass spectrometers to measure gas-phase molecules and ions, but most commonly used particle counters only work for aerosol particles larger than about 3 nm. There is thus a gap in our knowledge of sub-3 nm particles, just at the size range that is critical for new particle formation.

Here we concentrate on the detection methods for nanoparticles. Measuring nanoparticles is very challenging because of their small size and minuscule mass. Nanoparticles are also extremely diffusive, leading to substantial losses in the measurement system; they cannot be effectively charged, since their charging probability is small; they cannot be directly detected optically; and they need large supersaturations to grow. There are basically two different techniques available for the measurement of nanoparticles: condensation particle counting (Section 7.2.1) and electrostatic methods (Section 7.2.2).

7.2.1 Condensation Particle Counting

As nanoparticles are much smaller than the wavelength of visible light, they cannot be directly observed optically like larger aerosol particles. The working principle of a condensation particle counter (CPC; previously often called a condensation nuclei counter, CNC) is as follows: the aerosol is exposed to a supersaturated vapor, which condenses on the particles, making them large enough to be observed by optical methods. The supersaturation can be created by at least three different methods: (i) adiabatic expansion, (ii) thermal diffusion in a laminar flow, or (iii) mixing two flows with different temperatures or properties. These methods also serve as a classification of different CPC types.

CPCs have been used for aerosol research for more than a century. The first condensation nucleus counters, based on water vapor condensation on the surface of aerosol particles, were developed individually by M. Coulier and J. Aitken in the late nineteenth century (McMurry, 2000). Practically all early CNCs were expansion-type instruments that used water as the condensing vapor. Modern CPCs with single-particle counting optics, which usually have either alcohol or water as the working fluid, were developed in the 1970s. Their continuous flow rate and fast response time made them suitable for field measurements and connectable to, for example, size spectrometers. The first CPCs designed specifically for nanoparticles (ultrafine CPCs) were developed in the 1990s (Stolzenburg and McMurry, 1991). The history of CPCs is reviewed in more detail by McMurry (2000). CPCs are now commonly used in aerosol research, air-quality monitoring, and so on, both as standalone instruments and as counters in different measurement set-ups.

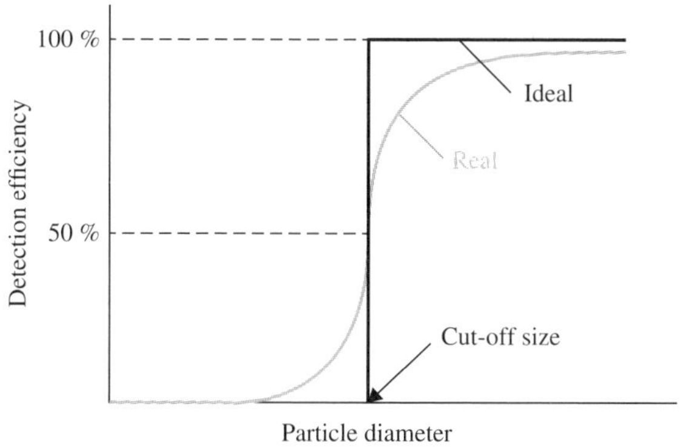

Figure 7.3 *Definition of the cut-off size of a condensation particle counter (CPC).*

The lower size limit of a CPC's measurement range is described by its cut-off size. The cut-off size is defined as the lowest particle diameter at which the instrument still detects at least half of the particles (Figure 7.3). Liu and Kim (1977) were the first to notice that the detection efficiency of a CPC drops when it approaches smaller sizes, even though ideally the counting efficiency curve would be a step function. The cut-off diameter and the steepness of the cut-off curve are determined both by the activation probability of particles and by diffusion losses of particles inside the instrument. The activation probability – that is, the fraction of particles which start growing inside the CPC – in turn depends on both the supersaturation profile created in the instrument and the physicochemical properties of the particles, such as their composition, electric charge, and shape.

A lot of effort has recently been put into developing CPCs that can measure particles down to molecular sizes (e.g. Mordas *et al.*, 2005; Kulmala *et al.*, 2007b; Sipilä *et al.*, 2008, 2009, 2010; Vanhanen *et al.*, 2011; Iida, Stolzenburg, and McMurry, 2009; Jiang *et al.*, 2011). The cut-off size of a CPC can be lowered by minimizing the diffusion losses inside the instrument and/or by increasing the supersaturation achieved. However, when the supersaturation is too high, homogenous nucleation of the working fluid commences, leading to false background counts. Iida, Stolzenburg, and McMurry (2009) evaluated different working fluids and concluded that by selecting a working fluid with a high surface tension but sufficiently low saturation vapor pressure, one can achieve a high supersaturation without considerable homogeneous nucleation. Based on that study, the fluid chosen by several different research groups has been diethylene glycol (DEG; Vanhanen *et al.*, 2011; Jiang *et al.*, 2011). However, DEG-based instruments require a second stage – a growth booster using water or alcohol – to reach optically detectable sizes.

In comparison workshops of different CPCs, mixing-type instruments have been reported to have very low cut-off sizes (Bartz *et al.*, 1985; Ankilov *et al.*, 2002). The work by Fernandez de la Mora and his group especially has shown the superiority of the mixing-type instruments in detecting small particles (e.g. Seto *et al.*, 1997; Gamero-Castaño and Fernandez de la Mora, 2000). The first commercially available instrument with a cut-off size of 1.5 nm that could activate a substantial amount of particles even at 1 nm was the Particle Size Magnifier (Airmodus A09; Vanhanen *et al.*, 2011), which is a mixing-type CPC that uses DEG as the working fluid.

A CPC, by default, measures the total number concentration of particles above a certain size. The differences between the concentration readings of two or more CPCs with different cut-off sizes (e.g. Brock *et al.*, 2000; Kulmala *et al.*, 2007a) – or among the readings of one CPC whose cut-off size can be quickly varied (Gamero-Castaño and Fernandez de la Mora, 2000; Vanhanen *et al.*, 2011) – can give information about the particle concentration in the small size range between the cut-off sizes. The Particle Size Magnifier can be used in scanning mode, continuously changing the flow mixing ratio and therefore the supersaturation, which gives size information on particles between about 1 and 2 nm in diameter (Vanhanen *et al.*, 2011).

Another method for getting size information using only a CPC is the pulse-height analysis technique (e.g. Saros *et al.*, 1996). This technique relies on detecting the intensity of light scattered by particles after their condensational growth in the CPC, which, for white light, is directly proportional to the scattering cross-section of the particle, and thus to its diameter. Due to the axial supersaturation gradient through which particles pass inside the instrument's condenser tube, different-sized particles are activated at different axial positions and have different amounts of time to grow before detection. The smaller the particle, the later it will be activated, leading to smaller final droplet sizes. The pulse-height CPC is sensitive to size only at sizes below about 15 nm, since all larger particles activate almost immediately, leading to equal-sized final droplets whose size information is lost in the growth process. Pulse-height analysis has been used to obtain information on particles between 3 and 10 nm (Weber *et al.*, 1996, 1997) and even below 3 nm (Sipilä *et al.*, 2008, 2009, 2010; Lehtipalo *et al.*, 2009, 2010).

Development of CPC applications that can both measure concentration and provide size information is advantageous compared to electrical measurement techniques, especially at small sizes, because a CPC detects both neutral and charged particles without the need to neutralize the particle population and without the additional diffusion losses that a separate size-selection unit would bring. Also, the signal-to-noise ratio at low concentrations is better, due to single-particle counting. However, the charge and chemical properties of nanoparticles affect their activation probability in a CPC and can affect measurements near the cut-off size of the instrument. The cut-off size of a CPC for material that is soluble in its working liquid is lower than that for insoluble material (e.g. Ankilov, *et al.*, 2002; Hering and Stolzenburg, 2005; Petäjä *et al.*, 2006; Kulmala *et al.*, 2007a). According to Winkler *et al.* (2008), negative ions are activated at smaller sizes than positive ones, and both polarities are activated before neutral particles.

By exploiting the material dependence of the cut-off sizes, CPCs can also be used to obtain indirect information on the chemical composition of nanoparticles. By comparing the counting of two CPCs (or several pairs) calibrated to the same cut-off size using insoluble test particles (e.g. silver), but with different working fluids (a so-called CPC battery; Kulmala *et al.*, 2007a; Riipinen *et al.*, 2009), one can detect changes in the particle composition in the size range near the cut-off size. Also, pulse-height analysis has been used to retrieve composition information (O'Dowd *et al.*, 2002a, 2004).

7.2.2 Electrostatic Methods

Even the tiniest particle can be handled and detected easily if it happens to possess an electric charge. The motion of charged particles (ions) in an electric field is determined by their electrical mobility, which primarily depends on their size and charging state. Particles can thus be size-selected by electric methods, giving a so-called 'electrical mobility equivalent diameter'. The concentration of charged particles can be determined by measuring the current delivered by a flow of ions to an electrometer. The limitation of electrometers is their sensitivity, since the currents are very small, especially when the ion concentration is low.

An important instrument used to produce aerosol samples of known mobility is the differential mobility analyzer (DMA; Knutson and Whitby, 1975; Liu and Pui, 1974; Reischl, Mäkelä, and Necid, 1997 Winklmayr *et al.* 1991, Stolzenburg and McMurry, 2008). There are numerous different DMA designs and configurations, but the diffusional broadening of the transfer function degrades the resolution of most of them at sizes below 3 nm (Stolzenburg, 1988). The demand for measurements below 2 nm has given rise to several high-resolution DMAs that use high flow rates and are thus capable of size-selecting ions even at 1 nm (e.g. Chen *et al.*, 1998; Rosser and Fernandez de la Mora, 2005; Steiner *et al.*, 2010).

When electric measurements are used to determine the total aerosol size distribution, it is necessary to bring the sample to a known charge distribution using forced charging/neutralizing. This can be achieved by artificial unipolar or bipolar charging (Flagan, 1998), such as corona discharges, a radioactive source, or an X-ray charger. The charging efficiency of small particles is very low, about 1% at 3 nm (Hoppel and Frick, 1986), which leads to very poor counting statistics at sizes below that, especially when accounting for diffusion losses in the system. Also, the ions produced inside the neutralizer can extend close to 2 nm depending on polarity and air composition, which makes it hard to separate between the signal and the charger ions in the overlapping area (e.g. Asmi *et al.*, 2009; Manninen *et al.*, 2011). Charging can also be selective to particle composition, since certain chemical compounds, like acids, are more prone to taking the charge than others.

A common tool for measuring the total size distribution of atmospheric aerosol particles is the DMPS or scanning mobility particle size (SMPS) (Hoppel, 1981). This consists of a neutralizer, used to bring the aerosol into charge equilibrium, a DMA for size-selection of the particles, and usually a CPC, used to count the concentration of particles in each size class. There are different configurations of DMPS/SMPS system calibration and operation protocols and data inversion methods. The basic principle is to select one mobility size at a time and to measure its concentration and scan through the desired size range by changing the DMA classification voltage.

Harmonization of DMPS measurements has been suggested, but different instruments still show large deviations, especially for particles below 20 nm (Wiedensohler *et al.*, 2012). New types of SMPS using high-resolution DMAs and low-cut-off CPCs are starting to emerge at the time of writing (Jiang *et al.*, 2011), improving the detection of nanoparticles. However, as a combined effect of low charging efficiency, losses in the sampling lines, DMA penetration efficiency, and CPC detection efficiency, the counting statistics of such instruments drop drastically at sizes below 2 nm.

Ion spectrometry has been used for more than a century for aerosol research (e.g. Hirsikko *et al.*, 2011 and references therein). Ion spectrometers use the electric charge of particles to measure their size distribution. Size selection of the ions is usually carried out based on their mobility in an electric field, and their concentration can be determined from the current they carry to an electrometer. This technique is also applicable to measuring the total particle size distribution, if it is first brought to charge equilibrium. Ion spectrometers are widely used in atmospheric nucleation studies in sub-3 nm size ranges, due to their ability to measure down to molecular sizes.

Lately, the AIS (Air Ion Spectrometer; Mirme *et al.*, 2007) and NAIS (Neutral cluster and Air Ion Spectrometer; Manninen *et al.*, 2009), manufactured by Airel, have been used extensively around the world in new-particle formation studies (e.g. Kulmala *et al.*, 2007b; Manninen *et al.*, 2010). Both measure the mobility distribution of positive and negative ions simultaneously with two parallel DMAs, but in NAIS there is also a measurement mode for total size distribution, in which the sample is charged by corona-discharge-generated ions before size selection is carried out. An electric post-filter section is used to remove the ions produced in the charger unit. The lower size limit of the AIS and NAIS ion mode is about 0.8 nm, whereas the lower size limit of the NAIS in total mode is around 2–3 nm depending on the corona ion distribution and post-filtering (Asmi *et al.*, 2009; Manninen *et al.*, 2011). A second-generation NAIS, the Airborne Neutral cluster and Air Ion Spectrometer (A-NAIS), was introduced in 2010 (Mirme *et al.*, 2010) and has been used for measurements onboard aircraft.

7.2.3 Mass Spectrometric Methods for Cluster Detection

Very recently the development of mass spectrometric methods has made atmospheric measurements of ionic clusters possible, and some success in detecting neutral clusters has been achieved. Only mass spectrometric methods can provide direct information on the chemical composition of ion and neutral clusters in the sub-3 nm diameter range. Two types of mass spectrometer are used in atmospheric cluster detection: quadrupole mass spectrometers and time-of-flight (TOF) mass spectrometers. A common feature is the atmospheric pressure interface (APi) required to guide the sample from ambient pressure conditions to the high vacuum of the mass analyzer.

Mass spectrometers can be used to characterize the natural ion cluster population or can be equipped with a chemical ionization source to charge the neutral clusters. The first mass spectrometric studies on atmospheric molecular ions and ion clusters date back to the 1980s and Fred Eisele's (see Eisele, 1989 and references therein) outstanding dual–collision-chamber tandem triple-quadrupole mass spectrometer. Eisele was the first to indentify the true nature of tropospheric ions. His methods were so progressive that it took more than 20 years before substantial advances in measurement technology allowed other research groups to confirm his findings (Ehn *et al.*, 2010).

The quantum leap in the mass spectrometry was the development of the atmospheric pressure interface–time-of-flight (APi-TOF) mass spectrometer (Junninen *et al.*, 2010), manufactured by Tofwerk AG., Switzerland. In APi-TOF, the sample at atmospheric pressure is guided through three differentially pumped chambers to a TOF analyzer. The first and second chambers contain short segmented quadrupoles used in ion-guide mode and the third chamber contains an ion lens assembly. The electric fields in quadrupoles are largely responsible for determining the ion transmission as a function of mass-to-charge ratio and affect the fragmentation of clusters to core ions. This combination produces a beam of ions that is directed to the TOF chamber. In the TOF system, the ions interact with a transverse electric field that gives them kinetic energy K according to the number z of elementary charges e carried by the ion (atmospheric ions typically have $z = 1$), $K = Uze$. Kinetic energy determines the velocity of the ion, $K = \frac{1}{2}mv^2$, and the velocity determines the time it takes for the ion to travel the distance s to the detector, $t = s/v = s(2Ue)^{-1/2}(\mathrm{m/z})^{1/2}$. Thus, TOF t is a direct measure of the mass-to-charge ratio. The constant $s(2Ue)^{-1/2}$ is determined by calibration using known ions. As a result of real-world effects, the relationship between time and m/z is rarely exactly linear, so calibration with more than just two masses is often desirable.

Mass resolution of a TOF analyzer depends on the distance s that ions travel before hitting the detector. The longer the distance, the more separated in time the two masses. To maximize the distance in a compact instrument, the ions must not travel a linear path in the APi-TOF but rather follow a V- or W-shaped trajectory. Such trajectories are achieved by means of electric fields inside the flight chamber, analogous to what happens when a bowling ball is thrown uphill. APi-TOF mass resolution is approximately 3000 Th/Th in V-mode, meaning that at $m/z = 100$ Th, for example, the peak width at half maximum is 0.033 Th (if $z = 1$, the peak from the 100 Da (= 100 amu) ion is 0.033 Da wide). High resolution is crucial, as there can be several constituents with the same integer mass in the atmosphere, but their actual masses will differe slightly due to their different atomic compositions. For example, the mass of a bisulfate ion–sulfuric acid cluster, which is frequently present in an atmospheric ion cluster population, is 194 927 Da, and the mass of the hexafluorobutanoate, the artificial compound that can originate from, for example, the Teflon tubing often used in various instruments, is 194 989 Da – a 0.06 Da difference (Figure 7.4). With insufficient mass resolution, the two compounds would not be able to be separated and other methods, such as preceding ion mobility measurement, would be required for the identification and quantification of the ions at an integer mass of 195 (Jokinen *et al.*, 2012).

Besides the resolution, the mass accuracy must also be high in order to identify compounds according to their exact mass. If properly calibrated, APi-TOF achieves a mass accuracy of 20 ppm or better, meaning that at 100 Th the error in m/z is less than 0.002 Th.

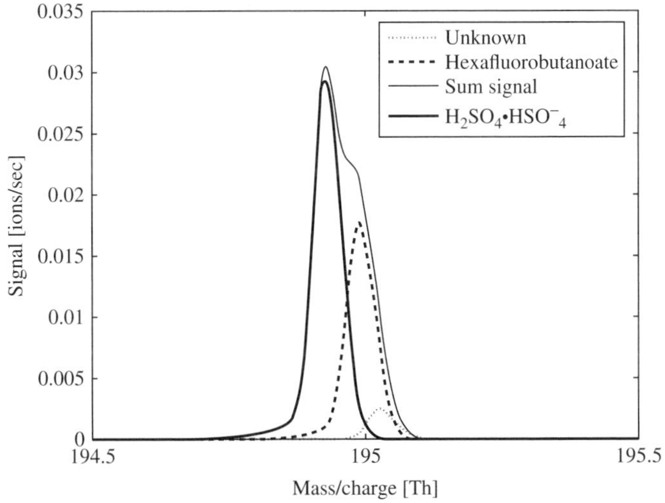

Figure 7.4 *High mass resolution is required for the separation of a bisulfate ion–sulfuric acid cluster of interest from the hexafluorobutanoate that originates from manmade materials, which has nothing to do with atmospheric composition.*

Along with the exact mass, isotopic distributions are also routinely used for peak identification. High resolution and mass accuracy can also be achieved with, for example, a triple quadrupole mass spectrometer and are not unique for TOF. The main advantage of a TOF mass spectrometer in comparison to quadrupole systems is that with TOF the whole mass spectrum is measured simultaneously. In order to achieve a reasonable time resolution with quadrupole mass spectrometry, which is capable of measuring only one *m/z* at time, the measured masses would need to be preselected, so the information obtained would be very limited.

APi-TOF is not a plug-and-play instrument; its efficient use requires highly automated data-processing software, such as the recently developed TofTools (Junninen *et al.*, 2010). Processing software should indeed be considered one of the crucial components of APi-TOF.

Despite APi-TOF's great performance, it only detects ions. In order to detect neutral constituents and clusters, they would need to be charged. Selective ion chemical ionization under atmospheric pressure has proven to be a highly efficient method for charging molecular sulfuric acid for subsequent mass spectrometric detection (Eisele and Tanner, 1993). The Chemical Ionization (quadrupole) Mass Spectrometer (CI-MS) has been widely used to detect atmospheric sulfuric acid, with an outstanding sensitivity on the order of a few 10^4 molecules per cm^{-3} (Eisele and Tanner, 1993; Petäjä *et al.*, 2009). The instrument uses nitrate ions and their clusters with nitric acid for the chemical ionization of sulfuric acid via proton transfer from sulfuric acid to nitrate ion, forming a bisulfate ion–nitric acid cluster. If necessary, this cluster can be dissociated to bare bisulfate ion prior to mass filtration by means of electric-field-induced collisions in a collision dissociation chamber.

$$H_2SO_4 + NO_3^- \cdot (HNO_3)_n \rightarrow HSO_4^- \cdot HNO_3 + n(HNO_3)$$

The same chemistry has been applied to the detection of neutral sulfuric acid containing clusters, with the assumption that they are charged in collisions with a nitrate ion (Eisele and Hanson, 2000; Hanson and Eisele, 2002; Hanson and Lovejoy, 2006; Zhao *et al.*, 2010). This is not necessarily always a valid assumption, as theoretically shown by Kurtén *et al.* (2011).

Although in laboratory nucleation flow tubes with relatively high concentrations of neutral clusters, nitrate ion-based CI-MS systems have been proven to work to some extent for the detection of neutral clusters (Hanson and Lovejoy, 2006; Petäjä *et al.*, 2009), the problems related to their application in cluster detection in ambient atmosphere are numerous (Jokinen *et al.*, 2012). First, a number of different compounds can become charged upon collision with nitrate ion, via either charge transfer or clustering with the nitrate ion. Most of the resulting signal does not come from neutral clusters but can have an almost identical *m/z* to the ions expected to result from the charging of the neutral clusters. Most of the compounds also have a similar diurnal cycle to that expected for sulfuric acid clusters, so the time behavior cannot be utilized straightforwardly. A high mass resolution and a low concentration of constituents that form overlapping signal are thus required. Second, naturally charged ions can penetrate the ionization system and be detected by a mass spectrometer. Since the calibration factor for the conversion of the measured signal to a neutral concentration is usually large, even miniscule concentrations of ions can affect the measurement. Third, ion-induced clustering of, for example, sulfuric acid readily takes place inside the ion source after the ionization of molecular sulfuric acid. If the neutral-cluster concentrations are low, the clusters formed by ion–molecule attachment can overwhelm the signal from the neutrals. Any experiment must be designed so that ion-induced clustering can be separated from neutral-cluster charging (Zhao *et al.*, 2010). Fourth, there are no reliable experimental data on the charging probabilities of clusters with different chemical compositions; they might become charged, but with unknown probability (Kurtén *et al.*, 2011).

The fundamental problem underlying these challenges is that the concentrations of clusters with a defined composition are expected to be very low and the concentration of neutral clusters with a certain composition is expected to decrease with increasing cluster size and complexity. CI-APi-TOF is probably the most sensitive mass spectrometer available for cluster detection. Assuming no interference from other gas-phase constituents or ions and a collision-limited cluster charging probability, a concentration of about 10^4 clusters/cm^{-3} per composition would be required to allow sufficient signal to be quantified by the CI-APi-TOF (Jokinen *et al.*, 2012).

7.3 Atmospheric Observations of New Particle Formation

7.3.1 Nucleation

The developer of the first condensation nucleus counter, J. Aitken, was also the first to understand that particles can be formed in the atmosphere as a result of photochemical processes (Aitken, 1911). Since then, hundreds of articles have revealed that new particle formation is a common phenomenon that happens frequently almost everywhere in the troposphere (Kulmala *et al.*, 2004a). Observations of new particle formation have been made both in pristine environments and in areas with heavy anthropogenic influence, including boreal forest (Mäkelä *et al.*, 1997; Vehkamäki *et al.*, 2004), rural areas (Weber *et al.*, 1997; Birmili *et al.*, 2003), African savannah (Vakkari *et al.*, 2011), Antarctica (Asmi *et al.*, 2010), and large cities (Wehner *et al.*, 2004; Mönkkönen *et al.*, 2005). The longest continuous time series of aerosol size distribution measurements and statistics on new particle formation events comes from the Hyytiälä SMEAR II measurement station in southern Finland (Hari and Kulmala, 2005), beginning in 1996. As well as the boundary layer, observations have been made of new particle formation in the free troposphere from high-altitude stations and airplane measurements (e.g. Venzac, Sellegri, and Laj, 2007; Mirme *et al.*, 2010).

New particle formation is mainly a daytime phenomenon, with only a few exceptions (Wiedensohler *et al.*, 1997; Vehkamäki *et al.*, 2004; Lee *et al.*, 2008; Suni *et al.*, 2008), although the concentration of sub-3 nm ions and particles can sometimes increase during night (Junninen *et al.*, 2008; Lehtipalo *et al.*, 2011). This indicates the importance of photochemistry in some crucial stage of the formation process. Other features common to conditions favorable to new particle formation are low condensation sink (preexisting aerosol concentration), low relative humidity, and high vapor source rate (Kulmala and Kerminen, 2008).

It is commonly agreed that sulfuric acid is the key compound responsible for atmospheric new particle formation (e.g. Weber *et al.*, 1997; Hanson and Eisele, 2002; Sihto *et al.*, 2006; Kuang *et al.*, 2008; Petäjä, Kerminen, and Kulmala, 2009; Nieminen *et al.*, 2009, Sipilä *et al.*, 2010). However, many studies show that binary sulfuric acid–water nucleation alone cannot explain observations of new particle formation in the troposphere (e.g. Weber *et al.*, 1996, Kulmala *et al.*, 1998; Fiedler *et al.*, 2005; Boy *et al.*, 2005; Wehner *et al.*, 2005; Kirkby *et al.*, 2011). Both quantum chemical calculations and experimental data show that bases such as ammonia or amines might be needed to stabilize the sulfuric acid clusters (e.g. Hanson and Eisele, 2002; Torpo *et al.*, 2007; Kurtén *et al.*, 2008, Ehn *et al.*, 2010, Yu *et al.*, 2012).

Indirect and direct measurements of chemical composition have shown the presence of organic compounds in ultrafine particles (O'Dowd *et al.*, 2002a; Smith *et al.*, 2008; Riipinen *et al.*, 2009). However, discerning nucleation from initial growth was impossible until very recently, as most measurements started at sizes at which growth had already taken place. Some claim that organics affect only growth, while others say that they have an effect at the nucleation step (Metzger *et al.*, 2010; Paasonen *et al.*, 2010). The ability of volatile organic compounds (VOCs) to form particles has been demonstrated in many laboratory and chamber experiments (e.g. Mentel *et al.*, 2009). VOCs are a large group of different compounds emitted by both anthropogenic sources and vegetation. They can become oxidized in the atmosphere to create less-volatile products, which can thus condense into particle phase. As the particles grow, the supersaturation needed for condensation decreases as a result of the Kelvin effect, and even vapors with higher saturation vapor pressure can start to condense. The connection between different compounds and particle formation is complex, as some VOCs can also inhibit particle formation, possibly by acting as OH scavengers (Kiendler-Scharr *et al.*, 2009).

Marine environments are a special case regarding new particle formation, and nucleation is rarely observed above the ocean. On coastal sites, however, strong local new-particle-formation bursts frequently occur during daytime low tide (O'Dowd *et al.*, 2002b; Yoon *et al*, 2006). Coastal nucleation is strongly connected to the biological emissions of iodine-containing compounds, which are processed by photochemical reactions when algae are exposed to sunlight (O'Dowd *et al.*, 2002b,c).

There is still controversy regarding the importance of ions in new particle formation. Electric charge can in principle help both the nucleation and growth steps (Laakso *et al.*, 2003; Kirkby *et al.*, 2011). Since the amount of electric charge, and thus of small ions in the atmosphere, is limited by the ionization rate, the relative role of ion-induced nucleation seems to decrease with increasing total particle formation rate (Gagné *et al.*, 2010; Manninen *et al.*, 2010). On average, the ion-induced fraction in the boundary layer seems to be mostly around or below 10% (Iida *et al.*, 2006; Gagné *et al.*, 2008; Manninen *et al.*, 2010; Kulmala *et al.*, 2010), although higher estimations can be found in literature (Yu and Turco, 2008; Yu *et al.*, 2008). However, ions may be more important at the beginning of particle formation (Laakso *et al.*, 2007a,b) and ion-mediated nucleation is likely to be important under some conditions and in environments – for example, in the upper troposphere (Curtius *et al.*, 2006; Yu, 2010).

The magnitude of secondary particle formation as a source of atmospheric aerosols and cloud condensation nuclei (CCN) can be very sensitive to the first steps of the particle formation and growth process. A reason for the lack of knowledge about the nucleation process has been the inability to detect or chemically identify the initial clusters and their precursor vapors at atmospheric concentration levels. The recent development of measurement techniques (Chapter 2) has provided us with tools to reach these sizes. While the continuous existence of small ions in the atmosphere has been known for at least a century, the

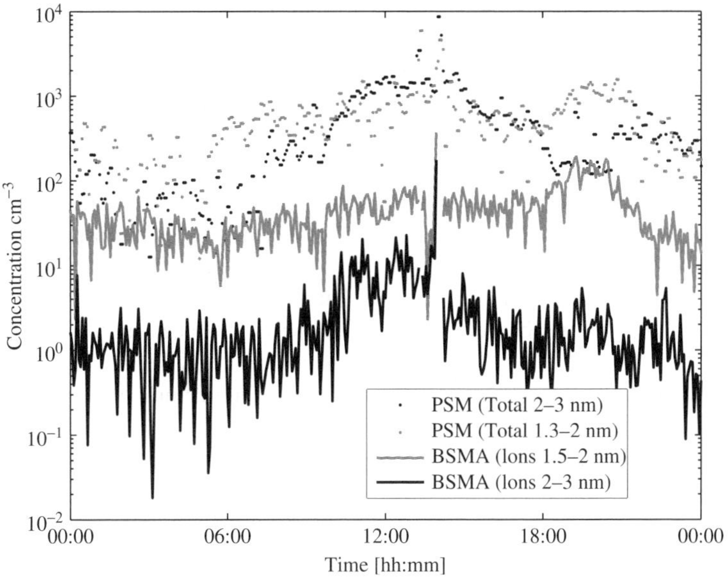

Figure 7.5 *Total concentration of particles between 1.3–2.0 nm and 2.0–3.0 nm mobility diameter, measured by the Airmodus A09 PSM in Hyytiälä SMEAR II station 16.8.2010, and ion concentration in corresponding size classes, measured by the BSMA ion spectrometer. See plate section for colour version.*

existence of their neutral counterparts was uncertain, as they could not be detected. Using model calculations, Kulmala *et al.* (2000, 2005) predicted a pool of neutral clusters as the explanation for the observed new-particle-formation events, and neutral prenucleation clusters have been measured in laboratory experiments (Hanson and Eisele, 2002). Weber *et al.* (1995) detected large molecular clusters during nucleation events at Mauna Loa, Hawaii. The continuous existence of neutral clusters in the atmosphere was first measured by Kulmala *et al.* (2007b) and Sipilä *et al.* (2008). Recently, particle size magnifier (PSM) technology has enabled a more quantitative measurement of neutral sub-3 nm particles and clusters (Figure 7.5). However, the size ranges of molecules, clusters (of several molecules), and tiny aerosol particles overlap, and it is usually not possible to distinguish between them using physical measurements. Also, the terminology around sub-3 nm atmospheric objects is ambiguous. Therefore, the concept of 'nano-CN' (nanometer-sized condensation nuclei) was introduced (Zhao *et al.*, 2010; Lehtipalo *et al.*, 2010; McMurry *et al.*, 2011). The dynamics and concentrations of sub-3 nm nano-CN in the atmosphere have been investigated by Lehtipalo *et al.* (2009, 2010) and Jiang *et al.* (2011) but the role of nano-CN and ion clusters in particle formation still needs to be studied and quantified.

7.3.2 Growth

The growth rate of newly formed particles finally determines what fraction of them reaches a size where they can scatter light and act as CCN and thus affect the climate before they are scavenged by different loss processes. The growth rate can also provide information about the amount and properties of condensing vapors in the atmosphere.

The spatial area over which new particle formation takes place affects the shape and time evolution of the measured size distributions. If the particle source is a relatively small area (point or line source) close to the measurement site, the shift (growth) of particles toward larger sizes cannot usually be observed. However, when the formation occurs over a larger area, the apparent growth of particles can be followed from the size distribution measurements at a fixed site. Thus, for regional new-particle-formation events, the growth rate of newly formed particles can be determined from the time evolution of the measured size distribution. There are several methods for doing this, such as following the geometric mean diameter of the nucleation mode (Dal Maso *et al.*, 2005) or finding the time of the maximum concentration for each size bin (Hirsikko *et al.*, 2005). Iida *et al.* (2008) used the size-dependency of the charged fraction of the particles to determine their growth rate. The time lag between the rise in sulfuric acid concentration and the smallest detectable particles has also been used to estimate the growth rate at the beginning of particle formation (Weber *et al.*, 1997; Sihto *et al.*, 2006).

The growth rate of nucleation-mode particles seems typically to vary between 1 and 20 nm in the atmosphere (Yli-Juuti *et al.*, 2011 and references therein). Higher values have been observed; for example, in polluted (Iida *et al.*, 2008) and coastal (O'Dowd *et al.*, 2007) areas, where sources of condensable vapors are numerous.

Observations on the size dependency of growth are rare, since often the growth rates have been integrated over large size/time windows. There is an indication that the growth rate increases with size (Hirsikko *et al.*, 2005; Yli-Juuti *et al.*, 2011; Kuang *et al.* 2012). Neglecting the size dependency might lead to a significant overestimation of the survival probability of particles in climate models (Kuang *et al.*, 2012). Also, the growth rate below 3 nm shows a different seasonal variation to the growth rate at larger sizes in boreal forest (Hirsikko *et al.*, 2005; Yli-Juuti *et al.*, 2011), indicating that the main condensing vapor(s) can change during the growth process. The fraction of growth that sulfuric acid can explain also seems to diminish with increasing particle size (Fiedler *et al.*, 2005; Kuang *et al.*, 2012), and usually other vapors are responsible for the main part of the growth (e.g. Birmili *et al.*, 2003; Boy *et al.*, 2005; Nieminen *et al.*, 2010; Riipinen *et al.*, 2011).

For technical reasons discussed in Chapter 2, size distribution measurements extending to molecular sizes have so far been possible only for naturally charged particles, and thus growth information in sub-3 nm size range has also been limited to ions. Yli-Juuti *et al.* (2011) compared different methods and instruments for calculating growth rates and concluded that the relative uncertainty between them increased toward smaller sizes, being about 25% below 3 nm. Due to differences in their dynamics, small ions may exhibit a different growth rate to the neutral particle population (Leppä *et al.*, 2011). New methods and instruments are beginning to bring information on the initial growth of the total particle population (Kuang *et al.*, 2012; Yu *et al.*, 2012).

7.4 Laboratory Experiments

7.4.1 Sulfuric Acid Nucleation

As already discussed, nucleation – or at least formation – of new particles in the atmosphere readily takes place in the presence of sulfuric acid at concentrations on the order of

10^6 molecules cm^{-3}. For this reason, nucleation of sulfuric acid with water and with ternary species has been widely investigated, both theoretically using classical (e.g. Mirabel and Katz, 1974; Vehkamäki *et al*, 2002) and quantum chemical (e.g. Kurtén *et al.*, 2008, 2010, 2011) approaches and experimentally in various laboratories, typically utilizing flow tubes or chambers.

Laboratory experiments performed in systems apparently containing only sulfuric acid and water have yielded contradictory information, with nucleation rates and onset sulfuric acid concentrations (the concentration required to produce a nucleation rate of 1 particle $cm^{-3}\,s^{-1}$) varying by several orders of magnitude (Figure 7.6). Some of the disagreement has been shown to result from technical problems in the experimental design (Sipilä *et al.*, 2010). The growth rate of newly formed particles by condensation of sulfuric acid is too slow: approximately only 1 nm/h for $[H_2SO_4] = 10^7$ molecules cm^{-3} (Nieminen *et al.*, 2010). Therefore, without either a sufficiently long residence time or a particle detector capable of measuring particles well below 3 or 2 nm in diameter, the detector's size-dependent detection efficiency falsifies the true concentration of nucleated particles (Sipilä *et al.*, 2010). This effect is visualized in Figure 7.7. Most of the experiments performed in flow tubes prior to that of Sipilä *et al.* (2010) necessarily suffered from a short residence time (from seconds to a few minutes), rapid loss of sulfuric acid to the tube walls, and poor counting efficiency of the CPC used to detect the nucleated particles. Flow-tube experiments performed using a cluster mass spectrometer as a detector (Eisele and Hanson, 2000; Hanson and Eisele, 2002; Hanson and Lovejoy, 2006, Petäjä *et al.*, 2011) do not, of course, suffer similar counting problems to those of CPCs, but the counting efficiency of a selective CI-MS might be affected by different reasons (Kurtén *et al.*, 2011).

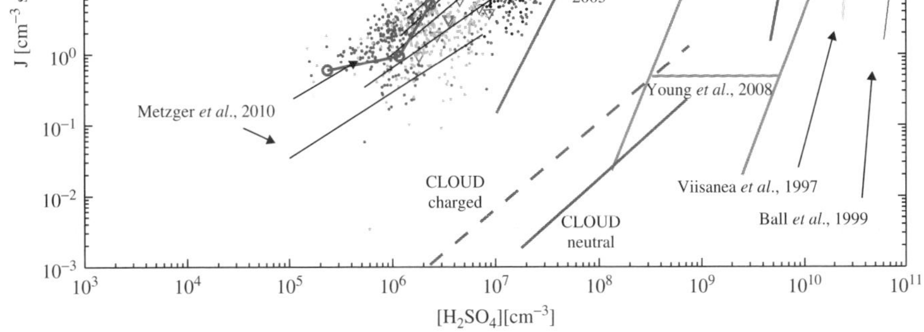

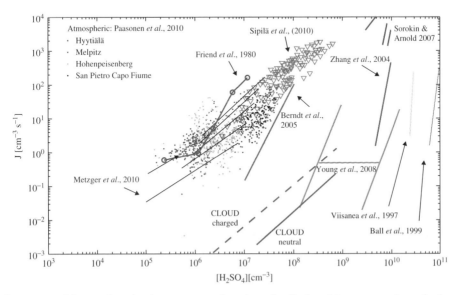

Figure 7.6 *Observed nucleation rates as a function of sulfuric acid concentration, obtained from a selection of different laboratory experiments. See plate section for colour version.*

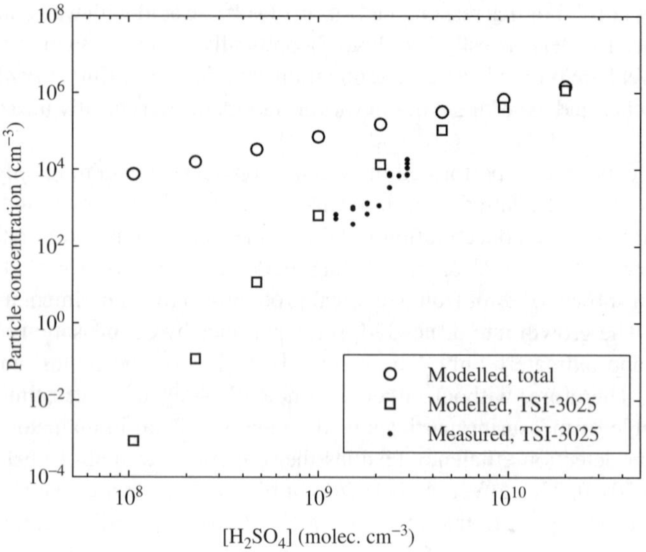

Figure 7.7 *Simulated total and simulated 'detected' particle number concentration, assuming a size-dependent detection limit of TSI-3025A CPC at the output of an IfT-LFT (Institute for Tropospheric Research – Laminar Flow Tube). In this simulation, sulfuric acid is injected into the beginning of the 5.05 m long flow tube and a total residence time of 379 seconds is assumed. For comparison, the concentration measured by TSI-3025A CPC is shown. The nucleation rate is expressed as* $J = 3 \cdot 10^{-6} [H_2SO_4]^1$*, the mechanism describing the data obtained by Sipilä* et al. *(2010) in the IfT-LFT, and growth is assumed to result from pure sulfuric acid condensation. Depending on the initial sulfuric acid concentration, the apparent particle concentration is skewed due to insufficient growth. With sulfuric acid concentration approaching the atmospheric range* $< 10^8$ *molecules cm*$^{-3}$*, no particles can be observed any longer.*

7.4.2 Hunt for Compound X

Qualitative explanation of the disagreement between many earlier flow-tube experiments was given by Sipilä *et al.* (2010), who indeed succeeded in producing an atmospherically relevant relation between nucleation rate and sulfuric acid concentration in the apparent absence of ternary species (i.e. without adding any ternary species to the system). In the proceeding study, performed in the same system using a mass spectrometer for detection of sulfuric acid clusters, it was found that the sulfuric acid dimers formed in the system in an initial nucleation process step are so strongly bound that the presence of some stabilizing compound is required (Petäjä *et al.*, 2011). Quantum chemical simulations accompanied by dynamical modeling of cluster formation suggested that a stabilizing compound with a similar stabilization capability to that of dimethyl amine would explain the observations if present in the system at a concentration as low as a few 10^7 molecules cm^{-3}. Such a compound might be responsible for the particle formation rates observed in the experiment by Sipilä *et al.* (2010), as well as in some other experiments (including Berndt *et al.*, 2005) and in the ambient atmosphere. Due to this observation and previous field observations by Mäkelä *et al.* (2001) and Smith *et al.* (2010), as well as theoretical findings by Kurtén *et al.*

(2008), several laboratory studies on enhancing the effect of different organic amines have been performed within the last year or two (Berndt *et al.*, 2010; Benson *et al.*, 2011; Zhao *et al.*, 2011). All of these studies report an increased nucleation rate upon amine injection. Similar enhancing effects on nucleation rates have been reported for ammonia. However, most recent works suggest that the required concentrations are too high for the stabilizing effect of ammonia to explain the atmospheric boundary layer nucleation rates (Kirkby *et al.*, 2011).

Organics are a plausible candidate for explaining atmospheric nucleation rates as most of the atmospheric trace constituents are organics, at least in the boundary layer. Obviously, most organics are not capable of nucleation. Organic amines, discussed earlier, are often separated colloquially from the rest of the organic compounds, which contain only carbon, hydrogen, and oxygen. The role of oxidized organics in nucleation has been investigated in a chamber study by Metzger *et al.* (2010), who suggested based on the results that atmospheric nucleation might be initiated by the formation of a cluster containing one molecule of sulfuric acid and one molecule of nucleating oxidized organic vapor. This result gets support from the flow-tube experiment by Zhang *et al.* (2004), who showed that certain organics are able to contribute to the initial steps of particle formation, if not nucleation, at least in early growth. That experiment, however, was performed with relatively high concentrations of sulfuric acid and suffers from the short residence time and size-sensitive particle counting discussed earlier. Thus, still more experimental work is required before non-nitrogen-containing organics can be declared to participate in nucleation and their total role in secondary particle formation can be assessed.

7.5 Concluding Remarks and Future Challenges

Work toward solving the mechanisms of atmospheric nucleation and early growth continues. One of the keys to a deeper understanding of these processes is the continuous development of new methods and instruments for direct measurement of the concentration, size, and chemical composition of newly formed clusters and their potential precursors. Better tools for detecting amines from the atmosphere and from experimental laboratory systems are needed to demonstrate their role in particle formation, for example. Knowledge of particle formation at a molecular level, and of the sources and sinks of precursor gases, is required before its global impact can be projected. Recent findings by Welz *et al.* (2012) indicate that not even the cycles of sulfuric acid are fully understood. Amine sources, sinks, and concentrations, and the factors determining them, are still poorly understood; the same holds for the other, non-nitrogen-containing organics.

While work toward understanding atmospheric nucleation mechanisms in detail continues, one thing is clear: sulfuric acid plays a crucial role in a number of atmospheric environments (Sipilä *et al.*, 2010; Kirkby *et al.*, 2011; Zhao *et al.*, 2010) and is one of the key compounds in atmospheric nucleation and new particle formation. Sulfuric acid mostly forms via photo-oxidation of sulfur dioxide, SO_2. SO_2 has three major sources in the atmosphere: volcanoes, photo-oxidation of dimethyl sulfide (DMS) emitted by phytoplankton in the seas (Charlson *et al.*, 1987), and human emissions; the latter make the dominant contribution to the total atmospheric burden (Smith *et al.*, 2001).

Global modeling studies show that secondary aerosol particle formation initiated by nucleation significantly increases aerosol number concentrations in the boundary layer. Increased particle number concentrations have a major effect on CCN (and cloud droplet number concentrations) and, therefore, on cloud properties (Makkonen *et al.*, 2009). The present-day contribution of boundary layer nucleation to CCN is estimated to range from a few percent up to about 50%, depending on the nucleation mechanism and assumed supersaturation required for CCN activation (e.g. Spracklen *et al.*, 2006; Merikanto *et al.*, 2010). These models often apply an activation, $J = A[H_2SO_4]$ (Kulmala *et al.*, 2006), or kinetic, $J = K[H_2SO_4]^2$ (McMurry, 1980), nucleation mechanism in the planetary boundary layer and binary H_2SO_4–H_2O nucleation (Vehkamäki *et al.*, 2002) in the upper troposphere. The models are sensitive to the correct description of nucleation mechanism (besides the correct description of growth processes and of CCN activation), however, and therefore the remaining open questions in atmospheric nucleation must be answered before the present-day anthropogenic aerosol effect on radiative forcing and climate can be reliably assessed. The latest global-climate-model-based estimate of the contribution of atmospheric nucleation to radiative forcing is about $-0.6\,W\,m^{-2}$ (Makkonen *et al.*, 2012), but further work is required to define this value and estimate the range of uncertainty.

Even though we cannot yet properly quantify the magnitude of the present-day anthropogenic aerosol effect, the future can be projected, at least quantitatively. Sulfur dioxide plays a major role in atmospheric nucleation and new particle formation but the concentration of SO_2 is in a state of change. Global anthropogenic SO_2 emissions, while still projected to increase for the next decade or two, are predicted to significantly decrease by the year 2100 (IPCC, 2000). Emissions in Europe and North America had already started to decrease by the 1970s, but that decrease is still globally overwhelmed by increasing emissions in East Asia and other strongly developing regions of the world (Smith *et al.*, 2011). This situation is expected to change in near future. Since SO_2 is the precursor of sulfuric acid, decreasing trends in H_2SO_4 levels and subsequently in new particle formation rates are also expected. The aerosol indirect effect on climate is cooling, and therefore air-pollution control can result in an acceleration of climate warming (e.g. Arneth *et al.*, 2009). Due to the decrease of SO_2 emissions and the subsequent suppression of aerosol particle formation rates, together with the suppression of primary aerosol emissions, the total anthropogenic aerosol forcing, $-1.61\,W\,m^{-2}$ in the year 2000, is predicted to be greatly reduced in the future, down to $-0.23\,W\,m^{-2}$ in 2100 (Makkonen *et al.*, 2012). Such a change in radiative forcing corresponds roughly to a doubling of the present-day total anthropogenic radiative forcing ($+0.6 \ldots +2.4\,W\,m^{-2}$ in 2005) and would certainly accelerate climate warming. However, as discussed before, climate models are sensitive to a proper description of particle dynamics, including nucleation, and therefore definite conclusions about future changes are pending.

References

Aalto, P., Hämeri, K., Becker, E. *et al.* (2001) Physical characterization of aerosol particles in boreal forests. *Tellus*, **53B**, 344–358.

Aitken, J. (1911) The sun as a fog producer. *Proceedings of the Royal Society of Edinburgh*, **32**, 138–215.

Ankilov, A., Baklanov, A., Colhoun, M. *et al.* (2002) Particle size dependent response of aerosol counters. *Atmospheric Research*, **63**, 209–237.

Arneth, A., Unger, N., Kulmala, M. and Andreae, M.O. (2009) Clean the air, heat the planet? *Science*, **326**, 672–673.

Asmi, E., Sipilä, M., Manninen, H.E. *et al.* (2009) Results of the first air ion spectrometer calibration and intercomparison workshop. *Atmospheric Chemistry and Physics*, **9**, 141–154.

Asmi, E., Frey, A., Virkkula, A., Ehn, M., Manninen, H.E., Timonen, H., O., Tolonen-Kivimäki, Aurela, M., Hillamo, R., and Kulmala, M. (2010) Hygroscopicity and chemical composition of Antarctic sub-micrometre aerosol particles and observations of new particle formation, *Atmospheric Chemistry and Physics*, **10**, 4253–4271, doi: 10.5194/acp-10-4253-2010.

Ball, S.M., Hanson, D.R., Eisele, F.L. and McMurry, P.H. (1999) Laboratory studies of particle nucleation: initial results for H_2SO_4, H_2O, and NH_3 vapors. *Journal of Geophysical Research*, **104** (D19), 23709–23718.

Bartz, H., Fissan, H., Helsper, C. *et al.* (1985) Response characteristics for four different condensation nucleus counters to particles in the 3-50 nm diameter range. *Journal of Aerosol Science*, **16**, 443–456.

Benson, D.R., Yu, J.H., Markovich, A. and Lee, S.H. (2011) Ternary homogeneous nucleation of H 2 SO 4, NH 3, and H 2 O under conditions relevant to the lower troposphere. *Atmospheric Chemistry and Physics*, **11**, 4755–4766.

Berndt, T., Boge, O., Stratmann, F. *et al.* (2005) Rapid formation of sulphuric acid particles at near atmospheric conditions. *Science*, **307**, 698–700.

Berndt, T., Stratmann, F., Sipilä, M. *et al.* (2010) Laboratory study on new particle formation from the reaction $OH + SO_2$: influence of experimental conditions, H_2O vapour, NH_3 and the amine tert-butylamine on the overall process. *Atmospheric Chemistry and Physics*, **10**, 7101–7116.

Birmili, W., Berresheim, H., Plass-Dülmer, C. *et al.* (2003) The Hohenpeissenberg aerosol formation experiment (HAFEX): a long-term study including size-resolved aerosol, H_2SO_4, OH, and monoterpenes measurements. *Atmospheric Chemistry and Physics*, **3**, 361–376.

Boy, M., Kulmala, M., Ruuskanen, T.M. *et al.* (2005) Sulphuric acid closure and contribution to nucleation mode particle growth. *Atmospheric Chemistry and Physics*, **5**, 863–878.

Brock, C.A., Schröder, F., Kärcher, B. *et al.* (2000) Ultrafine particle size distributions measured in aircraft exhaust plumes. *Journal of Geophysical Research*, **105**, 26555–26567.

Charlson, R.J., Lovelock, J.E., Andreae, M.O. and Warren, S.G. (1987) Oceanic phytoplankton, atmospheric sulphur, cloud albedo and climate. *Nature*, **326**, 655–661.

Chen, D.-P., Pui, D.Y.H., Hummes, D. *et al.* (1998) Design and evaluation of a nanometer aerosol differential mobility analyzer (Nano-DMA). *Journal of Aerosol Science*, **29**, 497–509.

Curtius, J., Lovejoy, E.R. and Froyd, K.D. (2006) Atmospheric ion-induced aerosol nucleation. *Space Science Reviews*, **125**, 159–167.

Dal Maso, M., Kulmala, M., Riipinen, I. *et al.* (2005) Formation and growth of fresh atmospheric aerosols: eight years of aerosol size distribution data from SMEAR II, Hyytiälä Finland. *Boreal Environment Research*, **10**, 323–336.

Ehn, M., Junninen, H., Petäjä, T. *et al.* (2010) Composition and temporal behavior of ambient ions in the boreal forest. *Atmospheric Chemistry and Physics*, **10**, 8513–8530. doi: 10.5194/acp-10-8513-2010.

Eisele, F.L. (1989) Natural and anthropogenic negative-ions in the troposphere. *Journal of Geophysical Research*, **94**, 2183–2196.

Eisele, F. and Tanner, D. (1993) Measurement of the gas phase concentration of H_2SO_4 and methane sulfonic acid and estimates of H_2SO_4 production and loss in the atmosphere. *Journal of Geophysical Research*, **98**, 9001–9010.

Eisele, F.L. and Hanson, D.R. (2000) First measurement of prenucleation molecular clusters. *Journal of Physical Chemistry A*, **104**, 830–836.

Fiedler, V., Dal Maso, M., Boy, M. *et al.* (2005) The contribution of sulphuric acid to atmospheric particle formation and growth: a comparison between boundary layers in Northern and Central Europe. *Atmospheric Chemistry and Physics*, **5**, 1773–1785.

Flagan, R.C. (1998) History of electrical aerosol measurements. *Aerosol Science and Technology*, **28**, 301–380.

Friend, J.P., Barnes, R.A. and Vasta, R.M. (1980) Nucleation by free radicals from the photooxidation of sulfur dioxide in air. *The Journal of Physical Chemistry*, **84**, 2423–2436.

Gagné, S., Laakso, L., Petäjä, T. *et al.* (2008) Analysis of one year of Ion-DMPS data from the SMEAR II station, Finland. *Tellus B*, **60**, 318–329.

Gagné, S., Nieminen, T., Kurtén, T. *et al.* (2010) Factors influencing the contribution of ion-induced nucleation in a boreal forest, Finland. *Atmospheric Chemistry and Physics*, **10**, 3743–3757. doi: 10.5194/acp-10-3743-2010.

Gamero-Castaño, M. and Fernández de la Mora, J. (2000) A condensation nucleus counter (CPC) sensitive to singly charged sub-nanometer particles. *Journal of Aerosol Science*, **31**, 757–772.

Hanson, D.R. and Eisele, F.L. (2002) Measurement of prenucleation molecular clusters in the NH_3, H_2SO_4, H_2O system. *Journal of Geophysical Research*, **107**, 4158–4174.

Hanson, D.R. and Lovejoy, E.R. (2006) Measurement of the thermodynamics of the hydrated dimer and trimer of sulfuric acid. *Journal of Physical Chemistry A*, **110**, 9525–9528.

Hari, P. and Kulmala, M. (2005) Station for measuring ecosystems – atmosphere relations (SMEAR II). *Boreal Environment Research*, **10**, 315–322.

Hering, S. and Stolzenburg, M. (2005) A method for particle size amplification by water condensation in a laminar, thermally diffusive flow. *Aerosol Science and Technology*, **39**, 484–436.

Hinds, W.C. (1999) *Aerosol Technology*, John Wiley & Sons, Inc., New York.

Hirsikko, A., Laakso, L., Hõrrak, U. *et al.* (2005) Annual and size dependent variation of growth rates and ion concentrations in boreal forest. *Boreal Environment Research*, **10**, 357–369.

Hirsikko, A., Nieminen, T., Gagné, S. *et al.* (2011) Atmospheric ions and nucleation: a review of observations. *Atmospheric Chemistry and Physics*, **11**, 767–798.

Hoppel, W.A. (1981) The use of differential mobility analyzers of second order in determining the aerosol size distribution. *Journal of Aerosol Science*, **12**, 55–57.

Hoppel, W.A. and Frick, G.M. (1986) Ion attachment coefficients and the steady state charge distribution on aerosols in a bipolar ion environment. *Aerosol Science and Technology*, **5**, 1–21.

Iida, K., Stolzenburg M., McMurry, P., Dunn, M.J., Smith, J.N., Eisele, F., and Keady, P. (2006) Contribution of ion-induced nucleation to new particle formation: methodology and its application to atmospheric observations in Boulder, Colorado, *Journal of Geophysical Research*, **111**, D23201, doi: 10.1029/2006JD007167.

Iida, K., Stolzenburg, M.R., McMurry, P.H., and Smith, J.N. (2008) Estimating nanoparticle growth rates from size-dependent charged fractions: analysis of new particle formation events in Mexico City, *Journal of Geophysical Research*, **113**, D05207.

Iida, K., Stolzenburg, M.R. and McMurry, P.H. (2009) Effect of working fluid on sub-2 nm particle detection with a laminar flow ultrafine condensation particle counter. *Aerosol Science and Technology*, **43**, 81–96.

IPCC (the Intergovernmental Panel on Climate Change) (2000) *Emissions Scenarios: A Special Report of Working Group II of the Intergovernmental Panel on Climate Change*, Cambridge University Press, Cambridge.

IPCC (the Intergovernmental Panel on Climate Change) (2007) Climate Change 2007: The Physical Science Basis. Summary for Policymakers, IPCC Secretariat, Geneva, http://www.ipcc.ch/publications_and_data/ar4/wg1/en/spm.html (last accessed August 2, 2013).

Jiang, J., Chen, M., Kuang, C. *et al.* (2011) Electrical mobility spectrometer using diethylene glycol condensation particle counter for measurement of aerosol size distributions down to 1 nm. *Aerosol Science and Technology*, **45**, 510–521.

Jokinen, T., Sipilä, M., Junninen, H. *et al.* (2012) Atmospheric sulphuric acid and neutral cluster measurements using CI-APi-TOF. *Atmospheric Chemistry and Physics*, **12**(9), 4117–4125.

Junninen, H., Hulkkonen, M., Riipinen, I. *et al.* (2008) Observations on nocturnal growth of atmospheric clusters. *Tellus*, **60B**, 365–371.

Junninen, H., Ehn, M., Petäjä, T. *et al.* (2010) A high-resolution mass spectrometer to measure atmospheric ion composition. *Aerosol Science and Technology*, **3**, 1039–1053.

Kiendler-Scharr, A., Wildt, J., DalMaso, M. *et al.* (2009) Isoprene emissions inhibit new particle formation in forests. *Nature*, **461**, 381–384. doi: 10.1038/nature08292.

Kirkby, J., Curtius, J., Almeida, J. *et al.* (2011) Role of sulphuric acid, ammonia and galactic cosmic rays in atmospheric aerosol nucleation. *Nature*, **476**, 429–433.

Knutson, E.O. and Whitby, K.T. (1975) Aerosol classification by electric mobility: apparatus, theory and applications. *Journal of Aerosol Science*, **6**, 443.

Kuang C., McMurry, P.H., McCormick, A.V., and Eisele, F.L. (2008) Dependence of nucleation rates on sulfuric acid vapor concentration in diverse atmospheric locations, *Journal of Geophysical Research*, **113**, D10209.

Kuang, C., Chen, M., Zhao, J. *et al.* (2012) Size and time-resolved growth rate measurements of 1 to 5 nm freshly formed atmospheric nuclei. *Atmospheric Chemistry and Physics*, **12**, 3573–3589. doi: 10.5194/acp-12-3573-2012.

Kulmala, M., Toivonen, A., Mäkelä, J. and Laaksonen, A. (1998) Analysis of the growth of nucleation mode particles observed in Boreal forest. *Tellus*, **50B**, 449–462.

Kulmala, M., Pirjola, L. and Mäkelä, J.M. (2000) Stable sulphate clusters as a source of new atmospheric particles. *Nature*, **404**, 66–69.

Kulmala, M. (2003) How particles nucleate and grow? *Science*, **302**, 1000–1001.

Kulmala, M., Vehkamäki, H., Petäjä, T. *et al.* (2004) Formation and growth rates of ultrafine atmospheric particles: a review of observations. *Journal of Aerosol Science*, **35**, 143–176.

Kulmala, M., Lehtinen, K.E.J., Laakso, L. *et al.* (2005) On the existence of neutral atmospheric clusters. *Boreal Environment Research*, **10**, 79–87.

Kulmala, M., Lehtinen, K.E.J. and Laaksonen, A. (2006) Cluster activation theory as an explanation of the linear dependence between formation rate of 3 nm particles and sulphuric acid concentration. *Atmospheric Chemistry and Physics*, **6**, 787–793.

Kulmala, M., Mordas, G., Petäjä, T. *et al.* (2007) The condensation particle counter battery (CPCB): a new tool to investigate the activation properties of nanoparticles. *Journal of Aerosol Science*, **38**, 289–304.

Kulmala, M., Riipinen, I., Sipilä, M. *et al.* (2007) Towards direct measurement of atmospheric nucleation. *Science*, **318**, 89–92.

Kulmala, M. and Kerminen, V.-M. (2008) On the formation and growth of atmospheric nanoparticles. *Atmospheric Research*, **90**, 132–150.

Kulmala, M., Riipinen, I., Nieminen, T. *et al.* (2010) Atmospheric data over a solar cycle: no connection between galactic cosmic rays and new particle formation. *Atmospheric Chemistry and Physics*, **10**, 1885–1898.

Kurtén, T., Loukonen, V., Vehkamäki, H. and Kulmala, M. (2008) Amines are likely to enhance neutral and ion-induced sulfuric acid-water nucleation in the atmosphere more effectively than ammonia. *Atmospheric Chemistry and Physics*, **8**, 4095–4103.

Kurtén, T., Kuang, C., Gomez, P. *et al.* (2010) The role of cluster energy nonaccommodation in atmospheric sulfuric acid nucleation. *Journal of Chemical Physics*, **132**, 024304. doi: 10.1063/1.3291213.

Kurtén, T., Petäjä, T., Smith, J. *et al.* (2011) The effect of H_2SO_4–amine clustering on chemical ionization mass spectrometry (CIMS) measurements of gas-phase sulfuric acid. *Atmospheric Chemistry and Physics*, **11**, 3007–3019.

Laakso, L., Kulmala, M. and Lehtinen, K.E.J. (2003) Effect of condensation rate enhancement factor on 3-nm (diameter) particle formation in binary ion-induced and homogenous nucleation. *Journal of Geophysical Research*, **108**, 4574.

Laakso, L., Gagné, S., Petäjä, T. *et al.* (2007) Detecting charging state of ultra-fine particles: instrumental development and ambient measurement. *Atmospheric Chemistry and Physics*, **7**, 1333–1345.

Laakso, L., Grönholm, T., Kulmala, L. *et al.* (2007) Hot-air balloon as a platform for boundary layer profile measurements during particle formation. *Boreal Environment Research*, **12**, 279–294.

Lee, S.-H., Young, L.-H., Benson, D.R., Suni, T., Kulmala, M., Junninen, H., Campos, T.L., Rogers, D.C., and Jensen J. (2008) Observations of night-time new particle formation in the troposphere. *Journal of Geophysical Research* **113**, D10210, doi: 10.1029/2007JD009351.

Lehtipalo, K., Sipilä, M., Riipinen, I. *et al.* (2009) Analysis of atmospheric neutral and charged molecular clusters in boreal forest using pulse-height CPC. *Atmospheric Chemistry and Physics*, **9**, 4177–4184.

Lehtipalo, K., Kulmala, M., Sipilä, M. *et al.* (2010) Nanoparticles in boreal forest and coastal environment: a comparison of observations and implications of the nucleation mechanism. *Atmospheric Chemistry and Physics*, **10**, 7009–7016.

Lehtipalo, K., Sipilä, M., Junninen, J. *et al.* (2011) Observations of nano-CN in the nocturnal boreal forest. *Aerosol Science and Technology*, **45**, 499–509.

Leppä, J., Anttila, T., Kerminen, V.-M. *et al.* (2011) Atmospheric new particle formation: real and apparent growth of neutral and charged particles. *Atmospheric Chemistry and Physics*, **11**, 4939–4955. doi: 10.5194/acp-11-4939-2011.

Liu, B.Y.H. and Pui, D.Y.H. (1974) A submicron aerosol standard and the primary, absolute calibration of the condensation nuclei counter. *Journal of Colloid and Interface Science*, **47**, 155–171.

Liu, B.Y.H. and Kim, C.S. (1977) On the counting efficiency of condensation nuclei counters. *Atmospheric Environment*, **11**, 1097–1100.

Mäkelä, J.M., Yli-Koivisto, S., Hiltunen, V. *et al.* (2001) Chemical composition of aerosol during particle formation events in boreal forest. *Tellus B*, **53**, 380–393.

Makkonen, R., Asmi, A., Korhonen, H. *et al.* (2009) Sensitivity of aerosol concentrations and cloud properties to nucleation and secondary organic distribution in ECHAM5-HAM global circulation model. *Atmospheric Chemistry and Physics*, **9**, 1747–1766.

Makkonen, R., Asmi, A., Kerminen,V.-M., Boy, M., Arneth, A., Hari, P., and Kulmala, M. (2012) Air pollution control and decreasing new particle formation lead to strong climate warming. *Atmospheric Chemistry and Physics*, **12**, 1515–1524.

Manninen, H.E., Nieminen, T., Riipinen1, I. *et al.* (2009) Charged and total particle formation and growth rates during EUCAARI 2007 campaign in Hyytiälä. *Atmospheric Chemistry and Physics*, **9**, 4077–4089.

Manninen, H.E., Nieminen, T., Asmi, E. *et al.* (2010) EUCAARI ion spectrometer measurements at 12 European sites – analysis of new particle formation events. *Atmospheric Chemistry and Physics*, **10**, 7907–7927.

Manninen, H.E., Franchin, A., Schobesberger, S. *et al.* (2011) Characterisation of corona-generated ions used in a Neutral cluster and Air Ion Spectrometer (NAIS). *Atmospheric Measurement Techniques Discussions*, **4**, 2099–2125.

McMurry, P.H. (1980) Photochemical aerosol formation from SO_2: a theoretical analysis of smog chamber data. *Journal of Colloid and Interface Science*, **78**, 513–527.

McMurry, P.H. (2000) The history of CPCs. *Aerosol Science and Technology*, **33**, 297–322.

McMurry, P.H., Kulmala, M. and Worsnop, D.R. (2011) Special Issue on aerosol measurements in the 1 nm range. *Aerosol Science and Technology*, **45**, 1.

Mentel, T.F., Wildt, J., Kiendler-Scharr, A. *et al.* (2009) Photochemical production of aerosols from real plant emissions. *Atmospheric Chemistry and Physics*, **9**, 4387–4406.

Merikanto, J., Spracklen, D.V., Pringle, K.J. and Carslaw, K.S. (2010) Effects of boundary layer particle formation on cloud droplet number and changes in cloud albedo from 1850 to 2000. *Atmospheric Chemistry and Physics*, **10**, 695–705.

Metzger, A., Verheggen, B., Dommen, J. *et al.* (2010) Evidence for the role of organics in aerosol particle formation under atmospheric conditions. *Proceedings of the National Academy of Sciences*. doi: 10.1073/pnas.0911330107.

Mirabel, P. and Katz, J.L. (1974) Binary homogeneous nucleation as a mechanism for the formation of aerosols. *Journal of Chemical Physics*, **60**, 1138–1144.

Mirme, A., Tamm, E., Mordas, G. *et al.* (2007) A wide-range multi-channel air ion spectrometer. *Boreal Environment Research*, **12**, 247–264.

Mirme, S., Mirme, A., Minikin, A. *et al.* (2010) Atmospheric sub-3 nm particles at high altitudes. *Atmospheric Chemistry and Physics*, **10**, 437–451.

Mordas, G., Kulmala, M., Petäjä, T., Aalto, P.P., Matulevicius,V., Grigoraitis,V., Ulevicius, V., Grauslys, V., Ukkonen, A., and Hämeri, K. (2005) Design and performance characteristics of a condensation particle counter UF-02proto, *Boreal Environment Research*, **10**, 543–552.

Mäkelä, J.M., Aalto, P.P., Jokinen, V. *et al.* (1997) Observations of ultrafine aerosol particle formation and growth in boreal forest. *Geophysical Research Letters*, **24**, 1219–1222.

Mönkkönen, P., Koponen, I.K., Lehtinen, K. *et al.* (2005) Measurements in a highly polluted Asian mega city: observations of aerosol number size distributions, modal parameters and nucleation events. *Atmospheric Chemistry and Physics*, **5**, 57–66.

Nieminen, T., Manninen, H., Sihto, S.-L. *et al.* (2009) Connection of sulphuric acid to atmospheric nucleation in boreal forest. *Environmental Science and Technology*, **43**, 4715–4721.

Nieminen, T., Lehtinen, K.E.J. and Kulmala, M. (2010) On condensational growth of clusters and nanoparticles in sub-10 nm size range. *Atmospheric Chemistry and Physics Discussions*, **10**, 1693–1717.

O'Dowd, C.D., Aalto, P., Hämeri, K. *et al.* (2002) Atmospheric particles from organic vapours. *Nature*, **416**, 497–498.

O'Dowd, C.D., Hämeri, K., Mäkelä, J.M. *et al.* (2002) Coastal new particle formation: environmental conditions and aerosol physicochemical characteristics during nucleation bursts. *Journal of Geophysical Research*, **107**(D19), 8107. doi: 10.1029/2000JD000206.

O'Dowd, C.D., Jimenez, J.L., Bahreini, R. *et al.* (2002) Marine aerosol formation from biogenic iodine emissions. *Nature*, **417**, 632–636.

O'Dowd, C.D., Aalto, P.P., Yoon, Y.J. and Hämeri, K. (2004) The use of the pulse height analyzer ultrafine condensation particle counter (PHA-UCPC) technique applied to sizing of nucleation mode particles of differing chemical composition. *Journal of Aerosol Science*, **35**, 205–216.

O'Dowd, C.D., Yoon, Y.J., Junkerman, W. *et al.* (2007) Airborne measurements of nucleation mode particles I: coastal nucleation and growth rates. *Atmospheric Chemistry and Physics*, **7**, 1491–1501.

Paasonen, P., Nieminen, T., Asmi, E. *et al.* (2010) On the role of sulphuric acid and low-volatility organic vapours in new particle formation at four European measurement sites. *Atmospheric Chemistry and Physics*, **10**, 11223–11242.

Penner, J.E., Quaas, J., Storelvmo, T. *et al.* (2006) Model intercomparison of indirect aerosol effects. *Atmospheric Chemistry and Physics*, **6**, 3391–3405.

Petäjä, T., Mordas, G., Manninen, H. *et al.* (2006) Detection efficiency of a water-based TSI condensation particle counter 3785. *Aerosol Science and Technology*, **40**, 1090–1097.

Petäjä, T., Mauldin, R.L. III,, Kosciuch, E. *et al.* (2009) Sulfuric acid and OH concentrations in a boreal forest site. *Atmospheric Chemistry and Physics*, **9**, 7435–7448.

Petäjä, T., Sipilä, M., Paasonen, P. *et al.* (2011) Experimental observation of strongly bound dimers of sulfuric acid: application to nucleation in the atmosphere. *Physical Review Letters*, **106**, 228302.

Ramanathan, V., Crutzen, P.J., Kiehl, J.T. and Rosenfeld, D. (2001) Aerosol, climate and the hydrologigal cycle. *Science*, **294**, 2119–2124.

Reischl, G.P., Mäkelä, J.M. and Necid, J. (1997) Performance of Vienna type differential mobility analyzer at 1.2–20 nanometer. *Aerosol Science and Technology*, **27**, 651–672.

Riipinen, I., Manninen, H., Yli-Juuti, T. *et al.* (2009) Applying the Condensation Particle Counter Battery (CPCB) to study the water-affinity of freshly-formed 2–9 nm particles in boreal forest. *Atmospheric Chemistry and Physics*, **9**, 3317–3330.

Riipinen, I., Pierce, J.R., Yli-Juuti, T., *et al.* (2011) Organic condensation: a vital link connecting aerosol formation to cloud condensation nuclei (CCN) concentrations. *Atmospheric Chemistry and Physics*, **11**(8), 3865–3878. doi: 10.5194/acp-11-3865-2011.

Rosser, S. and Fernández de la Mora, J. (2005) Vienna-type DMA of high resolution and high flow rate. *Aerosol Science and Technology*, **39**, 1191–1200.

Rosenfeld, D. (2000) Suppression of rain and snow by urban and industrial air pollution. *Science*, **287**, 1793–1796.

Rosenfeld, D., Lohmann, U., Raga, G.B. *et al.* (2008) Flood or drought: how do aerosols affect precipitation? *Science*, **321**, 1309–1313.

Saros, M., Weber, R.J., Marti, J. and McMurry, P.H. (1996) Ultra fine aerosol measurement using a condensation nucleus counter with pulse height analysis. *Aerosol Science and Technology*, **25**, 200–213.

Seinfeld, J.H. and Pandis, S.N. (2006) *Atmospheric Chemistry and Physics: From Air Pollution to Climate Change*, John Wiley & Sons, Inc., New York.

Seto, T., Okuyama, K., de Juan, L. and Fernendez de la Mora, J. (1997) Condensation of supersaturated vapors on monovalent and divalent ions of varying size. *Journal of Chemical Physics*, **107**, 1576–1585.

Sihto, S.-L., Kulmala, M., Kerminen, V.-M. *et al.* (2006) Atmospheric sulphuric acid and aerosol formation: implications from atmospheric measurements for nucleation and early growth mechanisms. *Atmospheric Chemistry and Physics*, **6**, 4079–4091.

Sipilä, M., Lehtipalo, K., Kulmala, M. *et al.* (2008) Applicability of condensation particle counters to detect atmospheric clusters. *Atmospheric Chemistry and Physics*, **8**, 4049–4060.

Sipilä, M., Lehtipalo, K., Attoui, M. *et al.* (2009) Laboratory verification of PH-CPC's ability to monitor atmospheric sub-3 nm clusters. *Aerosol Science and Technology*, **43**, 126–135.

Sipilä, M., Berndt, T., Petäjä, T. *et al.* (2010) The role of sulphuric acid in atmospheric nucleation. *Science*, **327**, 1243–1246.

Smith, J.N., Dunn, M.J., VanReken, T.M., Iida, K., Stolzenburg, M.R., McMurry, P.H., and Huye, L.G. (2008) Chemical composition of atmospheric nanoparticles formed from nucleation in Tecamac, Mexico: evidence for an important role for organic species in nanoparticle growth, *Geophysical Research Letters*, **35**, L04808, doi: 10.1029/2007GL032523.

Smith, J.N., Barsanti, K.C., Friedli, H.R. *et al.* (2010) Observations of aminium salts in atmospheric nanoparticles and possible climatic implications. *Proceedings of the National academy of Sciences*, **107**, 6634–6639.

Smith, S.J., Pitcher, H. and Wigley, T.M.L. (2001) Global and regional anthropogenic sulfur dioxide emissions. *Global and Planetary Change*, **29**, 99–119.

Smith, S.J., van Aardenne, J., Klimont, Z. *et al.* (2011) Anthropogenic sulfur dioxide emissions: 1850–2005. *Atmospheric Chemistry and Physics*, **11**, 1101–1116.

Sorokin, A. and Arnold, F. (2007) Laboratory study of cluster ions formation in H2SO4-H2O system: implications for threshold concentration of gaseous H2SO4 and ion-induced nucleation kinetics. *Atmospheric Environment*, **41**, 3740–3747.

Spracklen, D.V., Carslaw, K.S., Kulmala, M. *et al.* (2006) The contribution of boundary layer nucleation events to total particle concentrations on regional and global scales. *Atmospheric Chemistry and Physics*, **6**, 5631–5648.

Steiner, G., Attoui, M., Wimmer, D. and Reischl, G.P. (2010) A medium flow, high-resolution Vienna DMA running in recirculating mode. *Aerosol Science and Technology*, **44**, 308–315.

Stolzenburg, M.R. (1988) An ultrafine aerosol size distribution measuring system. PhD Thesis. Mechanical Engineering Department, University of Minnesota, Minneapolis.

Stolzenburg, M.R. and McMurry, P.H. (1991) An ultrafine aerosol condensation nucleus counter. *Aerosol Science and Technology*, **14**, 48–65.

Stolzenburg, M.R. and McMurry, P.H. (2008) Equations governing single and tandem DMA configurations and a new lognormal approximation to the transfer function. *Aerosol Science and Technology*, **42**, 421–432.

Suni, T., Kulmala, M., Hirsikko, A. *et al.* (2008) Formation and characteristics of ions and charged aerosol particles in a native Australian Eucalypt forest. *Atmospheric Chemistry and Physics*, **8**, 129–139.

Torpo, L., Kurtén, T., Vehkamäki, H. *et al.* (2007) Significance of ammonia in atmospheric nanoclusters. *Journal of Physical Chemistry A*, **111**, 10671–10674.

Twomey, S. (1974) Pollution and the planetary albedo. *Atmospheric Environment*, **8**, 1251–1256.

Vakkari, V., Laakso, H., Kulmala, M. *et al.* (2011) New particle formation events in semi-clean South African savannah. *Atmospheric Chemistry and Physics*, **11**, 3333–3346. doi: 10.5194/acp-11-3333-2011.

Vanhanen, J., Mikkilä, J., Lehtipalo, K. *et al.* (2011) Particle size magnifier for nano-CN detection. *Aerosol Science and Technology*, **45**, 533–542.

Vehkamäki, H., Kulmala, M., Napari, I. *et al.* (2002) An improved parameterization for sulfuric acid-water nucleation rates for tropospheric and stratospheric conditions. *Journal of Geophysical Research*, **107**, 4622–4632.

Vehkamäki, H., Dal Maso, M., Hussein, T. *et al.* (2004) Atmospheric particle formation events at Värriö measurement station in Finnish Lapland 1998–2002. *Atmospheric Chemistry and Physics*, **4**, 2015–2023.

Venzac, H., Sellegri, K. and Laj, P. (2007) Nucleation events detected at the high altitude site of the Puy de Dome Research Station, France. *Boreal Environment Research*, **12**, 345–359.

Viisanen, Y., Kulmala, M. and Laaksonen, A. (1997) Experiments on gas–liquid nucleation of sulphuric acid and water. *Journal of Chemical Physics*, **107**, 920–926.

Weber, R.J., McMurry, P.H., Eisele, F.L. and Tanner, D.J. (1995) Measurement of expected nucleation precursor species and 3-500-nm diameter particles at Mauna Loa Observatory Hawaii. *Journal of Atmospheric Science*, **52**, 2242–2257.

Weber, R.J., Marti, J.J., McMurry, P.H. *et al.* (1996) Measured atmospheric new particle formation rates: implications for nucleation mechanisms. *Chemical Engineering Communications*, **151**, 53–64.

Weber, R.J., Marti, J.J., McMurry, P.H. *et al.* (1997) Measurements of new particle formation and ultrafine particle growth rates at a clean continental site. *Journal of Geophysical Research*, **102**, 4375–4385.

Wehner, B., Wiedensohler, A., Tuch, T.M., Wu, Z.J., Hu, M., Slanina, J., Kiang, C.S. (2004) Variability of the aerosol number size distribution in Beijing, China: new particle formation, dust storms, and high continental background, *Geophysical Research Letters*, **31**, L22109, doi: 10.1029/2004GL021596.

Wehner, B., Petäjä, T., Boy, M., Engler, C., Birmili, W., Tuch, T., Wiedensohler, A., and Kulmala, M. (2005) The contribution of sulfuric acid and non-volatile compounds on the growth of freshly formed atmospheric aerosols, *Geophysical Research Letters*, **32**, L17810, doi: 10.1029/2005GL023827.

Welz, O., Savee, J.D., Osborn, D.L. *et al.* (2012) Direct kinetic measurements of Criegee Intermediate (CH_2OO) formed by reaction of CH_2I with O_2. *Science*, **335**, 204–207.

Wiedensohler, A., Hansson, H.-C., Orsini, D. *et al.* (1997) Night-time formation and occurrence of new particles associated with orographic clouds. *Atmospheric Environment*, **16**, 2545–2559.

Wiedensohler, A., Birmili, W., Nowak, A. *et al.* (2012) Mobility particle size spectrometers: harmonization of technical standards and data structure to facilitate high quality long-term observations of atmospheric particle number size distributions. *Atmospheric Measurement Techniques*, **5**, 657–685. doi: 10.5194/amt-5-657-2012.

Winkler, P.M., Steiner, G., Vrtala, A. *et al.* (2008) Heterogeneous nucleation experiments bridging scale from molecular ion clusters to nanoparticles. *Science*, **319**, 1374–1377.

Winklmayr, W., Reischl, G.P., Lindner, A.O. and Berner, A. (1991) A new electromobility spectrometer for the measurement of aerosol size distributions in the size range from 1 to 1000 nm. *Journal of Aerosol Science*, **22**, 289–296.

Yli-Juuti, T., Nieminen, T., Hirsikko, A. *et al.* (2011) Growth rates of nucleation mode particles in Hyytiälä during 2003–2009: variation with particle size, season, data analysis method and ambient conditions. *Atmospheric Chemistry and Physics*, **11**, 12865–12886. doi: 10.5194/acp-11-12865-2011.

Yoon, Y.J., O'Dowd C.D., Jennings, S.G., and Lee, S.H. (2006) Statistical characteristics and predictability of particle formation events at Mace Head, *Journal of Geophysical Research*, **111**, D13204, doi: 10.1029/2005JD006284.

Young, L.H., Benson, D.R., Kameel, F.R. *et al.* (2008) Laboratory studies of H_2SO_4/H_2O binary homogeneous nucleation from the SO_2+OH reaction: evaluation of the experimental setup and preliminary results. *Atmospheric Chemistry and Physics*, **8**, 4997–5016.

Yu, F. and Turco, R.P. (2008) Case studies of particle formation events observed in boreal forests: implications for nucleation mechanisms. *Atmospheric Chemistry and Physics*, **8**, 6085–6102.

Yu, F., Wang, Z., Luo, G. and Turco, R.P. (2008) Ion-mediated nucleation as an important source of tropospheric aerosols. *Atmospheric Chemistry and Physics*, **8**, 2537–2554.

Yu, F. (2010) Ion-mediated nucleation in the atmosphere: key controlling parameters, implications and look-up table, *Journal of Geophysical Research*, **115**, D03206.

Yu, H., McGraw, R., and Lee, S.-H. (2012) Effects of amines on formation of sub3 nm particles and their subsequent growth, *Geophysical Research Letters*, **39**, L02807, doi: 10.1029/2011GL050099.

Zhang, R., Suh, I., Zhao, J. *et al.* (2004) Atmospheric new particle formation enhanced by organic acids. *Science*, **304**, 1487–1490.

Zhao, J., Eisele, F.L., Titcombe, M., Kuang, C.G., and McMurry, P.H. (2010) Chemical ionization mass spectrometric measurements of atmospheric neutral clusters using the cluster-CIMS. *Journal of Geophysical Research* **115**, D08205, doi: 10.1029/2009JD012606.

Zhao, J., Smith, J.N., Eisele, F.L. *et al*. (2011) Observation of neutral sulfuric acid-amine containing clusters in laboratory and ambient measurements. *Atmospheric Chemistry and Physics*, **11**, 10823–10836.

8

Atmospheric Aerosols and Climate Impacts

Maria Kanakidou
Environmental Chemical Processes Laboratory, Department of Chemistry, University of Crete, Greece

8.1 Introduction

Atmospheric aerosols are complex ensembles of solid or liquid atmospheric materials of varying chemical composition and size that are associated with variable amounts of water depending on the atmospheric conditions and their hygroscopicity. Aerosols compromise human and ecosystem health, influence visibility and thus human activities, affect ozone and the global radiation budget, modify cloud properties, and lead to feedbacks in the hydrological cycle and climate perturbation, as they interact both with light (in the visible and near-infrared spectrum) and with atmospheric water.

8.2 Global Aerosol Distributions

Aerosol atmospheric concentrations result from emission to, formation in, and removal from the atmosphere, as well as transport and physicochemical transformation. Aerosol sources are both natural and anthropogenic. Anthropogenic sources of aerosols include fossil fuel and biofuel burning for energy production, industrial activities and transportation (ground, maritime, and aviation), dust resuspension from ground transportation, and other human activities. Submicron aerosols can be directly emitted into the atmosphere, mainly by anthropogenic sources (mostly combustion sources). Sources of large uncertainties are

Aerosol Science: Technology and Applications, First Edition. Edited by Ian Colbeck and Mihalis Lazaridis.

the natural aerosols sea salt, dust, marine organic aerosol, plant debris, and primary biogenic particles, including spores, viruses, and microbes. These are mainly coarse aerosols (with diameters larger than 2.5 μm). Sporadic volcanic eruptions are also significant sources of aerosols or their gaseous precursors in the atmosphere. Natural aerosols from the marine and the terrestrial biosphere may potentially represent a feedback mechanism coupling the marine and terrestrial biosphere and the atmosphere.

Secondary aerosols can be formed in the atmosphere from nucleation of vapors and grow to the Aitken (diameters 0.01–0.1 μm) and accumulation (0.1–1.0 μm) sizes by **coagulation** with other particles or/and **condensation** of vapors on their surfaces. There is competition between coagulation and condensation for specific conditions for the control of aerosol turnover time (see details in Kanakidou *et al.*, 2005a); that is, the time required for e-fold reduction of aerosol concentrations in the atmosphere. Thus, there are regions in which aerosol turnover time is controlled by coagulation and others in which it is controlled by condensation of soluble vapors. **Cloud processing** also affects aerosols. Cloud droplets can absorb soluble gases like SO_2 and glyoxal (CHOCHO) which are oxidized in the droplets, thereby increasing the 'residual' aerosol mass and size when droplets evaporate (Myriokefalitakis *et al.*, 2011). In turn, cloud droplets can be formed from aerosols (**condensation nuclei**, CNs) above a critical size threshold that are activated (follow unconstrained growth via water condensation) under conditions of high relative humidity (RH). Understanding and simulating secondary aerosol formation in the atmosphere and their interactions with atmospheric water remains challenging for the scientific community and for atmospheric chemistry transport modeling.

Over the past 15 years, focused field campaigns have provided detailed characterizations of regional aerosol, chemical, microphysical, and radiative properties, along with relevant surface and atmospheric conditions. Studies from these campaigns provide highly reliable characterizations of submicrometer spherical particles such as sulfate aerosol. *In situ* characterization of larger particles, like dust, is much less reliable (Chin *et al.*, 2009). Figures 8.1a,b provide a global view of surface concentrations of carbonaceous aerosols (black carbon, BC, from Koch *et al.* (2009) and organic carbon, OC, from Tsigaridis *et al.* (2013)). Figure 8.1c shows the locations of CN concentration observations compiled by Heintzenberg, Covert, and Van Dingenen (2000) and depicted by Spracklen *et al.* (2011). The compiled latitudinal distribution of observed CN levels is given in Figure 8.1d. These figures clearly show higher levels recorded over polluted continental locations. They also show that the observational network is biased toward northern hemisphere (NH) locations. There is a clear need for additional data to be acquired in the tropics, southern hemisphere (SH) continental locations, and remote locations in order to complement the global view of these aerosol components and their number distribution. Therefore, at present, global models are unique tools that provide a concise picture of the global distribution of aerosol components and characteristics and thus allow an evaluation of their impacts.

8.3 Aerosol Climate Impacts

The extinction of light from atmospheric aerosols resulting from multiple scattering and/or absorption of solar radiation reduces atmospheric visibility and causes serious transportation problems, but also induces the so-called aerosol direct climate effect. As such, we

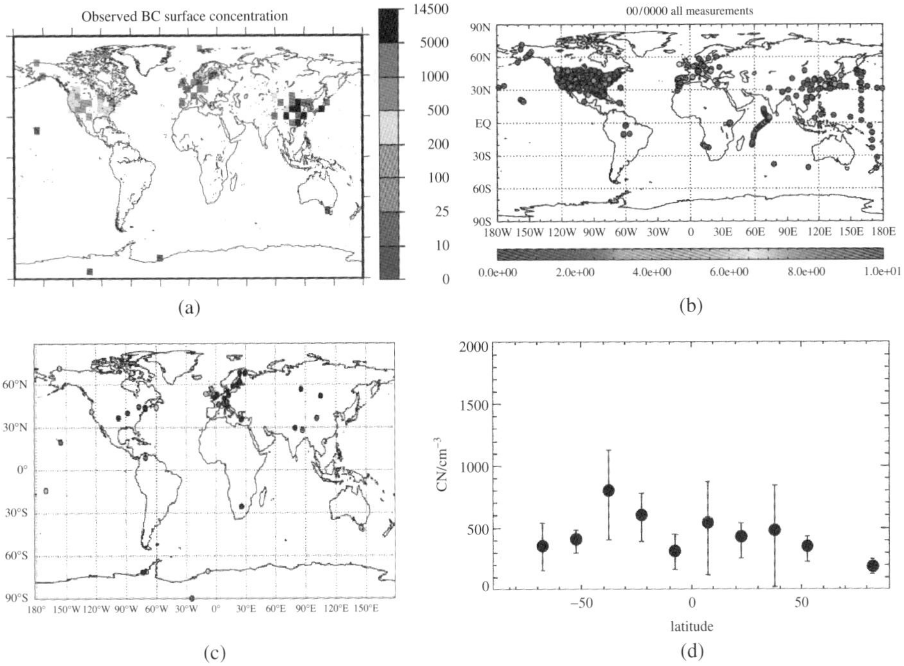

Figure 8.1 *(a) Surface BC observation nanograms per cubic meter. Reproduced with permission from Koch* et al., *2009. Copyright © 2009, Koch* et al. *(b) Surface OC observations in micrograms per cubic meter. Reproduced with permission from Tsigaridis* et al. *2013. Copyright © 2013, Tsigardis* et al. *and references therein. (c) Locations of CN observations. (d) Latitudinal distribution of CN concentration observations in cubic centimeters. Reproduced with permission from Spracklen* et al., *2010. Copyright © 2010, Spracklen* et al. *Figures (c) and (d) adapted from Spracklen* et al. *(2010) based on data compiled by Heintzenberg* et al. *(2000). See plate section for colour version.*

define the change in the energy balance of the atmosphere due to extinction by aerosols as compared to a 'non-aerosol' atmosphere.

The earth/atmosphere system is constantly receiving solar radiation (short-wave and near-infrared). A small fraction of this incoming solar radiation is reflected at the earth's surface and is thus directly emitted into space; the remainder is absorbed by the earth. In addition, long-wave (infrared) radiation is emitted from the earth's surface and escapes into space from the top of the atmosphere (Figure 8.2a, assuming an absence of any absorbing or scattering material in the atmosphere). The balance between incoming and outgoing radiation controls the earth's temperature, and absorbing and scattering material in the atmosphere affects this balance. Multiple scattering of radiation by atmospheric aerosols reduces the incoming solar radiation that reaches the earth's surface and increases the radiation that escapes from the top of the atmosphere, thus cooling the earth/atmosphere system (Figure 8.2b). Thus, scattering aerosol components exert an effect opposite to that of greenhouse gases on the global atmospheric temperature by decelerating the warming of the climate (Forster *et al.*, 2007). On the other hand, absorbing aerosols like BC and dust capture radiation and warm the atmospheric layer in which they reside, modifying the dynamic stability of the atmosphere. All types of aerosol reduce the radiation that reaches the surface, leading to a surface cooling.

Atmospheric aerosols also indirectly influence climate by modifying cloud amounts, microphysical and radiative properties (the so-called cloud albedo effect or first indirect effect), and precipitation efficiency (the so-called cloud lifetime effect or second indirect effect), and overall the hydrological cycle (Figure 8.2c). Thus, for fixed liquid water content, an increase in aerosols results in more droplets but of smaller size (Twomey, 1974). Furthermore, the increase in number of cloud droplets and the reduction in their size increases the cloud thickness and the amount of droplet surface that is in contact with the surrounding air.

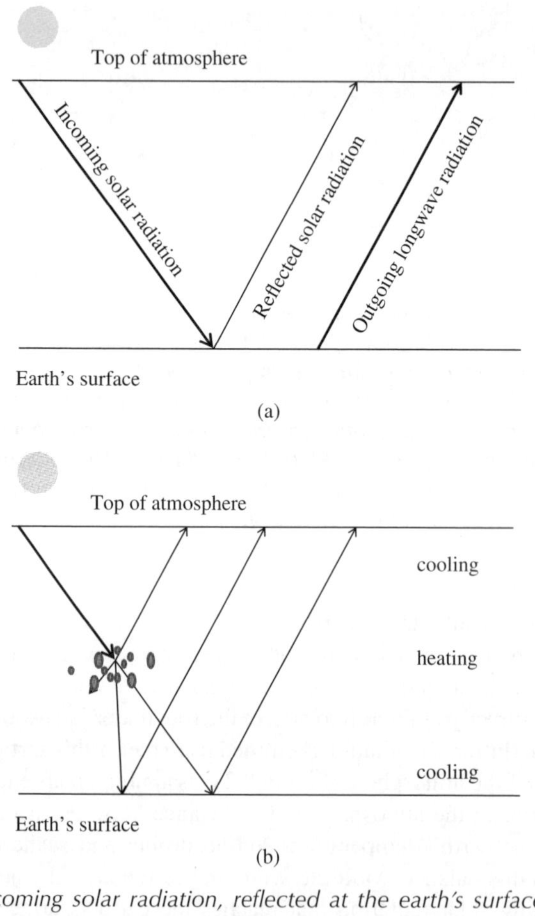

Figure 8.2 *(a) Incoming solar radiation, reflected at the earth's surface as outgoing solar radiation, and outgoing long-wave radiation from the earth, in the absence of aerosols and gases in the atmosphere (transparent atmosphere). (b) Atmosphere with aerosols and their direct impact on solar radiation. (c) Indirect climate effect of aerosols in clean and polluted atmospheres. (d) Radiative forcing estimates (1750–2000), as derived by Forster* et al. *(2007). Reproduced with permission from IPCC 2007. Copyright © 2007, IPCC. See plate section for colour version.*

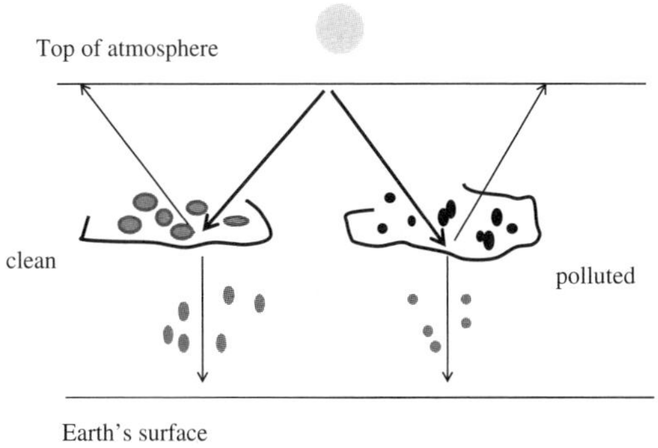

(c)

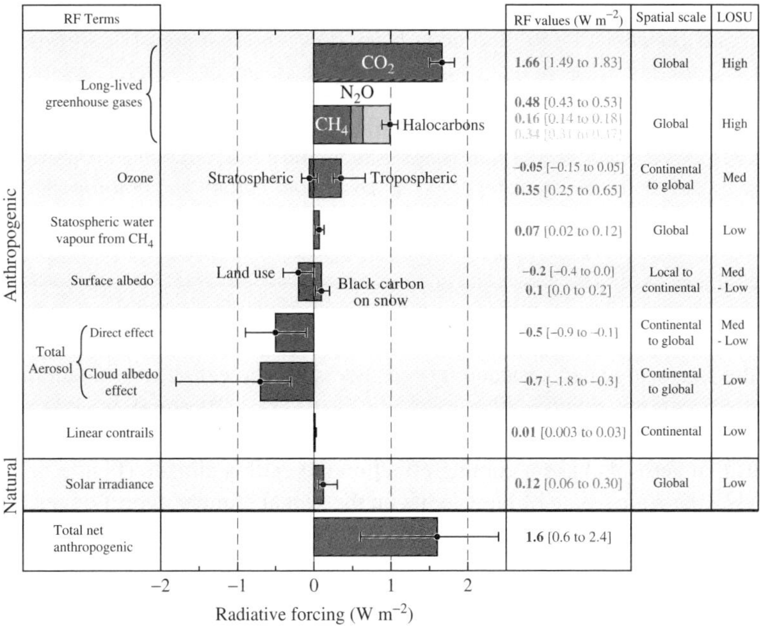

(d)

Figure 8.2 *(continued)*

It thus enables evaporation and increases the number of evaporation/condensation cycles that a cloud goes through before it finally precipitates, changing the cloud lifetime (Albrecht, 1989) (Figure 8.3). Furthermore, heating of the troposphere due to absorption of short-wave radiation by tropospheric aerosols (e.g. BC, Figure 8.2b) changes the RH and the stability of the troposphere and thereby influences cloud formation and lifetime, inducing the so-called 'semidirect' effect of aerosols (Figure 8.3). Aerosols also influence climate by affecting the amount of ozone that is a greenhouse gas, because heterogeneous reactions take place on the aerosol surface and in liquid aerosol particles, and photolysis rates are affected by the changes in radiation caused by the presence of aerosols.

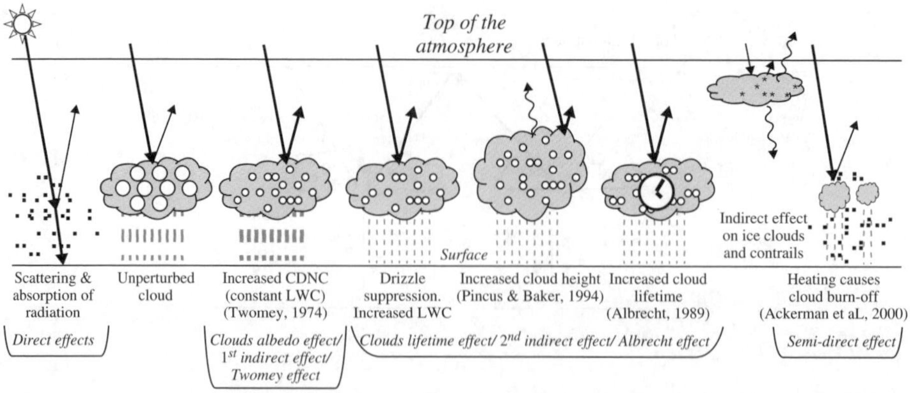

Figure 8.3 *Schematic diagram showing the various radiative mechanisms associated with cloud effects that have been identified as significant in relation to aerosols. Modified from Haywood and Boucher (2000). The small black dots represent aerosol particles; the larger open circles, cloud droplets. Straight lines represent the incident and reflected solar radiation and wavy lines represent terrestrial radiation. The filled white circles indicate cloud droplet number concentration (CDNC). The unperturbed cloud contains larger cloud drops as only natural aerosols are available as cloud condensation nuclei (CCN), while the perturbed cloud contains a greater number of smaller cloud drops as both natural and anthropogenic aerosols are available as CCNs. The vertical gray dashes represent rainfall and 'LWC' refers to the liquid water content. Reproduced with permission from IPCC 2007. Copyright © 2007, IPCC. Fig 2.10 from Forster* et al. *(2007).*

From the size-segregated concentration of aerosols, the extinction of radiation can be calculated, as can the cloud condensation nuclei (CCNs) concentrations that result from the growth of small fine atmospheric aerosols to critical sizes. However, although there is consensus that aerosols have a cooling effect on the earth's climate (Figure 8.2d; Forster *et al.*, 2007), the actual state of knowledge on the global climate impact of aerosol is very uncertain, as depicted by the bars associated with the direct and indirect aerosol effect estimates in Figure 8.2d.

8.4 Simulations of Global Aerosol Distributions

Global three-dimensional models have to account for the atmospheric processes that affect aerosols in the atmosphere, namely emissions, nucleation, condensation, evaporation, coagulation, cloud processing, atmospheric transport, dry and wet deposition, and chemistry/climate feedback mechanisms. Aerosol removal processes and climate impacts are greatly influenced by the size distribution. Moreover, anthropogenic emissions favor the transformation of insoluble to mixed particles by coating them with secondary products formed in the atmosphere. This increases the number of CNs that can grow to critical sizes to become CCNs, affecting the microphysical properties of aerosol, with impacts on aerosol lifetimes, global distributions, and radiative properties. Simplifications are required to describe this complex atmospheric system, which result in differences in the model simulations of the budgets and the properties of the tropospheric aerosols and of their climatic impact.

It is only in the last 2 decades that our understanding of aerosol has sufficiently advanced to successfully integrate all information on its atmospheric occurrence, fate, and impact on global atmospheric chemistry transport and general circulation models (GCMs). The climate impact of atmospheric aerosols depends on both their concentrations and their properties. The first modeling studies of the global aerosol system simulated the aerosol mass of individual aerosol components: the so-called 'bulk approach'. These studies represented aerosol either by an **external** mixture, in which each particle is made of solely one compound, or by an **internal** mixture, in which each particle consists of a uniform aggregate of all the individual components. They also assumed a constant aerosol **size distribution**. However, observations show that the mixing state and size distribution of the global aerosol system are highly variable, with both externally and internally mixed contributions and varying ratios among constituents.

Figure 8.4 presents a global view of the annual mean atmospheric loadings (atmospheric column amounts) of the major aerosol components and of CN number as computed by the global climate–chemistry model GISS-TOMAS (Lee and Adams, 2010). This figure depicts the high loadings of sulfate (SO_4^{2-}) aerosols over and downwind of industrialized areas of the NH, where anthropogenic activities are strong emitters of sulfur dioxide (SO_2), the precursor of sulfate aerosols. Oxidation of SO_2 occurs in both the gaseous and the aqueous phases of the atmosphere (clouds and water associated with the aerosol) to produce sulfate (SO_4^{2-}), which also has a non-negligible contribution (about one-fourth of the total *S* emissions) from natural marine sources. This secondary marine source, from the oxidation of dimethylsulfide (DMS), which is naturally emitted from the oceans, is spread across the oceanic areas, in contrast to the land sources of S, which are concentrated over regions subject to important human activities. A reduction of *S* content in fuels and 'clean atmosphere' measures for industrial exhausts have already led to a reduction in sulfate levels in the atmosphere. This trend is expected to continue in the coming years. Therefore, sulfate is expected to lose importance relative to the other aerosol components, such as nitrate and organics (Kanakidou *et al.*, 2005b; Tsigaridis and Kanakidou, 2007; Im and Kanakidou, 2012; Im *et al.*, 2012; Megaritis *et al.*, 2013). The elemental carbon (EC) aerosol component is often termed 'black carbon' (BC), which refers to the bulk absorbing material of carbonaceous aerosol, since EC is very often associated with some organic material. EC freshly emitted to the atmosphere (Figure 8.4d) or aged (mixed with other constituents, Figure 8.4c) shows high loading over areas subject to high fossil fuel emissions from human activities, mostly in the NH. It also shows high loadings over open fire-emission regions from naturally ignited biomass burning or human-driven fires (deforestation, agricultural waste burning). These activities and their impact maximize in the tropics. Particulate organic matter (OA or OM; primary or secondary in origin) also has major sources from the combustion of fossil fuel and biofuel and from open fires. In addition, OA has important natural sources from terrestrial biosphere, which can be both primary (plant debris, viruses, bacteria, microbes) and secondary (oxidation of biogenic volatile organic compounds (VOCs) such as isoprene, monoterpenes, sesquiterpenes, etc.), from soil dust, and from the ocean (OA is mainly associated with sea-salt emissions) (Kanakidou *et al.*, 2012). OA is also formed during oxidation of anthropogenic VOC. The annual mean loading of OA shows high levels in the NH and maximizes in the continental tropics.

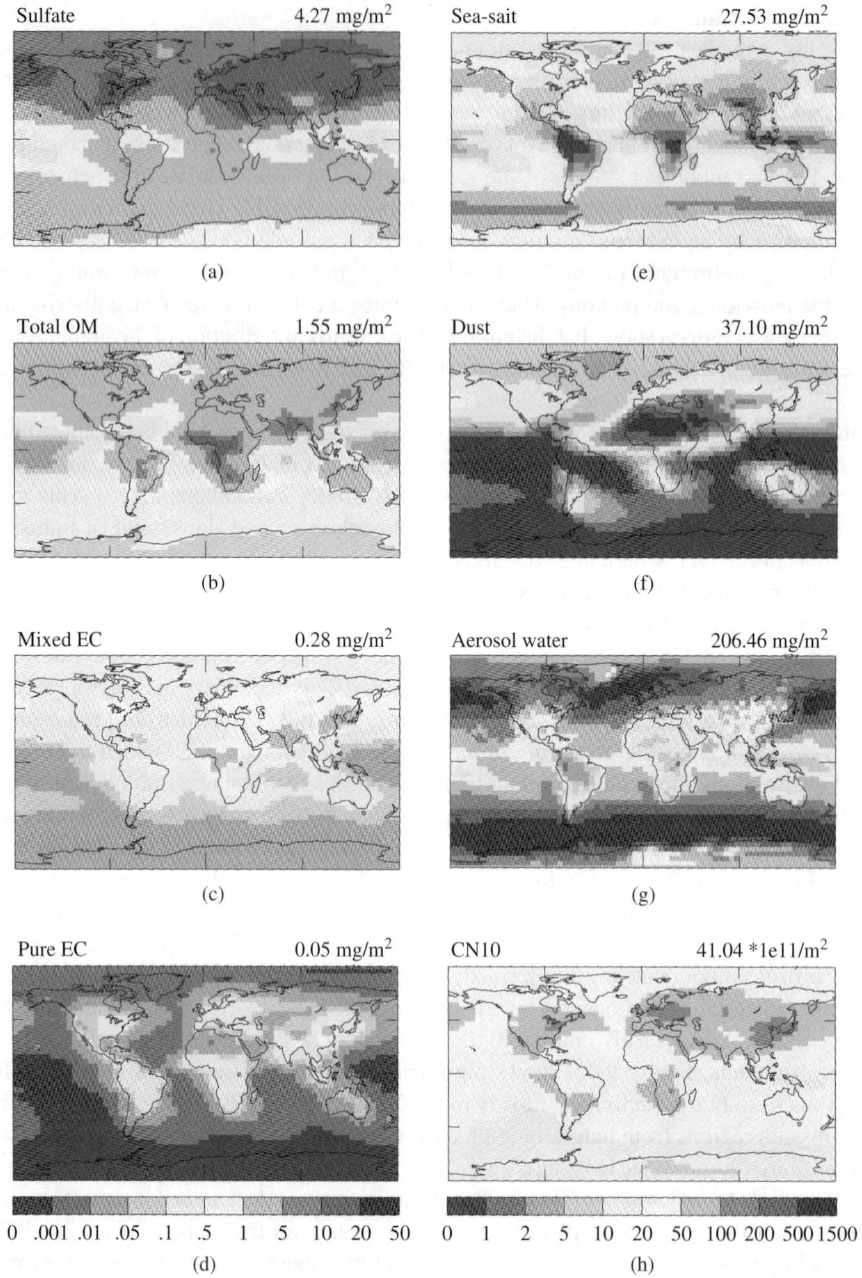

Figure 8.4 *(a–h) Annual mean aerosol component atmospheric loadings (mg.m^{-2}) and number concentration (CN10 in 10^{11}m^{-2}) as calculated by the GISS-TOMAS global model. Reproduced with permission from Lee and Adams, 2010. Copyright © 2010, Lee and Adams. See plate section for colour version.*

It presents strong seasonality, however, since boreal forests make a significant contribution to OA levels during the summer, when biogenic emissions of organics maximize, and during the boreal forest burning periods. The strong naturally driven sources of OA support the potential existence of significant feedback mechanisms between climate/biosphere and atmospheric chemistry.

Sea-salt emissions are wind- and temperature-driven and maximize in the SH west-wind zone and in the NH high latitudes (Figure 8.4e). Natural desert-dust aerosol loadings (Figure 8.4f) show by far the largest maxima compared to other aerosol components, particularly over Africa and the outbreaks of African deserts. Natural contributions of dust, sea salt and secondary organic aerosol (SOA) to the total aerosol mass are shown to be significant over regions as the Mediterranean, the tropical Atlantic and the outflow of Asia.

Differences between aerosol model simulations are documented in the frame of the international aerosol model intercomparison exercise AEROCOM (AEROsol model inter COMparison project: http://aerocom.met.no/Welcome.html; Textor *et al.*, 2006). Different assumptions concerning the size distributions of emissions and the resulting residence times in the atmosphere (Figure 8.5), driven by wet and dry deposition, yield discrepancies between models in simulating aerosol as a whole and in the aerosol composition (Figure 8.6).

Major factors of uncertainty in the model simulations are related to the water associated with the particulate phase, which can equal or even exceed the dry aerosol mass (Figures 8.6a and 8.4g) (Metzger *et al.*, 2002; Kanakidou *et al.*, 2005b; Lee and Adams, 2010).

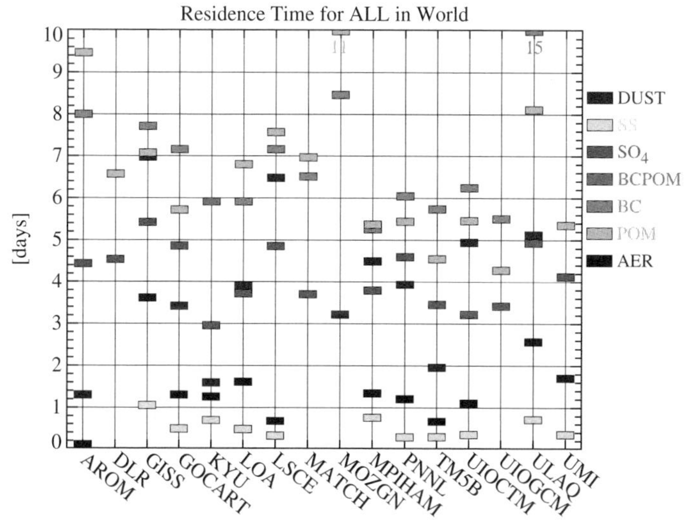

Figure 8.5 *Residence times of the major aerosol components as computed by the AEROCOM models during phase I experiment. Mean (and % standard deviation) values are: DUST: 4.14 days (±43%); SS (sea salt): 0.48 days (±58%); SO_4^{2-}: 4.12 days (±18%); black carbon (BC): 7.12 days (±33%); organic aerosol (POM): 6.54 days (±27%); total dry aerosol mass (AER): 1.42 days (±65%). Reproduced with permission from Textor* et al.*, 2006. Copyright © 2006, Textor* et al. *See plate section for colour version.*

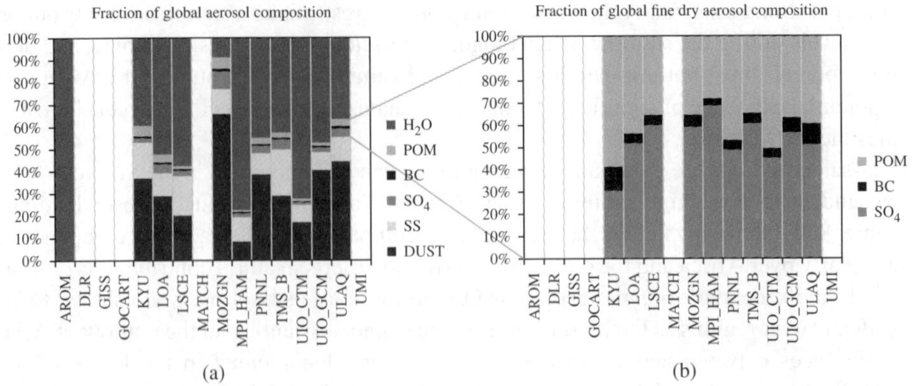

Figure 8.6 *(a) Global aerosol composition, expressed as percent mass fraction of the major aerosol components in global chemistry transport models that participated in the AEROCOM I exercise: H_2O: aerosol water; POM: organic aerosol; BC: black carbon; SO_4: sulfate; SS: sea salt; DUST: dust. (b) Relative contributions of POM, BC, and SO_4 in the fine dry aerosol in these models. Reproduced with permission from Textor* et al. *(2006). Copyright © 2006, Textor* et al. *See plate section for colour version.*

SOAs, chemically formed in the atmosphere, are another major source of uncertainties in the models. Although SOA modeling is progressing fast, it is still based on oversimplifications, due to the gaps in our understanding of the SOA occurrence and fate in the atmosphere (Kanakidou *et al.*, 2005a; Fuzzi *et al.*, 2006; Hallquist *et al.*, 2009). The global models seem to capture the general pattern of the spatial distribution of the major aerosol components. However, they significantly fail to reproduce the observed profiles (for instance, for BC see Koch *et al.*, 2009, 2010, and for organic aerosols, see Heald *et al.*, 2005, 2011).

Recently, CN-free tropospheric observations became available through intensive aircraft campaigns (e.g. the EUCAARI-LONGREX campaign in May 2008; Reddington *et al.*, 2011). In parallel, microphysical model developments enable simulations of the particle number concentrations that can improve estimations of aerosol impacts on health and climate (Reddington *et al.*, 2011; Fountoukis *et al.*, 2012). Comparison between the global model results and observations during the LONGREX campaign (Figure 8.7) showed considerable uncertainty associated with treatments of particle formation (nucleation) in the boundary layer and in the size distribution of emitted primary particles, leading to uncertainties in predicted CCN concentrations.

8.5 Extinction of Radiation by Aerosols (Direct Effect)

The extinction of light by atmospheric aerosol depends both on the amount of aerosol in the atmospheric path of the light and on the aerosol's optical properties, which vary as a function of wavelength, RH, and scattering or absorption characteristics. Scattering and absorption of light by atmospheric aerosols, assuming they are spherical particles, can be described by the Mie theory. This requires the relative refractive index, which is

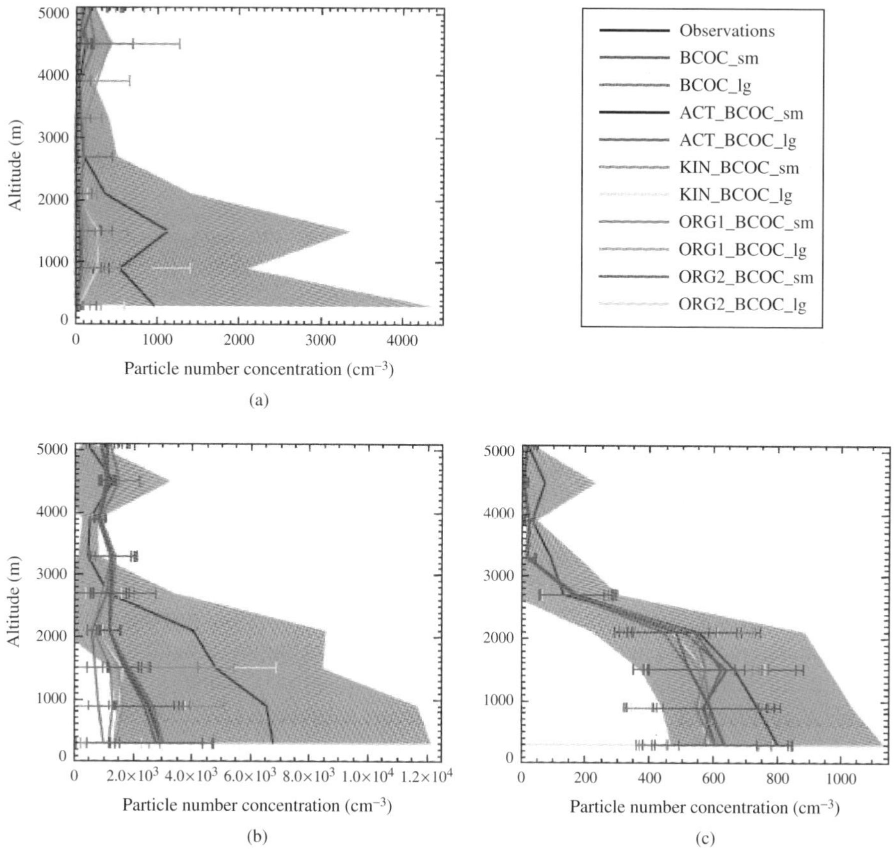

Figure 8.7 *(a–c) Black line: vertical distribution of CN concentrations as observed during the LONGREX campaign over Europe in May 2008; shaded lines: comparison with model results. The error bars and shading represent the standard deviation of the model and observations, respectively. Reproduced with permission from Reddington* et al. *(2011). Copyright © 2011, Reddington* et al. *See plate section for colour version.*

the refractive index of a particle (i.e. of the material of which the aerosol is composed) divided by the refractive index of the air in which the aerosol is present. The refractive index of a material describes how radiation propagates through it. Homogeneous particles are characterized by a single refractive index at a given wavelength. For a single particle, the complex refractive index:

$$N(\lambda) = n(\lambda) + i\,k(\lambda) \tag{8.1}$$

has two terms: $k(\lambda)$ is the wavelength-dependent absorption index component and $n(\lambda)$ is the corresponding scattering index component. This, together with the aerosol size and composition, determines the single scattering albedo (SSA or ω)

$$SSA = [scattering]/([absorption + scattering]) \tag{8.2}$$

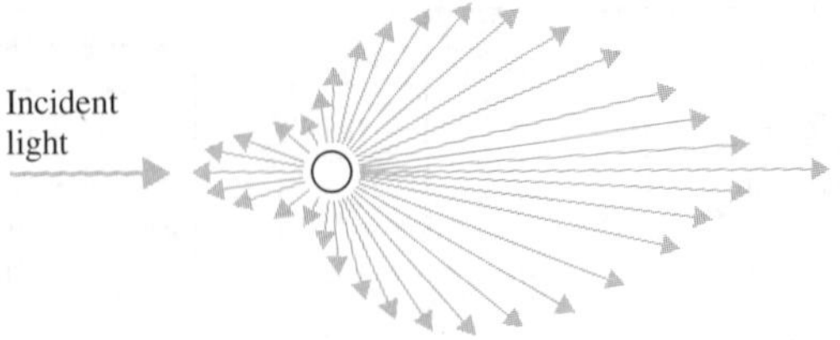

Figure 8.8 *The phase function provides information on the directions under which the incident light is scattered by a particle. The larger the particle, the larger the forward scattering. Reproduced with permission from http://www.iup.uni-bremen.de/E-Learning/at2-els_NO2/index.htm.*

This is defined as the ratio of the scattering to the total extinction (scattering plus absorption) efficiencies of the aerosol. The interpretation of the SSA is that for SSA = 1, the aerosol investigated is totally scattering and will lower the global mean surface temperature; very low values of ω indicate that an aerosol will increase the global mean surface temperature.

Atmospheric particles with more than one component require the effective refractive index that accounts for their inhomogeneity. Based on Mie theory and assuming a single spherical particle, the scattering phase function (Figure 8.8) that provides the fraction of the scattered radiation in the various directions (different angles) around the particle can be calculated. According to Mie theory, the largest extinction of radiation by atmospheric aerosols occurs for wavelengths similar in size to the particles.

The **Mie parameter** or **size parameter** (a), which is calculated from the ratio of the perimeter of the cross-section of the scattering particle ($2\pi r$, where r is the particle radius) to the wavelength (λ) of the incident light ($a = 2\pi r/\lambda$), determines the scattering mechanism of the light from the particle. For $a \approx 1$, Mie scattering (elastic scattering on aerosols and droplets) takes place.

Absorption of light by an optical mean is described by the Beer–Lambert law based on the absorption coefficient of the medium (b_a), which is the integral of the absorption cross-section (σ_a) of the medium over the optical path. Extinction of light by scattering is treated similarly using the scattering cross-section (σ_s) of the medium and the scattering coefficient (b_s). Integration of extinction by scattering and absorption for the ensemble of aerosol particles in an atmospheric layer (along a vertical path, e.g. altitude) per wavelength provides the wavelength-dependent effective aerosol optical depth (AOD) in this layer. Note that the overall effective (total) optical depth of an atmospheric layer also accounts for the extinction of radiation by clouds and aerosol water, as well as for absorption by gases and molecular (Rayleigh) scattering. Like the SSA of a single particle, for an atmospheric layer, the SSA gives the fraction of extinction of radiation that results from scattering of light (Equation 8.1) and the effective asymmetry factor (g), which provides information on the distribution of the radiation scattered by nonspherical aerosols in various directions around them. The asymmetry factor is equal to the mean value of the cosine of the scattering angle weighted by the scattering phase function and is usually used in two-stream radiative transfer calculations. When the asymmetry factor (g) approaches +1, scattering strongly peaks in the forward direction, and when g is −1, scattering strongly peaks in the backward direction. In general, $g = 0$ indicates scattering directions evenly distributed between forward and backward directions; for example, isotropic scattering (like scattering from

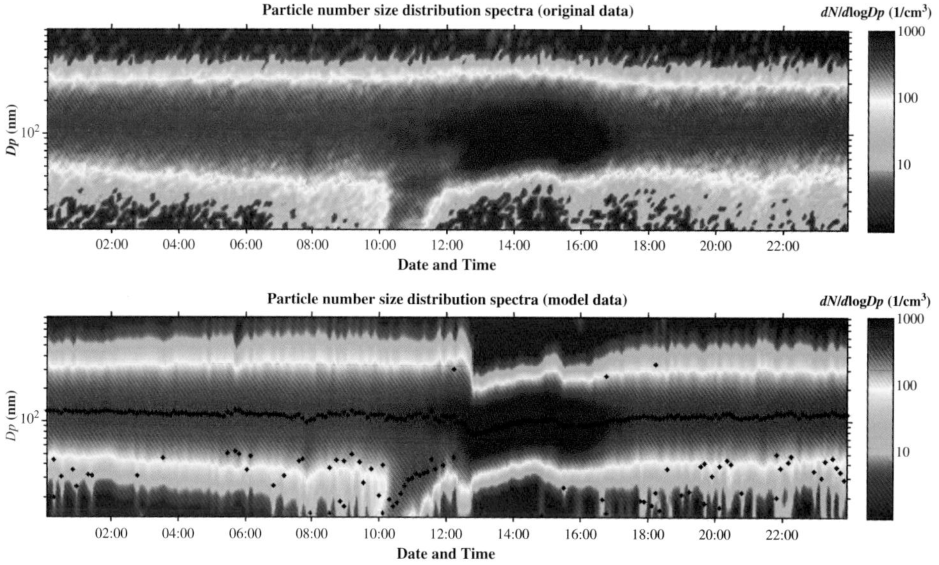

Figure 2.3 *New particle formation event observed during the summer of 2009 (7 July) at the Akrotiri research station (observations and modelled data), Crete, Greece.*

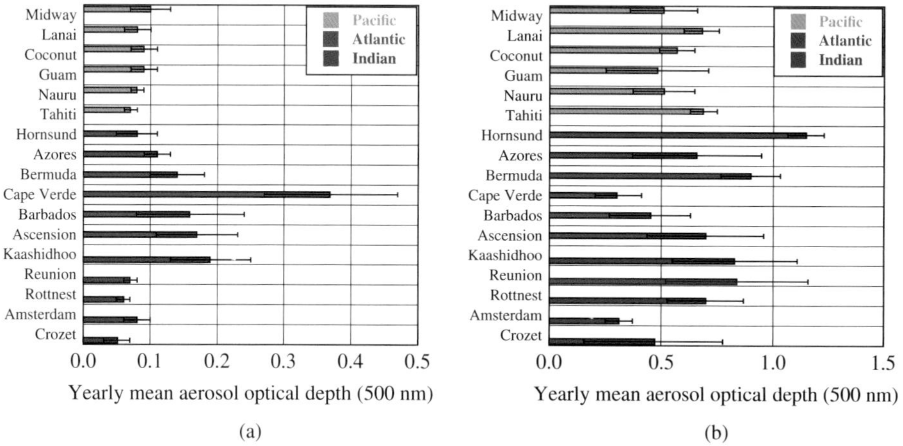

Figure 6.1 *Mean annual (a) AOD at 500 nm and (b) Angstrom parameter, at various island-based AERONET sites. Reproduced with permission from Smirnov* et al. *(2009). Copyright (2009), John Wiley and Sons Ltd.*

Aerosol Science: Technology and Applications, First Edition. Edited by Ian Colbeck and Mihalis Lazaridis.

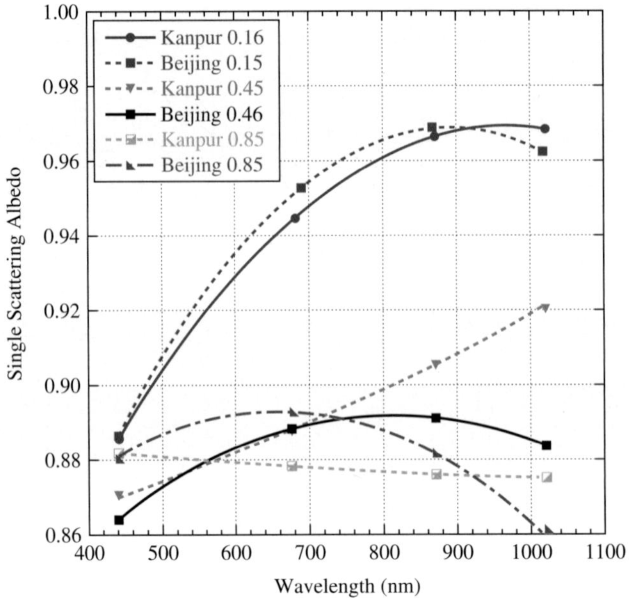

Figure 6.2 *Comparison of spectral SSAs at Kanpur (26.5°N, 80.2°E) and Beijing (39.9°N, 116.4°E) for low (~0.15), medium (~0.45), and high (~0.85) fine-mode fractions. Reproduced with permission from Eck* et al. *(2010). Copyright (2010), American Geophysical Union.*

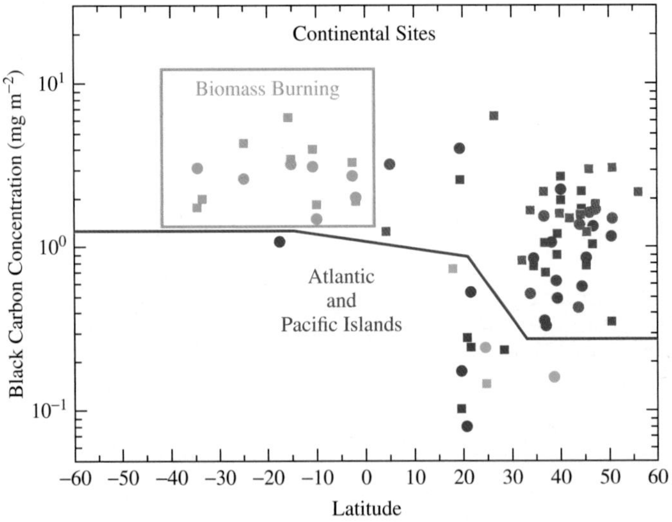

Figure 6.3 *Black carbon (BC) concentration at various AERONET sites in North America (red), Europe (gray), Asia (purple), South America (green), Africa (orange), Atlantic Islands (light blue), and Pacific Islands (dark blue) for the year 2000 (circles) and 2001 (squares). Reproduced with permission from Schuster* et al. *(2005). Copyright (2005), American Geophysical Union.*

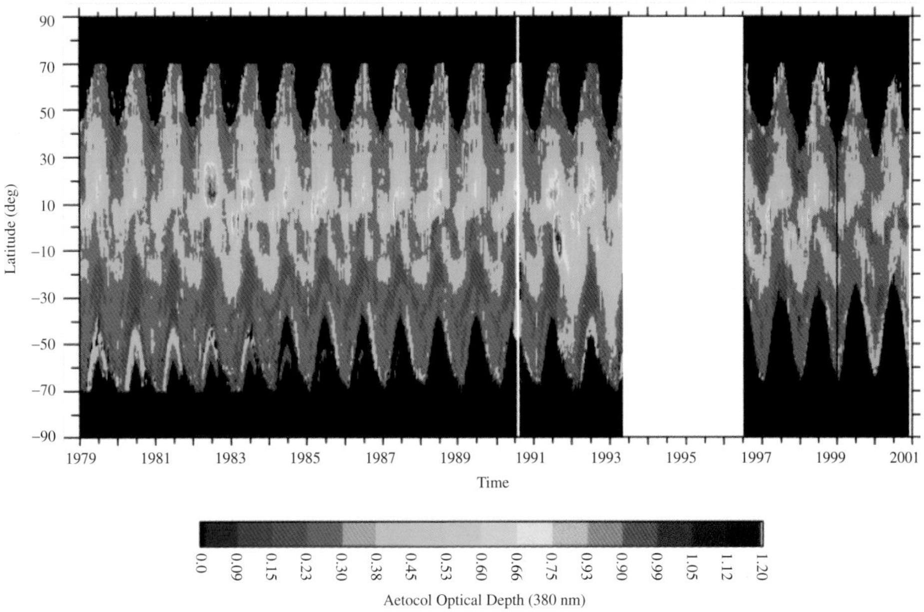

Figure 6.7 *Time series of the 380 nm aerosol optical depth during the period of operation of the TOMS instruments onboard the Nimbus 7 (1979–1992) and Earth Probe satellites (1996–2001). Weekly averages plotted over a 1 × 1° geographical grid. Figure adapted from Torres* et al. *(2002) (with permission from American Meteorological Society).*

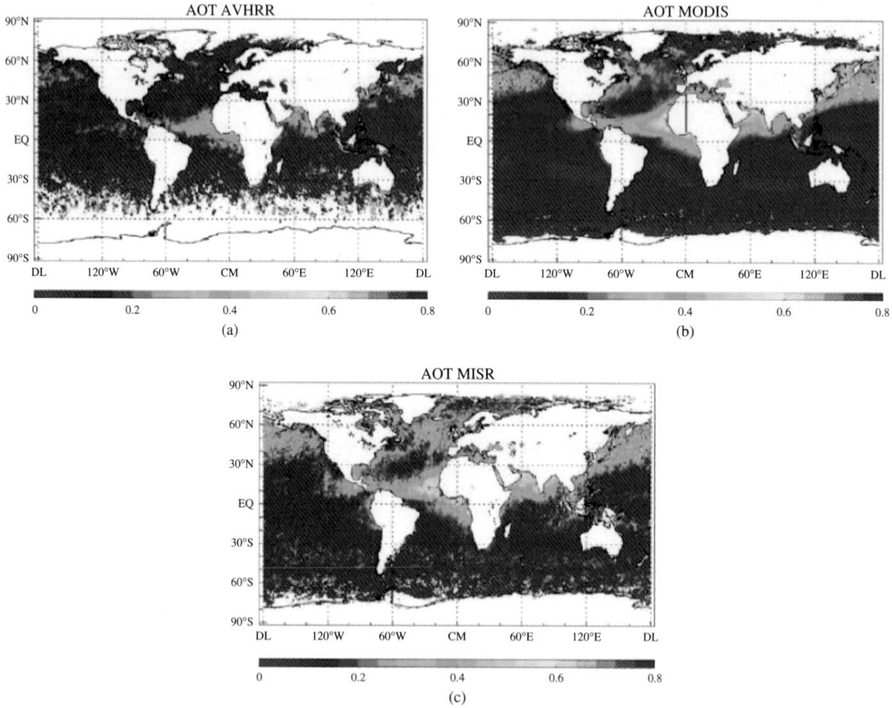

Figure 6.8 *Global distribution of aerosol optical depth for the period March to July 2000 from three passive sensors: AVHRR, MODIS, and MISR. Reproduced with permission from Geogdzhayev* et al. *(2002). Copyright (2004), Elsevier Ltd.*

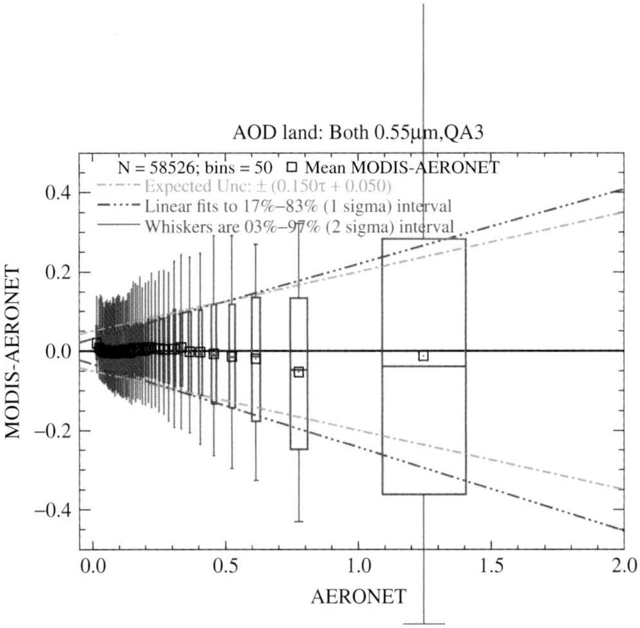

Figure 6.9 *Absolute error of AOD (MODIS-AERONET) vs. AERONET-AOD at 0.55* μm, *for QAC = 3. Data sorted by AERONET AOD and grouped into 50 equal bins. Each box plot represents the statistics of the MODIS-AERONET differences in the bin. The means and standard deviations of the AERONET AOD are the centers and half-widths in the horizontal (red). The mean, medians, and 66% ($1-\sigma$) interval of the MODIS-AERONET differences are the black squares, with the center and top–bottom red intervals in the vertical (also red). The blue whiskers are the 96% ($2-\sigma$) intervals. The red dashed curves are linear best fits to $1-\sigma$ intervals and the green dashed curves are the over-land expected errors for total AOD ± (0.05 + 0.15). Reproduced with permission from Levy* et al. *(2010). Copyright (2010), Levy* et al.

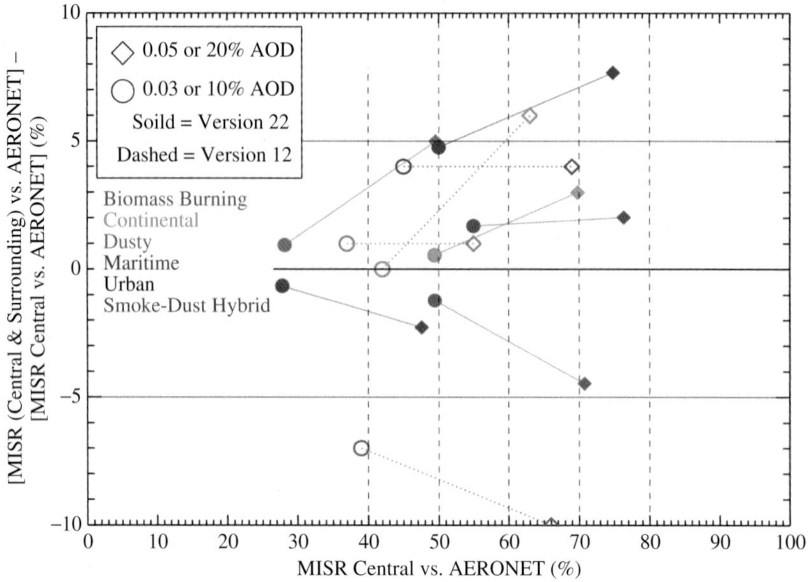

Figure 6.10 *MISR AOD bias relative to AERONET (in %) for all coincident samples stratified according to the dominant aerosol air-mass type. Filled diamonds and circles represent the class-averaged percentage meeting regular (0.05 or 20% × AOD) and more stringent (0.03 or 10% × AOD) criteria, respectively. Open symbols represent class average results for the earlier version of the product from Kahn* et al. *(2005). Figure adapted from Kahn* et al. *(2010) (with permission from AGU).*

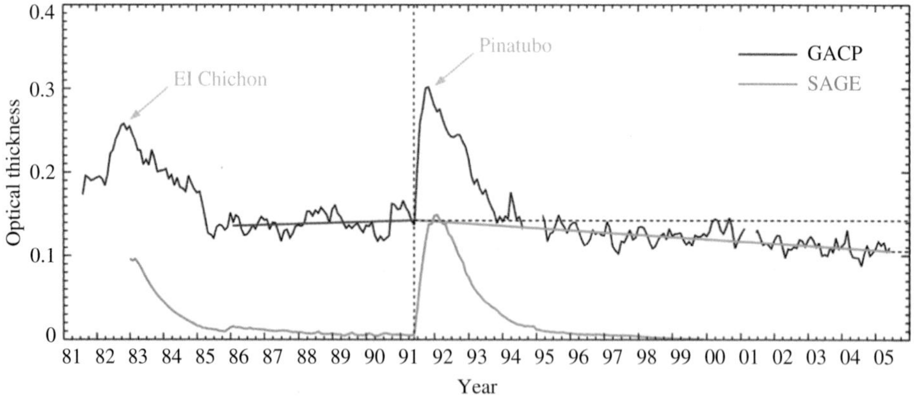

Figure 6.11 *GACP record of the globally averaged column AOD over the oceans and SAGE record of the globally averaged stratospheric AOD. Reproduced with permission from Mishchenko* et al. *(2007b). Copyright (2007), American Association for Advancement of Science.*

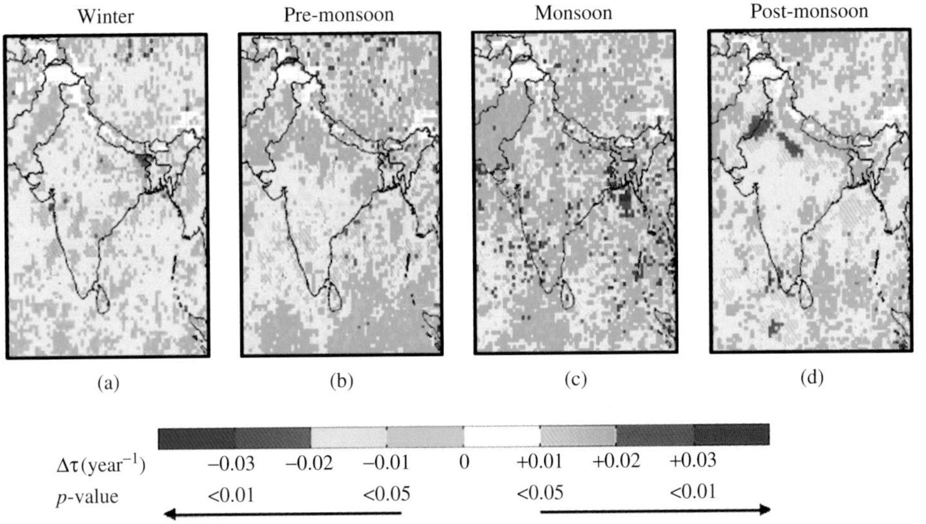

Figure 6.12 *Seasonal aerosol hotspots (significant rate of increase of AOD per year) over the Indian subcontinent for the period 2000–2010. Reproduced with permission from Dey and Di Girolamo (2011). Copyright (2011), John Wiley and Sons Ltd.*

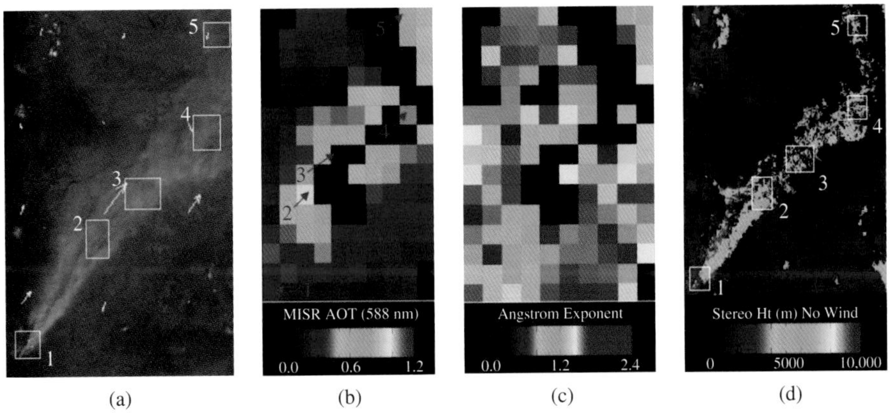

Figure 6.13 *Oregon fire, 4 September 2003, orbit 19 753, path 044, blocks 53–55, 19:00 UTC. (a) MISR nadir view of the fire plume, with five study site (patch) locations indicated as numbered white boxes, and MISR stereo-derived wind vectors superposed in yellow. (b) MISR mid-visible column AOD (version 17) retrieved at 17.6 km spatial resolution, with study site locations indicated by red arrows. There are no retrieval results for the black pixels, in most cases due to the high AOD and AOD variability of the plume core. (c) MISR-derived, column-average Angstrom exponent for the plume and surrounding area. (d) MISR stereo height product (version 13), without wind correction (labeled 'no wind' in the figure), for the same region. Reproduced with permission from Kahn* et al. *(2007). Copyright (2007), John Wiley and Sons Ltd.*

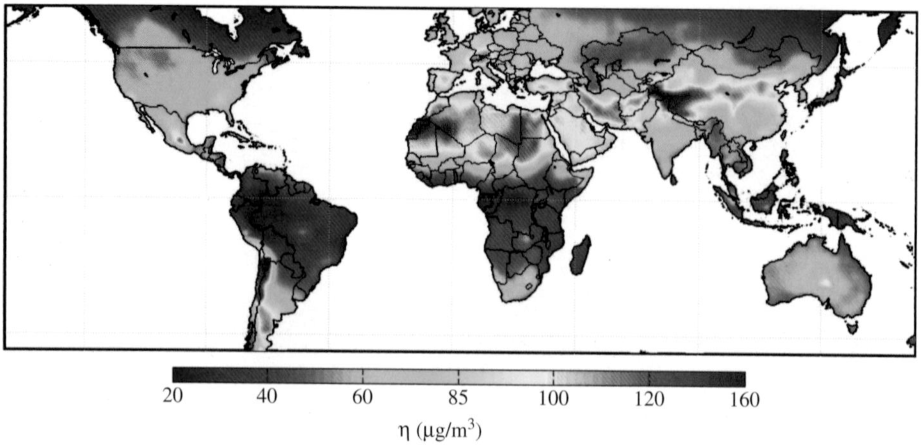

Figure 6.14 *Global view of satellite-derived $PM_{2.5}$ over land for the period 2001–2006. Reproduced from van Donkelaar* et al. *(2010). Copyright © 2010. This is an Open Access article.*

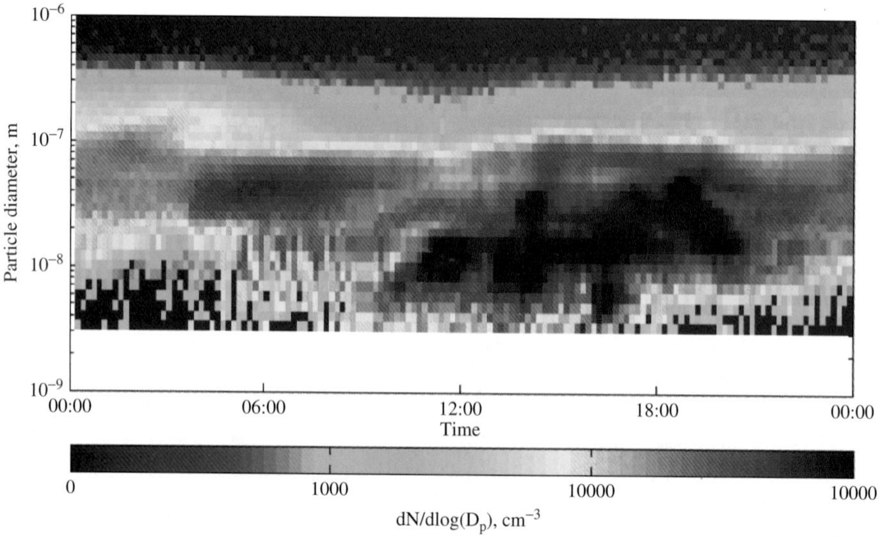

Figure 7.1 *Twenty-four-hour surface plot showing the development of the particle size distribution during a nucleation event day at the Hyytiälä SMEAR II forest field station, Finland. Shading depicts the number of particles in the (logarithmic) size channel. These data were obtained with an ordinary differential mobility particle sizer (DMPS) system, so no information on sub-3 nm concentrations can be obtained.*

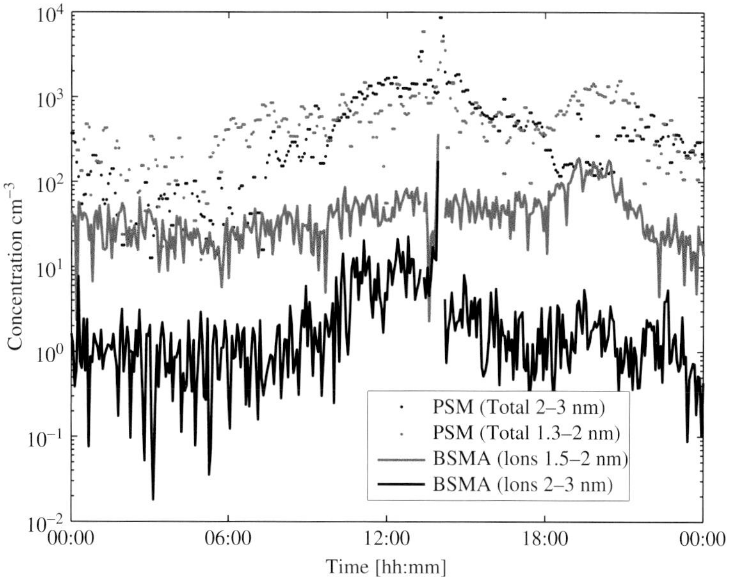

Figure 7.5 *Total concentration of particles between 1.3–2.0 nm and 2.0–3.0 nm mobility diameter, measured by the Airmodus A09 PSM in Hyytiälä SMEAR II station 16.8.2010, and ion concentration in corresponding size classes, measured by the BSMA ion spectrometer.*

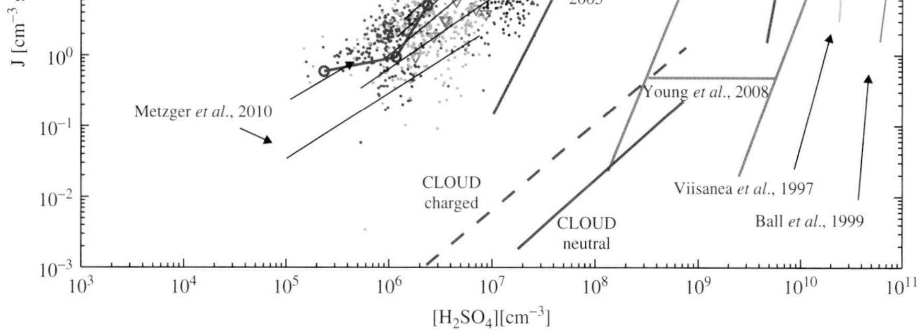

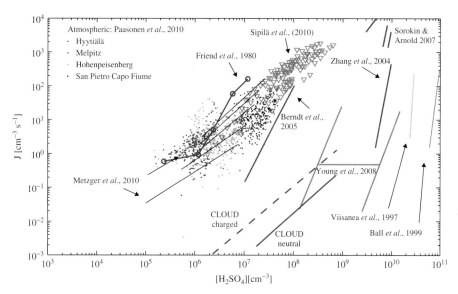

Figure 7.6 *Observed nucleation rates as a function of sulfuric acid concentration, obtained from a selection of different laboratory experiments.*

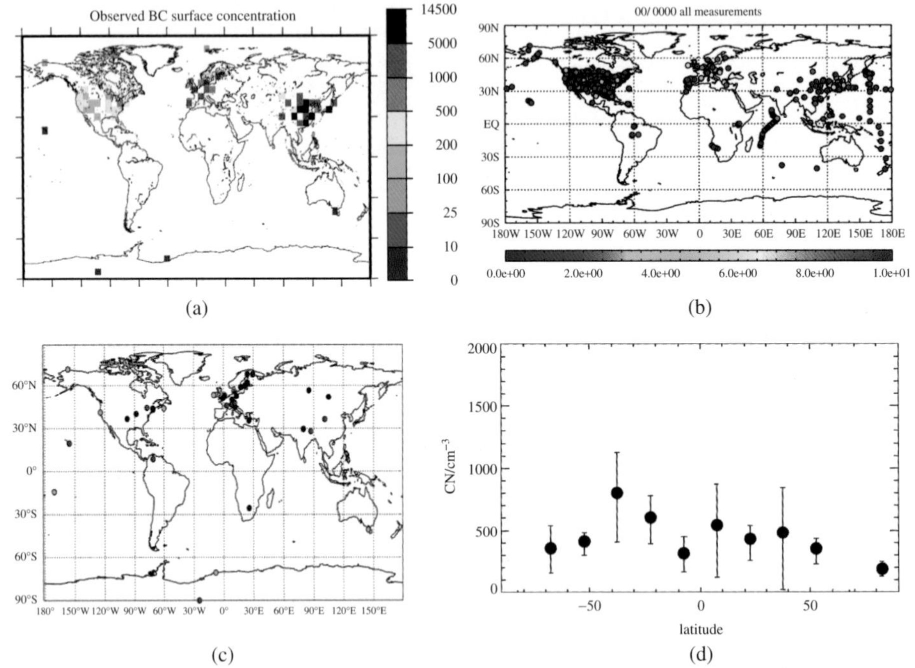

Figure 8.1 *(a) Surface BC observation nanograms per cubic meter. Reproduced with permission from Koch* et al.*, 2009. Copyright © 2009, Koch* et al. *(b) Surface OC observations in micrograms per cubic meter. Reproduced with permission from Tsigaridis* et al. *2013. Copyright © 2013, Tsigardis* et al. *and references therein. (c) Locations of CN observations. (d) Latitudinal distribution of CN concentration observations in cubic centimeters. Reproduced with permission from Spracklen* et al.*, 2010. Copyright © 2010, Spracklen* et al. *Figures (c) and (d) adapted from Spracklen* et al. *(2010) based on data compiled by Heintzenberg* et al. *(2000).*

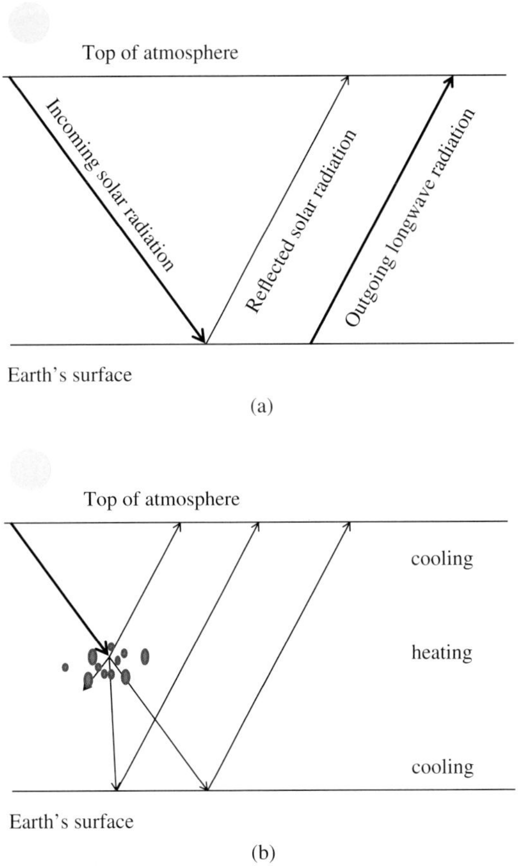

Figure 8.2 *(a) Incoming solar radiation, reflected at the earth's surface as outgoing solar radiation, and outgoing long-wave radiation from the earth, in the absence of aerosols and gases in the atmosphere (transparent atmosphere). (b) Atmosphere with aerosols and their direct impact on solar radiation. (c) Indirect climate effect of aerosols in clean and polluted atmospheres. (d) Radiative forcing estimates (1750–2000), as derived by Forster* et al. *(2007). Reproduced with permission from IPCC 2007. Copyright © 2007, IPCC.*

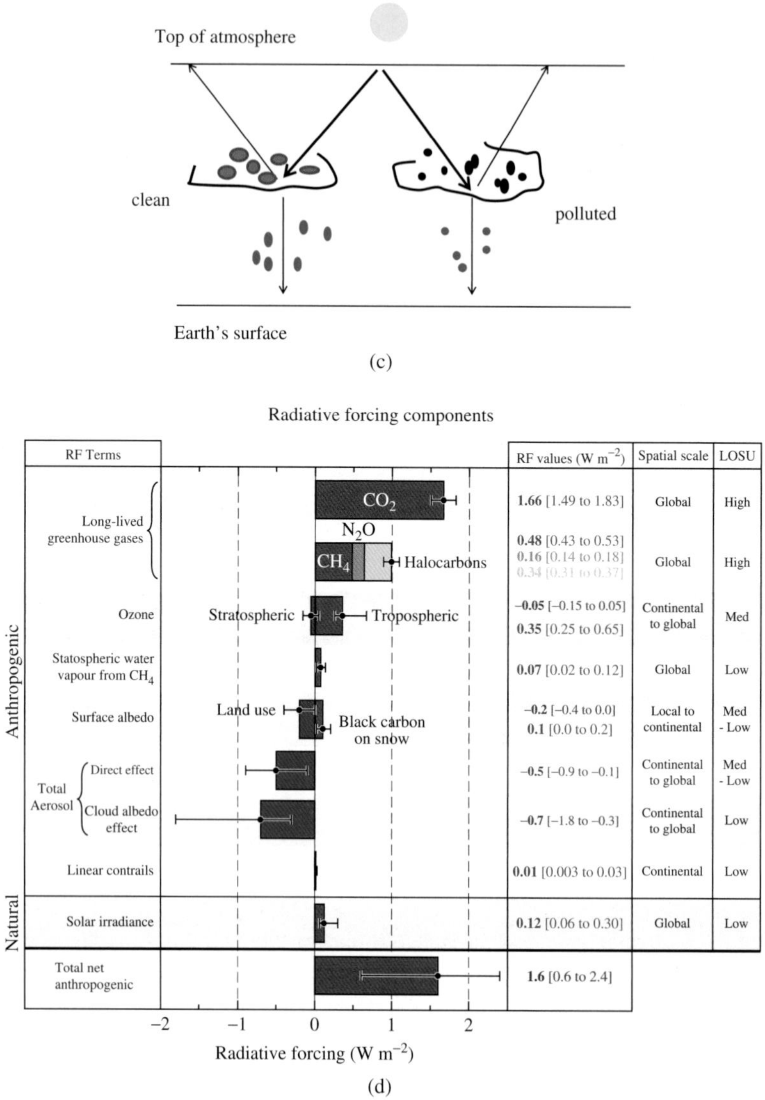

Figure 8.2 *(continued)*

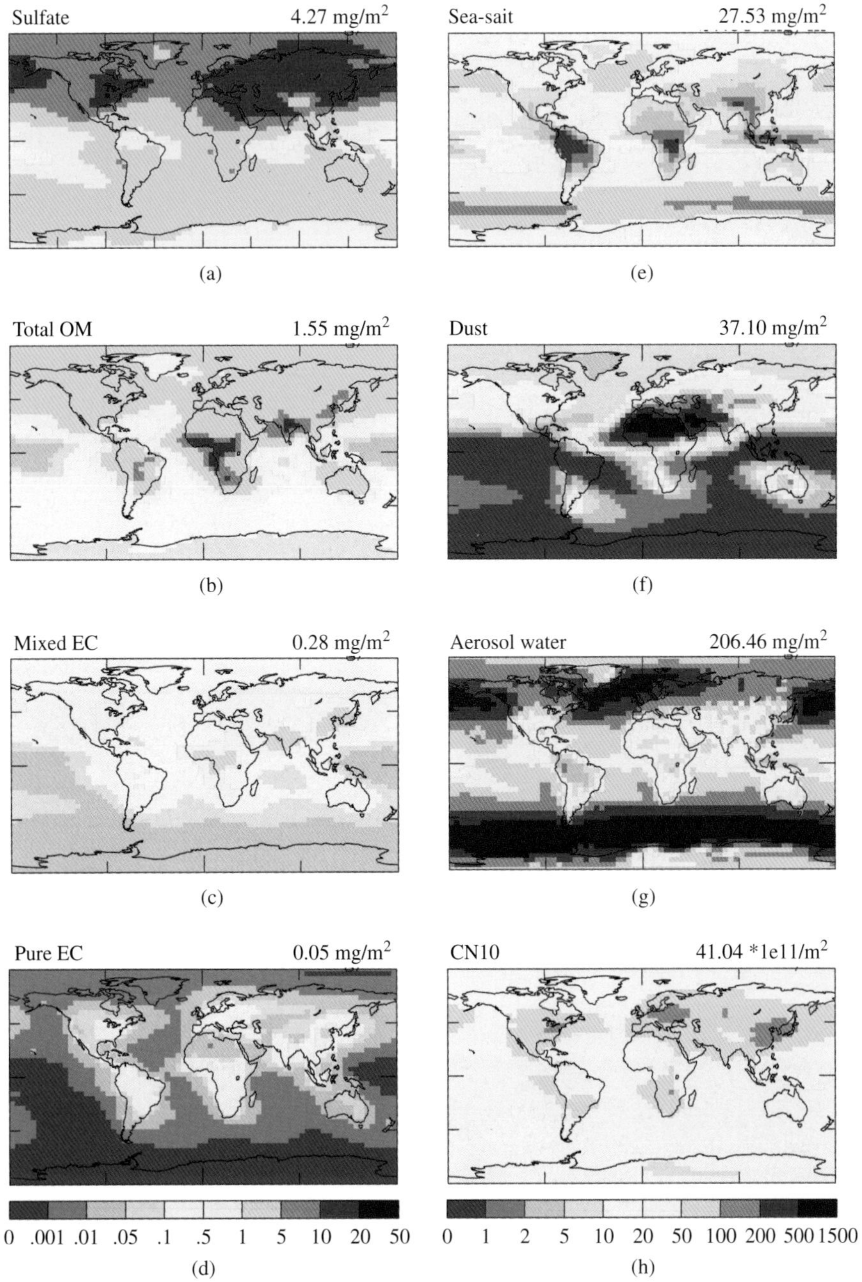

Figure 8.4 *(a–h) Annual mean aerosol component atmospheric loadings (mg.m^{-2}) and number concentration (CN10 in m^{-2}) as calculated by the GISS-TOMAS global model. Reproduced with permission from Lee and Adams, 2010. Copyright (2010), Lee and Adams.*

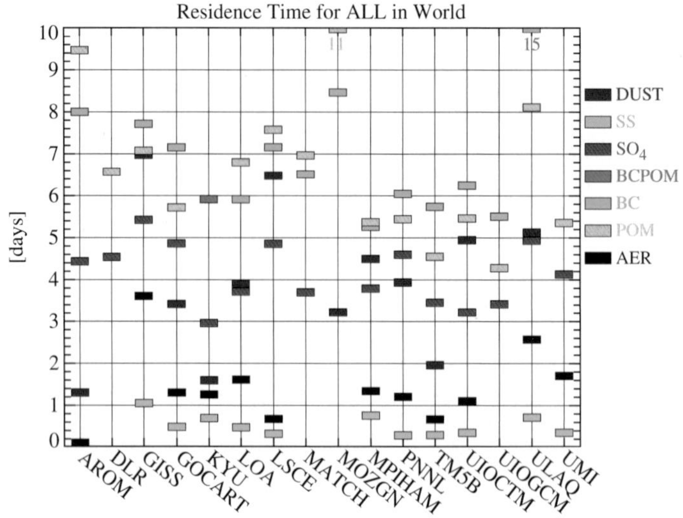

Figure 8.5 *Residence times of the major aerosol components as computed by the AEROCOM models during phase I experiment. Mean (and % standard deviation) values are: DUST: 4.14 days (±43%); SS (sea salt): 0.48 days (±58%); SO_4^-: 4.12 days (±18%); black carbon (BC): 7.12 days (±33%); organic aerosol (POM): 6.54 days (±27%); total dry aerosol mass (AER): 1.42 days (±65%). Reproduced with permission from Textor* et al. *(2006). Copyright (2006) Textor* et al.

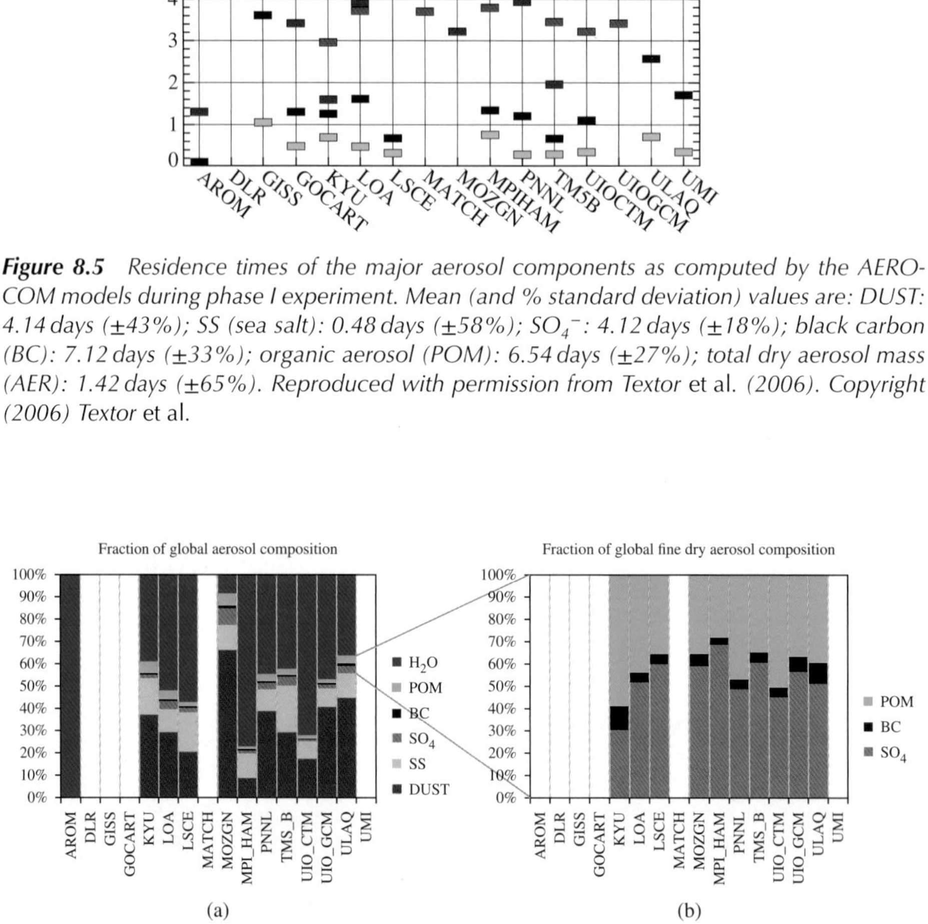

Figure 8.6 *(a) Global aerosol composition, expressed as percent mass fraction of the major aerosol components in global chemistry transport models that participated in the AEROCOM I exercise: H_2O: aerosol water; POM: organic aerosol; BC: black carbon; SO_4: sulfate; SS: sea salt; DUST: dust. (b) Relative contributions of POM, BC, and SO_4 in the fine dry aerosol in these models. Reproduced with permission from Textor* et al. *(2006). Copyright © 2006, Textor* et al.

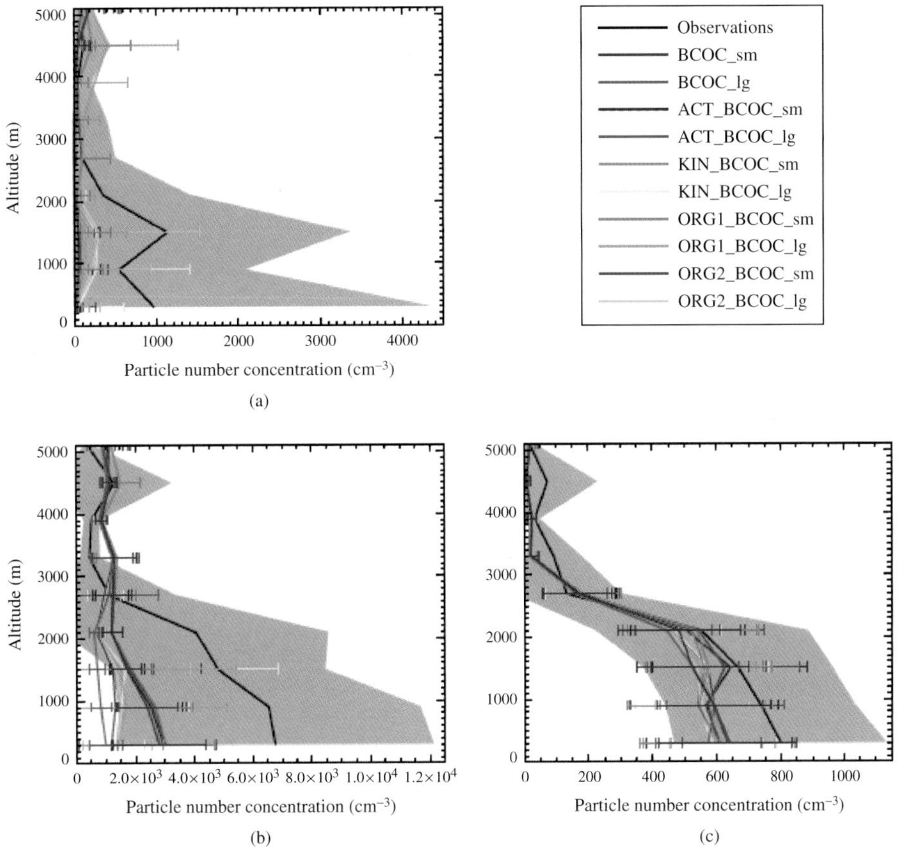

Figure 8.7 *(a–c) Black line: vertical distribution of CN concentrations as observed during the LONGREX campaign over Europe in May 2008; shaded lines: comparison with model results. The error bars and shading represent the standard deviation of the model and observations, respectively. Reproduced with permission from Reddington* et al. *(2011). Copyright © 2011, Reddington* et al.

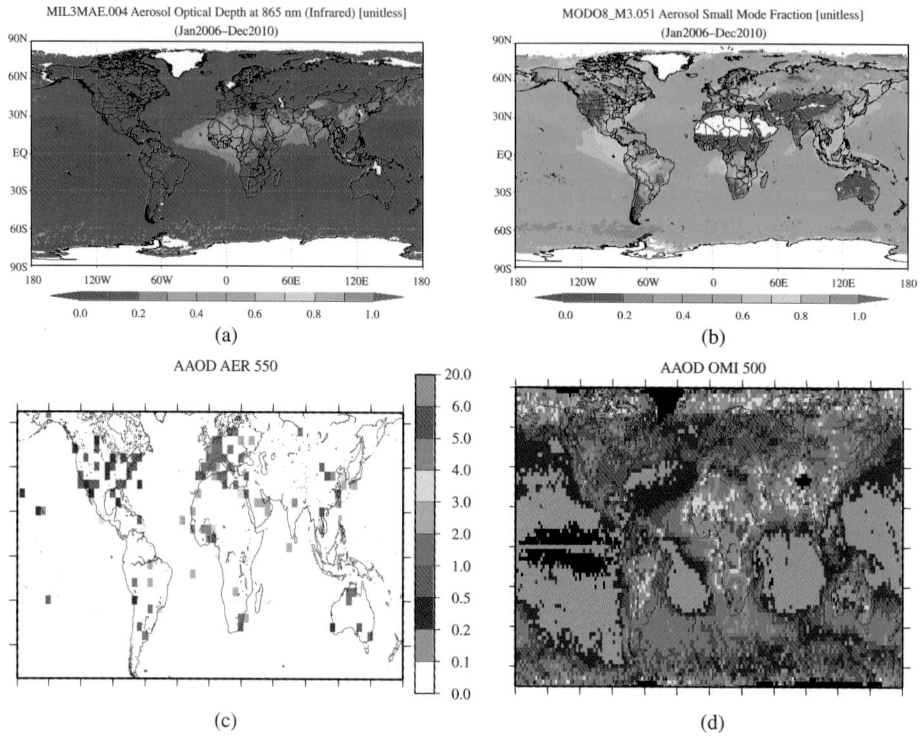

Figure 8.9 *(a) Aerosol optical depth at 550nm as retrieved from MODIS satellite-based observations. (b) Fraction of small-mode AOD as seen by MODIS (averaged over the 5 years from January 2005 to December 2010). Figures (a) and (b) produced using the GiovanniNASA Web-based tool. Reproduced with permission from Acker and Lepkouch, 2007. (c) Absorption AOD from AERONET. (d) Annual absorption AOD from OMI based on daily products from 2005 to 2007. Figures (c) and (d) adapted with permission from Koch* et al., *2009. Copyright © 2009, Koch* et al.

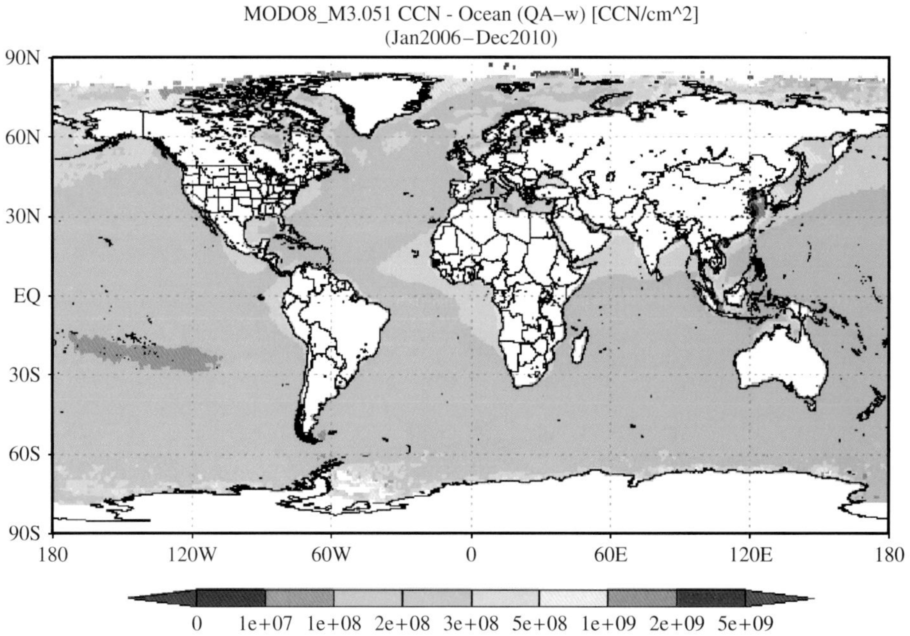

Figure 8.10 *Cloud condensation nuclei (CCNs) concentrations over the ocean, as seen by MODIS, averaged over the 5 years from January 2005 to December 2010. Figure produced using the Giovanni NASA web-based tool (http://disc.sci.gsfc.nasa.gov/giovanni /overview/index.html)*

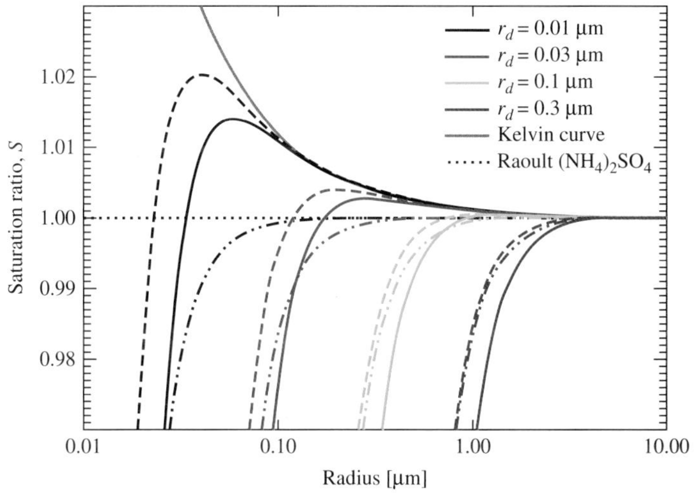

Figure 8.12 *Koehler curves for ammonium sulfate (dashed) and sodium chloride (solid) for various aerosol radii, separating the Kelvin from the Raoult effects. Reproduced with permission from Lohmann, 2009. Copyright © 2009, Lohmann.*

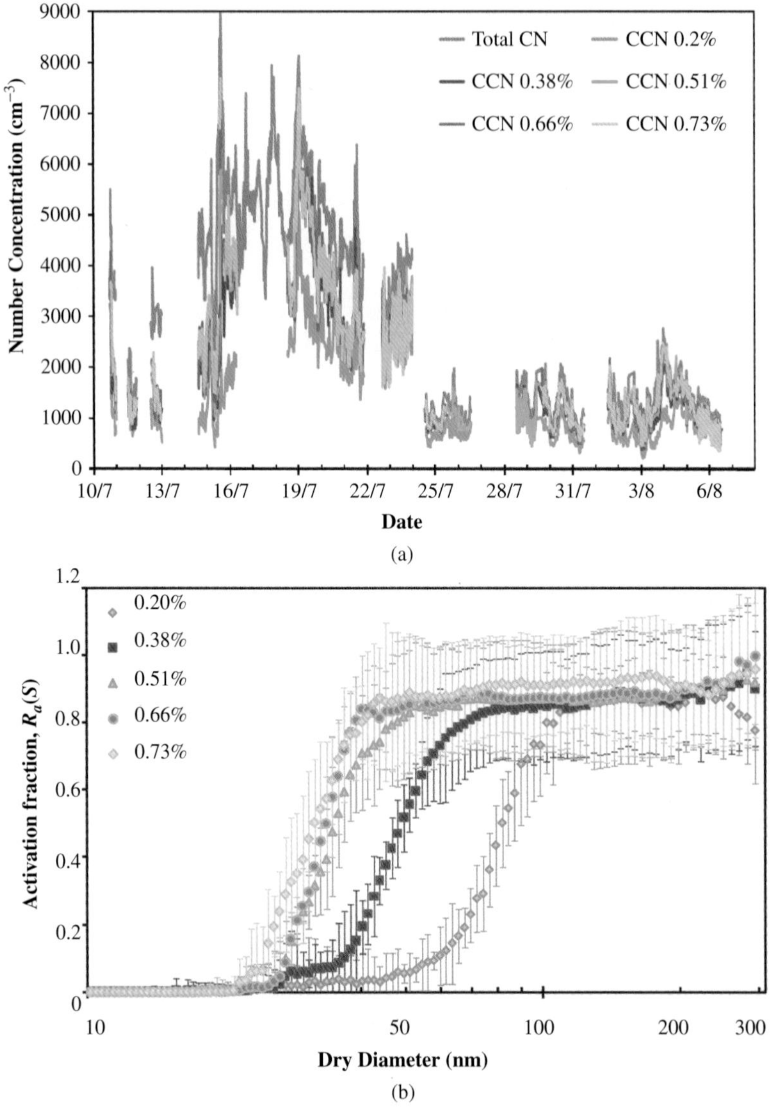

Figure 8.13 *(a) Time series of CN and CCN measurements at Finokalia, Crete, Greece during the FAME campaign. (b) Aerosol size-resolved CCN spectra at different levels of supersaturation (0.2–0.73%), showing the activated aerosol fraction against particle dry diameter. Reproduced with permission from Bougiatioti* et al. *(2011). Copyright © 2011, Bougiatioti.*

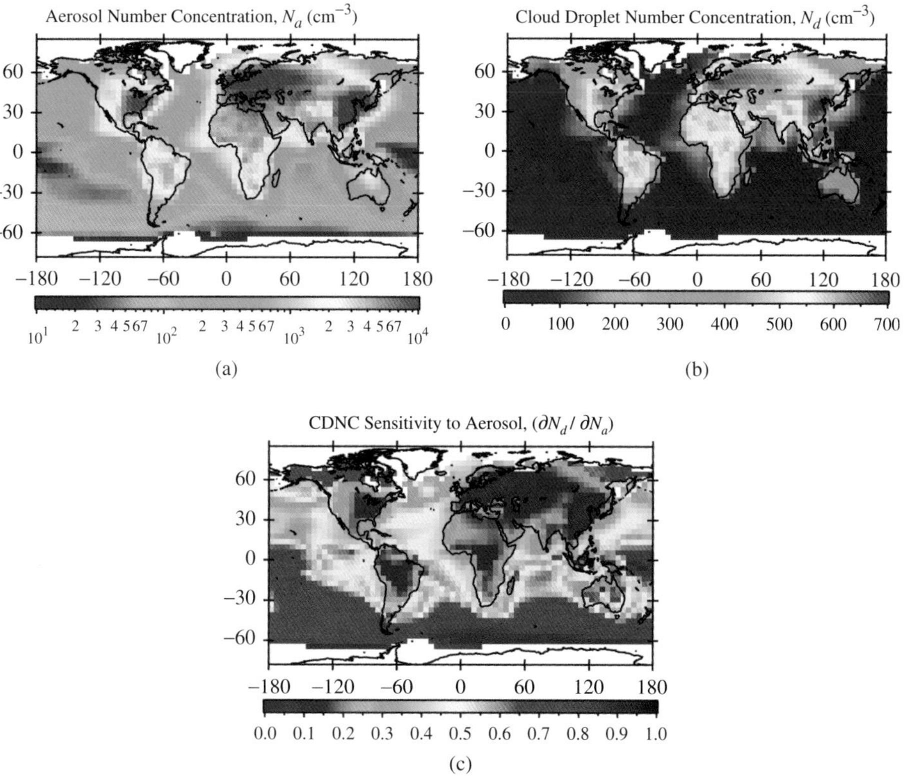

Figure 8.14 *Simulated annual mean (a) condensation nuclei (aerosol number) concentration, (b) CDNC concentration, and (c) CDNC sensitivity to aerosol changes maximizes over the ocean and remote locations. Reproduced with permission fromMoore* et al. *(2013). Copyright © 2013, Moore* et al.

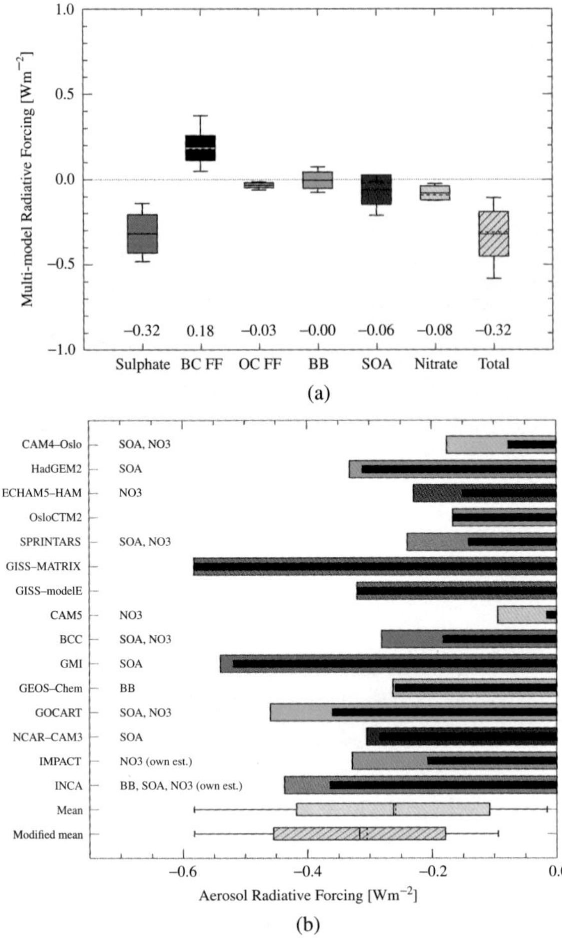

Figure 8.15 *(a) Direct radiative forcing per aerosol component (internal bars) for the period 1750–2010 – mean of AEROCOM models. (b) Total aerosol direct radiative forcing computed by 15 global models. Black bars show the bare modeled forcing, shaded bars show the forcing modified by the model components. Light bars show the AEROCOM mean of the total radiative forcing of direct aerosol effect. The light shaded bar shows the AEROCOM mean when an aerosol component adjustment is made for the missing aerosol components. The boxes indicate one standard deviation, while the whiskers indicate the maximum and minimum distribution. Reproduced with permission with Myhre* et al. *(2013). Copyright © 2013, Myhre* et al.

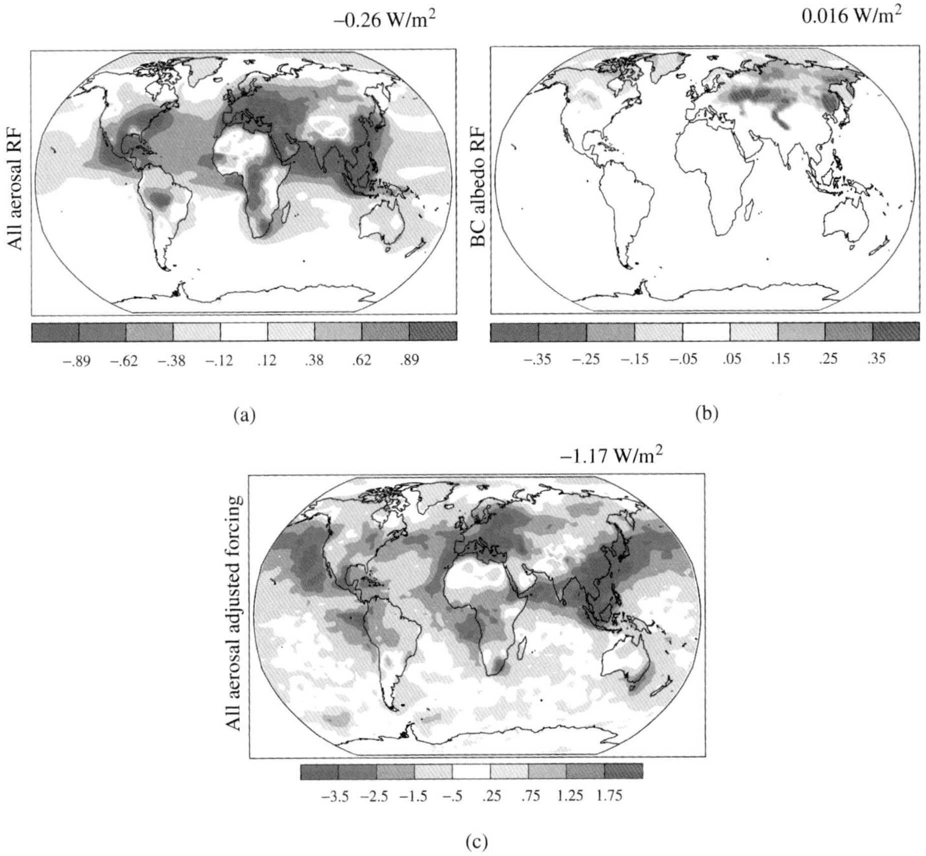

Figure 8.16 *(a) Direct RF from all aerosols (mean of 10 models). (b) BC radiative forcing due to changes in the surface albedo. (c) Adjusted (direct and indirect) RF from all aerosols. Results from the ACMIP model exercise. Adopted with permission from Shindell* et al. *(2013). Copyright © 2013, Shindell* et al.

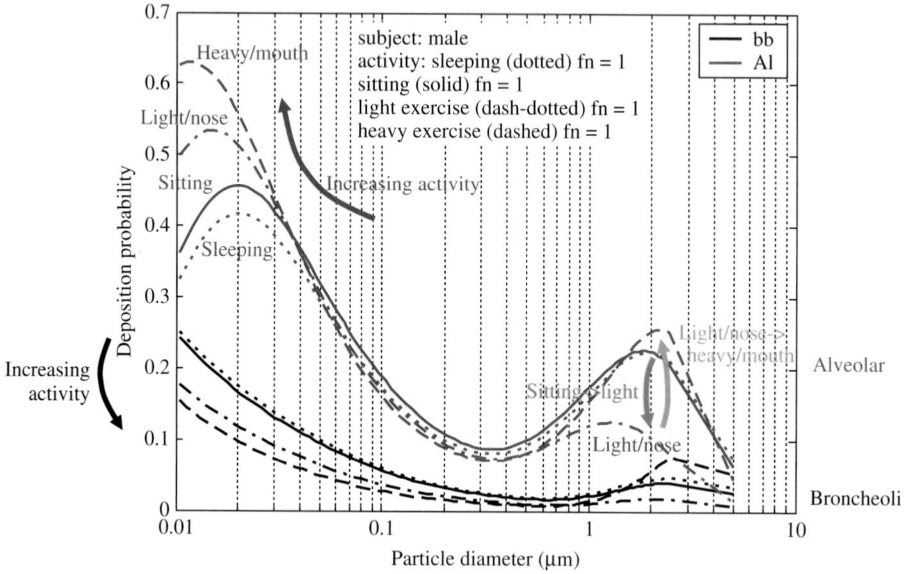

Figure 9.4 *Dependence of the alveolar and bronchiolar deposition efficiency of aerosol particles by size on the physical activity level, as estimated by the ICRP (1994) model. Aerosol density: 1.5 g cm^{-3} corresponding ambient aerosols (Sorjamaa and Hänninen, 2011).*

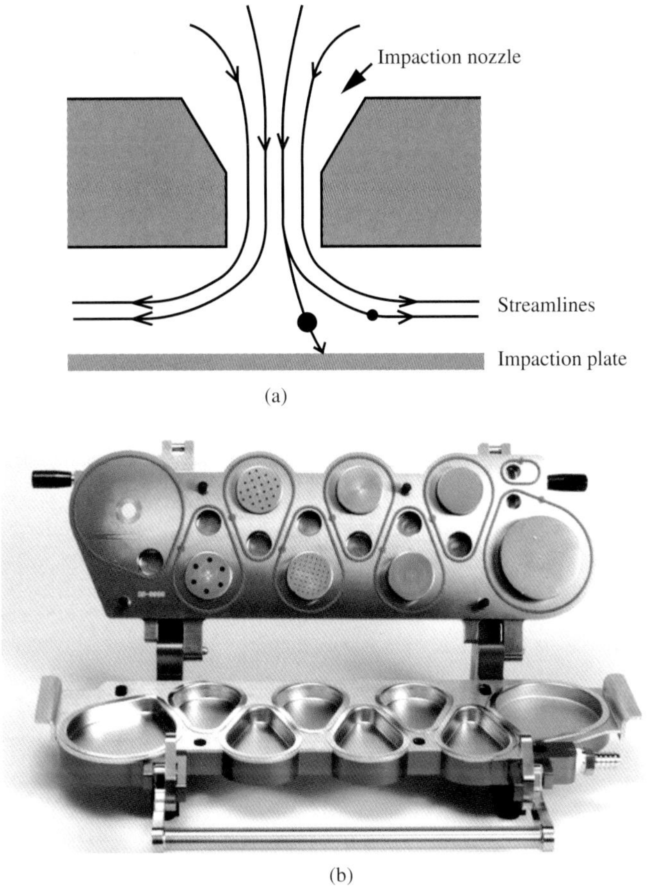

Figure 10.9 *Schematic of (a) a single nozzle of an inertial impactor stage and (b) the next–generation pharmaceutical impactor (NGI).*

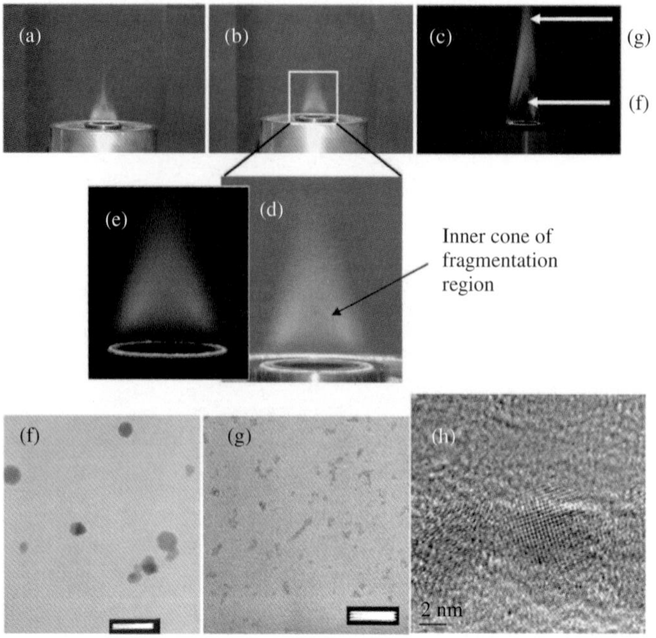

Figure 12.1 *Photographic images of flame cones and transmission electron microscopy (TEM) images for iron oxide nanoparticles. (a) Appearance of the flame cone in the case of no fragmentation at a hydrogen : oxygen ratio of 3 : 4 at 1345 °C. (b) Appearance of the flame cone in the case of fragmentation at a hydrogen : oxygen ratio of 4 : 5 at 1650 °C. (c) Appearance of the flame cone in the case of fragmentation at a hydrogen : oxygen ratio of 14 at 1100 °C (considerably reduced). (d,e) Magnified part of (c) taken with and without a flash light, respectively. The fragmentation region is clearly distinguished. (f) A TEM image of iron-oxide nanoparticles collected inside the cone in (c) from the region at z = 20 mm. (g) Outside the cone region at z = 100 mm (sampling regions are shown by arrows in (c), scale bar is 50 nm). (h) High-resolution TEM image of fragmented particles from (g). Reprinted with permission from Pikhitsa* et al. *(2007). Copyright © 2007, American Institute of Physics.*

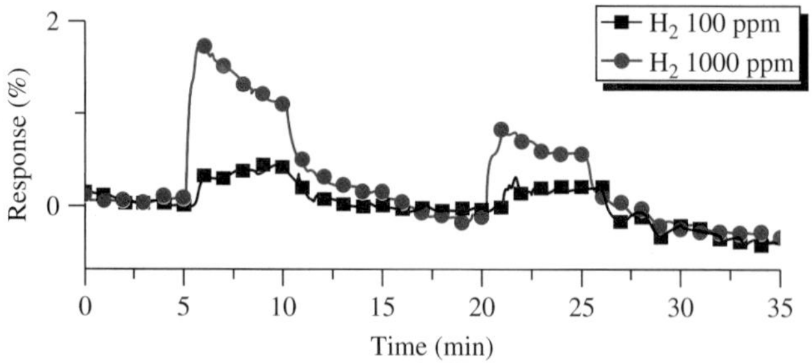

Figure 12.16 *Sensing performance of an acid-treated soot-particle sensor at room temperature. The response to hydrogen was obtained in nitrogen atmosphere in order to check the recovery mechanism. From Kim* et al. *(2011). © IOP Publishing. Reproduced by permission of IOP Publishing. All rights reserved.*

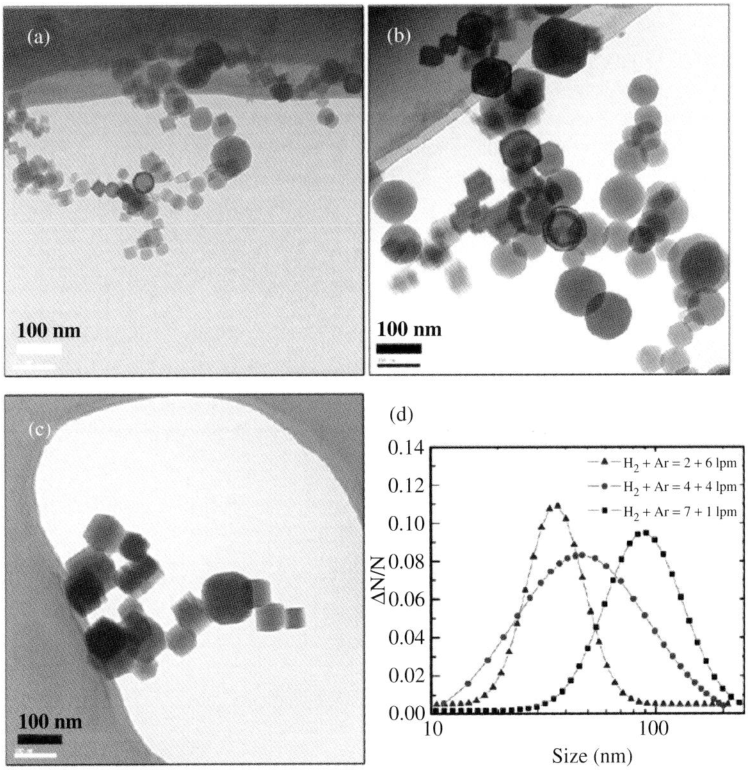

Figure 12.18 *TEM images of magnesium oxide nanoparticles prepared at different hydrogen flow rates. (a) 2 lpm hydrogen gas. (b) 4 lpm hydrogen gas. (c) 7 lpm hydrogen gas. (d) Particle size distributions for all three cases. Reprinted from Yang* et al. *(2010). Copright © 2010, Elsevier.*

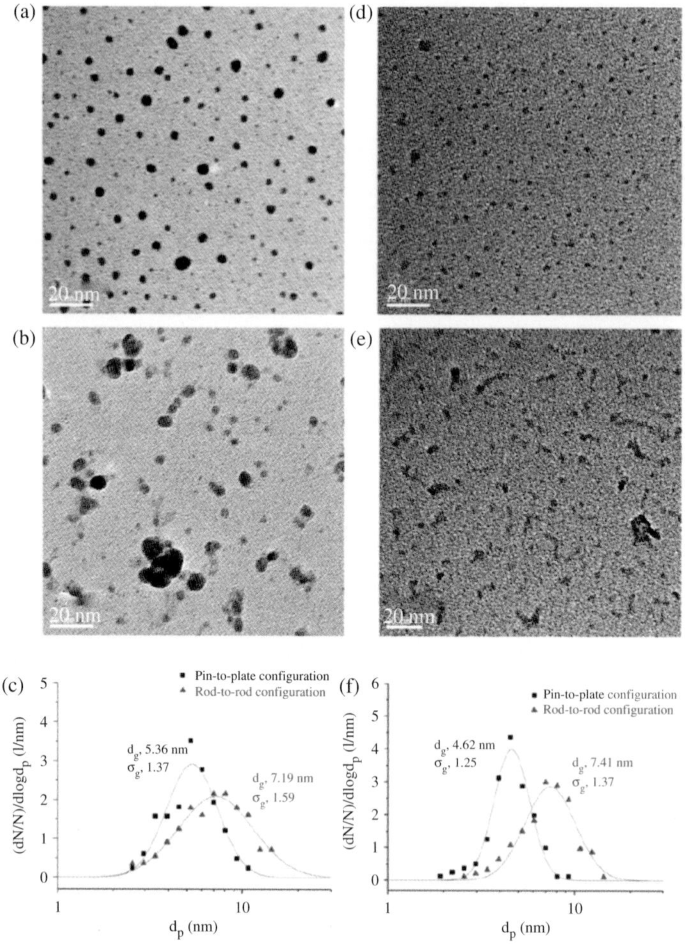

Figure 12.22 *Silver and copper nanoparticles generated by spark discharge with a gap distance of 2.5 mm, argon gas flow rate of 3.5 lpm, and applied positive potential of 5 kV. (a) TEM image of silver particles for pin-to-plate-type SDG. (b) TEM image of silver particles for RR-SDG. (c) Size distributions of silver particles. (d) TEM image of copper particles for pin-to-plate-type SDG. (e) TEM image of copper particles for RR-SDG. (f) Size distributions of copper particles. Reprinted with permission from Han* et al. *(2012). Copyright © 2012, Elsevier.*

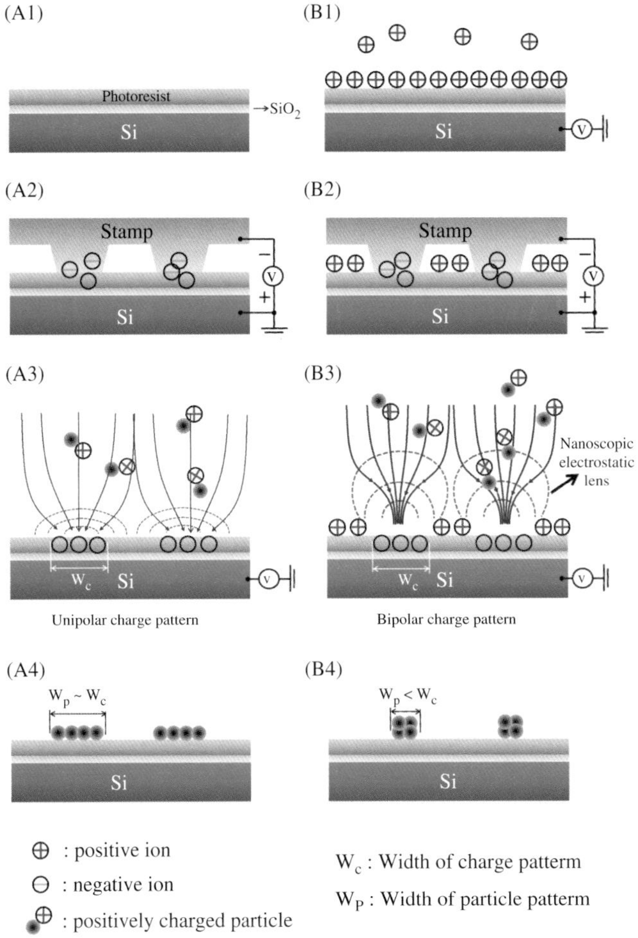

Figure 12.24 *Schematic diagram of conventional and contemporary nanoxerography. (A1) A nonconductive polymer-coated substrate is prepared. (A2) A conductive stamp comes in contact with a PR surface and negative charges are transferred on to the PR. (A3,A4) Due to the Coulombic force, nanoparticles are attracted to the negative surface charge pattern area. (B1) Positive ions are deposited on the PR surface. (B2) A prepatterned metal-coated stamp is contacted with the PR surface and transfers negative charges on to the PR, which replace the positive ions on the contacting area. (B3) Positively charged particles are focused toward the center of the negatively charged surface via a nanoscopic electrostatic lens, induced by alternating bipolar surface charges. (B4) Nanoparticle pattern sizes are much reduced compared to the original negative charge patterns. Reprinted with permission from Lim* et al. *(2012). Copyright © 2012, American Institute of Physics.*

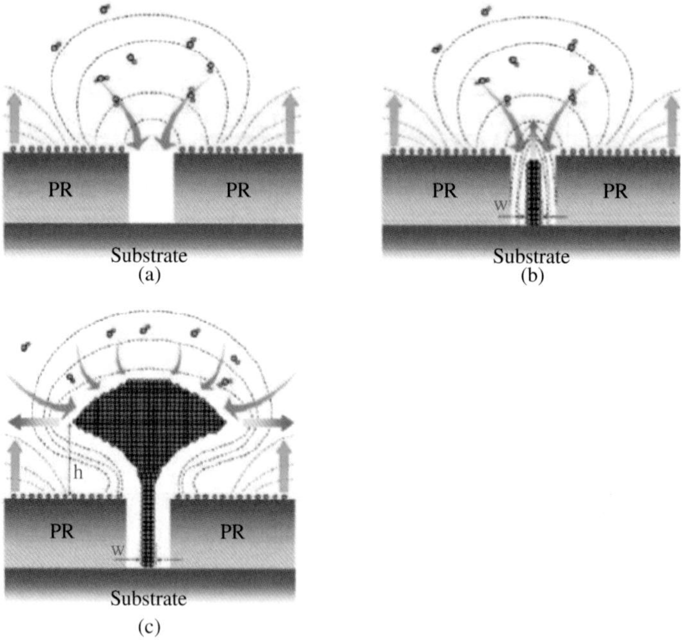

Figure 12.25 *Scheme for the 3D assembly of nanoparticles. (a) Ion-induced electrodynamic focusing guides charged nanoparticles into the center region of the opened Si substrate. (b) Vertical growth of nanostructure within the PR pattern is caused by ion-induced focusing and the antenna effect. (c) Further growth in the lateral direction results from an enhanced antenna effect near the side of the 3D nanoparticle structure. The repelling field is shown on the PR surface. Reprinted with permission from Lee* et al. *(2011). Copyright © 2011, American Chemical Society.*

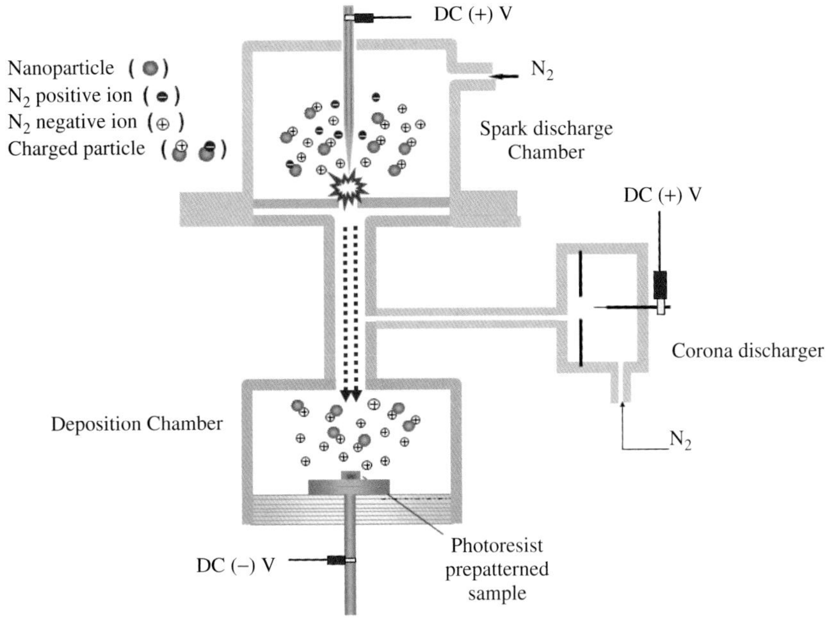

Figure 12.26 *Experimental set-up for the generation and deposition of charged nanoparticles by spark discharge. Charged nanoparticles and ions with mostly positive charge are generated when a positive voltage is applied to the pin electrode and the plane-type electrode is grounded. These are fed into the deposition chamber under an N2 flow rate of 4 l/min. 4 kV negative potential is applied to the substrate in order to attract ions and charged nanoparticles toward it. Therefore, only positively charged ions and nanoparticles move toward the substrate, and a small portion of ions and particles with negative charge cannot be deposited on to it. A corona discharger provides N2 positive ions, which are accumulated on the PR surface prior to deposition. The ions accumulate on the PR surface along the electric field line and form ion-induced electrostatic lens. Charged nanoparticles are convergently guided by the lens and placed at the desired position upon the substrate. As the deposition time increases, various 3D nanoparticle structures, such as hemispheres, pillars, mushrooms, and four-leaf clovers, are obtained from the originally 500 nm-wide square PR pattern (Figure 12.27a) and the 3D nanoparticle structures, such as 3D crosses and flowers, are obtained from originally cross-shaped PR patterns with one side of 500 nm (Figure 12.27b). As observed early at 4 minutes, nanoparticles are convergently guided only within the center region of the PR pattern, due to the ion-induced focusing mentioned earlier (see Figure 12.27a). Reprinted with permission from Lee* et al. *(2011). Copyright © 2011, American Chemical Society.*

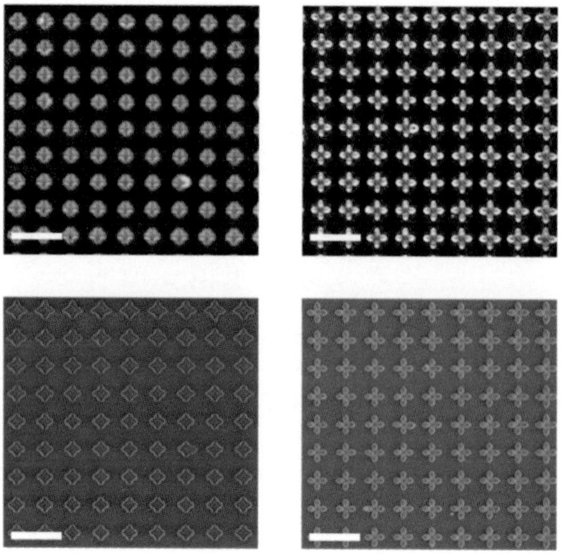

(a) (b)

Figure 12.29 *Dark-field scattering image (top) and scanning electron microscopy (SEM) image (bottom) of (a) silver-nanoparticle-covered and (b) gold-film-covered 3D copper nanoparticle structure arrays. The scale bar is 5 μm. Reprinted with permission from Lee* et al. *(2011). Copyright © 2011, American Chemical Society.*

Figure 17.7 *A Collison nebulizer.*

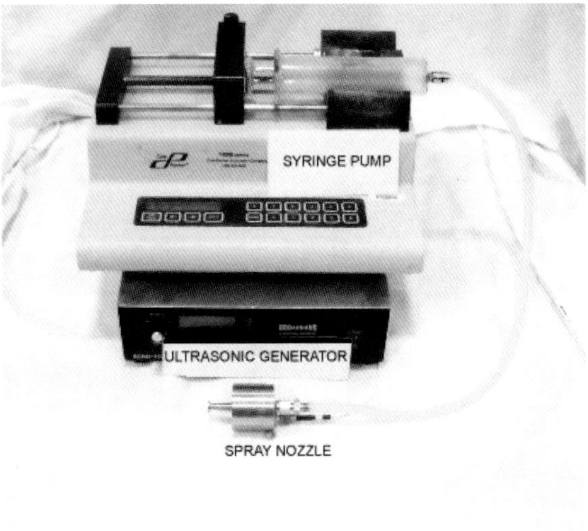

Figure 17.8 *Sono-Tek aerosol generators.*

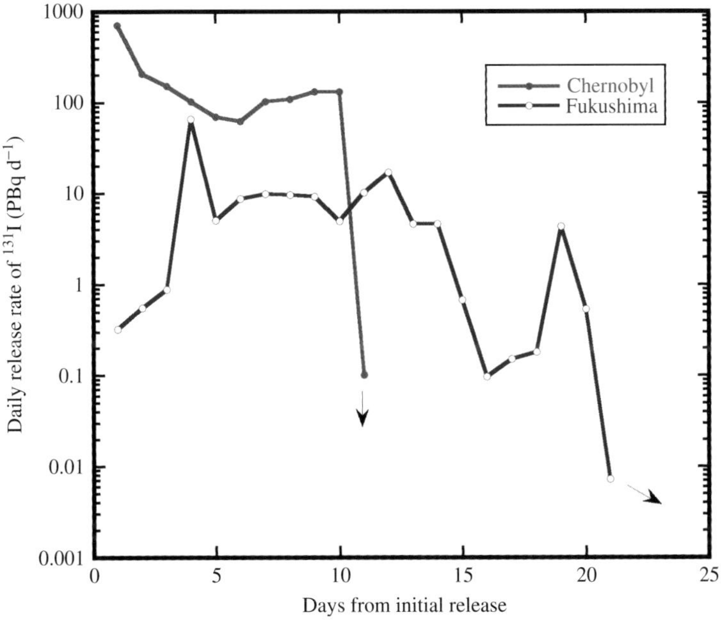

Figure 18.1 *Temporal changes in daily ^{131}I emission rates from the nuclear reactors at Chernobyl and Fukushima Daiichi NPP.*

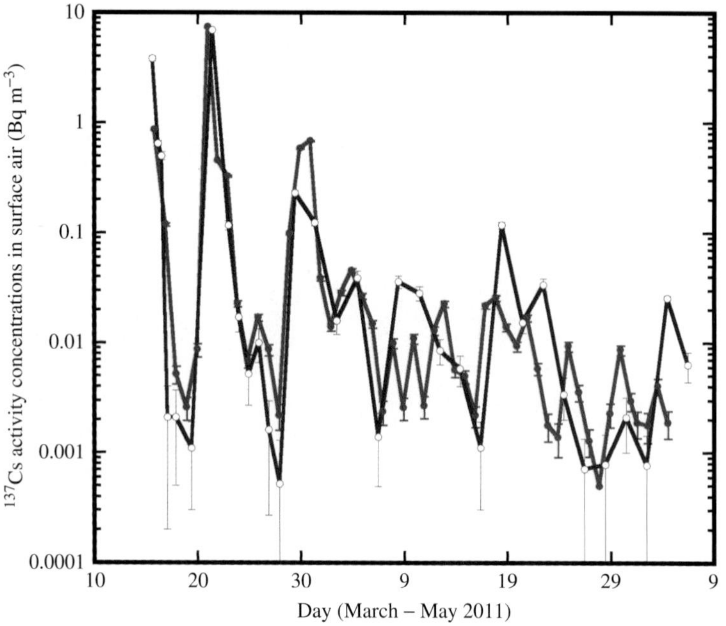

Figure 18.2 *Temporal changes in Fukushima-derived ^{137}Cs concentrations in surface air at Tsukuba and Inage, Japan. Data cited from Amano* et al. *(2012) and Doi* et al. *(2013).*

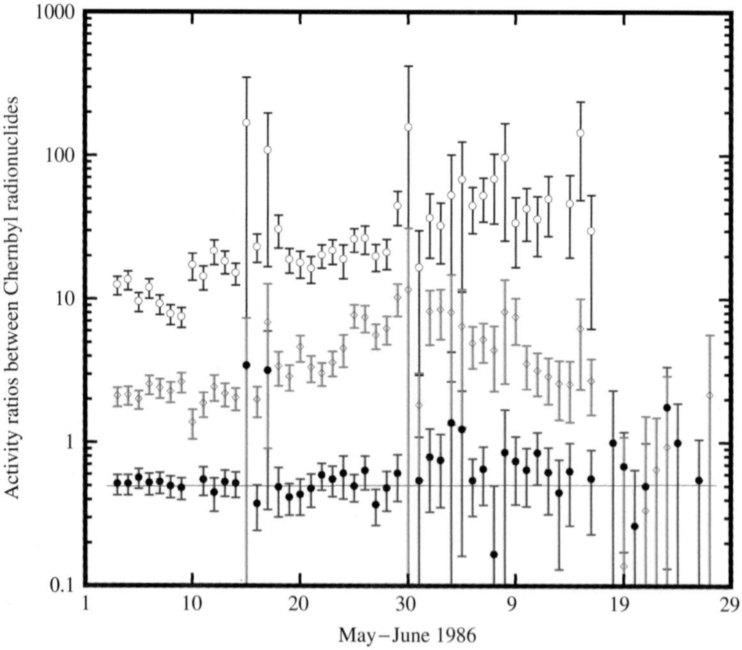

Figure 18.4 *Temporal variations of ^{131}I/^{137}Cs (open circle), ^{103}Ru/^{137}Cs (open rhombic), and ^{134}Cs/^{137}Cs (closed circle) activity ratios in surface air samples observed at Tsukuba, Japan.*

small spherical particles). These parameters are most frequently determined as a function of wavelength.

8.5.1 Aerosol Optical Depth and Direct Radiative Forcing of Aerosol Components

Although size-segregated aerosol mass concentration is the basic measure of aerosol loading in the models, this quantity is translated to AOD via the extinction efficiency of aerosol components in order to compare with remote-sensing (ground-based or satellite-based) observations and then estimate aerosol direct radiative forcing (RF). Note that each model employs its own extinction efficiencies based on limited knowledge of the optical and physical properties of each aerosol type. Thus, it is possible for the models to produce different distributions of aerosol loading as mass concentrations but to agree in their distributions of AOD, and vice versa (Chin *et al.*, 2009).

The AOD of the atmosphere retrieved from satellite observations shows large aerosol extinction due to desert dust and biomass burning in the tropics and to pollution aerosol over South Asia (Figure 8.9a). Due to rapidly increasing retrieval capabilities, satellite observations now provide information on the size of the aerosols, separating the

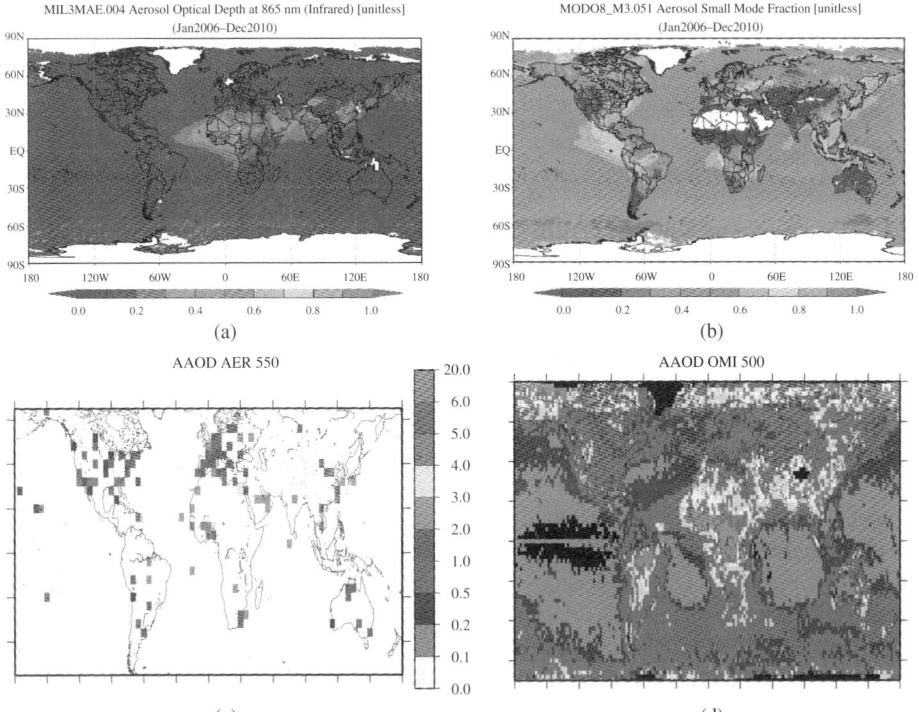

Figure 8.9 *(a) Aerosol optical depth at 550 nm as retrieved from MODIS satellite-based observations. (b) Fraction of small-mode AOD as seen by MODIS (averaged over the 5 years from January 2005 to December 2010). Figures (a) and (b) produced using the Giovanni NASA Web-based tool. Reproduced with permission from Acker and Lepkouch, 2007. (c) Absorption AOD from AERONET. (d) Annual absorption AOD from OMI based on daily products from 2005 to 2007. Figures (c) and (d) adapted with permission from Koch* et al.*, 2009. Copyright © 2009, Koch* et al. *See plate section for colour version.*

fine-mode fraction of AOD (as depicted in Figure 8.9b) from the MODIS observations (January 2006 to December 2010). This figure shows the presence of small aerosols downwind of industrial and biomass-burning pollution regions in the NH and the tropics. The pictures were made using the Giovanni NASA Web-based tool (http://disc.sci.gsfc.nasa.gov/giovanni/overview/index.html) and are complemented/validated by data from the global network of sun photometers (http://gcmd.nasa.gov/records/GCMD_AERONET_NASA.html). Figure 8.9c shows data on absorption AOD (which is mainly driven by BC levels) from the AERONET station, as compiled by Koch *et al.* (2009), while Figure 8.9d shows the annual absorption AOD as derived from the Ozone Monitoring Instrument (OMI).

The RF of aerosols in the atmosphere is therefore determined based on the spatiotemporal distribution of the studied aerosol components and their optical properties. In addition, changes in ambient RH affect the water that is taken up by atmospheric aerosols and the amount of soluble aerosol constituents dissolved in the aerosol water phase, thereby changing the aerosols' chemical composition, the size of the particles and the particles' optical properties. Radiative forcing depends upon the underlying surface reflectance and the insolation. For scattering aerosol, the strongest radiative forcing occurs when the surface reflectance is low, while for absorbing aerosol, radiative forcing is strongest when the surface reflectance is high. For the same reasons, the RF of an aerosol depends on its location relative to any clouds, since clouds increase radiation above their layer and reduce it below (Figure 8.3). Thus, aerosols below clouds will receive less radiation than those in cloud-free atmospheric conditions, while those above clouds will receive even more.

8.6 Aerosols and Clouds (Indirect Effect)

Aerosols serve as CNs; that is, surfaces on which water vapor and semi- or low-volatility gases can condense, growing the aerosol until it reaches a critical size and becomes a CCN. Under favorable low-temperature ambient conditions, aerosols can also serve as ice nuclei (IN). Figures 8.1 and 8.5 depict observed distributions of CNs in the global atmosphere. Figure 8.10 depicts the distribution of CCNs over the ocean, as retrieved from MODIS observations. An increase in aerosol concentration and/or changes in aerosol properties can therefore modify the microphysical and radiative properties of clouds, as well as the precipitation efficiency and thus the cloud lifetime (Haywood and Boucher, 2000). However, the cause–effect relationship between aerosols and clouds/precipitation is not evidenced in all studied conditions. Despite decades of research, the establishment of climatically meaningful relationships among aerosol, clouds, and precipitation remains a challenge. Stevens and Feingold (2009) proposed that this difficulty was due to unaccounted processes that buffer cloud and precipitation responses to aerosol perturbations. In that thorough review of existing data, they proposed the existence of **microphysical** (on the scale of the cloud droplet) and **macrophysical** (on the scale of the cloud and its surrounding atmosphere) buffering mechanisms in the atmosphere. Microphysical buffers are related, for example, to changes in the size distribution or composition of cloud-active aerosol. In the absence of other effects, such changes lead to fewer cloud droplets but also result in locally higher supersaturation. This tends to partially compensate for, or even counterbalance, the initial changes, by allowing smaller CNs to be activated. Macrophysical buffers are associated with, for example, the deepening of clouds, caused by a reduction in local precipitation due

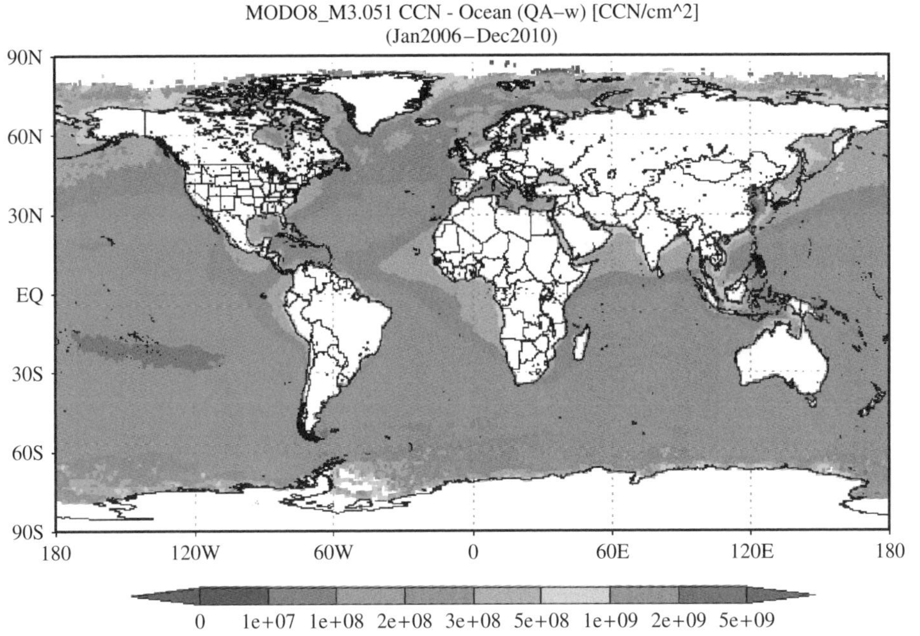

Figure 8.10 *Cloud condensation nuclei (CCNs) concentrations over the ocean, as seen by MODIS, averaged over the 5 years from January 2005 to December 2010. Figure produced using the Giovanni NASA web-based tool (http://disc.sci.gsfc.nasa.gov/giovanni/overview/index.html). See plate section for colour version.*

to the higher number of (smaller) CCNs. The deeper clouds produce heavier rain, which compensates for the initial inhibition of precipitation due to pollution aerosol.

The first modeling studies to investigate aerosol–cloud interactions applied empirical observation-based relationships between the sulfate aerosol mass concentration (used as a proxy for anthropogenic hygroscopic aerosol) and cloud droplet number concentration (CDNC) concentrations, separating observations over land from those over ocean (Figure 8.11 from Haywood and Boucher, 2000). Since then, Koehler-based parameterizations accounting for aerosol/cloud microphysics have been adopted in global models. These parameterizations are of varying complexity and describe the water uptake and growth of aerosols to form cloud droplets and their impact on precipitation.

8.6.1 How Aerosols Become CCNs and Grow into Cloud Droplets

The first step in simulating aerosol–cloud interactions is to simulate which aerosols activate and become cloud droplets, and when. The hygroscopic growth of inorganic particles is described by the Koehler equation, which refers to the equilibrium conditions of an aerosol with its surrounding atmosphere. Koehler theory provides the water vapor saturation needed to form a droplet as a function of the dry radius (or diameter) and chemical composition of the aerosol, often expressed in terms of the equilibrium ambient water saturation ratio, S_{eq}, and the droplet size (details in Seinfeld and Pandis, 1998 and the appendix in Kanakidou *et al.*, 2005a). The Koehler theory accounts for both the Kelvin (thermodynamic effect) and the Raoult (chemical composition – solution term) effects. This curve

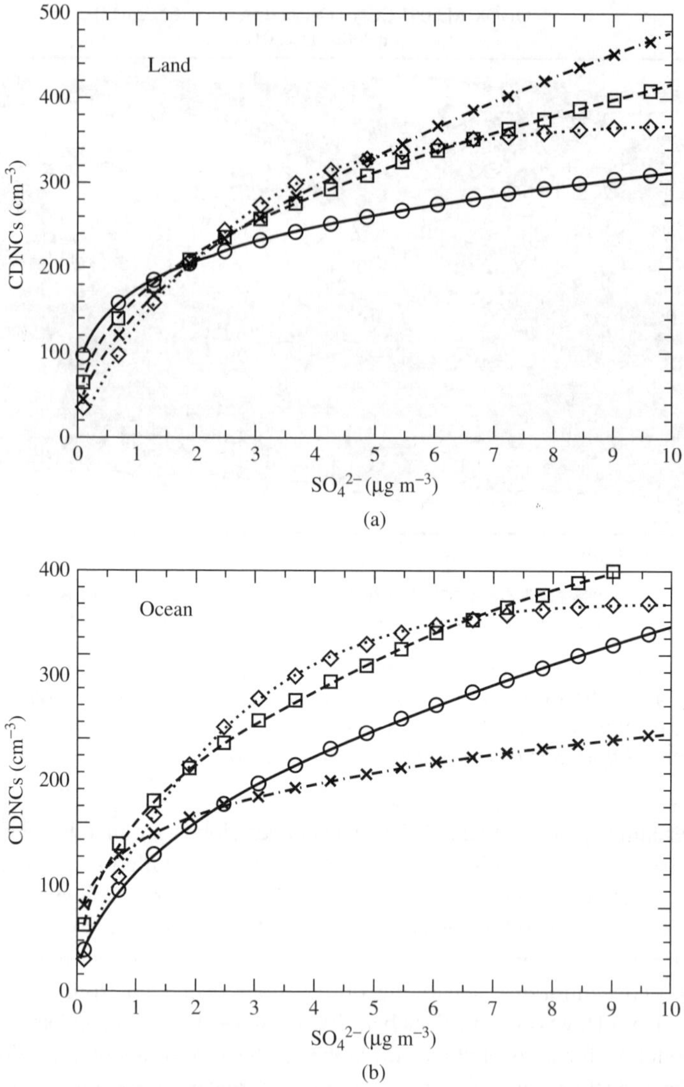

Figure 8.11 *Observed correlations between cloud droplet number concentrations (CDNCs) and sulfate aerosol mass concentration. Symbols show different parameterisations. (a) over land, (b) over ocean. Figure modified from the original Haywood and Boucher (2000) where more details can be found.*

(Figure 8.12) provides the water vapor pressure needed to form a droplet over equilibrium vapor pressure (saturation ratio) as a function of the droplet size (radius). The Kelvin term $(1 + a/r)$ depends on the size of the particle and is proportional to the inverse of the particle radius r ($\sim 1/r$). It describes the so-called curvature effect, which is the increase in the saturation vapor pressure over a droplet of size r compared to a plane surface. Opposing the Kelvin effect, the presence of solute (chemical composition effect – the Raoult term of the

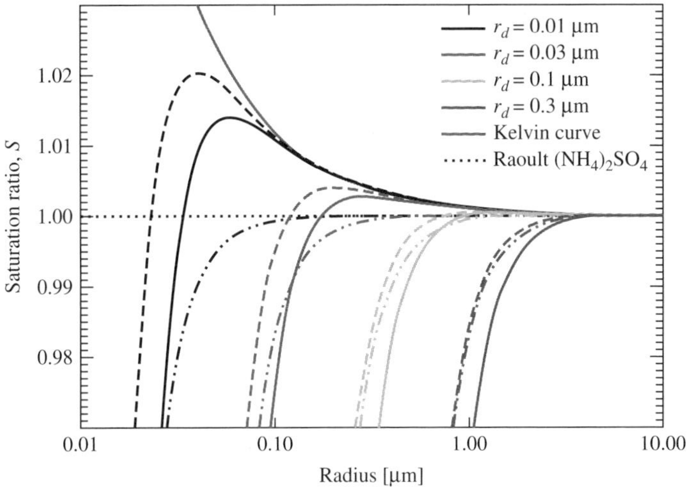

Figure 8.12 *Koehler curves for ammonium sulfate (dashed) and sodium chloride (solid) for various aerosol radii, separating the Kelvin from the Raoult effects. Reproduced with permission from Lohmann, 2009. Copyright © 2009, Lohmann. See plate section for colour version.*

equation) acts to lower the saturation vapor pressure. This effect depends on the volume of the aerosol $(1 - b/r^3)$ and is therefore proportional to the inverse of the cube of the radius, $1/r^3$. At small sizes, the Raoult effect is more important than the Kelvin effect.

The Koehler curve (an example is given in Figure 8.12) increases steeply when $S_{eq} < 1$ and goes through a maximum at some S_{eq} above unity, which is the so-called 'critical supersaturation', S_c. If the ambient S exceeds S_c, water vapor diffuses toward and condenses on the droplet, resulting in unconstrained growth of the droplet, which is thus 'activated', and in atmospheric cloud formation. Due to the competing Kelvin and Raoult terms in the Koehler equation, small particles must acquire more soluble material relative to their mass than larger ones. However, the required soluble mass per particle is relatively insensitive to the diameter of the initial insoluble particle.

Inside a cloud, the driving force for the increase in S is cooling, usually caused by expansion of rising air, mixing, or radiation. Depletion of water vapor and release of latent heat by water condensation slow the increase in S, and at some point S reaches a maximum. In principle, all droplets with critical supersaturations below the maximum value of S can activate to cloud droplets, although the diffusional growth of droplets may be sufficiently slow that all such droplets will not have time to activate (Nenes *et al.*, 2001).

Competition between the ambient cooling rate in rising air and condensational depletion of water vapor from the growing CCNs determines the maximum value of S as well as the number of activated droplets. Thus, anything that can affect the growth rate of droplets will potentially affect cloud droplet number. The maximum value of S is affected by the number concentration, size distribution, and composition of the aerosol particles present. Compositional effects are not only expressed in S_{eq} but can also affect the mass-transfer coefficient of water vapor to the droplet. This is because the size of CCNs is comparable to the mean free path of air molecules; thus, the probability of a water vapor molecule 'sticking' upon the growing droplet will influence the mass-transfer coefficient of water vapor to the droplet (Seinfeld and Pandis, 1998).

Experimentally, the efficiency of dry aerosols in capturing atmospheric water and grow is determined by the aerosol hygroscopicity, which is commonly expressed by the growth factor (GF). Aerosol GF can be measured in the atmosphere as the ratio between the humidified and the dry particle diameter at a reference RH that in most studies is 90% (or the results are converted to 90% RH). Recently, Petters and Kreidenweis (2007) presented a method for describing the relationship between particle dry diameter and CCN activity using a single hygroscopicity parameter, κ, that is defined through its effect on the water activity, a_W, of the solution; that is, the water vapor pressure ratio that the particles senses:

$$1/a_w = 1 + \kappa V_s/V_w \tag{8.3}$$

where V_s is the volume of the dry particulate matter and V_w is the volume of the water. Values of κ are between 0.5 and 1.4 for highly CCN-active salts such as sodium chloride, between 0.01 and 0.5 for organic species of intermediate hygroscopicity, and 0 for nonhygroscopic components. Observations indicate that atmospheric particles are typically characterized by $0.1 < \kappa < 0.9$. A multicomponent hygroscopicity parameter can be computed by weighting component hygroscopicity parameters by their volume fractions in the aerosol mixture, for known chemical composition of the aerosols and hygroscopicity parameters of each aerosol component. κ is then used in the Koehler theory to simulate cloud droplet activation. Figure 8.13a shows summertime observations of CNs and CCNs in the Mediterranean (Bougiatioti *et al.*, 2011), which clearly indicate the different activation fractions of aerosol (Figure 8.13b) and that not all aerosols are activating.

Generally, aerosol–cloud interaction studies assume that CCNs are formed from aerosols that contain significant amounts of soluble material and thus that their growth can be described by the Koehler theory. However, recent studies suggest that mineral dust aerosols can also act as good CCNs, since they can absorb water on to their surfaces (DeMott *et al.*, 2003; Levin *et al.*, 2005; Solomos *et al.*, 2011). The dependence of CCNs on dry aerosol diameter for aerosols composed mainly of soluble material is different from that of those that are mainly composed of insoluble material. Kumar, Sokolik, and Nenes (2009), assuming an external mixture of soluble and insoluble aerosols, developed a parameterization for CCN activity that follows the Koehler theory for soluble materials and accounts for the absorption activation theory for insoluble materials. Improvement of this parameterization enables treatment of aerosols that contain both soluble and insoluble material (Kumar *et al.*, 2011).

Karydis *et al.* (2011) have further applied this concept in a GCM accounting for activated CCNs from (i) particles that follow the Koehler theory (mainly soluble material), (ii) particles that follow the Frenkel–Halsey–Hill absorption theory, (iii) giant CCNs whose activation behavior represents the border between particles that experience significant growth after activation and those that are strongly kinetically limited, and (iv) giant CCNs that follow the absorption activation theory and do not significantly grow after activation. They have thus found that a coating of dust by hygroscopic salts increases the contribution of dust to CCNs by a factor of two, while substantially depleting the in-cloud supersaturation during the initial stages of cloud formation, and eventually reduces the CDNC.

Furthermore, Moore *et al.* (2013), investigating the sensitivity of CDNC to CCN-active aerosol number concentrations, found little dependence over the continents (10–30%), but

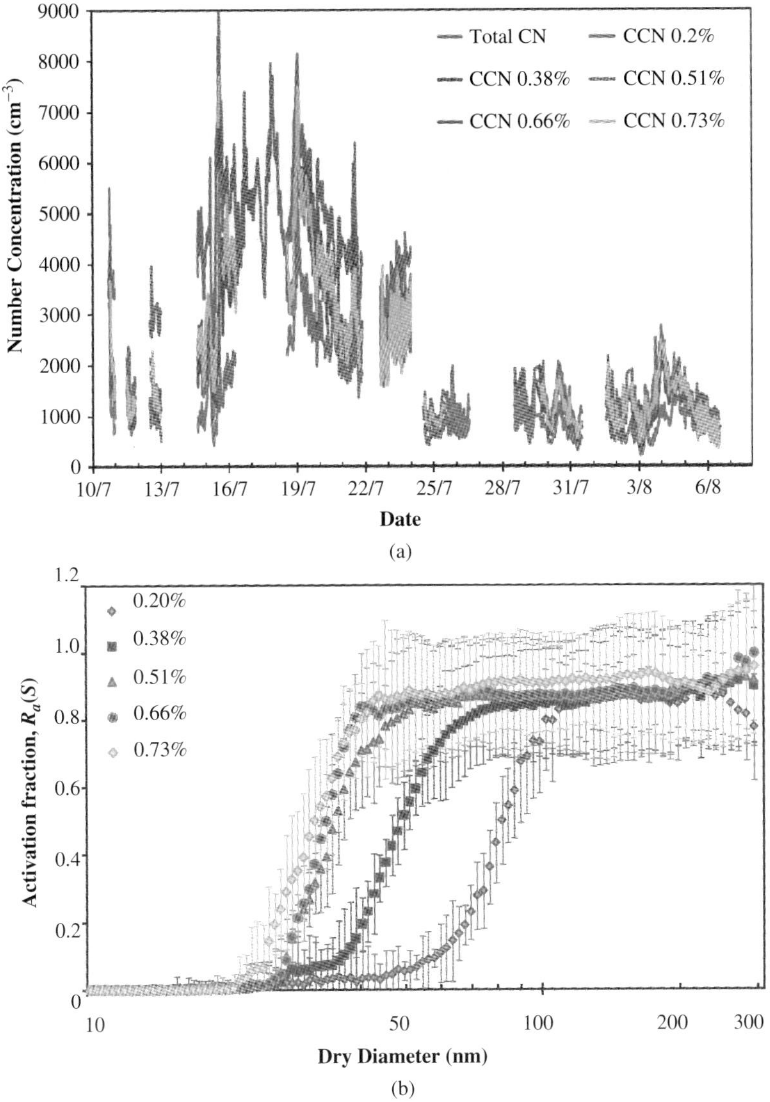

Figure 8.13 (a) Time series of CN and CCN measurements at Finokalia, Crete, Greece during the FAME campaign. (b) Aerosol size-resolved CCN spectra at different levels of supersaturation (0.2–0.73%), showing the activated aerosol fraction against particle dry diameter. Reproduced with permission from Bougiatioti et al. (2011). Copyright © 2011, Bougiatioti. See plate section for colour version.

that sensitivities exceed 70% in pristine regions (Figure 8.14). This result suggests that an anthropogenically driven aerosol indirect effect can be expected over remote regions. These findings are in line with the modeling and satellite synergistic investigations by Sorooshian *et al.* (2009), who found that for clouds with an intermediate liquid water path, aerosol effectively suppresses precipitation.

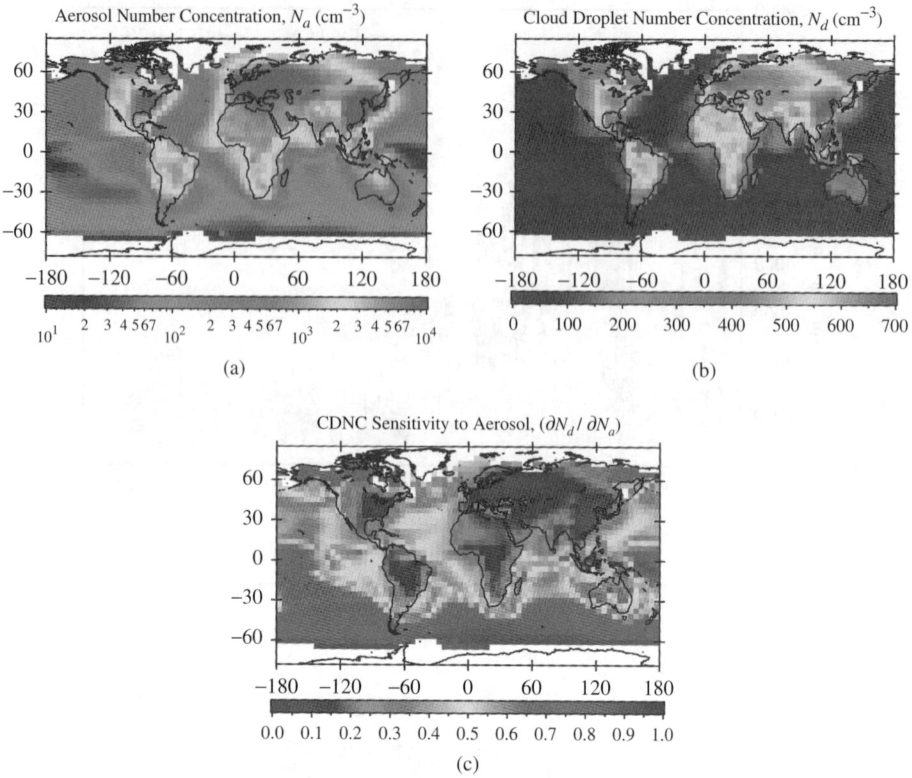

Figure 8.14 *Simulated annual mean (a) condensation nuclei (aerosol number) concentration, (b) CDNC concentration, and (c) CDNC sensitivity to aerosol changes maximizes over the ocean and remote locations. Reproduced with permission from Moore* et al. *(2013). Copyright © 2013, Moore* et al. *See plate section for colour version.*

8.7 Radiative Forcing Estimates

'Radiative forcing' is a measure of changes in the energy balance of the earth–atmosphere system resulting from changes in factors (e.g. aerosols) that affect climate; that is, the change in the energy balance between the incoming solar radiation and the outgoing infrared radiation within the earth's atmosphere that forces the system away from its normal state (Forster *et al.*, 2007). Positive radiative forcing leads to a surplus of energy in the system compared to its normal state and thus to global warming, whereas negative radiative forcing reduces the earth's temperature. On a global scale, radiative forcing can be linked to changes in the global mean temperature by a global mean parameter called the **climate sensitivity** ($\lambda = \Delta RF/\Delta T$), which is a model-dependent parameter. However, this is not the case when investigating the regional impacts of atmospheric climate forcers that are short-lived, such as aerosols, and thus present significant spatial heterogeneity.

Because of the large uncertainties in the model estimates of aerosol radiative forcing, the modeling community seeks robustness in the results of model intercomparison exercises and calculates the 'mean' model behavior. For this reason, Myhre *et al.* (2013) evaluated 16 global models with regard to aerosol direct radiative forcing calculations and found significant spread. The models calculate strong positive BC radiative forcing but also strong negative forcing from scattering aerosol components (sulfate or organic aerosol; Figure 8.15).

Overall, the errors in individual aerosol component forcing are higher than those in the total aerosol forcing. Diversities in aerosol burden, mass extinction coefficient, and

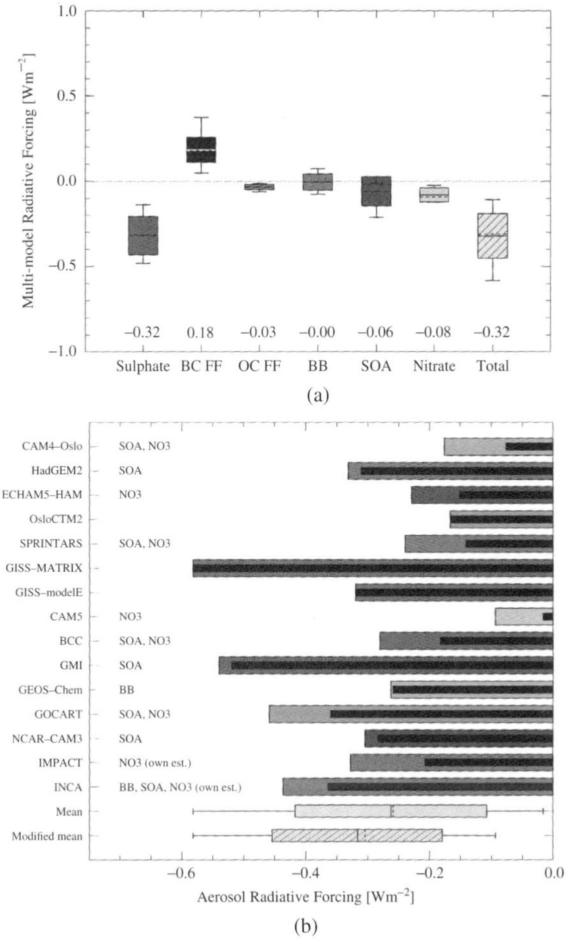

Figure 8.15 *(a) Direct radiative forcing per aerosol component (internal bars) for the period 1750–2010 – mean of AEROCOM models. (b) Total aerosol direct radiative forcing computed by 16 global models. Black bars show the bare modeled forcing, shaded bars show the forcing modified by the model components. Light bars show the AEROCOM mean of the total radiative forcing of direct aerosol effect. The light shaded bar shows the AEROCOM mean when an aerosol component adjustment is made for the missing aerosol components. The boxes indicate one standard deviation, while the whiskers indicate the maximum and minimum distribution. Reproduced with permission with Myhre* et al. *(2013). Copyright © 2013, Myhre* et al. *See plate section for colour version.*

normalized radiative forcing to the AOD have been identified as the major reasons for this, making similar contributions to the spread. The computed mean direct aerosol radiative forcing for the years from 1750 to 2010 is calculated to be $-0.39\,W\,m^{-2}$.

Similar results were obtained in the Atmospheric Chemistry and Climate Model Intercomparison Project (ACCMIP), which examined both conventional direct radiative forcing at the tropopause and the forcing, including rapid adjustments (adjusted forcing, including direct and indirect effects, Forster *et al.* (2007)), for the period 1850–2000 (Shindell *et al.*, 2013). The direct aerosol forcing, found to be $-0.26\,W\,m^{-2}$ (Figure 8.16a), increased to $-0.39\,W\,m^{-2}$ when results were corrected for missing aerosol components in some models. Shindell *et al.* (2013) also calculated the mean aerosol adjusted forcing during this period to be $-1.17\,W\,m^{-2}$ (Figure 8.16b, range -0.71 to $-1.44\,W\,m^{-2}$). This value, which is almost three times the direct radiative forcing, indicates that adjustments to aerosols, which include cloud, water vapor, and temperature, lead to stronger forcing than the aerosol direct radiative forcing. Both forcings are strongest over North America, Europe, and South-East

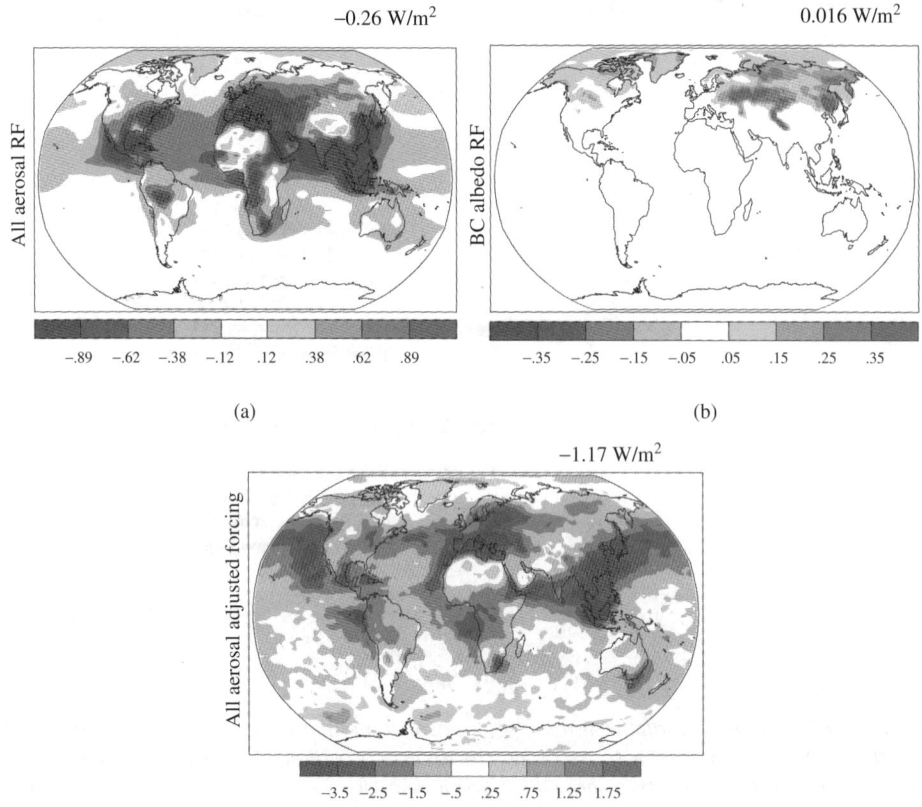

Figure 8.16 *(a) Direct RF from all aerosols (mean of 10 models). (b) BC radiative forcing due to changes in the surface albedo. (c) Adjusted (direct and indirect) RF from all aerosols. Results from the ACMIP model exercise. Adopted with permission from Shindell* et al. *(2013). Copyright © 2013, Shindell* et al. *See plate section for colour version.*

Asia. Adjusted aerosol forcing is positive over the polar regions. Figure 8.16c shows the computed radiative forcing induced by BC deposition that changes the snow albedo, leading to a globally small but regionally important positive forcing (warming effect).

8.8 The Way Forward

The spatiotemporal variability of aerosol properties and the gaps in our knowledge of the processes that change aerosol properties in the atmosphere and link aerosols to clouds prohibit a simple method for evaluating the climate impacts of aerosols. Therefore, further process understanding is needed, particularly with regard to the role of BC and organics. Recently, BC reduction has been proposed as a feasible measure for slowing down global warming while improving air quality (UNEP, 2011). However, the efficiency of such potential measures in slowing down global warming depends on the extent of the impact of BC on the climate, which is very uncertain (Koch *et al.*, 2009; Vignati *et al.*, 2010). There is also increasing interest in the climatic role of natural aerosol components and how it changes when combined with pollution. As discussed earlier, pollution has been shown to be more efficient in affecting climate over 'clean' regions.

Stevens and Feingold (2009) suggest that the effect that aerosol has on clouds and precipitation is almost certainly regime-dependent. They further state that more research is needed to determine how the trace of aerosol–cloud–precipitation interactions depends on the state of the system and to improve the representation of such cloud regimes in models.

Maintaining and increasing observational capabilities can enhance our understanding of aerosol climate impacts. In this respect, ground-based and vertical profiles of size-segregated aerosol chemical composition, CN and CCN concentrations, AOD, and SSA observations remain invaluable. Satellite observations with a continuously increasing quality of retrieval algorithms provide global views of aerosols and radiation in the atmosphere.

In parallel, modeling capabilities keep improving, with several models now including aerosol microphysics with online calculation of oxidant chemistry and parameterizations of aerosol–cloud interactions that can accommodate new knowledge from large experimental campaigns, where available.

References

Acker, J.G. and Leptoukh, G. (2007) Online analysis enhances use of NASA earth science data. *Eos, Transactions American Geophysical Union*, **88**(2), 14–17.

Ackerman, A.S., Toon, O.B., Stevens, D.E., *et al.* (2000) Reduction of tropical cloudiness by soot. *Science*, **288**, 1042–1047, doi: 10.1126/science.288.5468.1042.

Albrecht, B.A. (1989) Aerosols, cloud microphysics, and fractional cloudiness. *Science*, **245**, 1227–1230.

Bougiatioti, A., Nenes, A., Fountoukis, C. *et al.* (2011) Size-resolved CCN distributions and activation kinetics of aged continental and marine aerosol. *Atmospheric Chemistry and Physics*, **11**, 8791–8808. doi: 10.5194/acp-11-8791-2011.

Chin, M., Kahn, R.A., Remer, L.A. *et al.* (2009) *Atmospheric Aerosol Properties and Climate Impacts, Synthesis and Assessment Product 2.3 Report by the U.S. Climate Change Science Program and the Subcommittee on Global Change Research*, p. 115.

DeMott, P.J., Sassen, K., Poellot, M.R. *et al.* (2003) African dust aerosols as atmospheric ice nuclei. *Geophysical Research Letters*, **30**(14), 1732. doi: 10.1029/2003GL017410.

Fountoukis, C., Riipinen, I., Denier van der Gon, H.A.C. *et al.* (2012) Simulating ultrafine particle formation in Europe using a regional CTM: contribution of primary emissions versus secondary formation to aerosol number concentrations. *Atmospheric Chemistry and Physics*, **12**, 8663–8677. doi: 10.5194/acp-12-8663-2012.

Fuzzi, S., Andreae, M.O., Huebert, B.J. *et al.* (2006) Critical assessment of the current state of scientific knowledge, terminology, and research needs concerning the role of organic aerosols in the atmosphere, climate, and global change. *Atmospheric Chemistry and Physics*, **6**, 2017–2038. doi: 10.5194/acp-6-2017-2006.

Hallquist, M., Wenger, J.C., Baltensperger, U. *et al.* (2009) The formation, properties and impact of secondary organic aerosol: current and emerging issues. *Atmospheric Chemistry and Physics*, **9**, 5155–5236. doi: 10.5194/acp-9-5155-2009.

Haywood, J. and Boucher, O. (2000) Estimates of the direct and indirect radiative forcing due to tropospheric aerosols: a review. *Reviews of Geophysics*, **38**, 513–543.

Heald, C.L., Coe, H., Jimenez, J.L. *et al.* (2011) Exploring the vertical profile of atmospheric organic aerosol: comparing 17 aircraft field campaigns with a global model. *Atmospheric Chemistry and Physics*, **11**, 12673–12696. doi: 10.5194/acp-11-12673-2011.

Heald, C.L., Jacob, D.J., Park, R.J. *et al.* (2005) A large organic aerosol source in the free troposphere missing from current models. *Geophysical Research Letters*, **32**, L18809. doi: 10.1029/2005GL023831.

Heintzenberg, J., Covert, D. and Van Dingenen, R. (2000) Size distribution and chemical composition of marine aerosols: a compilation and review. *Tellus B*, **52**, 1104–1122.

Im, U. and Kanakidou, M. (2012) Impacts of East Mediterranean megacity emissions on air quality. *Atmospheric Chemistry and Physics*, **12**, 6335–6355. doi: 10.5194/acp-12-6335-2012.

Im, U., Markakis, K., Koçak, M. *et al.* (2012) Summertime aerosol chemical composition in the Eastern Mediterranean and its sensitivity to temperature. *Atmospheric Environment*, **50**, 164–173. doi: 10.1016/j.atmosenv.2011.12.044.

Forster, P., Ramaswamy, V., Artaxo, P. *et al.* (2007) Changes in atmospheric constituents and in radiative forcing, in *Climate Change 2007: The Physical Science Basis. Contribution of Working Group I to the Fourth Assessment Report of the Intergovernmental Panel on Climate, Change* (eds S. Solomon, D. Qin, M. Manning *et al.*), Cambridge University Press, Cambridge, New York, pp. 129–134.

Kanakidou, M., Seinfeld, J.H., Pandis, S.N. *et al.* (2005a) Organic aerosol and global climate modelling: a review. *Atmospheric Chemistry and Physics*, **5**, 1053–1123.

Kanakidou, M., Tsigaridis, K., Dentener, F.J. *et al.* (2005b) *Particles of Human Origin Extinguishing Natural Solar Radiation in Climate Systems – PHOENICS Synthesis and Integration Report*, University of Crete, Greece, ISBN 960-88712-0-4, 2005.

Kanakidou, M., Duce, R.A., Prospero, J., Baker, A.R., *et al.* (2012) Atmospheric fluxes of organic N and P to the global ocean. Global Biogeochemical Cycles, 10.10.1029/2011GB004277.

Karydis, V.A., Kumar, P., Barahona, D. *et al.* (2011) On the effect of dust particles on global cloud condensation nuclei and cloud droplet number. *Journal of Geophysical Research*, **116**, D23204. doi: 10.1029/2011JD016283.

Koch, D., Schulz, M., Kinne, S. *et al.* (2009) Corrigendum to 'Evaluation of black carbon estimations in global aerosol models'. *Atmospheric Chemistry and Physics*, **10**, 79–81.

Koch, D., Schulz, M., Kinne, S. *et al.* (2010) Corrigendum to 'Evaluation of black carbon estimations in global aerosol models'. *Atmospheric Chemistry and Physics*, **9**, 9001–9026. doi: 10.5194/acp-10-79-2010.

Kumar, P., Sokolik, I.N. and Nenes, A. (2009) Parameterization of cloud droplet formation for global and regional models: including adsorption activation from insoluble CCN. *Atmospheric Chemistry and Physics*, **9**(7), 2517–2532. doi: 10.5194/acp-9-2517-2009.

Kumar, P., Sokolik, I.N. and Nenes, A. (2011) Cloud condensation nuclei activity and droplet activation kinetics of fresh unprocessed regional dust samples and minerals. *Atmospheric Chemistry and Physics*, **11**(7), 3527–3541. doi: 10.5194/acp-11-3527-2011.

Lee, Y.H. and Adams, P.J. (2010) Evaluation of aerosol distributions in the GISS-TOMAS global aerosol microphysics model with remote sensing observations. *Atmospheric Chemistry and Physics*, **10**, 2129–2144. doi: 10.5194/acp-10-2129-2010.

Levin, Z., Teller, A., Ganor, E. and Yin, Y. (2005) On the interactions of mineral dust, sea-salt particles, and clouds: a measurement and modeling study from the Mediterranean Israeli Dust Experiment campaign. *Journal of Geophysical Research*, **110**, D20202. doi: 10.1029/2005JD005810.

Lohmann, U. (2009) Marine boundary layer clouds, in *Surface Ocean – Lower Atmosphere Processes*, Geophysical Monograph Series, Vol. **187** (eds C. Le Quéré and E.S. Saltzman), Hardbound, p. 350. ISBN: 978-0-87590-477-1., AGU Code GM1874771.

Megaritis, A.G., Fountoukis, C., Charalampidis, P.E. *et al.* (2013) Response of fine particulate matter concentrations to changes of emissions and temperature in Europe. *Atmospheric Chemistry and Physics*, **13**, 3423–3443, doi: 10.5194/acp-13-3423-2013.

Metzger, S.M., Dentener, F.J., Jeuken, A. *et al.* (2002) Gas/aerosol partitioning II: global modeling results. *Journal of Geophysical Research*, **107**, D16. doi: 10.1029/2001JD001103.

Moore, R.H., Karydis, V.A., Capps, S.L. *et al.* (2013) Droplet number prediction uncertainties from CCN: an integrated assessment using observations and a global adjoint model. *Atmospheric Chemistry and Physics*, **13**, 4235–4251, doi:10.5194/acp-13-4235-2013.

Myhre, G., Samset, B.H., Schulz, M. *et al.* (2013) Radiative forcing of the direct aerosol effect from AeroCom Phase II simulations. *Atmospheric Chemistry and Physics Discussions*, **13**, 1853–1877, doi:10.5194/acp-13-1853-2013.

Myriokefalitakis, S., Tsigaridis, K., Mihalopoulos, N., Sciare, J., Nenes, A., Kawamura, K., Segers, A., Kanakidou, M. (2011) In-cloud oxalate formation in the global troposphere: A 3-D modeling study. *Atmos. Chem. Phys.*, **11**, 5761–5782, doi:10.5194/acp-11-5761-2011.

Nenes, A., Ghan, S., Abdul-Razzak, H. *et al.* (2001) Kinetic limitations on cloud droplet formation and impact on cloud albedo. *Tellus B*, **53**, 133–149.

Petters, M.D. and Kreidenweis, S.M. (2007) A single parameter representation of hygroscopic growth and cloud condensation nucleus activity. *Atmospheric Chemistry and Physics*, **7**, 1961–1971. doi: 10.5194/acp-7-1961-2007.

Pincus, R. and Baker, M.B. (1994) Effect of precipitation on the albedo susceptibility of clouds in the marine boundary layer Nature, **372**, 250–252. doi:10.1038/372250a0.

Reddington, C.L., Carslaw, K.S., Spracklen, D.V. *et al.* (2011) Primary versus secondary contributions to particle number concentrations in the European boundary layer. *Atmospheric Chemistry and Physics*, **11**, 12007–12036. doi: 10.5194/acp-11-12007-2011.

Seinfeld, J.H. and Pandis, S.N. (1998) *Atmospheric Chemistry and Physics: From Air Pollution to Climate Change*, John Wiley & Sons, Inc., New York.

Shindell, D.T., Lamarque, J.-F., Schulz, M. *et al.* (2013) Radiative forcing in the ACCMIP historical and future climate simulations. *Atmospheric Chemistry and Physics*, **13**, 2939–2974. doi: 10.5194/acp-13-2939-2013.

Solomos, S., Kallos, G., Kushta, J. *et al.* (2011) An integrated modeling study on the effects of mineral dust and sea salt particles on clouds and precipitation. *Atmospheric Chemistry and Physics*, **11**, 873–892. doi: 10.5194/acp-11-873-2011.

Sorooshian, A., Feingold, G., Lebsock, M.D. *et al.* (2009) On the precipitation susceptibility of clouds to aerosol perturbations. *Geophysical Research Letters*, **36**, L13803. doi: 10.1029/2009GL038993.

Spracklen, D.V., Carslaw, K.S., Merikanto, J. *et al.* (2010) Explaining global surface aerosol number concentrations in terms of primary emissions and particle formation. *Atmospheric Chemistry and Physics*, **10**, 4775–4793. doi: 10.5194/acp-10-4775-2010.

Spracklen, D.V., Carslaw, K.S., Pöschl, U. *et al.* (2011) Global cloud condensation nuclei influenced by carbonaceous combustion aerosol. *Atmospheric Chemistry and Physics*, **11**, 9067–9087. doi: 10.5194/acp-11-9067-2011.

Stevens, B. and Feingold, G. (2009) Untangling aerosol effects on clouds and precipitation in a buffered system. *Nature*, **461**(1) 607–612. doi: 10.1038/nature08281.

Textor, C., Schulz, M., Guibert, S. *et al.* (2006) Analysis and quantification of the diversities of aerosol life cycles within AeroCom. *Atmospheric Chemistry and Physics*, **6**, 1777–1813. doi: 10.5194/acp-6-1777-2006.

Tsigaridis, K. and Kanakidou, M. (2007) Secondary organic aerosol importance in the future atmosphere. *Atmospheric Environment*, **41**, 4682–4692, doi: 10.1016/j.atmosenv.2007.03.045.

Tsigaridis, K., Daskalakis, N., Kanakidou, M., *et al.* (2013) An AeroCom intercomparison exercise on state-of-the-art organic aerosol global modeling, to be submitted to *Atmos. Chem. Phys.*, 2013.

Twomey, S. (1974) Pollution and the planetary albedo. *Atmospheric Environment*, **8**, 1251–1256.

UNEP and WMO Report (2011) Integrated Assessment of Black Carbon and Tropospheric Ozone – Summary for Decision Makers, UNEP/GC/26/INF/20.

Vignati, E., Karl, M., Krol, M. *et al.* (2010) Sources of uncertainties in modelling black carbon at the global scale. *Atmospheric Chemistry and Physics*, **10**, 2595–2611. doi: 10.5194/acp-10-2595-2010.

9

Air Pollution and Health and the Role of Aerosols

Pat Goodman[1] and Otto Hänninen[2]

[1]School of Physics, Environmental Health Sciences Institute, Dublin Institute of Technology, Ireland

[2]Department of Environmental Sciences, University of Eastern Finland, Finland

9.1 Background

This chapter investigates the literature for evidence of adverse health effects associated with exposure to air pollution, and specifically to the size range of pollution particle associated with this. Specifically, we discuss the evidence linking exposure to aerosols with adverse health outcomes. Although there is significant evidence pertaining to adverse health effects in employees in certain types of industry, this chapter concentrates on effects in the general public. Presenting a comprehensive review of all papers on aerosols and health is not possible; the vast number of original papers reflects both the importance as well as challenges in this research area. We aim at giving an overview of the main findings and current understanding on the relationship of aerosols, especially ultrafine particles, and health of general populations. There may well be other papers that are not cited but which provide similar results: it would be impossible to cite all relevant publications within this chapter. The Health Effects Institute (HEI) has conducted an expert review of the state of knowledge surrounding ultrafine particles (UFPs) and health, which by its nature is more detailed than the overview presented here (HEI, 2013).

The role of particulate pollution in the air and its association with adverse health effects such as increased morbidity and mortality is well documented. The extreme pollution events in Donora, Pennsylvania in 1948 (Snyder, 1994) and in London, UK

Aerosol Science: Technology and Applications, First Edition. Edited by Ian Colbeck and Mihalis Lazaridis.

in 1952 (HMSO, 1954) were associated with many thousands of excess deaths and led to the realisation that action was needed to reduce air pollution levels. In fact, there are references throughout history to air pollution events associated with adverse health outcomes. In his excellent book on the history of air pollution, Brimblecombe (1988) refers to the documentation of adverse health effects throughout the ages, most of which are attributable to coal burning. Brimblecombe very clearly illustrates the evidence in many different countries and over the past few millennia.

Current research shows that even at low levels of pollution, adverse health effects can be detected in the general population. One key paper is that known as the 'Harvard Six Cities Study' (Dockery *et al.*, 1993), which showed that life expectancy was lowest in the most polluted city and highest in the least polluted. Long exposure to particulate air pollution was associated with chronic health outcomes, while the events in London and Pennsylvania, and more recently Dublin (Kelly and Clancy, 1984), show immediate or acute health effects. Interestingly, when a pollution event occurs and the pollution then dissipates, there is evidence that the health effects of that exposure continue on (Bell and Davis, 2001; Goodman, Clancy and Dockery, 2004).

Over the past few decades, particulate air pollution was measured as the 'blackness' of material deposited on a filter (British Standards Institute, 1969). This system was ideally suited to measuring pollution from coal burning and from diesel emissions. It had a size cut-off of ~4.5 μm (McFarland, Ortiz and Rodes, 1982). Although based on light refelectometry, this system produced a pollution level in micrograms per cubic metre, whereby a conversion factor was applied to the reflectometry reading. The European Union (EEC, 1980) has adopted a similar measurement system for particulate matter (PM) and sulfur dioxide.

This type of measurement has now been superseded by techniques that measure particle mass based on various size cut-offs, with air drawn through a filter and the mass of material collected and measured. The parameters measured are referred to as PM_{10} and $PM_{2.5}$; these are both now legislated for by the European Union (EU, 2008; USEPA, 2012), and the World Health Organization (WHO) has defined guideline values for them (WHO, 1987, 2000, 2006). These limit or guideline values are set in order to protect human health.

The WHO guidelines for air quality are based on a systematic review of the scientific evidence on the association between ambient levels and health outcomes in large populations. Over the last decade, the strength of such evidence has increased, and more health end points have become associated with PM exposures in particular (WHO, 2013).

9.2 Size Fractions

In health studies, particles in the size range 2.5–10.0 μm (measured as the difference $PM_{10} - PM_{2.5}$) are referred to as the *coarse fraction*, particles smaller than 2.5 μm are referred to as the *fine fraction* and particles smaller than 0.1 μm are termed the *ultrafine fraction*. It should be noted that PM_{10} actually also contains the $PM_{2.5}$ and UFPs, and likewise $PM_{2.5}$ also contains the ultrafine fraction. All these particulate-matter fractions in the atmosphere consist of aerosol.

From an environmental physics perspective, when one gets down to these very-small-size particles, particle mass may not be the most appropriate metric to use, as there may be many thousands of small particles that collectively have the same mass as one large $PM_{2.5}$ particle, so particle number count or surface area might be a more relevant metric. However, as already mentioned, current air-quality legislation only specifies the use of $PM_{2.5}$ and PM_{10} mass as approved metrics.

The availability of particle number counts for use in health studies is still quite limited, and the choice of monitoring site is very important and can show greater variability than PM measurements in a given urban area (Aalto *et al.*, 2005). Pekkanen and Kulmala (2004) also report that central site monitoring may give a somewhat worse proxy for human exposure to UFPs than for exposure to $PM_{2.5}$.

9.3 Which Pollution Particle Sizes Are Important?

Air pollution can comprise a complex mix of gases and particles (i.e. complex aerosols), with the mix very much depending on the sources, be they industrial, traffic or domestic. The mixture of particles in the air can range in size over a number of orders of magnitude, typically from a few nanometres to tens of micrometres in aerodynamic diameter, 3–4 orders of magnitude in diameter and thus 9–12 orders of magnitude in volume and mass. This raises the question as to which particle sizes are most strongly associated with adverse health outcomes: the smaller aerosol particles or the coarse fraction? The fact that particle pollution is currently controlled in terms of mass concentration is based on the abundant evidence for the association of mass concentration and health. However, the associations between different size fractions in the particulate pollution mix can vary quite significantly; for example, Boogaard *et al.* (2010) report that mean concentrations of UFP (number concentration) are poorly associated with PM_{10} and soot. In their extensive review, Pope and Dockery (2006) report that the vast majority of the literature suggests that it is the fine and potentially also UFPs that are most associated with adverse health effects. They report that people more exposed to ambient air pollution experience more respiratory and cardiovascular health outcomes.

9.4 What Health Outcomes Are Associated with Exposure to Air Pollution?

Initially it was assumed that respiratory problems would be found to be the major health of effect of air pollution, and while this is true, there is significant evidence in the literature that exposure to air pollution is also associated with adverse cardiovascular health outcomes, such as stroke, heart attacks and so on. There is also some evidence that cardiovascular effects are more immediate, occurring within hours to days, while respiratory effects follow longer exposures of days to weeks. The literature also suggests that those who already have health conditions are more susceptible to exposure to pollution. Brook *et al.* (2010) provide a detailed review of some of the health effects associated with exposure to particulate pollution. As we progress through this chapter, we will consider both the respiratory and cardiovascular evidence from exposure to aerosols in the air and the different response times associated with each health outcome.

9.5 Sources of Atmospheric Aerosols

The production of aerosols is often from combustions processes, be they in motor vehicles, domestic heating and cooking or even natural sources.

Particles can be released into the atmosphere directly (e.g., soot emissions from combustion processes, desert sandstorms) but a substantial fraction of the ambient mass concentration is formed in the atmosphere from gaseous precursors in photo- and other chemical reactions. The three most abundant components affecting these processes are sulfates, nitrates and ammonia, although volatile organic compounds play a significant role in the formation of secondary organic aerosols.

Most of our routine measurement techniques cannot distinguish between the sources of aerosols, so the majority of health-related studies have focussed on reporting aerosol mass and number concentrations. Franck *et al.* (2006) report that in urban areas, UFPs originate primarily from rapidly increasing traffic, which is the dominating source at many urban sites.

9.6 Particle Deposition in the Lungs

If we take the example of tobacco smoke, we know that the smoke is predominantly in the aerosol size range, and it is well known that people who smoke tobacco and other illegal substances rapidly experience a 'high', indicating that these aerosols have passed through the lungs and into the bloodstream.

A lot of the early aerosol research was conducted in Dublin, where Nolan and Pollak (Nolan, 1972) developed the condensation nucleus counter (CNC). Burke and Nolan (1955) reported that the number of aerosols exhaled by a person is lower than the number they inhale; some of this difference is accounted for by coagulation processes but some aerosols remain 'trapped' in the body.[1]

Some researchers have studied the deposition of aerosols in the healthy and diseased lung. For example, Wiebert *et al.* (2006) report a high retention of UFPs, with little difference between healthy and diseased lungs. Moller *et al.* (2008) showed negligible clearance of UFP 24 hours after exposure; that is, the particles are retained in the body. There is some evidence that UFP can translocate or travel to other organs via the blood (Geiser and Kreyling, 2010), but this has not been observed for larger particles, although the mechanisms are not currently well understood.

If we focus on the area of tobacco smoking as an example of aerosol exposure, we now know that this is a major cause of disease and death across the world, with 95% of lung cancers attributable to it and tobacco smokers having a significantly greater risk of many aspects of cardiovascular disease as compared to nonsmokers (US Surgeon General, 2006). This illustrates that the inhalation of aerosols – in this case the products of tobacco smoke combustion – leads to the deposition of a fraction of their number in the lungs and bloodstream, where over time they give rise to adverse health effects. Certainly not all aerosols will be associated with such adverse health outcomes, but this illustrates the efficiency of aerosols to enter deep into the human body.

[1] While mentioning the Dublin group, it is also interesting to note that they developed the ionisation 'smoke alarm', which responds to aerosols in the air: clearly a positive health outcome that has saved countless lives.

There is also significant evidence in the research literature to show that nonsmokers exposed to environmental tobacco smoke (ETS) also show adverse health effects that are greater than those found in people not so exposed (US Surgeon General, 2006).

In their recent review of smoking bans, Goodman *et al.* (2009) showed that the health of the general population improved very soon after implementation of the bans, mostly in association with reduced cardiovascular health events. This review looked at bans across many continents and found consistent results, suggesting that the smoking bans removed the exposure of the general population when they were out socialising. Goodman *et al.* (2009) also reported that a number of studies observed improved health among workers.

When a person inhales air containing aerosols, a number of things must happen for these aerosols to be associated with adverse health effects: the aerosols must be able to enter into the lungs, and once in there they must either be deposited or else be absorbed into the bloodstream. Significant work has been done by the International Commission on Radiation Protection (ICRP) on the deposition of aerosols in the human lung, and in fact the deposition models it has developed are considered the definitive models in this regard. Although the ICRP was initially interested in the deposition of radioactive particles, the results are directly transferrable to any type of aerosol.

The ICRP model for particle-size-dependent uptake was first published in 1994. This model divides the human respiratory tract into three regions (extrathoracic, thoracic and alveolar) and five subregions (Figure 9.1). As demonstrated in Figures 9.2–9.4, coarser particles have higher probabilities of depositing in the nasal region and UFPs in the alveolar region. However, even supermicron particles have substantial deposition efficiencies in the alveolar region and, due to their huge masses in comparison with ultrafines, easily dominate mass-based particle uptake even in the alveolar region. (The mass of a 1 μm particle is a million times greater than that of a 10 nm particle, assuming the same density. Often the UFPs have lower densities due to their fractal shapes.) UFPs dominate the particle numbers, as well as the deposited fractions in all respiratory tract regions.

This preferential deposition by particle size also has beneficial applications in the delivery of medications, as covered later in this chapter.

Werner Hofmann (2011) reviewed different models developed to characterise the uptake of particles in the human respiratory tract and showed that despite major differences in model formulation, the differences between them are relatively modest in comparison with the overall variability in particle-size-specific deposition efficiency.

9.7 Aerosol Interaction Mechanisms in the Human Body

The mechanisms by which aerosol particles interact and give rise to adverse health effects are not clearly understood. Brown *et al.* (2001) report that low-toxicity particles such as polystyrene give rise to proinflammatory activity as a consequence of their large surface area, and they suggest that this is one of the ways in which particles cause adverse health effects. Araujo *et al.* (2008) report that in animal studies, UFPs result in the inhibition of the antiinflammatory capacity of high-density lipoprotein and in greater systemic oxidative stress, as evidenced by a significant increase in hepatic malondialdehyde levels and upregulation of Nrf2-regulated antioxidant genes, from which they conclude that UFPs

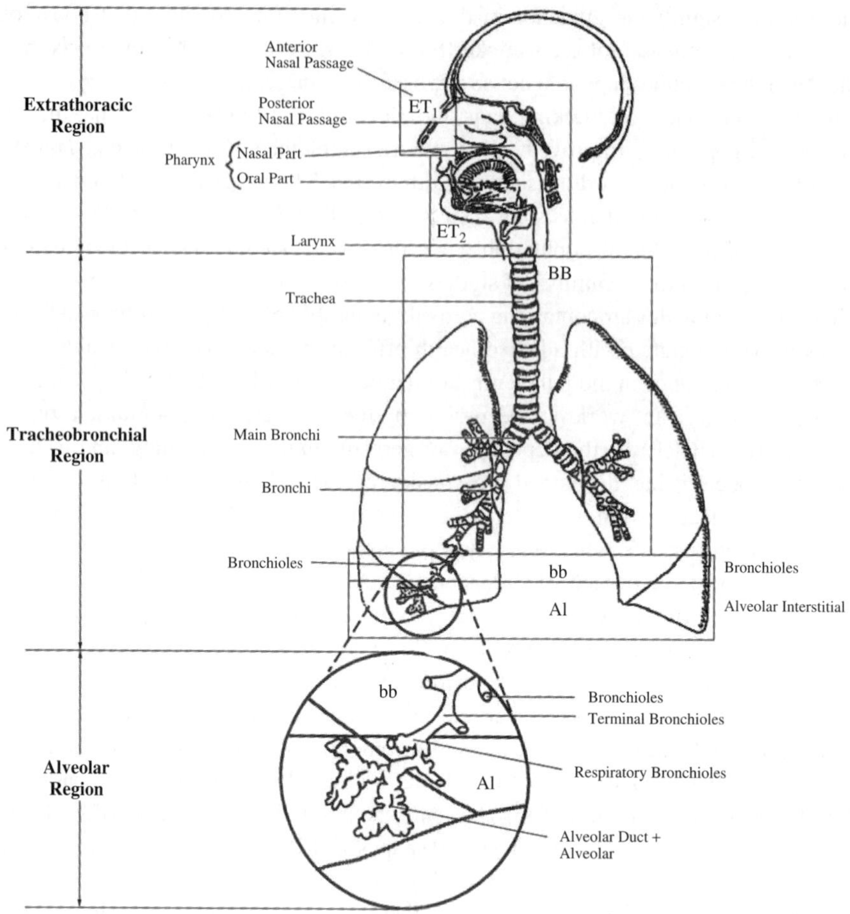

Figure 9.1 *Illustration of the major anatomical regions of the human respiratory tract (ICRP, 1994). ET1, anterior nasal passages; ET2, posterior nasal passages, naso-oropharynx and larynx; BB, bronchial region, including trachea and bronchi; bb, bronchiolar region, consisting of bronchioles and terminal bronchioles; AI, alveolar–interstitial region, consisting of respiratory bronchioles and alveolar ducts and sacs surrounded by alveoli. Reprinted from Hofmann (2011). Copyright © 2011, with permission from Elsevier.*

concentrate the proatherogenic effects of ambient PM and may constitute a significant cardiovascular risk factor.

Song *et al.* (2011) report significant associations between concentrations of UFP and itchiness symptoms in children with atopic dermatitis. Sannolo, Lamberti and Pedata (2010), in their review of the literature relating to UFP and cellular interactions, report that *in vitro* toxicological research has shown that UFPs induces several types of adverse cellular effect, including cytotoxicity, mutagenicity, DNA oxidative damage and stimulation of proinflammatory cytokine production. Eder *et al.* (2009) also suggest some mechanisms that might account for some of the effects of inhaled particles, such as their activation and/or the

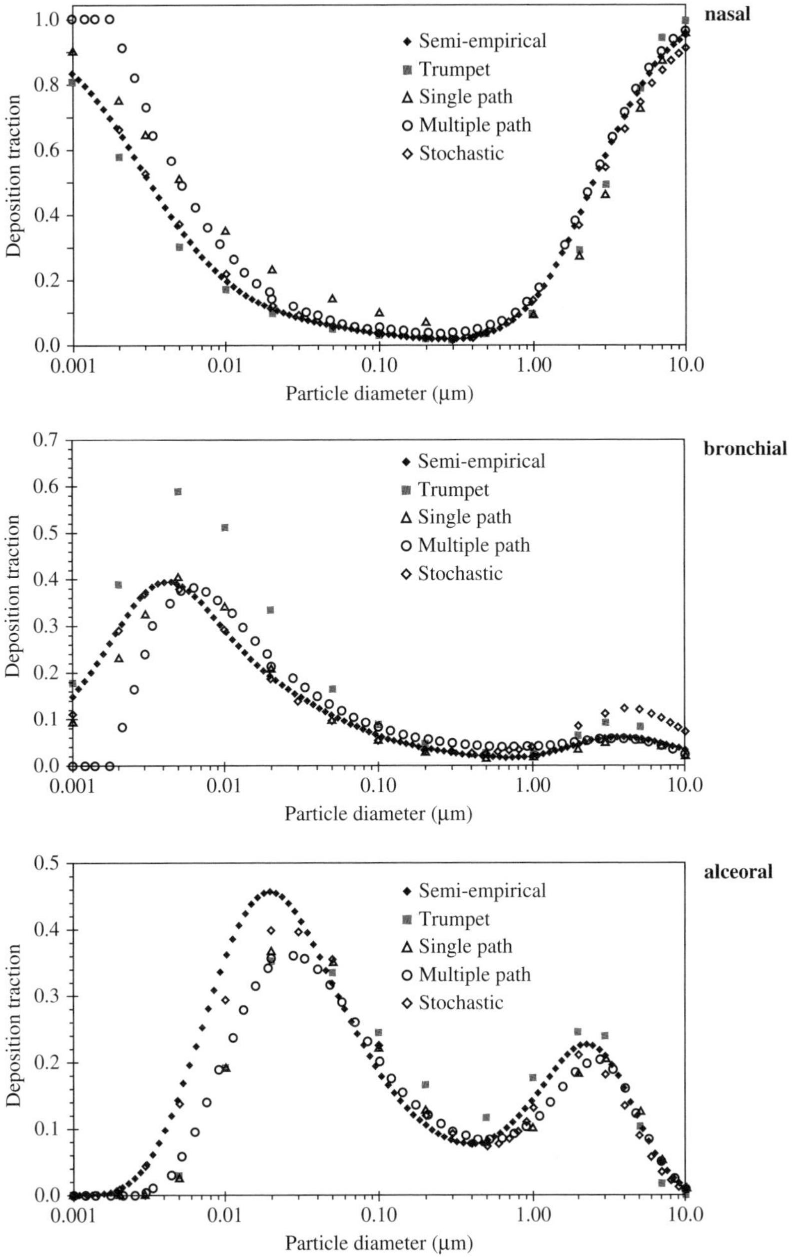

Figure 9.2 *Comparison of five mainstream modelling approaches for estimation of the respiratory tract uptake of particles as a function of particle size. Reprinted from Hofmann (2011). Copyright © 2011, with permission from Elsevier.*

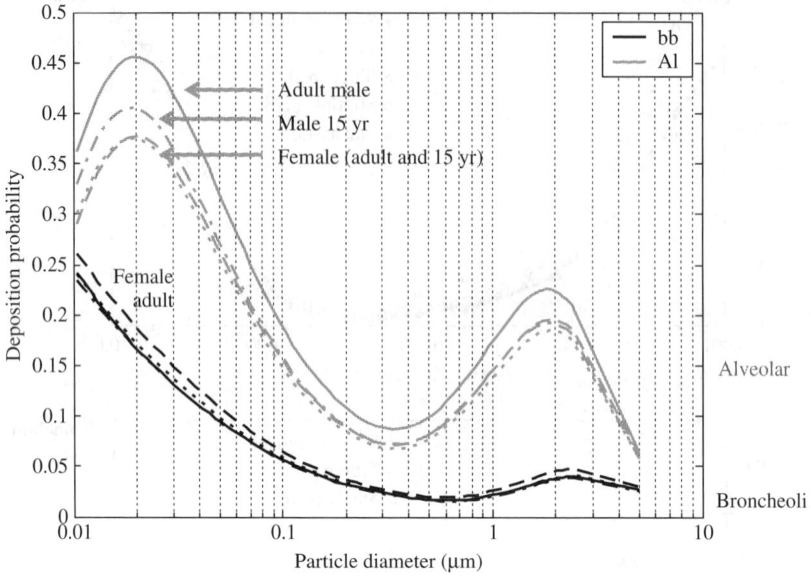

Figure 9.3 *Comparison of the alveolar and broncheoliar deposition efficiencies for subjects of different ages according to the ICRP (1994) model.*

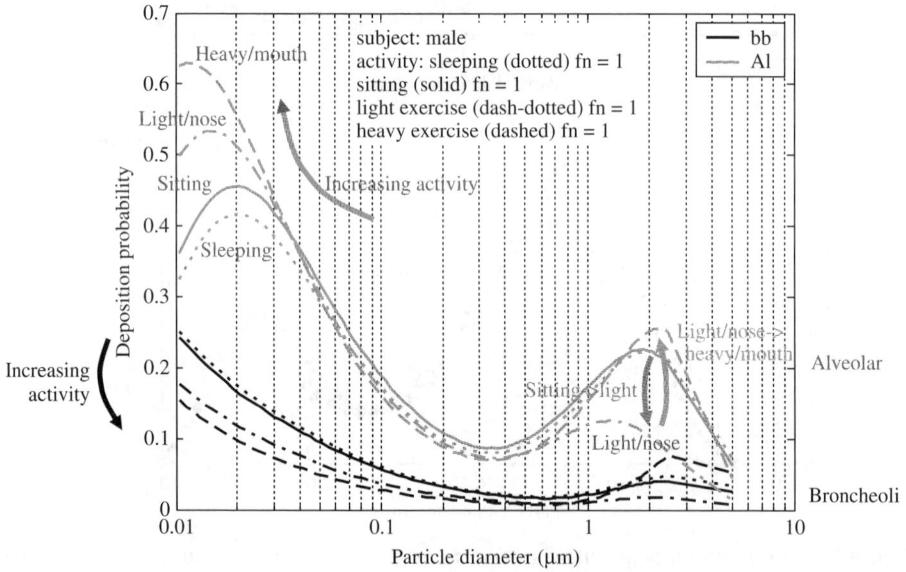

Figure 9.4 *Dependence of the alveolar and bronchiolar deposition efficiency of aerosol particles by size on the physical activity level, as estimated by the ICRP (1994) model. Aerosol density: 1.5 g cm^{-3} corresponding ambient aerosols (Sorjamaa and Hänninen, 2011). See plate section for colour version.*

detoxification capabilities of inhaled toxic compounds. Alessandrini *et al.* (2006), having conducted exposure studies, report that allergen-sensitised individuals may be more susceptible to the detrimental effects of UFPs. Pope (2000) summarises the evidence as follows: particle-induced pulmonary inflammation, cytokine release and altered cardiac autonomic function may be part of the pathophysiological mechanisms or pathways linking particulate pollution with cardiopulmonary disease.

9.8 Human Respiratory Outcomes and Aerosol Exposure

In this section we investigate the literature relating to the evidence or otherwise of respiratory effects caused by exposure to aerosol particles. Oberdorster *et al.* (1995) conclude based on animal studies that since UFPs are always present in the urban atmosphere, they play a role in causing acute lung injury in sensitive subgroups of the population. However, in contrast, Pekkanen *et al.* (1997) found UFPs to be no different to PM_{10} or black smoke in their effects on respiratory function in asthmatic children, while Peters *et al.* (1997) reported that UFPs were more strongly associated with adverse changes in pulmonary function in asthmatic adults.

Another study providing contradictory evidence comes from Iskandar *et al.* (2012), who report that coarse and fine particles, but not UFPs, are a trigger for hospital admissions for asthma in children.

In a multicity study, Karakatsani *et al.* (2012) reported that no consistent association was observed between fine-particle concentrations and respiratory health effects.

Overall, the evidence for respiratory health effects from aerosol exposure in relation to pulmonary function is inconsistent. It may be that there is no effect, or if there is an effect, that the current studies have not been able to detect it. Another factor in some of these studies is that a significant percentage of the study populations may have been taking respiratory medications for their asthma.

9.9 Cardiovascular Outcomes and Aerosol Exposure

In a study of hypertensive crisis, Franck *et al.* (2011) reported that significant effects were detected for UFP, with two days' lag after exposure, but that no consistent effects were detected for $PM_{2.5}$ and PM_{10}. Weichenthal (2012) reports that the evidence to date suggests that UFPs have a measurable impact on physiological measures known to be altered in cases of acute cardiovascular morbidity.

Amatullah *et al.* (2012) report that their findings 'indicate that coarse and fine PM influence lung function and airways responsiveness, while ultrafine PM can perturb cardiac function. This study supports the hypothesis that coarse and fine PM exerts its predominant physiologic effects at the site of deposition in the airways, whereas ultrafine PM likely crosses the alveolar epithelial barrier into the systemic circulation to affect cardiovascular function.'

In a study in Bejing, Breitner *et al.* (2011) reported an increase in cardiovascular mortality associated with UFP concentrations, with a 2 day lag, and that unlike the other particle measures, the association with UFP number count was not modified by air mass origin.

It is most likely that UFP number count is a better indicator of local sources. Cho *et al.* (2008) found similar results for cardiovascular and respiratory mortality in a study in Seoul, especially in the elderly population.

Stolzel *et al.* (2007) found an association between UFP number concentration and increases in cardiorespiratory mortality in Eufurt; the lag in this case was 4 days. They did not observe any effect for UFP mass.

9.10 Conclusions and Recommendations

The evidence shows that exposure to ambient UFPs and aerosols are associated with adverse health effects. These effects are more pronounced in those with preexisting respiratory and cardiovascular disease.

In the case of cardiovascular outcomes, the evidence suggests that acute adverse effects are detected within 2 days of exposure. There is a large body of evidence in the literature on the adverse health effects associated with long term exposure to air pollution, which are too extensive to present here. The Dockery *et al* paper from 1993 is one of the key indicators that living in a location with higher air pollution reduces life expectancy. Recently in Europe, Hannien and Knol (2011) suggest that exposure to airborne particles are the dominant source of environmental health risk to the european population. For respirator health outcomes, the evidence is less clear. The exposure metric that seems to be most relevant when considering adverse health effects is the particle number concentration.

Because there is currently no legislative requirement to measure UFP/aerosol number concentrations in the ambient air, the number of health-related studies is somewhat limited, with none available for large population groups, particularly compared to the number of studies available in relation to both $PM_{2.5}$ and PM_{10}.

Currently neither EU nor USEPA air quality guidelines cover exposure to UFPs/aerosols, and neither gives guideline values or any requirements for the measurement of such particles in the ambient air.

We recommend that monitoring of the number concentrations of aerosols in ambient air be undertaken, ideally at the same time as other measurements. This would allow the development of a database on the concentrations of aerosols in ambient air, and would facilitate future research into the health effects of such exposures. It would also allow the development of guideline limit values.

References

Aalto, P., Hämeri, K., Paatero, P. *et al.* (2005) Aerosol particle number concentration measurements in five European cities using TSI-3022 condensation particle counter over a three-year period during health effects of air pollution on susceptible subpopulations. *Journal of the Air & Waste Management Association*, **55**(8), 1064–1076.

Alessandrini, F., Schulz, H., Takenaka, S. *et al.* (2006) Effects of ultrafine carbon particle inhalation on allergic inflammation of the lung. *Journal of Allergy and Clinical Immunology*, **117**(4), 824–830.

Amatullah, H., North, M.L., Akhtar, U.S. *et al.* (2012) Comparative cardiopulmonary effects of size-fractionated airborne particulate matter. *Inhalation Toxicology*, **24**(3), 161–171.

Araujo, J.A., Barajas, B., Kleinman, M. *et al.* (2008) Ambient particulate pollutants in the ultrafine range promote early atherosclerosis and systemic oxidative stress. *Circulation Research*, **102**, 589–596.

Bell, M.L. and Davis, D.L. (2001) Reassessment of the lethal London fog of 1952: novel indicators of acute and chronic consequences of acute exposure to air pollution. *Environmental Health Perspectives*, **109**(Suppl. 3), 389–394.

Boogaard, H., Montagne, D.R., Brandenburg, A.P. *et al.* (2010) Comparison of short-term exposure to particle number, PM10 and soot concentrations on three (sub) urban locations. *Science of the Total Environment*, **408**(20), 4403–4411.

Breitner, S., Liu, L., Cyrys, J. *et al.* (2011) Sub-micrometer particulate air pollution and cardiovascular mortality in Beijing, China. *Science of the Total Environment*, **409**(24), 5196–5204.

Brimblecombe, P. (1988) *The Big Smoke*, Routlddge, London, New York.

British Standards Institute 1969 Standard No. 1747. *Methods for the Measurement of Air Pollution. Part 2, Determination of Concentrations of Suspended Matter*, British Standards Institute, London.

Brook, R.D., Rajagopalan, S., Pope, C.A. III, *et al.* (2010) Particulate matter air pollution and cardiovascular disease: an update to the scientific statement from the American Heart Association. *Circulation*, **121**, 2331–2378.

Brown, D.M., Wilson, M.R., MacNee, W. *et al.* (2001) Size-dependent proinflammatory effects of ultrafine polystyrene particles: a role for surface area and oxidative stress in the enhanced activity of ultrafines. *Toxicology and Applied Pharmacology*, **175**(3), 191–199.

Burke, T.P. and Nolan, P.J. (1955) Nucleus content of air in occupied rooms. *Geofisica Pura e Applicata*, **31**, 191–196.

Cho, Y.S., Lee, J.T., Jung, C.H. *et al.* (2008) Relationship between particulate matter measured by optical particle counter and mortality in Seoul, Korea, during 2001. *Journal of Environmental Health*, **71**(2), 37–43.

Dockery, D.W., Pope, C.A. III, Xu, X. *et al.* (1993) An association between air pollution and mortality in six U.S. cities. *New England Journal of Medicine*, **329**, 1753–1759.

Eder, C., Frankenberger, M., Stanzel, F. *et al.* (2009) Ultrafine carbon particles downregulate CYP1B1 expression in human monocytes. *Particle and Fibre Toxicology*, **6**, 27.

EEC 1980. Air Quality Limit Values and Guide Values for Sulphur Dioxide and Suspended Particulates Directives 80/779/EEC.

EU 2008. Directive 2008/50/EC. On Ambient Air Quality and Clean Air for Europe.

Franck, U., Odeh, S., Wiedensohler, A. *et al.* (2011) The effect of particle size on cardiovascular disorders–the smaller the worse. *Science of the Total Environment*, **409**(20), 4217–4221.

Franck, U., Tuch, T., Manjarrez, M. *et al.* (2006) Indoor and outdoor submicrometer particles: exposure and epidemiologic relevance ('the 3 indoor Ls'). *Environmental Toxicology*, **21**(6), 606–613.

Geiser, M. and Kreyling, W.G. (2010) Deposition and biokinetics of inhaled nanoparticles. *Particle and Fibre Toxicology*, **7**, 2.

Goodman, P.G., Clancy, L. and Dockery, D.W. (2004) Cause-specific mortality and the extended effects of particulate pollution and temperature exposure. *Environmental Health Perspectives*, **112**(2), 179–185.

Goodman, P.G., Haw, S., Kabir, Z. and Clancy, L. (2009) Are there health benefits associated with comprehensive smoke-free laws: a review. *International Journal of Public Health*, **54**, 367–378.

Otto Hänninen and Anne Knol (Eds.). EBoDE-Report. Environmental Perspectives on Environmental Burden of Disease. Estimates for Nine Stressors in Six European Countries. National Institute for Health and Welfare (THL), Report 1/2011.

HEI (2013) Understanding the Health Effects of Ambient Ultrafine Particles HEI Review Panel on Ultrafine Particles, HEI, Boston, MA.

HMSO (1954) Her Majesty's Public Health Service. Mortality and Morbidity during the London Fog of December 1952. Public Health and Medical Subjects Report No. 95. Her Majesty's Stationery Office, London.

Hofmann, W. (2011) Modelling inhaled particle deposition in the human lung – a review. *Journal of Aerosol Science*, **42**, 693–724.

ICRP (International Commission on Radiological Protection) (1994) Human respiratory tract model. *Annals of ICRP*, **24**, 1–482 http://www.sciencedirect.com/science/journal /01466453 (last accessed 9 August 2013).

Iskandar, A., Andersen, Z.J., Bønnelykke, K. *et al.* (2012) Coarse and fine particles but not ultrafine particles in urban air trigger hospital admission for asthma in children. *Thorax*, **67**(3), 252–257.

Karakatsani, A., Analitis, A., Perifanou, D. *et al.* (2012) Particulate matter air pollution and respiratory symptoms in individuals having either asthma or chronic obstructive pulmonary disease: a European multicentre panel study. *Environmental Health*, **11**, 75.

Kelly, I. and Clancy, L. (1984) Mortality in a general hospital and urban air pollution. *Irish Journal of Medical Science*, **77**, 322–324.

McFarland, A.R., Ortiz, C.A. and Rodes, C.E. (1982) Wind tunnel evaluation of the British Smoke shade sampler. *Atmospheric Environment*, **16**(2), 325–328.

Möller, W., Felten, K., Sommerer, K. *et al.* (2008) Deposition, retention, and translocation of ultrafine particles from the central airways and lung periphery. *American Journal of Respiratory and Critical Care Medicine*, **177**, 426–432.

Nolan, P.J. (1972) The photoelectric nucleus counter. The Boyle medal lecture. *Proceedings of the Royal Dublin Society, Series A*, **4**, 161–180.

Oberdorster, G., Gelein, R.M., Ferin, J. and Weiss, B. (1995) Association of particulate air pollution and acute mortality: involvement of ultrafine particles? *Inhalation Toxicology*, **7**(1), 111–124.

Pekkanen, J. and Kulmala, M. (2004) Exposure assessment of ultrafine particles in epidemiologic time-series studies. *Scandinavian Journal of Work Environment & Health*, **30**(Suppl. 2), 9–18.

Pekkanen, J., Timonen, K.L., Ruuskanen, J. *et al.* (1997) Effects of ultrafine and fine particles in urban air on peak expiratory flow among children with asthmatic symptoms. *Environmental Research*, **74**(1), 24–33.

Peters, A., Wichmann, H.E., Tuch, T. *et al.* (1997) Respiratory effects are associated with the number of ultrafine particles. *American Journal of Respiratory and Critical Care Medicine*, **155**(4), 1376–1383.

Pope, C.A. III., (2000) What do epidemiologic findings tell us about health effects of environmental aerosols? *Journal of Aerosol Science*, **13**(4), 335–354.

Pope, C.A. and Dockery, D.W. (2006) Health effects of fine particulate air pollution: lines that connect. *Journal of the Air & Waste Management Association*, **56**, 709–742.

Sannolo, N., Lamberti, M. and Pedata, P. (2010) Human health effects of ultrafine particles. *Giornale Italiano di Medicina del Lavoro ed Ergonomia's*, **32**(4 Suppl), 348–351.

Snyder, L.P. (1994) 'The death-dealing smog over Donora, Pennsylvania': industrial air pollution, public health policy, and the politics of expertise. *Environmental History Review*, **18**(1), 117–139.

Song, S., Lee, K., Lee, Y.M. *et al.* (2011) Acute health effects of urban fine and ultrafine particles on children with atopic dermatitis. *Environmental Research*, **111**(3), 394–399.

Sorjamaa R Hänninen O. (2011) Modelling Particle Exposures and Doses: Development and Validation of a Size Segregated Micro-Environmental Model. TRANSPHORM Deliverable D2.5.2. Kuopio, Finland, December 2011, 71 pp.

Stölzel, M., Breitner, S., Cyrys, J. *et al.* (2007) Daily mortality and particulate matter in different size classes in Erfurt, Germany. *Journal of Exposure Science and Environmental Epidemiology*, **17**, 458–467.

USEPA (2012) National AirStandards for Particulate matter, http://www.epa.gov/air /criteria.html (last accessed 9 August 2013).

US Surgeon General (2006) The Health Consequences of Involuntary Exposure to Tobacco Smoke: A Report of the US Surgeon General, Department of Health and Human Services, CDC Office on Smoking and Health.

Wiebert, P., Sanchez-Crespo, A., Seitz, J. *et al.* (2006) Negligible clearance of ultrafine particles retained in healthy and affected human lungs. *European Respiratory Journal*, **28**(2), 286–290.

Weichenthal, S. (2012) Selected physiological effects of ultrafine particles in acute cardiovascular morbidity. *Environmental Research*, **115**, 26–36.

WHO (1987) *Air Quality Guidelines for Europe*, European Series, Vol. **23**, World Health Organization, Regional Office for Europe, Copenhagen.

WHO (2000) *Air quality guidelines for Europe*, European Series, Vol. **91**, 2nd edn, World Health Organization, Regional Office for Europe, Copenhagenhttp://www.euro.who.int /document/e71922.pdf (last accessed 9 August 2009).

WHO (2006) World Health Organisation Air Quality Guidelines, Global Update 2005, Copenhagen, 484 pp.

WHO (2013). WHO Regional Office for Europe. Review of Evidence on Health Aspects of Air Pollution –REVIHAAP: First Results, World Health Organization, Copenhagen, http://www.euro.who.int/__data/assets/pdf_file/0020/182432/e96762-final.pdf (last accessed 9 August 2013).

[illegible] Wichmann, [illegible] (2002) [illegible] Respiratory [illegible] associated with [illegible] outdoor [illegible] studies. *Journal of [illegible]* [illegible]

[illegible] (2005) [illegible] Epidemiology [illegible] and [illegible] Health [illegible] ... *Journal of [illegible]* [illegible]

[illegible] and [illegible] (2001) [illegible] effects of [illegible] ... *Journal of [illegible]* [illegible] 20, 73–[illegible]

[illegible] ... [illegible] (2001) [illegible] ... *American Journal of [illegible]* [illegible] 143–[illegible]

[illegible] (1994) The relationships between [illegible] pollution and health [illegible] ... [illegible] *Environmental [illegible]* [illegible] 1994, [illegible]

[illegible] and [illegible] (2011) [illegible] ... [illegible] 1141–[illegible]

[illegible] (1991) [illegible] ... The [illegible] ... TRC ... [illegible]

[illegible] ... *[illegible] and [illegible]*

[illegible]

Wichmann [illegible] (2000) [illegible] ... *Health [illegible]* [illegible]

[illegible] ... [illegible]

[illegible] ... [illegible]

WHO (2000) Air Quality Guidelines for Europe, 2nd edn, European Series, No. 91, World Health Organization, Regional Office for Europe, Copenhagen [illegible]

WHO (2006) World Health Organization Air Quality Guidelines. Global Update 2005. [illegible]

[illegible] ... World Health Organization [illegible] ... [illegible] (accessed 9 April 2013).

10

Pharmaceutical Aerosols and Pulmonary Drug Delivery

Darragh Murnane[1]*, Victoria Hutter*[1]*, and Marie Harang*[2]
[1]*Department of Pharmacy, University of Hertfordshire, UK*
[2]*Institute of Pharmaceutical Sciences, King's College London, UK*

10.1 Introduction

Drug delivery is the science of targeting drug administration to the required site of pharmacological action. The lungs are readily identifiable as a target organ for drug delivery, due to the (relative) ease of access during inhalation. Site-specific delivery of medicaments such as bronchodilators, corticosteroids and anti-infectives directly to their site of action in the airways is an attractive option. Such localised, topical drug delivery maximises a drug's pharmacological effect while limiting systemic exposure and the consequent side effects. Pulmonary drug delivery is by no means a recent phenomenon. As long as 4000 years ago, ancient Indian tribes inhaled vapours for the treatment of obstructed airways (Crompton, 2006). The lung has also long been employed as a systemic portal for drug molecules, albeit not always licitly or therapeutically, as demonstrated by the discovery of tobacco pipes dating back some 2000 years in South America and by Chinese opium inhalers (Sanders, 2007).

The primary physiological function of the lung is to achieve gaseous exchange across the alveoli into the blood circulation. The anatomic characteristics that have evolved to achieve gaseous exchange are the large surface area ($>100\,m^2$ in the adult human) and the thin epithelial barrier (typically $<1\,\mu m$) with extensive vascularisation (Matsukawa *et al.*, 1997; Scheuch *et al.*, 2006). It is precisely these properties that render the alveolar region of the lungs such a desirable route for drug administration, presenting an extensive and very rapid

Aerosol Science: Technology and Applications, First Edition. Edited by Ian Colbeck and Mihalis Lazaridis.

absorptive capacity for exogenous chemicals, including macromolecules such as insulin (Patton, Bukar and Eldon, 2004; Rave *et al.*, 2005). Aerosols of fine particulate matter are ubiquitous in the environment. In order to maintain gaseous exchange, the lung is equipped with protective mechanisms that exclude airborne contaminants such as dust and smoke. As initially proposed by Weibel (1963), it is now widely accepted that the airways, through which inhaled air is drawn, branch, change direction and decrease in calibre 23 times before the air reaches the alveolar epithelium (Figure 10.1). The branching of the airways and the high turbulence in the upper airways affect the deposition of inhaled aerosols on the airway walls, with subsequent particle removal achieved by defensive mechanisms such as dissolution (if soluble) and the cough reflex, and mucociliary clearance (Stahlhofen, 1980; Scheuch *et al.*, 2008) or macrophage activity (Brain, 1988) for poorly soluble aerosols. Although a barrier to achieving systemic drug delivery, particle deposition in the airways is an effective means of achieving localised therapy of lung diseases such as asthma, chronic obstructive pulmonary disease (COPD) and cystic fibrosis (CF).

Drug delivery to the lung is most effectively achieved through the administration of pharmaceutical aerosols. Significant advances have been made in inhalation products over the last 15 years and current research is attempting to unravel the effect of the physicochemical properties of inhaled aerosols on the interaction with the lung epithelial layer. It is still necessary to produce stable products that reproducibly generate aerosols suitable for

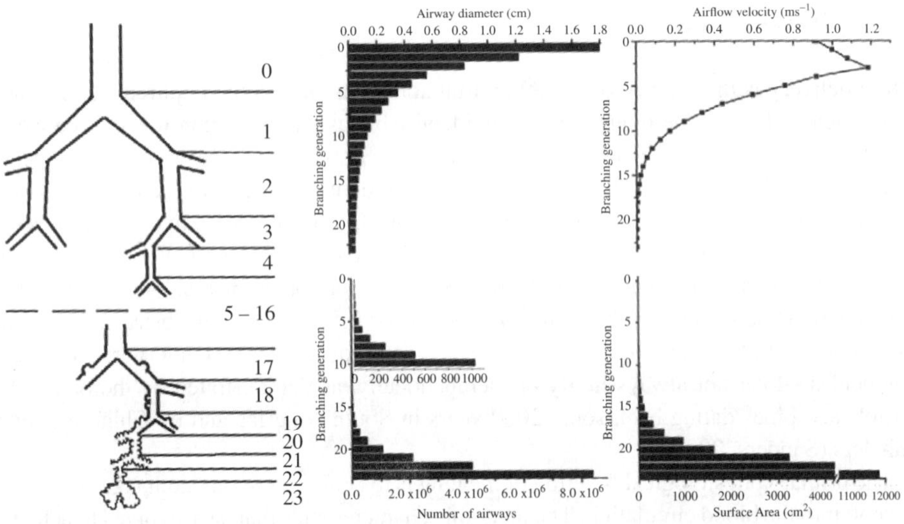

Figure 10.1 Schematic of the pulmonary anatomy according to the Weibel model of the lung. The cross-sectional diameter of the airways decreases (top-left graph) but the number of daughter airways (bottom-left graph) increases between the trachea and the alveoli. Consequently, the cross-sectional surface area of the lungs increases markedly beyond generation 17 (lower right), with concomitant decreases in the airflow velocity (upper right). Airflow velocity was calculated for sedentary conditions according to Martonen (1993) and morphometric data were taken from Swift et al. *(2007).*

achieving drug deposition. Therefore, it is important to consider the physicochemical properties of a pharmaceutical aerosol, which determine deposition, as well as the products used by patients to deliver that aerosol.

10.2 Pharmaceutical Aerosols in Disease Treatment

In order to achieve gaseous exchange at the alveolar surface, air is drawn into the lungs upon expansion of the chest cavity by contraction of the diaphragm and intercostal muscles. The upper respiratory tract serves to achieve particle filtration by inertial impaction. The lower airways consist of the central conducting airways (which warm and humidify the inhaled air), encompassing the bronchi and bronchioles, and the peripheral airways, comprising terminal and respiratory bronchioles and the alveoli. The conducting airways consist of ciliated epithelium, which contains mucus-secreting goblet cells and is coated by a two-layer barrier over the cells: a low-viscosity periciliary sol, which is covered by a highly viscoelastic mucus gel layer (Satir and Sleigh, 1990; Mathias, Yamashita and Lee, 1996). The alveoli and lower bronchioles are devoid of mucus and have a much flatter epithelium, consisting of squamous-type cells (Forbes, 2000) coated with a layer of pulmonary surfactant (Patton, 1996). Diseases of the airways therefore manifest as altered airway anatomies or pathophysiological changes such as altered mucus clearance and pulmonary surfactant production. Topical pulmonary administration directly to the airways is the gold-standard treatment option for obstructive lung diseases; it works by maximising a drug's pharmacological potential and limiting systemic exposure and the resultant side effects.

10.2.1 Asthma

Asthma is a chronic disease of the conducting airways, the most obvious symptom of which is breathlessness and wheezing. The disease is, in fact, a result of chronic inflammation involving cellular and noncellular inflammatory pathways that leads to bronchial hypersensitivity and ultimately airflow obstruction. Asthma is a significant worldwide cause of mortality and morbidity; it is estimated that 300 million individuals are affected by the disease, with all indications showing that global prevalence is rising (GINA, 2011). Indeed, worldwide 180 000 deaths annually are associated with asthma (Viegi, Annesi and Matteelli, 2003). In developed economies, despite increasing prevalence, pharmacological therapies are effective in reducing mortality. For example, in the UK it has been estimated that 5.2 million people live with asthma (Asthma, 2004) and that 250 million daily adjustable life years are lost to the disease (GINA, 2011).

The backbone of asthma therapy is the treatment of bronchoconstriction, which can be achieved readily by bronchodilators such as β-agonists (e.g., salbutamol) and antimuscarinics (e.g., ipratropium), which reduce breathlessness. However, it is also essential to address the causes of bronchoconstriction, and long-term inflammation and hyperresponsiveness are treated with corticosteroids (e.g., beclomethasone dipropionate) or leukotriene antagonists (e.g., montelukast). For many years, developments in asthma therapy revolved around altering the physicochemical properties of drug molecules in order to achieve higher potency (fluticasone propionate), longer action (formoterol and salmeterol) or specificity in lung activity (ciclesonide). It is accepted that inhaled therapy offers excellent control over asthma symptoms (Corrigan *et al.*, 2009), but decline in lung function is still observed in asthmatic patients due to the lack of effective treatment to alter lung remodelling, which results from long-term inflammation in the airways (Hanania and Donohue, 2007).

10.2.2 Chronic Obstructive Pulmonary Disease

COPD encompasses a broad range of diseases characterised by chronic airflow limitation that is not fully reversible. The true burden of COPD is hard to establish, due to the range of comorbidities and difficulties in definition, but approximately 3.28 million deaths were estimated in 2008 (World Health Organization, 2011). Inhalation of particulate matter (e.g., through smoking or pollution) has been related to biochemical changes associated with COPD pathology (Schwarze *et al.*, 2006) and it is estimated that COPD will be the third-leading cause of death by 2020, due to particle exposure in developing countries and the ageing population (Gartlehner *et al.*, 2006). COPD is progressive, with patients typically presenting with infective exacerbations of the disease that require hospitalisation. The combined direct and indirect cost of COPD to the UK's National Health Service was estimated at £1 billion per annum in 2005, with a substantial impact on the wider economy; for example, 21.9 million workdays were lost in 1994–1995 due to COPD (National Collaborating Centre for Chronic Conditions, 2004).

COPD is quite distinct from asthma in that the major site of airflow limitation is the small, peripheral airways (Burgel, 2011). COPD is characterised by obstructive bronchiolitis and inflammation identified by the accumulation of neutrophils and lymphocytes (Hogg, 2004). The associated release of oxygen radicals, elastase and several cytokines is responsible for increased mucus secretion, induction of extracellular matrix breakdown of parenchymal tissues and fibrosis of the small airways (Adcock, Caramori and Barnes, 2011). Current pharmacological therapies for COPD rely on the use of bronchodilators, which improves inspiratory capacity, and typically long-acting bronchodilators such as formoterol and salmeterol are employed. Bronchodilators are less effective in COPD than in asthma (Sturton, Persson and Barnes, 2008), but antimuscarinic agents such as ipratropium (O'Donnell, Lam and Webb, 1999) and the long-acting tiotropium (Cooper, 2006) have emerged as a pharmacological class that offers particular benefits in improving dyspnoea and exercise tolerance. The inflammatory pathways in COPD airways differ to those in asthma and the role of corticosteroid therapy is ambiguous (Gartlehner *et al.*, 2006), possibly due to phenotypic variance in disease. Because of the increasing global incidence of COPD, the coming years are likely to see the arrival of pharmacological agents with novel disease-modifying actions introduced into the clinic (Donnelly and Rogers, 2008; Adcock Caramori and Barnes, 2011).

10.2.3 Cystic Fibrosis

CF is the most common genetically inherited disease in white populations, with an occurrence of approximately 1 in 2500 newborns (Davies, Alton and Bush, 2007). Although CF is a complex multiorgan disease, lung disease accounts for approximately 85% of mortality (Flume *et al.*, 2007). CF is an autosomal recessive disease caused by mutations in the cystic fibrosis transmembrane conductance regulator (CFTR) gene, which is responsible for chloride transport across apical membranes of the lung epithelium (Rowe, Miller and Sorscher,

2005). As a result of impaired chloride transport, the volume of liquid lining the respiratory epithelium is reduced. Mucociliary clearance is impaired because the microcilia of the cell surfaces require the presence of sufficient periciliary fluid volume (Satir and Sleigh, 1990). Progressive lung destruction results from thickened lung secretions, bronchial infection and excessive inflammation in response to pathogens and allergens that cannot be removed.

Treatment of CF requires physiotherapy to remove airway mucus and strengthen respiratory muscle strength (Pryor, 1999). However, frequently antibiotic therapy is required to prevent infection and treat exacerbations in the early stages of disease (in childhood). In addition, mucolytic therapy is beneficial; in particular, aerosol therapy using recombinant human DNase has reached consensus acceptance in practice (Flume *et al.*, 2007). It has been demonstrated that a wide range of bacterial species are now involved in CF disease progression (Stressmann *et al.*, 2011), and the patient's environment appears to play a role in the infectious exacerbations. As CF disease progresses, the lung becomes colonised by bacteria and fungi (Davies, Alton and Bush, 2007). Thus the focus of disease therapy becomes eradication of disease-causing microflora, which can be achieved by using inhaled antibiotics such as tobramycin, as well as by systemic maintenance therapy. However, patients also require conventional obstructive-lung-disease therapies such as corticosteroids and bronchodilators to enhance lung function. Perhaps one of the most exciting developments is the entry into clinical trial of the long-awaited gene therapy. The UK CF Gene Therapy Consortium has developed a nebulised formulation containing a nonviral gene vector that delivers a functioning CFTR gene (Griesenbach and Alton, 2009). Although the current approach has certain limitations, surrogate markers have been identified and several patients have demonstrated CFTR function restoration (Davies *et al.*, 2011).

10.2.4 Respiratory Tract Infection

Respiratory infections tend to be relatively self-limiting and respond to short courses of orally administered antibiotics. However, several groups of patients experiencing long-standing lung colonisation (e.g., CF) or opportunistic infections (e.g., immune-compromised patients) would benefit from localised administration of antibiotic agents that are not well tolerated or available by systemic administration. It is often difficult to achieve sufficient localised drug concentrations for the treatment of tuberculosis infections, and localised delivery may offer a solution that achieves relatively high local concentrations (Smaldone, 2006). Available inhaled antibiotic therapies include nebulised tobramycin and colistin, which have achieved widescale use in the treatment of CF (Flume *et al.*, 2007). There is also much interest in aerosolised antibiotic delivery for the prevention and treatment of ventilator-associated pneumonia, again using tobramycin, colistin, ceftazidime and amikacin (Dhand and Sohal, 2008; Lu *et al.*, 2011). The renewed interest in the delivery of antibiotic agents for the treatment of lung infections has seen the addition of aztreonam nebulised solution and a dry-powder inhalation formulation of tobramycin to the marketed formulary (Geller, Weers and Heuerding, 2011).

10.2.5 Beyond the Lung: Systemic Drug Delivery

The high surface area, thin epithelial barrier and absence of mucociliary clearance make the terminal bronchioles and alveolar regions of the lung an attractive target site for drug absorption. Water-soluble aerosols achieve rapid dissolution in the lung-lining fluid, and subsequent extensive absorption results in the potential for a more rapid onset than is provided by traditional oral dosage forms. Systemic drug delivery is not a new phenomenon, as testified by the human ingenuity shown in identifying a pleasurable range of chemical agents and vapours to smoke or inhale. Two of the earliest pressurised metered-dose inhaler (pMDI) products were for systemic action – Riker (3M) Medihaler Nitro™ and Medihaler-Ergotamine™ – and an explosion of R&D interest in the 1990s and 2000s investigated and exploited the potential of the lungs as a portal for the systemic absorption of medicaments.

Inhaled systemic delivery is often considered in isolation from lung diseases, and the influence of lung disease on effective targeting is less understood, owing to the exclusion of such patients from many studies. However, even in healthy lungs the challenge of reproducibly delivering aerosols to the alveolar region is not trivial. In order to optimise systemic drug delivery, a range of bioengineering approaches have been developed with the aim of controlling the physicochemical properties of aerosols with respect to size, density, morphology and the solid state and thus achieving deep lung deposition (Edwards and Dunbar, 2002; Chow *et al.*, 2007). There has also been substantial investigation into delivery devices that minimise variability in breathing patterns and thus achieve reproducible, controlled drug delivery (Scheuch *et al.*, 2006). The pulmonary route offers substantial increases in the systemic bioavailability of macromolecules such as proteins and peptides compared to other noninvasive delivery routes (Patton, Fishburn and Weers, 2004). Indeed, the clinical potential of inhaled insulin has been identified by many companies, reaching the market as Exubera® (Nektar/Pfizer) (Mastrandrea and Quattrin, 2006). Since the withdrawal of Exubera®, there has been some rebalancing of the emphasis of pulmonary drug delivery toward applying the knowledge gained to improving localised therapy of airways disease. However, several companies maintain systemic drug delivery pipelines, including agents for improving pain management, erectile dysfunction, migraine and Parkinsonism, in all of which rapid onset of action is required.

10.3 Aerosol Physicochemical Properties of Importance in Lung Deposition

The challenge of pulmonary drug delivery is to achieve high lung deposition of inhaled aerosols in the target region for either local pharmacological response or for systemic absorption. Due to the dichotomous branching of the airways, the lung is divided up into generations of airways with decreasing calibre but which increase in number. There is a change in airflow direction and the velocity of airflow decreases from laminar to stagnant in the bronchio-alveolar region (Martonen, 1993). The inhaled aerosol is therefore exposed to a variety of stresses that prohibit aerosol penetration into the deep lung, and aerosol particles are mainly deposited in the airways by the processes of inertial impaction, sedimentation, diffusion, interception and electrostatic attraction (Figure 10.2). A variety of anatomical, physiological and physicochemical properties determine the deposition profile of inhaled aerosols (Table 10.1). It is possible to control the physicochemical properties of the inhaled aerosol in order to control particle deposition mechanisms. The principal properties for control are the particle size, density, shape, charge, solid state and hygroscopicity. It is useful to consider how the physicochemical properties of aerosols determine lung deposition.

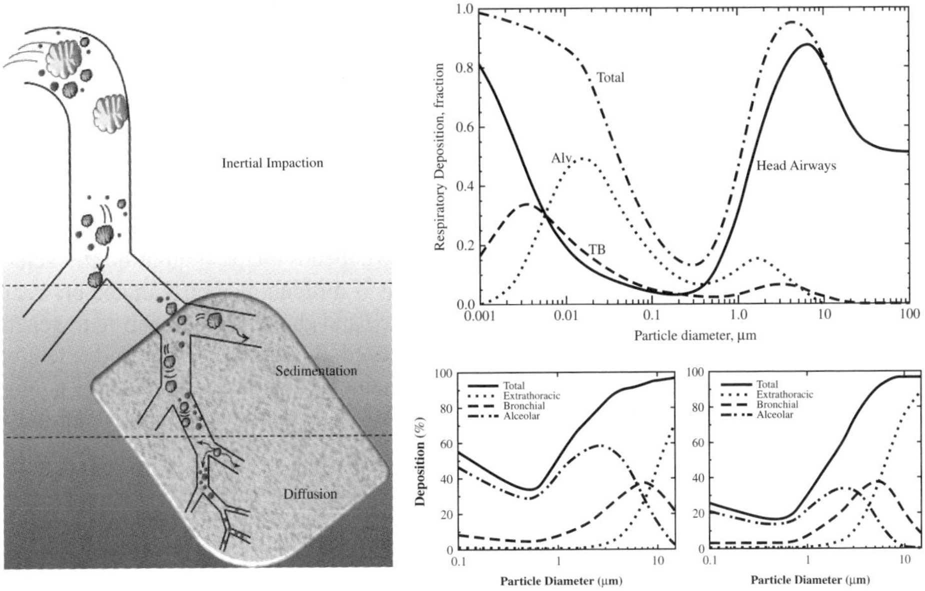

Figure 10.2 *The predominant deposition mechanisms in different lung regions and an indication of regional deposition as a function of particle size (top right). The site of deposition of smaller particles in the small airways decreases with a shift in the deposition profile to larger airways when the inhalation flow rate is increased (bottom right). Top right: Reprinted from Carvalho, Peters and Williams (2011). Copyright © 2011, with permission from Elsevier. Bottom right: Reprinted from Scheuch* et al. *(2006). Copyright © 2006, with permission from Elsevier.*

Table 10.1 *Physicochemical and physiological factors determining pulmonary deposition.*

Property classification	Variable affecting deposition
Particle properties	Density
	Diameter
	Shape
	Solubility/hygroscopicity
	Charge
Aerosol properties	Concentration
	Particle size range
	Bolus or continuous cloud
	Velocity of aerosol
Anatomical properties	Geometry
	Disease
	Humidity
Physiological properties	Inhalation flow rate
	Residence time
	Mouth/nasal breathing

A variety of mathematical and anatomical models of the lungs (Weibel, 1963; Martonen, 1993; Martonen and Katz, 1993) have now been substantiated by experimental evidence regarding the effect of particle properties on the site and extent of deposition in the lung airways (Pritchard, 2001; Rostami, 2009; Carvalho, Peters and Williams, 2011). An axiom generated from studies into aerosol deposition is that particles must possess an aerodynamic diameter (d_{ae}) <10 µm in order to promote the probability of their passage through the oropharynx (Heyder, 2004). In particular, a d_{ae} of 2–6 µm is required for topical aerosol therapy, while a size of 1–3 µm is preferable for targeting of deposition to the peripheral (smaller) airways. Aerosol deposition in the lung also depends on the ventilation characteristics with which the aerosol is inhaled, although this is difficult to control in and between patients. It is therefore the goal of formulation science and chemical engineering to generate aerosols with the physicochemical properties needed to optimise the site and extent of aerosol deposition in the lungs. The aerodynamic diameter, d_{ae}, is the diameter of a unit-density sphere that possesses the same settling velocity as the particle under consideration, and can be expressed as:

$$d_{ae} = d_g \sqrt{\frac{\rho}{\chi}}$$

where d_g is the geometric diameter, ρ is the particle density (g cm^{-3}) and χ is the dynamic shape correction factor. It has previously been suggested that hygroscopicity and water solubility are critical physicochemical properties, due to the potential to alter particle density, geometric diameter or shape as a result of water sorption (Hiller, 1991; Ferron, 1994). However, the influence of hygroscopicity, despite modelling assumptions, is far from understood in the context of dry-powder and propellant-based aerosols.

Particles of large aerodynamic size and particles inhaled at high velocity deposit principally by inertial impaction at airway bifurcations. Particles that are small enough to escape impaction are susceptible to deposition by sedimentation (which dominates deposition in peripheral airways) and diffusion. These latter mechanisms are enhanced by periods of breath-holding. There is a degree of overlap between particles that deposit by impaction and sedimentation (1–10 µm), sedimentation and diffusion (0.1–1.0 µm) and particles that deposit solely by diffusion (<0.1 µm) (Heyder, 2004; Mitsakou, Helmis and Housiadas, 2005). The role of aerosol charge in deposition is a matter of some interest due to the clear evidence of electrostatic charging of aerosols during generation from portable devices (Kwok and Chan, 2009). Effects of charge on deposition have been convincingly modelled (Bailey, 1997; Bailey, Hashish and Williams, 1998). However, the role of electrostatic charge in lung deposition has been questioned (Jeffers, 2007) as aerosol particles need to be located close to the airway wall for charge attraction to have any effect (Jeffers, 2005).

10.4 The Fate of Inhaled Aerosol Particles in the Lung

The fate of inhaled particles is governed by several factors, including physicochemical characteristics, deposition site in the airways and biological defences such as mucociliary transport and resident airway macrophages. Although in certain situations particles are taken up intact by cells of the respiratory tract, in general it is the molecules released from the deposited particles by dissolution or partition that are biologically active. Molecular weight, hydrophobicity, pH, charge, chemical structure and molecular arrangement all influence solubility and impact absorption through the airway epithelium. Molecules can traverse the airway epithelium by two main pathways: either by passing between adjacent cells (paracellular transport) or by partitioning through the phospholipid bilayer (transcellular transport). Transcellular transport may occur through simple partition and passive diffusion or via more complex active mechanisms, including endocytosis and carrier-mediated transport. In reality, inhaled substances can partition across the airway epithelium via multiple pathways governed by their physicochemical characteristics.

10.4.1 Paracellular Transport

Polarised airway epithelial cells have highly specialised junctional complexes between adjacent cells that act as a selectively permeable barrier separating the external surroundings from the internal environment of the lung. In airway epithelial cells, three intercellular membrane specialisations constitute the junctional complex, namely tight junctions, adherent junctions and desmosomes (Denker and Nigam, 1998). Tight junctions comprise over 40 different proteins and form a continuous intercellular barrier between adjacent epithelial cells. They regulate the selective movement of solutes across the epithelium and provide the major barrier to the permeation of substances via the paracellular route (Mitic, van Itallie and Anderson, 2000). The rate of paracellular transport is governed by Fick's Law, driven by a concentration gradient and inversely proportional to molecular weight. Small, hydrophilic molecules such as mannitol are able to passively diffuse through the junctional complexes between adjacent cells (Anderson and Vanitallie, 1995). Solute permeability through tight junctions is variable between different epithelia, but it has been established that macromolecules less than 40 kDa in size can be passively absorbed across the airway epithelium through tight junctions (Patton, 1996; Li *et al.*, 2006). Large macromolecules (>40 kDa) and more hydrophobic entities cannot traverse cells paracellularly and can only permeate biological membranes by transcellular means.

10.4.2 Transcellular Transport

Inhaled drug molecules have different physical barriers to drug absorption in the airways, according to their sites of deposition and action. In the bronchial epithelium, transcellular uptake of drug molecules from the airspace requires diffusion through the apical mucus layer and partition into the lipid cell membrane. The mucosa of the conducting airways comprises several cell layers, which any drug moiety needs to navigate before it can pass into the blood stream via capillaries in the lamina propria. Conversely, further down the respiratory tract toward the terminal bronchioles, mucus production ceases and is replaced by a pulmonary surfactant. Additionally, the alveoli are closely associated with a large capillary network, and in this part of the lung inhaled molecules need only partition through the thin alveolar epithelium and capillary endothelium before being absorbed into the blood stream.

Due to the lipophilic nature of the phospholipid bilayer, low-molecular-weight lipophilic drug molecules can be transported transcellularly by simple passive diffusion, as they are able to partition through the cellular lipid bilayer. The lipid solubility of a compound is determined by its partition coefficient in oil/water (log P); therefore, if log P is >0 then the compound is lipid-soluble, and for low-molecular-weight compounds the percentage absorbed increases with increasing log P value. However, if the log P value is too high (>6), molecules may not partition easily out of the lipid bilayer, and this can result in poor absorption. Larger-molecular-weight drugs (Mw >500–700), and in particular charged molecules, have also been shown to possess restricted movement across cell membranes via passive diffusion (Camenisch *et al.*, 1998). Ionisation reduces the lipophilicity of a compound and limits partitioning into the lipid bilayer, reducing the rate of passive transcellular diffusion. Other transcellular pathways by which larger ionised molecules can be transported across the epithelium are discussed later.

10.4.3 Carrier-Mediated Transport

Embedded within the phospholipid bilayer are several different classes of specialised membrane protein that are able to bind substrates and traffic them across the cell membrane in both absorptive and secretory directions (Figure 10.3). These transporter systems can transport substances by either facilitated diffusion (with the concentration gradient) or active transport (against the concentration gradient). In the lung epithelium, transporters are situated on the apical membrane (closest to the airway lumen) and on the basolateral surfaces. There are two main types of transporter: efflux transporters, which extrude molecules from the membrane or inner leaflet and out of the cell, and uptake transporters, which traffic molecules through the plasma membrane into the cell. Many transporters have the ability to transport a diverse range of structurally and chemically similar substrates, including therapeutic drug moieties (Zhou *et al.*, 2008). There is also evidence for the expression and activity of cation transporters in the airway epithelium, which are capable of transporting charged molecules, including inhaled drugs such as salbutamol and ipratropium (Bosquillon, 2010).

Larger molecules (>500 Da) can be absorbed via endocytic pathways in which part of the plasma membrane is internalised and pinched off to form an intracellular vesicle, containing some of the extracellular environment and capable of incorporating particulate matter. In some instances, these vesicles (endosomes) fuse with lysosomes, degrading the contents

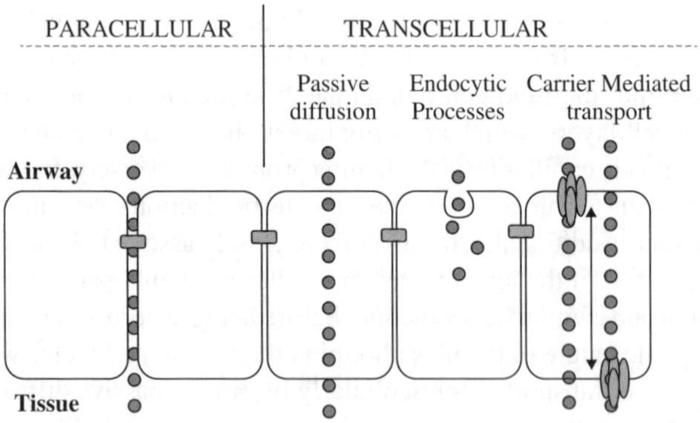

Figure 10.3 *Pathways for the permeation of substances across the airway epithelium.*

via enzymes in the lysosome and releasing them into the cells. However, transcytosis can also occur, whereby the endosome migrates through the cell and fuses with the basolateral membrane, releasing its contents into the extracellular environment.

10.4.4 Models for Determining the Fate of Inhaled Aerosols

There are several different types of biological model available for the study of the safety and efficacy of inhaled compounds. In addition to modelling the absorption and/or permeability of substances across the airway epithelium, these models can also be used to study metabolism and the toxicological and inflammatory responses of the airways to inhaled substances and to model respiratory disease.

10.4.4.1 In Vivo Models

Of the animal models available, the rat is the most commonly used as a result of its size, which is small enough to achieve economical dosing but large enough to manipulate for multiple administrations and blood sampling. Several methodologies exist, but in general the animal is first anaesthetised and then undergoes a tracheotomy or orotracheal intubation. The subject is kept in the supine position while either a drug solution is instilled or an aerosol formulation is administered via the trachea. While *in vivo* studies provide pharmacokinetic data regarding the fate of a drug and its metabolites in a whole-animal system, there is a high degree of interlaboratory variability, primarily caused by variation in methodology and regional distribution within the lung. In addition, the ethics of animal testing and the cost of *in vivo* experimentation bear consideration. Significant differences in the distribution of inhaled particles in the lungs of animals has been demonstrated, highlighting the need for confirmation of pharmacokinetic parameters using *ex vivo* and *in vitro* methods (Sakagami, 2006).

10.4.4.2 Ex Vivo Models

Ex vivo animal models include freshly extracted whole or part lung tissue (primarily rat, guinea pig or rabbit) and are commonly used in biopharmaceutical research when the mechanism of drug transport is not fully elucidated by *in vitro* and *in vivo* models. Thin sections of lung tissue have been used to assess drug metabolism, pharmacology and toxicology (Ressmeyer *et al.*, 2006), while tissue strips have been used in tissue baths and Ussing chambers to assess drug permeability (Widdicombe, 1997). One major limitation of lung-tissue-slice preparations is that due to the size of the tissue, they are only available for the large airways. Isolated perfused lung (IPL) models are a useful tool for allowing the investigation of lung pharmacokinetics without the systemic effects of *in vivo* models. In IPL models, animal lungs (commonly rat) are isolated from the systemic circulation and perfused via the pulmonary circulation. This experimental set-up has the advantage of maintaining tissue architecture (structure, permeability) and functionality (cell interactions, biochemical activity), being a closer representation of the *in vivo* state than is *in vitro* modelling. Additionally, ventilation and perfusion can be tightly controlled and sampling of the perfusate is easily accessible. However, the tissue only remains viable for 3–5 hours after isolation and this time restraint makes it impossible for slower pharmacokinetic processes to be investigated. Furthermore, while the pulmonary circulation is maintained, the bronchial circulation is likely to be severed during the tissue-isolation process. Therefore, absorption in the tracheo-bronchial region is unlikely to be represented in this model (Mehendale, Angevine and Ohmiya, 1981; Sakagami, 2006).

10.4.4.3 In Vitro Models

Respiratory epithelial *in vitro* models comprise homogeneous or heterogeneous populations of epithelial cells from either a primary source (isolated directly from the lungs of humans and other animals) or immortalised cell lines (derived from cancerous sources or transformed using viruses). The cells are typically cultured on suspended inserts, allowing differentiation of the cells at an air–liquid interface. Several of the cell models available form functional tight junctions in culture, creating polarised cell layers that separate the apical and basolateral epithelial surfaces and providing the opportunity to study the permeability of substances across the cells. However, cell lines that do not form polarised airway epithelial layers are still used to study metabolism, cellular response and other pharmacological aspects, using a high-throughput investigational approach (Sporty, Horalkova and Ehrhardt, 2008).

Established *in vitro* cell cultures used to model both the bronchial and the alveolar epithelial regions, as well as disease models, are summarised in Table 10.2. Although primary cells provide the closest native cultures of the airway epithelium *in vitro*, their

Table 10.2 *Summary of the biological characteristics of commonly used human* in vitro *models of airway epithelia (information adapted from Sporty, Horalkova and Ehrhardt, 2008).*

Model type	Origin	Phenotype	Polarised	Air–liquid interface culture
		Bronchial		
Primary cells	Healthy tissue	Mixed airway population, mucus-producing	Yes	Yes
Calu-3	Adenocarcinoma	Sparsely ciliated, mucus-producing	Yes	Yes
16HBE14o-	Transformed healthy tissue	Nonciliated, nonsecretory	Yes	No
BEAS-2B	Transformed healthy tissue	Nonciliated, nonsecretory	No	No
NuLi-1	Transformed healthy tissue	Sparsely ciliated, mucus-producing	Yes	Yes
		Alveolar		
Primary cells	Healthy tissue	Mixed airway population	Yes	Yes
A549	Adenocarcinoma	Alveolar type II cell-like morphology	No	Yes
H441	Adenocarcinoma	Alveolar type II cell-like and bronchiolar-like	Possibly	Yes
		Cystic fibrosis		
CFBE41o-	Transformed healthy tissue	Nonciliated, nonsecretory	Yes	No
CuFi-3	Transformed healthy tissue	Sparsely ciliated, mucus-producing	Yes	Yes

limited functional lifespan (two to three passages), donor variability and high cost to culture are major limitations (Forbes and Ehrhardt, 2005) that make them less economical and convenient than other *in vitro* models. Given their significantly lower culture costs, reduced variability and longer functional passage window, immortalised cell lines are often the *in vitro* models of choice. Whereas for other epithelia a single 'gold-standard' *in vitro* cell line has been established for permeability modelling, no agreement has been reached for the airway epithelium.

The development of *in silico* models for the prediction of adsorption, distribution, metabolism and excretion of oral drug candidates (e.g., GastroPlusTM) has spurred interest in the development of similar models to predict inhaled pharmacokinetics. Existing *in silico* models are best known for simulating and predicting the deposition of inhaled drug particles in the lung. Studies have shown that the physicochemical properties of drug molecules that are most influential on pulmonary absorption are lipophilicity, molecular polar surface area and hydrogen bonding potential. *In vitro–in vivo* correlations have also been established between *in vivo* inhaled drug absorption in the rat and human *in vitro* cell culture permeability (Mathias *et al.*, 2002; Tronde *et al.*, 2003; Manford *et al.*, 2005), allowing mathematical predictions of inhaled drug absorption. Recently, prediction of efficacious lung doses *in vivo* from quantitative structure–activity relationship models has also been possible (Cooper, Potter and Luker, 2010). The establishment of such models will enhance the understanding of inhaled biopharmaceutics and may potentially minimise the use of extensive *in vivo* testing in the future.

10.5 Production of Inhalable Particles

Exposure to environmental and atmospheric aerosols typically occurs in the form of pre-dispersed aerosol clouds. This in unfeasible for pharmaceutical formulations, in which an aerosol must be formed on demand with reproducible physicochemical properties for a patient to inhale as necessary. The three main platforms are techniques that produce droplets from bulk solutions or suspension (i.e., nebulisation), techniques that release dry-powder particle clouds (i.e., dry-powder inhalers, DPIs) and techniques that release condensation vapours (e.g., cigarettes). With the exception of solution-based nebulisers, propellant solution-based MDIs and developmental systems (e.g., the capillary aerosol generator; Longest, McLeskey and Hindle, 2010) the majority of orally inhaled medicinal products contain aerosolisable drug in particulate form. The advantages of the solid state are the potential for improved chemical stability compared to solution formulations and the possibility of controlling the particle properties that determine lung deposition. It is therefore necessary to produce particles in the low micron size range that can be formulated into drug delivery products. There are two classes of particle production method – destructive and constructive techniques – that are used for inhalation formulation.

10.5.1 Particle Attrition and Milling

The production of inhalable particles involves batch crystallisation of the active pharmaceutical ingredient, isolation and drying, all of which are performed while paying remarkable attention to control of the solid-state and particulate properties. Subsequently,

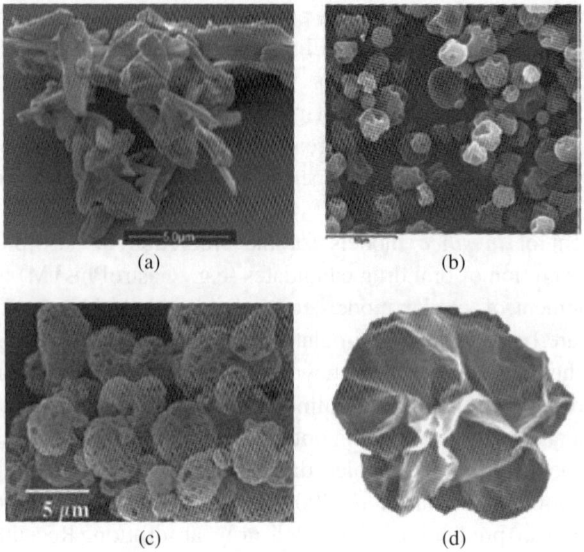

Figure 10.4 (a) Particles for inhalation produced by micronisation. Original image by D. Murnane. (b) Polymeric microspheres produced by conventional spray drying. Image from Li and Seville (2010). (c) Small porous PulmoSphere™ particles produced by spray drying with a perfluoroalkane blowing agent. Image from Duddu et al. (2002). (d) Alkermes AIR™ large porous particles, which can be produced by emulsification–evaporation or controlled spray drying. Image from Edwards (2002).

milling is carried out by a number of processes, which depend on the properties of the input material and the desired properties of the product. Fluid jet milling (also known as micronisation) is an attrition milling process that can produce material in the size range 1–10 μm and has traditionally been employed to reduce the size of crystalline materials for inhalation (Figure 10.4). Jet milling is a relatively old technique that was invented in the nineteenth century but has been employed extensively in inhaled drug delivery since the development of the first suspension pMDIs by Riker Laboratories in the 1960s. Particles are introduced into a grinding chamber using the fluid energy of a carrier gas. The particle stream meets with an opposing fluid jet, leading to particle impaction with high intensity. The carrier gas can be inert (e.g., nitrogen) and the entire process can be performed at low temperature in order to enable milling of thermolabile material. The particle size reduction arises from grinding and fracture upon impact of the feed material in the milling chamber (Midoux *et al.*, 1999).

The popularity of jet milling has been attributed to its relative simplicity and established scale-up, which depends on the size of the collection vessel and product hopper. The process is amenable to modelling (Teng *et al.*, 2009) (and particle size reduction can be optimised by altering the feed rates, carrier gas pressures and mill geometry (Zhao and Schurr, 2002; Teng *et al.*, 2009). Micronisation is a size-classification technique, which means that particles are subjected to many impaction events before reaching the appropriate size to exit the milling chamber. Classification is typically achieved through the use of sieves to guard the milling chamber or use of cyclone separators, which, depending on construction, ensure the collection of particles of respirable range.

Micronisation is not a universal solution to particle size reduction. Proteins and peptides are not generally amenable to milling, but the application of liquid-suspension bead milling at subambient temperatures has been shown to be acceptable (Irngartinger *et al.*, 2004). Microniser feed material requires good flow properties and particles must possess low tensile strength and elasticity. Even for materials that are suitable for jet milling, the efficiency of the process is low, with multiple passes of feed material required in order to achieve a suitable yield. Although particles in the respirable size range are produced, they show polydispersity of their particle size distribution (Midoux *et al.*, 1999), an undesirable property for therapy (Zanen, Go and Lammers, 1998). Destructive techniques also provide poor control over particle shape (Akbarieh and Tawashi, 1987) and porosity, both of which have substantial effects on aerodynamic diameters.

Micronisation subjects drug crystals to high-energy impaction events, leading to particle fracture, typically at the face of the smallest attachment energy (York *et al.*, 1998). Mechanical and tribomechanical forces from milling often induce solid-state transformations in crystalline drug particles, and the newly formed phases may be metastable upon removal of these forces. These solid-state transformations can be as severe as polymorphic transition (Bauer-Brandl, 1996; Chieng *et al.*, 2006; Chikhalia *et al.*, 2006) and extensive bulk amorphisation (Patterson *et al.*, 2005; Gaisford *et al.*, 2010). Total amorphisation of particles is, however, a rare occurrence in jet milling. Surface energisation is more common, in the form of amorphisation of surface regions (Elamin *et al.*, 1994; Brodka-Pfeiffer *et al.*, 2003). More typical is the creation of short-range disorder, such as crystal lattice disruption or dislocations (Perkins *et al.*, 2009). The main difficulty with micronisation is the unpredictability of fracture events and the uneven application of forces. Thus, in addition to particle size distribution, particles will also have a surface energy distribution (Tong *et al.*, 2005). Any alteration in the surface chemistry or energy as a result of milling can dictate formulation performance. For example, physicochemical instability and surface heterogeneity (Bérard *et al.*, 2002) will lead to alterations in powder characteristics over time (Joshi, Dwivedi and Ward, 2002). The ultimate product of micronisation is a particle with an electrostatic charge, high surface energy and physical and chemical instability, and which tends to be highly cohesive and to oppose aerodynamic dispersal in order to form an inhalable aerosol (Fukunaka *et al.*, 2005).

10.5.2 Constructive Particle Production

Constructive particle production techniques provide the opportunity to engineer properties into micron-sized particles that enhance deposition mechanisms and improve aerosolisation performance. Due to the limitations of micronisation with respect to controlling and optimising particle properties and its particularly poor ability to address the production of biopharmaceuticals, a range of constructive techniques have been applied to inhaled formulations. Thus far, the techniques have made little impact in the marketplace, and only two products have been marketed which use spray-drying. Alternative production methods can be broadly classed as droplet-to-particle (e.g., spray drying) or solution crystallisation processes (Figure 10.4). Droplet-to-particle techniques include spray-drying, spray freeze-drying and supercritical spray-drying processes, all of which have been extensively investigated and are amenable to biopharmaceuticals and small-molecular-weight entities (Chow *et al.*, 2007).

The typical process for spray-drying is preparation of a feed-drug solution (which may be a nonaqueous solvent), formation of a spray using an air-jet atomiser and surface contact with drying air, evaporation of solvent at elevated temperatures and finally separation of the dried product (typically using a cyclone collector). In spray freeze-drying, the atomised solution is collected in a cryogenic liquid (e.g., liquid nitrogen) and then lyophilised (Yu, Johnston and Williams, 2006). The dried powder resulting from spray-drying and spray freeze-drying can be obtained as a relatively free-flowing powder with a particle size range <5 μm, but this is not without difficulties in process optimisation and yield (Edwards and Dunbar, 2002). Solution-to-droplet and suspension-to-droplet techniques, including the spray-drying of nanoparticles to create micron-sized composites (Bhavna *et al.*, 2009), can be used to control the ultimate particle properties by controlling feed rate, atomisation conditions and drying rate (e.g., use of cosolvents). It is also possible to include excipients and cosolvents to alter the surface properties, density, surface area and shape of spray-dried particles. An example includes the briefly marketed Exubera® insulin formulation: low-density particles that provide for stabilisation of the protein using mannitol, glycine and sodium citrate (White *et al.*, 2005). An alternative approach involves the inclusion of a blowing agent to induce porosity into the powder, as achieved with the PulmoSphere™ technology. An emulsion of a fluorocarbon in water, stabilised by phospholipid surfactants (Tarara *et al.*, 2004), is used in the production of PulmoSphere™ Tobramycin Inhalation Powder™, the second marketed product to employ spray-drying techniques.

Solution-to-particle approaches include both conventional solvent crystallisation techniquesand precipitation processes involving unconventional solvents such as supercritical fluids (SCFs) (Shekunov and York, 2000) or the attractive environmentally benign approach of poly(ethylene glycol) (Murnane, Martin and Marriott, 2008a). Precipitation is a common industrial process consisting in the rapid crystallisation of drugs, usually at high supersaturation, through the alteration of solution conditions, which renders the drugs insoluble. Frequently, however, there is a dispersion in the crystallisation rates, which makes it difficult to generate particles reproducibly. Conventional antisolvent crystallisation processes involve the intimate mixing of organic and aqueous solvents and have been reported for the production of microcrystals (Larhrib *et al.*, 2003b). Precipitates generally undergo additional steps of agglomeration and ageing, which serve to broaden the particle size distribution, leading to coarser-sized crystals. The processes of agglomeration and ageing are preventable by the addition of polymeric additives (e.g., polyvinylpyrrolidone and poloxamer) that are undesirable for inhalation. Alternatively, controlled mixing techniques can be employed (D'Addio and Prud'homme, 2011) or ultrasound can be applied during crystallisation (Ruecroft *et al.*, 2005).

SCFs conveniently allow the alteration of solubility on the basis of pressure and temperature, which allows control not only of crystalline forms but also of the rate of desupersaturation. As most pharmaceuticals display poor solubility in the most convenient and important SCF medium (carbon dioxide) (Shekunov and York, 2000), SCFs can be used as an antisolvent (termed a 'gas antisolvent' (GAS) technique), with an organic or aqueous solution of the drug precipitated by the SCF. SCFs display diffusivities up to 2 orders of magnitude greater than conventional solvents (Reverchon, 1999) and demonstrate low interfacial tensions (Thiering, Dehghani and Foster, 2001). Upon mixing, therefore, the rapid diffusion into liquid solvents produces a controlled supersaturation and crystallisation, leading to small crystal sizes and narrow size distributions (Foster *et al.*, 2003).

A number of pharmaceuticals have been produced in a form suitable for inhalation by GAS techniques, including: salmeterol xinafoate (Beach *et al.*, 1999), terbutaline sulfate (Rehman *et al.*, 2004), salbutamol sulfate (Schiavone *et al.*, 2004) and flunisolide (Velaga, Berger and Carlfors, 2002).

Constructive techniques such as spray-drying and precipitation do have disadvantages, including leading to amorphous micron-sized particles (Chawla *et al.*, 1994) or metastable solid-state forms (Rehman *et al.*, 2004). Thus, alternative techniques may not always lead to inhaled drugs having improved solid-state characteristics and in many cases are not cost-efficient enough for low-value active pharmaceutical ingredients to allow replacement micronisation as the standard technique for aerosol particle production.

10.6 Aerosol Generation and Delivery Systems for Pulmonary Therapy

10.6.1 Nebulised Disease Therapies

Nebulisers are one of the oldest of the clinically employed aerosol-generation and drug-delivery systems, with the first developed in the 1860s by Siegle, using a pressurised steam spray (Sanders, 2007). Nebulisers use an external source of energy such as a jet of compressed air or ultrasonic vibrations to create aerosol droplets of drug solutions or suspensions, which the patient inhales as a cloud. Nebulisers only represent approximately 12% of all inhalation product sales (Oversteegen, 2008) but have an important place in acute care where inhalation is compromised or where infants and children are being treated. Nebulisers represent a relatively costly drug-delivery system for the treatment of airways diseases such as asthma and COPD, and alternative portable delivery systems offer improved outcomes for bronchodilator use (Leversha *et al.*, 2000). Nebulisers tend to be prescribed for patients whose lung function or competency at using portable devices is inadequate. They offer advantages in that the aerosol is delivered under conditions of normal tidal breathing, and nebulisation has an important place in diseases such as CF due to its ability to deliver high doses of antibiotics and mucolytics. Several developments have led to improvements in performance, such as miniaturisation through the use of vibrating-mesh nebulisers and improved dose output through the use of open-vent nebulisers or breath-coordinated nebulisers, which release aerosol during the inhalation phase.

10.6.1.1 Aerosol Formulation Considerations

Nebulised aerosol formulations are by definition liquid-based and the formulation choice as solutions and suspensions depend on the solubility profile and chemical stability of their active pharmaceutical ingredient. In suspension formulations, the smallest potential particle size of the nebulised aerosol cloud is dictated by the size of the primary micronised drug particles and hence it is necessary to include suspending agents such as the surface-active agent polysorbate 80 to prevent particle aggregation in Pulmicort™ budesonide and Flixotide™ fluticasone propionate. The majority of formulations, however, are aqueous solutions in which a water-soluble salt of the drug is employed to ensure high solubility (e.g., salbutamol sulfate) and prevent precipitation upon storage. Cosolvents such as ethanol and glycerol may be employed to solubilise poorly soluble drug compounds (e.g., Ventavis™ iloprost trometamol). Aqueous formulations are adjusted to isotonicity by the addition of sodium chloride or by pH-buffering systems (e.g., sodium citrate) in order to avoid potential bronchoconstriction in response to hypotonic and hypertonic aerosols (Eschenbacher, Boushey and Sheppard, 1984). The formulation of proteins and peptides is frequently required, due to the high doses needed for therapy, and this poses special formulation considerations. Polypeptides are susceptible to denaturation from exposure to shear and adsorption at the droplet–liquid interface (Patton, Fishburn and Weers, 2004), so surface-active agents must be included. In the case of recombinant human DNase (Pulmozyme™), prevention of deamidation requires adjustment to pH 6.3 and inclusion of $CaCl_2$ in order to maintain enzyme function (Gonda, 1996). Other typical excipients added to formulations include antimicrobial preservatives (e.g., benzalkonium chloride) and chelating agents (e.g., disodium edetate). The formulation components that are included in the nebuliser feed solution/suspension have quite a dramatic impact on the aerosolised mass output and particle size distribution, and excipients must be considered for their effects on surface tension, viscosity and the evaporation rate of the liquid dispersant.

10.6.1.2 Droplet Generation During Aerosolisation

The process of droplet generation and the subsequent aerosol properties depend on the combination of the feed solution/suspension properties as well as on the geometry and mechanical energy applied to aerosolisation (e.g., liquid jet or ultrasonic vibration). Droplets formed during nebulisation are spherical in shape, which renders them suitable for study by laser diffraction techniques (Mitchell *et al.*, 2006). There are two principal methods of nebulisation, which have been adapted in several configurations: air jet nebulisers, which operate on the Bernouilli principle and require a compressed air supply, and ultrasonic nebulisers, which supply mechanical energy produced by vibrations of a piezoelectric crystal or by vibration of a membrane.

In jet nebulisation, compressed air is passed through a small orifice, entraining the formulation liquid from one or more adjacent capillaries. The array of orifice and capillaries differs depending on the nebuliser design, but in all cases the liquid feed is dispersed into droplets in the airstream by primary shear and secondary droplet break-up (Le Brun *et al.*, 2000). The hydrodynamic break-up mechanism means that jet nebulisation is suitable for suspension and high-concentration solution aerosol generation (Hess, 2008). Because the droplets tend to be polydisperse and large in size, jet nebulisers also contain baffles, which provide an inertial barrier that effectively limits droplet output to small respirable droplets. Large droplets are returned to the bulk of the solution or suspension, whereas smaller droplets exit in the inhalation airstream, which may be augmented by a secondary airstream in a side vent. The respirable aerosol undergoes further conditioning during inhalation, which can include coalescence of droplets or solvent evaporation. The passage of air through the nebuliser chamber induces solvent evaporation, which concentrates dissolved solute in the chamber during the nebulisation cycle. Typically, volumes of 2–3 ml are employed in jet nebulisation, which can require 10–15 minutes to aerosolise (Dalby and Suman, 2003), but this time can be reduced when auxiliary airstreams are present, which reduce droplet coalescence and increase mass output rate (Barry and O'Callaghan, 1999). In addition to the jet configuration, the factors affecting the droplet size distribution are the surface tension of the liquid, the concentration of nonvolatile solutes, the liquid viscosity and the airflow rate through the device. It has been observed that there are significant differences in the aerosol output from devices of similar operating design, and although the devices are themselves relatively cheap, the compressors required to deliver the pressurised airstream are comparatively immobile and expensive.

Ultrasonic nebulisers generate droplets by the application of mechanical energy to the nebuliser fill volume produced by the vibration of a piezoelectric crystal. Application of the ultrasonic vibration to the liquid overcomes its surface tension, leading to the generation of droplets from the peaks of capillary waves on the bulk surface. During the application of ultrasonic vibration, the fill volume tends to increase in temperature, and although this is countered by the evaporation of the solvent, it may be necessary to interrupt nebulisation if the temperature rises excessively. Ultrasonic nebulisers atomise the liquid phase and as such are less suitable for nebulising suspensions (Nikander, Turpeinen and Wollmer, 1999), which tend to aerosolise as large-sized agglomerates. Like jet nebulisers, ultrasonic nebulisers contain baffles to retain large droplets; however, it is the patient's own inspiration which entrains the generated aerosol. Because no compressed air is required, ultrasonic nebulisers are smaller, quieter and more accepted by patients.

It is the frequency of vibration that determines the droplet formation, in conjunction with the properties of the nebulisation liquid. The frequency required to achieve capillary waves in the fluid increases as the viscosity of the fill liquid increases; the capillary wavelength is expressed as:

$$l = \left(\frac{8pg}{rf^2}\right)^{1/3}$$

where p is a proportionality constant, g is the surface tension of the solution, r is the density of the solution and f is the frequency of the ultrasound vibration. The size (number mean diameter) of the initial droplets formed can be expressed as:

$$D = 0.34l$$

However, as in jet nebulisation, the ultimate inhaled droplet diameter is also determined by recycling of the larger droplets, droplet coalescence and evaporation in the inhalation airstream. The drug output rate is proportional to the frequency of vibration, as well as the drug concentration and volume of nebuliser solution, but an optimum concentration may exist (Le Brun *et al.*, 2000). Because of the temperature generation, ultrasonic nebulisation may be unsuitable for thermolabile substances, including biopharmaceuticals, and the operation of an ultrasonic nebuliser requires moderate electrical power consumption, which limits device miniaturisation and portability.

In order to address the disadvantages of jet and ultrasonic nebulisers, the most recent class of portable nebuliser to be developed – vibrating mesh nebulisers – was introduced to the market in the first decade of the twenty-first century (Figure 10.5). The main advantages of vibrating mesh nebulisers are their high portability and low power requirements (AA batteries). Droplet formation is achieved by a mesh or membrane containing an array of perforated micron-sized apertures. The diameter of the apertures determines the droplet size of the primary aerosol, leading to sizes that are smaller than those produced by jet and ultrasonic nebulisers. The droplet size distributions are also controllable, in the size range 1–6 μm. The majority of the primary aerosol is inhalable and no baffles are required to achieve recirculation of larger droplets.

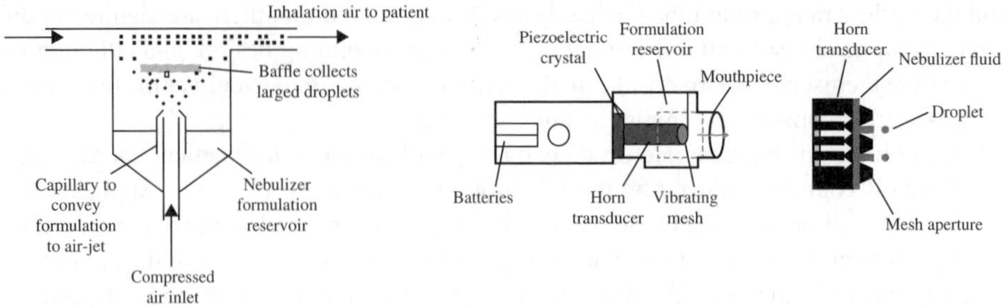

Figure 10.5 Schematics of (a) a traditional air-jet nebuliser and(b) a more modern passive vibrating-mesh nebuliser, including magnification of the vibrating mesh component. Reprinted from Ghazanfari et al., (2007). Copyright © 2007, with permission from Elsevier.

A selection of vibrating mesh devices are now available, which are broadly classed as passive and active vibrating mesh devices suitable for use with low fill volumes (Newman and Gee-Turner, 2005). In active devices, the mesh is vibrated by an electric current, which causes the liquid to be extruded from dome-shaped apertures. In passive devices, liquid is forced through the apertures by an ultrasonic horn, causing the mesh to vibrate in synchrony. Unlike ultrasonic nebulisers, vibrating mesh nebulisers have been shown to be suitable for delivery of suspension and nucleic acid formulations (Dhand and Sohal, 2008). Dhand and Sohal, 2008 report that droplet sizes decrease with increasing solution viscosity, but that excessive viscosity does impede aerosol mass output. Nebulisation is enhanced at higher concentrations of NaCl (or other conducting ions), which may arise from improvements in electrical conductivity (Deshpande *et al.*, 2002). The impact of surface tension on vibrating mesh nebulisation appears to be less important than the impact of viscosity, but surface tension appears to affect mass output, rather than the particle size of the aerosol (Ghazanfari *et al.*, 2007). In general, vibrating mesh nebulisers require low fill volumes and produce slow-moving aerosol clouds, the generation of which can be coordinated with patient inhalation. There are however issues with blockage of the apertures during use.

10.6.1.3 Aerosol Properties and Patient Inhalation Use

There are quite substantial differences between the performances of different marketed nebulisers for a given formulation (Barry and O'Callaghan, 1998), and this variability has led to the regulatory requirement to justify and detail the recommendation of a specific nebuliser for use with a specific licensed formulation. However, the particle sizing method must be considered carefully, because droplets evaporate at the lower relative humidity of laboratory environments. The effect of evaporation will depend on the mass output rate and the type of nebuliser employed (e.g., jet nebulisers cool solutions, while ultrasonic nebulisers heat them). This therefore decreases the measured *in vitro* size distribution compared to that which a patient would inhale (Berg, Svensson and Asking, 2007). It is also important to remember that the patient inhalation manoeuvre (i.e., tidal volume, frequency and flow rate) is crucial to the extent of lung deposition. Tidal breathing conditions are an important consideration for efficacy when nebulised drug delivery is used under conditions of mechanical ventilator support (Ceriana *et al.*, 2003). There is a poor correlation between *in vivo* deposition studies and *in vitro* measurements when only standing aerosol clouds are assessed, and it is thus important that aerosol sizing is measured under conditions of tidal breathing (Sangwan, Condos and Smaldone, 2003).

The major factors determining the aerosol performance of a nebuliser are the mass output rate, the aerosol size distribution and the aerosolisation period. Traditional nebulisers release aerosol across the entire nebulisation period, regardless of whether the patient is inhaling or not. Indeed, aerosol loss may occur if patients exhale into the device rather than removing the mouthpiece. In order to inhale the nebulised aerosol, a mouthpiece or facemask is required. Drug delivery from mouthpieces is higher than from nonsealed facemasks. However, not all patients can comply with mouthpiece use (e.g., infants) and when sealed masks are used with children, facial deposition can occur, leading to side effects or irritation (Brodie and Adalat, 2006).

The significant limit on lung deposition of inhaled aerosols is oropharyngeal deposition, which can be minimised by the use of aerosols with a small droplet diameter (achieved quite well by nebulisers) and by the patient inhaling slowly and tidally (Smaldone, 2002; Clark *et al.*, 2007). Vented nebulisers overcome the problem found in traditional jet nebulisers for which the airflow is not usually sufficient for tidal breathing by allowing extra make-up air to be inhaled through the nebuliser, thereby improving droplet entrainment and reducing coalescence (Le Brun *et al.*, 2000). The most advanced vented systems also contain directional flow valves, which allow exhaled air to vent away from the aerosolisation region. Breath-assisted vented nebulisers are constructed with two valves, which seal alternately upon inhalation and exhalation, such that when the patient does not inhale, no aerosol can vent the nebuliser device. As a result, output rates tend to be improved for breath-assisted open-vent nebuliser designs as compared to conventional jet nebulisers (Le Brun *et al.*, 1999).

One of the major advances in nebulised drug delivery has been the harnessing of the knowledge that regional drug deposition is determined by the volume and timing of aerosol inhalation during the breathing cycle (Brand *et al.*, 1999). The central conducting airways can be targeted with rapid inhalation rates (Clark *et al.*, 2007) and small inhalation volumes (Heyder, 2004), whereas low flow rates, early bolus inhalation and breath-holding promote alveolar deposition (Bennett *et al.*, 2002). It is possible not only to direct deposition to central/peripheral airways but also to target specific anatomical regions of the lungs (e.g., deposition in a specific lobe) by restricting expansion of the lung through supine, upright and reclining postures. Adaptive aerosol delivery systems use microcomputer monitoring, which can adapt drug delivery to the patient's inhalation cycle and pulses an aerosol bolus during the correct period of inhalation in order to achieve the desired drug deposition response. This provides the ability to target drug deposition, allowing systemic absorption to be maximised and improving drug deposition into poorly ventilated regions (Scheuch *et al.*, 2006). The incorporation of such technologies into portable mesh nebulisers provides an opportunity to advance development of regional targeting in pulmonary drug delivery.

10.6.2 Pressurised Metered-Dose Inhaler Systems

The development of the pMDI in the 1950s by Riker Laboratories represented a major leap in respiratory therapy, providing for the first time a compact, portable and cost-efficient drug-delivery device that is easy to use, albeit not always correctly. A pMDI product consists of an active pharmaceutical ingredient (in solution or suspension) and a liquefied propellant with excipients, filled into a canister and sealed with a metering valve. The metering valve limits the volume that can be dispensed upon actuation. The canister is placed in a plastic actuator sleeve containing a spray orifice that atomises the aerosol. Previously, one of the greatest challenges in aerosol medicine was the enforced phase-out of the ozone-depleting chlorofluorocarbon (CFC) propellants originally used in metered-dose inhalers (MDIs), but their replacement with hydrofluoroalkane (HFA) propellants has provided opportunities to improve portable aerosol therapies. pMDIs are still the most used devices worldwide, with their use increasing particularly in developing economies (Technology and Economic Assessment Panel, 2009). Despite the advantages of pMDI devices, a meta-analysis of disease outcomes has suggested that most patients fail to achieve the optimum benefit of pMDI therapy (Brocklebank, Ram and Wright, 2000).

10.6.2.1 Metered-Dose Inhaler Formulation Strategies

A formulation of drugs, excipients and propellants may be a solution or a suspension of micronised particles in liquid propellant. The propellant vaporises to fill the head space of the canister and rapid evaporation of the volatile propellant upon opening of the metering valve provides the force to expel the contents upon actuation. Traditionally marketed formulations were manufactured as suspensions for their ability to control the physicochemical properties of the particles and maximise the drug's chemical stability.

The transition from CFC to HFA propellants, enforced by environmental concerns, initially involved preparation of bioequivalent replacement 'branded generic' formulations, essentially replacing a CFC suspension with an HFA suspension (Keller, 1999). However, the physicochemical properties of HFA propellants differ from those of CFCs; in particular, the solubility profile of the excipients required to stabilise suspensions was not sufficiently high in HFA (Vervaet and Byron, 1999) and the propellants were incompatible with valve components. By altering the valve components, it has been possible to produce equivalent HFA suspensions for drugs such as fluticasone propionate (Kunka *et al.*, 2000). The solvent properties of HFA propellants have created both problems and opportunities to improve aerosol therapy.

10.6.2.2 Inhalation Aerosol Propellants

The propellant represents more than 80% of the formulation composition in most pMDI aerosols, and their high vapour pressure provides the energy necessary to achieve atomisation of the aerosol upon actuation of the pMDI device. The important properties of the propellant are therefore its toxicology, vapour pressure, solvency power, flammability and density (Noakes, 2002), all of which determine the formulation characteristics and performance. HFA propellants emerged from the search in other industries for a replacement for CFCs, for example as blowing agents and refrigerants. The physical properties of the HFA propellants selected (HFA134a and HFA227ea) are presented in Table 10.3, along with those of the CFC propellants they were designed to replace.

The obvious differences are the lower boiling points of both of the HFA propellants, which translate to a higher vapour pressure, and the inability to blend propellants in order to tune the vapour pressure. The low boiling point means that the cold-filling technique (where manufacture is achieved in open vessels and filled in a cooled, liquefied state) is expensive. The low density of the HFA propellants also creates difficulties in achieving stable

Table 10.3 *HFA properties.*

	Boiling point (°C)	Vapour pressure (kPa, at 20 °C)	Density (kg dm^{-3} at 20 °C)	Hildebrand solubility parameter	Dipole moment	Dielectric constant	Polarisability (m^3 mol^{-2} $\times 10^5$)
CFC 11	23.8	89	1.49	7.6	0.46	2.3	2.8
CFC 12	−29.8	566	1.33	6.1	0.51	2.1	2.3
CFC 114	3.6	182	1.47	6.4	0.50	2.3	3.2
HFA 134a	−25.8	572	1.23	6.6	2.06	9.5	6.1
HFA 227ea	−17.3	390	1.42	6.6	0.93	4.1	6.1

suspensions, where the denser suspended drug crystals will sediment rapidly in the lower density-liquefied propellant. In terms of solvency, HFA propellants possess a significantly large dipole, resulting in comparatively higher water and drug substance solubility (Vervaet and Byron, 1999; Hoye, Gupta and Myrdal, 2008), but conversely reduced solubility for suspension-stabilising surfactants. The consequences of the solvency changes are potential precipitation of suspended particles upon water ingress, irreversible Ostwald ripening of suspensions and, in the case of beclomethasone dipropionate, the slow formation with HFA134a of a solvate that rendered suspensions unstable (Vervaet and Byron, 1999).

Some strategies for addressing the formulation are to blend HFA134a and HFA227ea (Williams and Liu, 1998), dissolve cosolvents such as ethanol, glycerol and polyethylene glycol (Brambilla *et al.*, 1999) or develop novel surfactant classes (Peguin and da Rocha, 2008). The dissolution of additional excipient offers one of the most acceptable solutions, due to the alteration of the vapour pressure, density and viscosity of the formulation. Mixtures of HFA with cosolvents typically show deviation from ideal solutions (Williams and Liu, 1998; Vervaet and Byron, 1999). The addition of a cosolvent such as ethanol, however, offers the possibility of dissolving surfactants, and a suspension of salbutamol sulfate stabilised by oleic acid (replacing a salbutamol base suspension) could be formulated and marketed (Ross and Gabrio, 1999). The largest impact on therapy has been the use of cosolvents and other soluble additives to prepare solutions of drug molecules such as beclomethasone dipropionate or formoterol (Acerbi, Brambilla and Kottakis, 2007), which have the potential to improve the drug deposition profile in the lung or to match existing suspension products through formulation engineering.

10.6.2.3 Determinants of Product Aerosolisation

Upon actuation of a pMDI, the opening of the metering chamber to the atmospheric pressure across the spray orifice promotes flash evaporation of the propellant and propels a two-phase gas–liquid mixture through the spray orifice, resulting in the formation of an atomised aerosol cloud (Versteeg, Hargrave and Kirby, 2006). The volatile propellant then evaporates in the aerosol phase, leaving a residual drug-containing aerosol. The properties of the aerosol cloud are determined by the composition of the formulation and the 'hardware' of the metering valve design. In optimising a pMDI formulation, the first consideration is whether a suspension of a solution formulation should be manufactured. Suspensions are typically favoured because of the opportunity for enhanced chemical stability. The simplest formulation consists of propellant and dispersed drug. The vapour pressure of the propellant will be approximately equal to that of the pure propellant itself, leading to production of fast-moving aerosol clouds upon actuation and potentially to small aerosol droplet sizes. The presence of dissolved agents (e.g., stabilising agents or cosolvents) reduces the vapour pressure and can therefore increase the aerosol droplet diameter (Stein and Myrdal, 2004).

In the case of suspension pMDIs, factors governing atomisation are important, as are the volume concentration of the particles in the propellant, the original size distribution of those particles, the occurrence of particle aggregation and the homogeneity of particle dispersion (Gonda, 1985; Hak-Kim and Gonda, 1988). Formation of a stable suspension is crucial in determining the aerodynamic size distribution of the inhaled aerosol and ensuring reproducibility of the dose. The stability of a suspension is determined by the compatibility of the particles with the propellant, the lining of the canister (Michael *et al.*, 2000) and the valve components (Berry *et al.*, 2004, Peyron *et al.*, 2005). Particles can aggregate in the liquid propellant or during evaporation of droplets, resulting in a particle size distribution following atomisation that is different to that of the original particles. For example, the respirable fraction from salmeterol xinafoate and fluticasone propionate pMDIs is decreased as a result of the aggregation of the dispersed particles in suspension (Murnane, Martin and Marriott, 2008b).

Particles are formed from solution-type pMDI formulations, by evaporation of the propellant droplets after actuation. The size of the droplets in the spray plume is affected by the diameter of the atomisation orifice (Brambilla *et al.*, 1999; Berry *et al.*, 2003), the actuator design (Smyth, 2003), the volume metered by the valve and the vapour pressure driving the emission through the orifice (Brambilla *et al.*, 1999; Vervaet and Byron, 1999). The presence of nonvolatile components such as drugs, surfactants or cosolvents decreases the vapour pressure, and this can lead to an increase in the droplet size. However, a propellant with high vapour pressure can lead to droplet impaction on the device or in the throat upon atomisation. If the orifice diameter is increased, this ballistic effect can be more pronounced (Berry *et al.*, 2003). The droplet size of the aerosol cloud is therefore determined by an interaction between actuation, device and formulation parameters (Brambilla *et al.*, 1999; Berry *et al.*, 2003; Smyth *et al.*, 2006; Stein and Myrdal, 2006). The main effects of the discrete factors are summarised in Figure 10.6. However, it is generally true that the aerodynamic size distribution of a solution formulation is smaller than that of a suspension (Crampton, Kinnersley and Ayres, 2004). One particular advantage is the ability to 'tune' the droplet size to achieve peripheral lung deposition or to minimise variability in conducting airway deposition (Acerbi, Brambilla and Kottakis, 2007).

10.6.2.4 Aerosol Properties and Lung Deposition

A major benefit of pMDIs is that aerosol clouds are generated independently of the patient's respiratory effort, due to the atomising force of the highly volatile propellant. However, this benefit is also a major limitation of pMDI therapy as the aerosol cloud is produced in a short timeframe and moves at a high velocity, meaning that patients must be able to coordinate the actuation event and inhalation in order to maximise drug deposition in the lungs. Unfortunately, somewhere in the region of 50% of patients fail to use their pMDI correctly even with adequate training (Barnes, 2004); the most common error is failure to coordinate actuation–inhalation, with some patients terminating the inhalation.

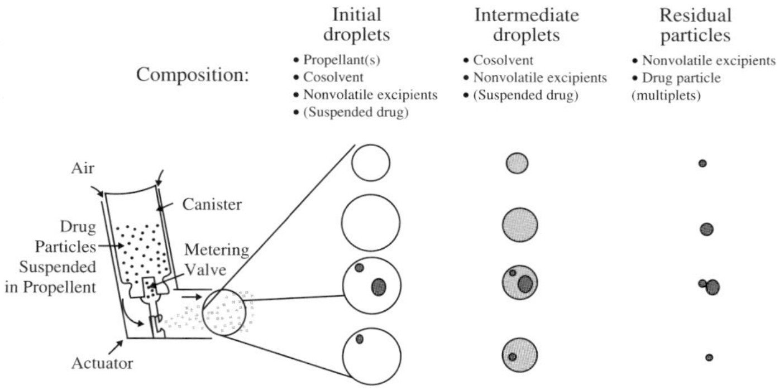

Factor	Increase infactor results in:
Orifice diameter	Larger droplet diameter
	Higher droplet momentum
	Higher throat deposition
Metering chamber	Larger droplet diameter (+/–)
	Higher plume velocity (+/–)
Expansion chamber size	Altered plume geometry
	Larger droplet diameter
Propellant vapour pressure	Faster evaporation rate
	Smaller droplet diameter
	Higher plume velocity
Non-volatile components	Lower the vapour pressure
	Larger droplet diameter

Figure 10.6 *Principles of operation of a pMDI, in which the formulation may consist of an active pharmaceutical ingredient dissolved or suspended in a propellant/cosolvent mixture.*

Using traditional CFC-based inhalers resulted in 10–20% lung deposition (Pauwels, Newman and Borgstrom, 1997) and even in patients who had undergone training and who showed correct technique, the lung deposition was less than one-third of the emitted dose (Newman *et al.*, 1998). The main factor contributing to the low lung deposition with pMDIs is the extensive oropharyngeal deposition that results from the ballistic nature of pMDI aerosols (Newman and Chan, 2008).

The introduction of HFA-based pMDI formulations has provided an opportunity to improve drug deposition in the lungs. In order to develop ways of improving on drug deposition, it is first necessary to consider the properties of the aerosol produced by pMDIs. Ballistic deposition is determined by the physical properties of the aerosol cloud, including particle size, plume geometry and plume velocity. Reducing particle aerodynamic diameter increases passage through the oropharynx and consequently lung deposition (Smaldone, 2006). It is generally stated that HFA formulations possess a smaller particle size than do CFC formulations and should therefore achieve low oropharyngeal and increased lung deposition. QVAR beclomethasone dipropionate, which is an HFA/ethanol solution product (mass median aerodynamic diameter, MMAD, 1.1 μm), improves lung deposition compared to the CFC-suspension it was designed to replace (MMAD 3.5 μm) (Leach, 1998; Leach *et al.*, 2005). However, many formulations were developed to match the MMAD of the CFC formulations they replaced (e.g., Flixotide™ fluticasone propionate, Ventolin™ salbutamol sulfate and Serevent™ salmeterol xinafoate), which show broadly similar lung deposition fractions. It is thus an inaccurate conclusion that HFA aerosols always have a smaller particle size and higher lung deposition – it is a matter of the formulations.

The major difficulty is that extra-fine HFA aerosols do still exhibit high oropharyngeal deposition, varying between roughly 30 and 56% of an actuated dose, depending on technique (Leach, 1998). The patient inhales a dynamic aerosol cloud of drug-containing propellant droplets that are undergoing evaporation competing with droplet coagulation and ageing in a turbulent plume. The size of the droplets is a function of the quality of the fluid flow in the expansion chamber of the pMDI and the pressure differential across the spray orifice. Therefore, the formulation vapour pressure and viscosity interact with the spray-orifice diameter to determine the discharge and atomisation event of the aerosol. The higher vapour pressure of HFA results in smaller droplets than are found in CFC propellant, but these droplets possess higher velocity (Dunbar, 1997; Dunbar and Miller, 1997). The difficulty with estimations of droplet velocity and size is the heterogeneity of both according to the time following actuation, with fluid ejection lasting up to approximately 200 ms (Dunbar, Watkins and Miller, 1997a). Initial velocities from HFA close to the spray orifice have been found to range from 16 to 58 $m\,s^{-1}$ (Crosland, Johnson and Matida, 2009). However, at a distance of 10 cm from the spray orifice, velocities range from 15 to 5 $m\,s^{-1}$ depending on the time during actuation (Dunbar, Watkins and Miller, 1997b). A range of marketed formulations demonstrate that several CFC-pMDIs have similar velocities (~6 $m\,s^{-1}$ at 10 cm distance), but several HFA suspensions do have a high droplet velocity (5.1–8.4 $m\,s^{-1}$) (Hochrainer *et al.*, 2005). The extended duration of a spray is advantageous for patients who struggle to coordinate actuation, provided they continue to inhale the aerosol cloud. It is important to note that the plume duration, velocity and impact force at 5 cm from the actuator mouthpiece can be altered by appropriate formulation manipulation and choice of spray orifice (Ross and Gabrio, 1999; Acerbi, Brambilla and Kottakis, 2007). Coordination is therefore not the full explanation for oropharyngeal deposition with pMDIs, and even with automated-firing breath-actuated inhalers or patients who show adequate technique, pMDIs prove to be exceedingly good at depositing drug in the mouth and throat.

10.6.2.5 Auxiliary Devices to Optimise Aerosol Delivery

In order to reduce oropharyngeal deposition, auxiliary devices termed 'spacers' were introduced in the late 1970s, which quickly took the form of valved holding chambers (VHCs). VHCs are an elegant solution to the high oropharyngeal deposition of pMDIs, temporally separating the actuation and inhalation events. It has generally been believed that the ballistic fraction is captured by deposition within the VHCs themselves. The fine-particle lung deposition (i.e., fine-particle mass) is therefore unchanged when a VHC is employed, but the extrathoracic deposition fraction is reduced (Newman and Newhouse, 1996). However, the situation may not be quite so straightforward; for example, substantial oropharyngeal deposition has been shown for budesonide (Turbuhaler® and pMDI plus spacer in asthmatic patients) (Hirst *et al.*, 2001) and relative lung bioavailability has been shown to increase when using a VHC for salmeterol/fluticasone propionate (Nair *et al.*, 2009). In attempting to understand drug delivery from spacers, it is important to consider the interaction between residence time and inhalation flow rate, which interact to determine the dosing particokinetics (Verbanck *et al.*, 2004).

The importance of dose deposition within a VHC lies in its relationship with the drug-deposition performance. Early studies reported values for dose retention in the VHC on the order of 50–80% for salbutamol (Dubus and Dolovich, 2000), with consequent decreases in deposited lung dose (Pierart *et al.*, 1999). The actuated dose from a pMDI is of droplets with large diameter and momentum, which undergo a process of evaporation and evolution of particle size distribution. There is a clear dependence on VHC design (Hall *et al.*, 2011): spray deposition within the VHC is likely to be higher for a small-volume chamber. Conversely, large-volume chambers dilute the aerosol concentration, requiring longer inhalation times, multiple inhalations for those with small lung volumes (Schultz *et al.*, 2010), potential sedimentation and a larger surface area for deposition. Therefore, the extent of spacer deposition depends on the residence time of the aerosol in the VHC, as well as the volume of inhalation and the flow rate (Barry and O'Callaghan, 1999).

Interestingly, dose retention in VHCs decreases following multiple actuations (Clark and Lipworth, 1996) and dose delivery from a spacer coated with an antistatic layer improves drug delivery and removes the effect of the duration of aerosol residence time in the VHC (O'Callaghan *et al.*, 1993; Clark and Lipworth, 1996). pMDI propellants and excipients can support charge. As a result of triboelectrification during high-shear atomisation, fine aerosol particles exhibit a charge (O'Leary *et al.*, 2008) that varies from drug product to drug product. Simply washing with detergent can provide an antistatic layer to the spacer; however, this relies on patient/carer cooperation and is subject to differences in washing and drying techniques. The use of antistatic plastics is a more rational approach to reducing dose variability when using VHCs. This removes the potential for variability in washing techniques and led to the introduction of electrically conducting VHCs such as the metal NebuChamber™ in the 1990s (Bisgaard *et al.*, 1995). Despite their advantages in achieving selective, targeted deposition in the lung, spacers are rarely used in children under paediatric care (35%) and they are the least preferred dosage form of adults (Mitchell, Coppolo and Nagel, 2007).

10.6.3 Dry-Powder Inhalation

The first DPI developed for the delivery of penicillin antibiotics was the Abbott Aerohaler™ in the 1940s. There were two major drivers for the development of DPIs in the twentieth century. Because the dose that can be delivered from pMDIs is limited by the volume of the metering valve (maximum of ~100 μl), there was a requirement for a portable dosage form that could deliver high doses of an active pharmaceutical ingredient. This led to the development of the Intal™ Spinhaler™, containing sodium cromoglicate, in 1967. The second major driver was the need to find alternatives to the use of CFC propellants following the Montreal Protocol, which led to the observation that drug-delivery performance can be enhanced using DPI technologies. Like nebuliser and pMDI drug-delivery systems, DPI systems require a formulation containing the inhalable drug and a device to disperse the formulation into an aerosol. DPIs may be either single-dose devices, in which the formulation is typically contained in a capsule, or multiple single-unit-dose devices, in which the formulation is contained in a series of unit-dose blisters on a strip. Multidose devices are also manufactured, in which the formulation is contained in a reservoir and metered by the patient prior to inhalation, much like in the pMDI. In the majority of DPIs, it is the patient's own inspiratory manoeuvre that aerosolises the inhalable drug particles; however, recent years have seen the development of active devices that use external energy (e.g., compressed air) to achieve aerosolisation. There are therefore three main components to consider: the inhalable drug particles, the powder formulation and the interaction between a patient and their device (Figure 10.7).

10.6.3.1 Physicochemical Properties and Behaviour of Respirable Powders

The aerosolisation of micron-sized powders from a DPI system is difficult to achieve. Micronised powders demonstrate a high surface-area-to-mass ratio and interparticulate forces arising from surface phenomena dominate gravitational forces. The main interactions between particles in the size range <20 μm are van der Waals forces and electrostatic

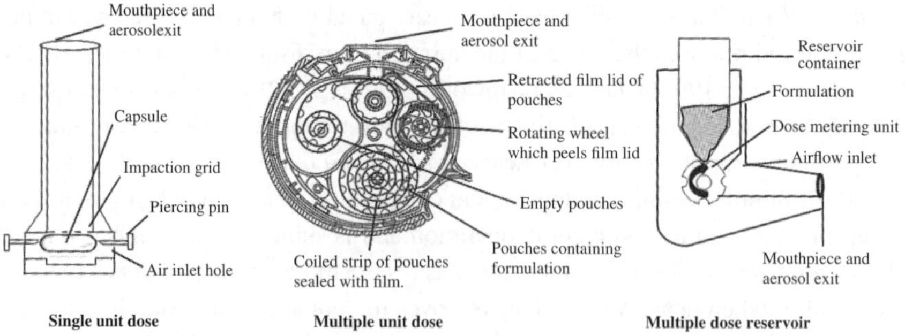

Figure 10.7 Examples of conventional passive dry-powder inhaler designs.

interactions (Schubert, 1981) (which are particularly relevant for particles that have undergone milling). In the presence of sufficient relative humidity, capillary forces also contribute extensively to interparticulate bridging forces (Price *et al.*, 2002). The result of the high interparticulate forces is that powders demonstrate high cohesivity and particles form strong powder agglomerates (Weiler *et al.*, 2010). The powders are also highly adhesive and stick to other surfaces (such as devices). As a consequence, powder flow is impaired, rendering the uniform filling of inhalable solid drug particles into devices, blisters or capsules difficult. Of equal concern is that the cohesive force of micron-sized particles must be overcome in order to achieve deaggregation of the powder upon inhalation, leading to low and highly variable respirable fractions of micron-sized drug particles.

During inhalation of powders in DPIs, there are three – often concurrent – processes that result in particle aerosolisation: powder-bed fluidisation, particle entrainment and deagglomeration. It has conventionally been thought that aerosolisation arises from deaggregation caused by aerodynamic shear forces (Zeng, Martin and Marriott, 2001). Aerosolisation requires both consideration of the forces of interaction between individual particles, which must be overcome by the dispersal forces, and an understanding of the bulk powder properties, such as inter-agglomerate forces, powder structure and air permeability. Agglomerates are weaker in strength when the primary particle size is greater, and particles with rough surfaces have reduced contact surface areas (Tang, Chan and Raper, 2004). In terms of the powder meso- and macroscale, agglomerates and powders with low bulk density are easier to entrain and particles with low surface energy promote better dispersion and deagglomeration in the airstream (Chow *et al.*, 2007). In recent years it has become evident that aerodynamic shear forces are not the only contributor to aerosolisation events: agglomerate impaction events within the DPI device are also important (Wong *et al.*, 2010).

In many ways, particles produced by micronisation are not ideal for dry-powder aerosols. Micronisation leads to the disruption of the surfaces of crystalline particles and the high surface energy of micronised particles contributes to their poor aerosolisation and dispersion. Regions of disorder also represent ‘hot spots’ for atmospheric sorption, which can lead to the recrystallisation of amorphous material (Ward and Schultz, 1995; Price and Young, 2004), with the potential for solid–solid particle bridging, which is a barrier to deagglomeration. The sensitivity of disordered sites to humidity provides a source of instability and variability in the agglomerate strength and powder cohesion. The real difficulty is in the intrabatch variability in properties, which means that batches of micronised powders comprise subpopulations with heterogeneous susceptibility to deagglomeration (Taki *et al.*, 2011). The result is that pure, micronised drug powders show agglomeration, which upon aerosolisation results in a large fraction of coarsely sized agglomerates (Figure 10.8).

In order to gain some control over the difficulties in the aerosolisation of dry powders, a variety of particle-engineering approaches have emerged, aimed at controlling particle surfaces, sizes, shapes and densities and thus manipulating the agglomeration forces and powder-blend structures (Chow *et al.*, 2007). Formulation strategies may also be employed, including the addition of small amounts of fine-sized sugar particles (Jones and Price, 2006), which alter the agglomerate or blend (macro)structure. Alternative strategies include coating with low-surface-energy components such as L-leucine or phospholipid, through either powder blending or constructive droplet-to-particle manufacture. Such formulation approaches attempt to alter particle–particle interaction forces (Lucas,

Anderson and Staniforth, 1998; Geller, Weers and Heuerding, 2011) and improve the deagglomeration of the fine particles (Figure 10.8).

10.6.3.2 Powder Formulations

The majority of respiratory therapies require the inhalation of low doses (typically <1 mg) at which it is almost impossible to achieve uniformity of filling and inhalable dose. It is necessary to mix the inhalable particles with a diluent such as lactose monohydrate particles in order to form an ordered mixture or granulation using an appropriate wetting liquid, with or without the addition of a bulking agent to aid in dose metering. By serendipity, the initial desire to triturate by blending with lactose also improved the flow properties and aerosolisation of the inhalable drug particles. The bulking agent most commonly employed is crystalline lactose monohydrate, which is usually termed a 'carrier particle' because its size is larger than that of the micronised particle. The use of the carrier particle introduces an additional level of interaction into the formulation: drug particles may adhere on the carrier surface directly or as agglomerates (Figure 10.8). Many drug particles form strong adhesive interactions with lactose carrier particles, resulting in poor redispersal upon aerosolisation of the blend (Hindle and Byron, 1995). Some of the carrier properties that have been identified as being important in formulating particles that can disperse adequately are: the surface roughness and shape of the carrier crystals (Larhrib *et al.*, 2003a, 2003b; Islam *et al.*, 2005), the size and grade of the carrier particles (Islam *et al.*, 2004a), the carrier crystallinity (Saleem, Smyth and Telko, 2008), the ratio of drug to carrier (Dickhoff *et al.*, 2003; Young *et al.*, 2005) and the blending intensity (Dickhoff *et al.*, 2003).

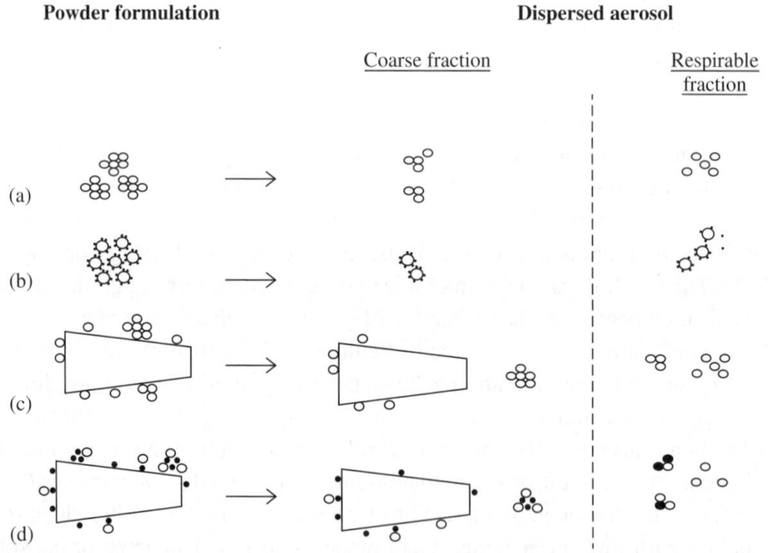

Figure 10.8 *Schematic of (a) pure-micronised-drug-particle, (b) micronised-drug-additive, (c) carrier-based and (d) carrier/additive-based DPI formulations.*

Mechanical activation of the drug particle surface by micronisation is a source of physicochemical instability and heterogeneity in carrier–drug and carrier–agglomerate interaction forces (Bérard *et al.*, 2002). Micronisation is responsible for alterations in adhesive/cohesive forces in DPI blends (Davies *et al.*, 2005), which affects blend uniformity. Worryingly, physical instability will lead to alteration in powder characteristics over time (Joshi, Dwivedi and Ward, 2002). Capillary forces of adhesion between drug and carrier (Price *et al.*, 2002) are increased when moisture condenses at disordered regions of the micronised particle surface. Thus, carrier-based formulations often show low aerosolisation performance for the inhalable drug. The poor efficiency of formulations can be accounted for, and even improved by, formulation engineering (e.g., using adhesion force control agents (Zhou *et al.*, 2010) or adding ternary dispersal agents (Islam *et al.*, 2004b)). It is unlikely that carrier-based formulations will disappear for the majority of DPI therapies, and hence it is necessary to optimise the blending performance and to gain more insight into the role of powder structure in aerosolisation, as well as to develop device strategies that mitigate heterogeneity in powder properties.

10.6.3.3 The Patient and the Device

Aerosolisation occurs when a patient inhales through the device, generating the pressure drop that results in powder fluidisation, entrainment and deagglomeration. In order to achieve effective deagglomeration of inhalable particles, a fast, forceful inhalation is required. High airflow through a low-resistance device (or lower airflow through a high-resistance device) produces the aerodynamic shear forces, turbulence and impaction required to produce high respirable fractions (Srichana, Martin and Marriott, 1998). Unfortunately, many patients would be unable to inhale effectively through a high-resistance device. *In vitro* aerosolisation studies show that micronised drug blended with carrier results in a relatively low-efficiency performance. Therefore, it is no surprise that traditional DPI formulations show poor pulmonary deposition ability – as little as 5–28% of a dose is deposited in the lung from DPIs (Pauwels, Newman and Borgstrom, 1997). Although this figure refers to relatively old devices and to formulations without the incorporation of ternary formulation agents, it is generally true to say that *in vivo* deposition underperforms compared to *in vitro* testing.

Patients with lung disease are unable to inhale with the necessary force through their DPI device (Cegla, 2004), leading to low DPI dose emission (Kamin *et al.*, 2002; Tarsin *et al.*, 2006). The fraction of drug available for lung deposition is suboptimal for many patients with lung disease (Bisgaard *et al.*, 1998; Burnell *et al.*, 2001; Kamin *et al.*, 2002). COPD patients (even after instruction) inhale with lower inspiratory flow rates than asthmatic patients (Broeders *et al.*, 2003), which is attributed to the patients' respiratory muscle strength and lung hyperinflation. In order to achieve deep lung penetration of aerosol, the patient must inhale with a low flow rate. The requirements for aerosolisation (a high flow rate) are thus incompatible with achieving enhanced peripheral deposition of inhaled particles in DPIs, as compared to other inhaler systems (Thorsson *et al.*, 1998). It is for this reason that the concept of improved drug-delivery devices that provide effective powder break-up at relatively low flow rates has been developed (e.g., Airmax™; Zeng *et al.*, 2002): devices that incorporate classifiers to minimise interpatient variability in aerosolised particle size (de Boer *et al.*, 2006) or active devices in which the patient's inhalation is not required to aerosolise the drug particles (Mastrandrea and Quattrin, 2006).

10.6.4 Advancing Drug-Delivery Strategies

It will be obvious to the reader that each of the drug-delivery systems examined here has disadvantages that provide ample opportunity for pulmonary drug-delivery research and development. Ten years ago, this section would have consisted of a discussion of the development of systemic protein and peptide delivery, which culminated in the marketing (and subsequent withdrawal) of Exubera® Inhaled Insulin. In addition, the almost ubiquitous nanotechnology 'revolution' has been notably absent from the discussion. Nanoparticles exhibit enhanced deposition fractions when below ~100 nm in size (Heyder *et al.*, 1986), but they pose difficulty in formulation and product manufacture. Nevertheless, several strategies have been proposed, including nanoparticle composite particles (Azarmi *et al.*, 2008) and dispersion in nebulised formulations. Indeed, a nanoparticle drug-delivery strategy involving nebulisation of an aqueous liposomal dispersion is now in clinical trial for the gene therapy of CF (Griesenbach *et al.*, 2011). The use of nanoparticles provides the opportunity to enhance dissolution rates, target uptake into macrophages and potentially avoid mucociliary clearance. 'Controlled-release technology' refers to a drug-delivery system that alters the drug release rate from a formulation beyond that achieved by dissolution alone. The field of controlled release in inhaled drug delivery has been slow to develop, due to the limited array of polymers that are approved for inhalation therapy on the basis of safety and toxicity profiles. One example is the use of mucoadhesive polymers (Sakagami *et al.*, 2001) or multicomponent microparticles and nanoparticles that extend residence time at the epithelium and promote cellular uptake (Alpar *et al.*, 2005). Additionally, a platform for the delivery of insulin has been developed by Mannkind Corporation, termed 'Technosphere™', which employs diketopiperazine derivatives to provide physical protection for the polypeptide but also rapid drug dissolution in the lungs (Cassidy *et al.*, 2011).

The development of active DPIs that aerosolise the drug independent of patient effort, breath-actuated MDI technologies to address coordination issues and miniaturisation and inhalation synchronisation of nebulisers has already been discussed. However, many of these techniques still exhibit high throat deposition, with the potential for side effects. The last 10 years have seen significant research into soft-mist inhalation technologies, in which slow-moving aerosol clouds of extended aerosol lifetime are produced for the patient to inhale. One example is the condensation aerosol generator under development by Chrysalis Technologies, which borrows from the age-old practice of smoking to evaporate thermostable drugs and subsequently submits them to controlled evaporation in order to produce aerosols in the size 0.25–2.0 μm (Hong, Hindle and Byron, 2002) with predicted reduced throat deposition (Worth Longest, Hindle and Das Choudhuri, 2009). Many of the newer device technologies require mechanical or electromechanical components, which can meet with regulatory objections (e.g., the impact of device failure during an emergency). However, one device that has achieved approval is the Respimat® Soft Mist™ inhaler, which has been marked by Boehringer Ingleheim with Ipratropium bromide. Respimat® is a mechanical device that contains a reservoir drug solution in which the mechanical energy of a compressed spring forces a metered volume of the solution at high pressures (250 bar) through a nozzle device, which causes two high-velocity liquid jets to collide, resulting in a slow-moving aerosol cloud. The Respimat® delivers a high respirable fraction in *in vitro* studies. The device has a number of recommendable features, include the absence of coordination issues (due to the low cloud velocity), low throat deposition and a high lung-deposition fraction (Brand *et al.*, 2008).

10.7 Product Performance Testing

In studying inhaled aerosols, a variety of techniques are employed that are universal to all branches of aerosol science. These include low-angle laser-light-scattering (laser diffraction) analysis, aerosol time-of-flight analysers, laser Doppler anonometers, particle image velocimetry and dynamic mobility analysers. The field of aerosol characterisation has been excellently reviewed by Mitchell *et al.* (2011). Unfortunately, many of the techniques that are well known to aerosol scientists achieve a use which is essentially confined to characterisation and development, perhaps with the exception of laser diffraction analysis, which has gained approval for use in nasal sprays and aqueous droplet systems. The main requirement for a technique in aerosol medicine is that it provide a (drug-specific) measure of the mass of aerosol of a given size, such that an indication of the pulmonary availability can be assessed. In essence, this refers to tests of emitted dose, and the aerodynamic particle size distribution of that emitted dose. The second component to these tests is that they must be performed in a way that is highly prescribed by the regulatory agencies in the form of national and international standards, such as the British and European Pharmacopoeias.

10.7.1 Total-Emitted-Dose Testing

Emitted-dose testing is performed for nebulisers, pMDIs and DPIs using a filter housing that is designed to capture the entire aerosol emitted from an inhaler. The aerosol is actuated into a moving airstream, which is intended to represent the conditions of use of the inhaler being tested. This poses several problems for nebuliser and DPI systems. In the case of nebulisers, patients inhale using tidal breathing, meaning that the dose emitted is affected by the tidal volume and the frequency of breathing (Nikander *et al.*, 2001). Therefore, it is recommended that breathing simulators be used to draw representative inhalation waveforms through the nebuliser filtration apparatus in order to mimic this situation *in vitro*. In the case of passive DPIs, the dose emission is achieved by the force provided by the patient's inhalation manoeuvre, and hence a means to standardise this process between inhalers is required (Hindle and Byron, 1995). This has been achieved by adjusting the airflow through the DPI to one that is equivalent to a 4 kPa pressure drop across the device as it draws in 4 L of air. However, this situation is less than ideal, as patients do not inhale at a constant flow rate through the device, but rather with an inspiratory profile that rises rapidly to a peak inspiratory flow rate and then gradually decreases (Kamin *et al.*, 2002). Improvements in standardised testing are therefore still required for DPIs.

10.7.2 Aerodynamic Particle Size Determination: Inertial Impaction Analysis

Inertial impaction is the separation into aerodynamic size fractions of particles on the basis of their differing inertias in a moving airstream. Particles resist changes in the direction of travel through their momentum (Figure 10.9). A particle with a large aerodynamic diameter has greater momentum and inertia than a smaller particle and thus greater probability of being deposited on the impaction plate. Increasing the airflow velocity increases the momentum of particles travelling within the airstream, thereby promoting particulate deposition. Thus, the diameter of the nozzle width at a given flow rate through the impactor determines the aerodynamic sizes of particles that deposit on the impactor stage. Particles

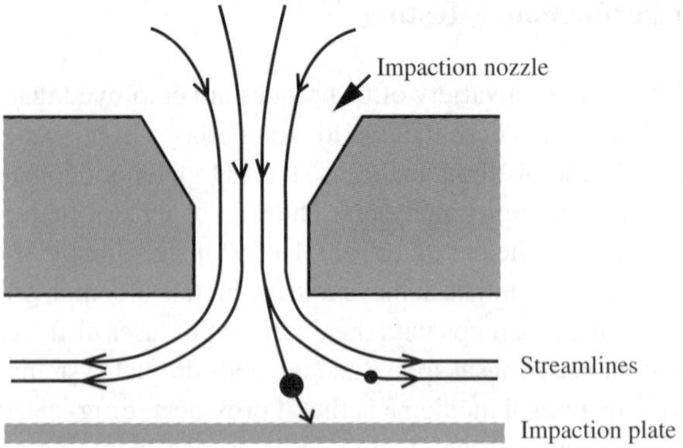

(a)

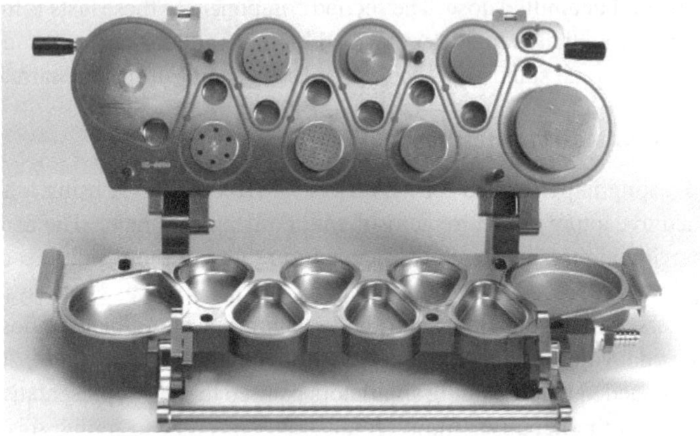

(b)

Figure 10.9 *Schematic of (a) a single nozzle of an inertial impactor stage and (b) the next-generation pharmaceutical impactor (NGI). See plate section for colour version.*

that have been aerosolised are size-fractionated to yield an aerodynamic particle size distribution of the emitted dose, and the fraction of the particles that can be respired is then determined; that is, the fine particle fraction/mass. Most modern pharmaceutical products are tested using the next-generation pharmaceutical impactor (NGI). The NGI was developed by an international consortium with the aim of producing an impactor suitable for high-throughput flow rates up to 100 L min^{-1}, used for the analysis of DPIs, and with a minimum of five stages, used to characterise particles in the size range 0.5–5.0 μm. The result is an impactor that is highly engineered and has been exhaustively validated as being suitable for inhalation testing. Impaction testing provides information on the nature of formulation interactions, the influence of the device on aerosolisation and ultimately the suitability of a product for use. The mass deposition on a given impaction stage is determined following recovery and chemical assay and is converted into a cumulative mass undersize particle size distribution on a logarithmic-probability basis (O'Shaughnessy and Raabe, 2003). Impaction testing is limited by the fact that deposition *in vivo* also depends on sedimentation and diffusion methods, rather than solely on impaction (de Boer *et al.*, 2002).

10.8 Conclusion and Outlook

This chapter aimed to discuss the applications of aerosol science to pharmaceutical therapies, and in particular to examine the methods of drug administration from the most commonly used dosage forms. There are many issues that we have deliberately not addressed, such as the pharmacological, pharmacokinetic and pharmacodynamic aspects of inhalation therapy. The reader will have noted, no doubt, an overlap between this and other chapters, including, for example, the physicochemical properties of aerosols, lung deposition of particles and exposure to environmental aerosols particles, while certain aspects of aerosol measurement have been introduced. Pharmaceutical aerosol scientists meet certain difficulties that do not exist in other fields of aerosol research, in that the major task is to turn what is, by definition, a disperse system into a compact, portable product. Therefore, much of the discussion has focussed on the different formulation approaches used to manufacture products that generate aerosols on demand in a way that is reproducible for easy patient use.

The difficulties in choosing the correct drug-delivery platform for a patient are evident, and there is no truly universal inhaler. Part of the difficulty for prescribers and patients alike is the linking of therapeutic agents to a proprietary device platform, rather than a truly addressing the needs of the patients. The progress of research into pulmonary drug delivery has been immense over the last 20 years, providing fundamental evidence of the factors determining formulation performance, the means of optimising regional drug deposition using aerosols and the mechanisms of drug absorption from and activity in the lungs. However, while some have notably been solved, many well-known problems still face pulmonary drug delivery in the second decade of the twenty-first century, including excessive throat deposition and interpatient differences in aerosol therapy performance. Many of the advances made in pharmaceutical aerosol therapy have emerged from the development of products for systemic drug delivery of biopharmaceuticals. However, the field is now in a period of consolidation, with the lessons from systemic drug delivery being applied to improving the therapy of respiratory diseases. The development of aerosol systems that achieve control over the physicochemical properties affecting drug deposition and the kinetics of pharmacological and toxicological effects remains a high priority for aerosol medicine. With the incidence of respiratory diseases increasing, there is no doubt that a concomitant development of aerosol medicines is required, which should provide sufficient impetus to continue pharmaceutical technological advances in the field of pulmonary drug delivery.

References

Acerbi, D., Brambilla, G. and Kottakis, I. (2007) Advances in asthma and COPD management: delivering CFC-free inhaled therapy using modulite® technology. *Pulmonary Pharmacology and Therapeutics*, **20**(3), 290–303.

Adcock, I.M., Caramori, G. and Barnes, P.J. (2011) Chronic obstructive pulmonary disease and lung cancer: new molecular insights. *Respiration*, **81**(4), 265–284.

Akbarieh, M. and Tawashi, R. (1987) Morphic features of solid particles after micronization in the fluid energy mill. *International Journal of Pharmaceutics*, **35**(1–2), 81–89.

Alpar, H.O., Somavarapu, S., Atuah, K.N. and Bramwell, V. (2005) Biodegradable mucoadhesive particulates for nasal and pulmonary antigen and DNA delivery. *Advanced Drug Delivery Reviews*, **57**(3), 411–430.

Anderson, J.M. and Vanitallie, C.M. (1995) Tight junctions and the molecular-basis for regulation of paracellular permeability. *American Journal of Physiology – Gastrointestinal and Liver Physiology*, **269**(4), G467–G475.

Asthma UK (2004) *Where Do We Stand? Asthma in the UK Today*, Asthma UK, London.

Azarmi, S., Loebenberg, R., Roa, W.H. *et al.* (2008) Formulation and in vivo evaluation of effervescent inhalable carrier particles for pulmonary delivery of nanoparticles. *Drug Development and Industrial Pharmacy*, **34**(9), 943–947.

Bailey, A.G. (1997) The inhalation and deposition of charged particles within the human lung. *Journal of Electrostatics*, **42**(1–2), 25–32.

Bailey, A.G., Hashish, A.H. and Williams, T.J. (1998) Drug delivery by inhalation of charged particles. *Journal of Electrostatics*, **44**(1–2), 3–10.

Barnes, P.J. (2004) Asthma guidelines: recommendations versus reality. *Respiratory Medicine*, **98**(Suppl. A), S1–S7.

Barry, P.W. and O'Callaghan, C. (1998) Drug output from nebulizers is dependent on the method of measurement. *European Respiratory Journal*, **12**(2), 463–466.

Barry, P.W. and O'Callaghan, C. (1999) A comparative analysis of the particle size output of beclomethasone diproprionate, salmeterol xinafoate and fluticasone propionate metered dose inhalers used with the Babyhaler, Volumatic and Aerochamber spacer devices. *British Journal of Clinical Pharmacology*, **47**(4), 357–360.

Bauer-Brandl, A. (1996) Polymorphic transitions of cimetidine during manufacture of solid dosage forms. *International Journal of Pharmaceutics*, **140**(2), 195–206.

Beach, S., Latham, D., Sidgwick, C. *et al.* (1999) Control of the physical form of salmeterol xinofoate. *Organic Process Research and Development*, **3**(5), 370–376.

Bennett, W.D., Brown, J.S., Zeman, K.L. *et al.* (2002) Targeting delivery of aerosols to different lung regions. *Journal of Aerosol Medicine*, **15**(2), 179–188.

Bérard, V., Lesniewska, E., Andrès, C. *et al.* (2002) Affinity scale between a carrier and a drug in DPI studied by atomic force microscopy. *International Journal of Pharmaceutics*, **247**(1–2), 127–137.

Berg, E., Svensson, J.O. and Asking, L. (2007) Determination of nebulizer droplet size distribution: a method based on impactor refrigeration. *Journal of Aerosol Medicine: Deposition Clearance and Effects in the Lung*, **20**(2), 104.

Berry, J., Heimbecher, S., Hart, J.L. and Sequeira, J. (2003) Influence of the metering chamber volume and actuator design on the aerodynamic particle size of a metered dose inhaler. *Drug Development and Industrial Pharmacy*, **29**(8), 865–876.

Berry, J., Kline, L., Naini, V. *et al.* (2004) Influence of the valve lubricant on the aerodynamic particle size of a metered dose inhaler. *Drug Development and Industrial Pharmacy*, **30**(3), 267–275.

Bhavna , Ahmad, F.J., Mittal, G. *et al.* (2009) Nano-salbutamol dry powder inhalation: a new approach for treating broncho-constrictive conditions. *European Journal of Pharmaceutics and Biopharmaceutics*, **71**(2), 282–291.

Bisgaard, H., Anhoj, J., Klug, B. and Berg, E. (1995) A nonelectrostatic spacer for aerosol delivery. *Archives of Disease in Childhood*, **73**(3), 226–230.

Bisgaard, H., Klug, B., Sumby, B.S. and Burnell, P.K.P. (1998) Fine particle mass from the Diskus inhaler and Turbuhaler inhaler in children with asthma. *European Respiratory Journal*, **11**(5), 1111–1115.

de Boer, A.H., Gjaltema, D., Hagedoorn, P. and Frijlink, H.W. (2002) Characterization of inhalation aerosols: a critical evaluation of cascade impactor analysis and laser diffraction technique. *International Journal of Pharmaceutics*, **249**(1–2), 219–231.

de Boer, A.H., Hagedoorn, P., Gjaltema, D. *et al.* (2006) Air classifier technology (ACT) in dry powder inhalation: part 3. Design and development of an air classifier family for the Novolizer® multi-dose dry powder inhaler. *International Journal of Pharmaceutics*, **310**(1–2), 72–80.

Bosquillon, C. (2010) Drug transporters in the lung-Do they play a role in the biopharmaceutics of inhaled drugs? *Journal of Pharmaceutical Sciences*, **99**(5), 2240–2255.

Brain, J.D. (1988) Lung macrophages: How many kinds are there? What do they do. *American Review of Respiratory Disease*, **137**(3), 507–509.

Brambilla, G., Ganderton, D., Garzia, R. *et al.* (1999) Modulation of aerosol clouds produced by pressurised inhalation aerosols. *International Journal of Pharmaceutics*, **186**(1), 53–61.

Brand, P., Haussinger, K., Meyer, T. *et al.* (1999) Intrapulmonary distribution of deposited particles. *Journal of Aerosol Medicine: Deposition Clearance and Effects in the Lung*, **12**(4), 275–284.

Brand, P., Hederer, B., Austen, G. *et al.* (2008) Higher lung deposition with Respimat Soft Mist inhaler than HFA-MDI in COPD patients with poor technique. *International Journal of Chronic Obstructive Pulmonary Disease*, **3**(4), 763–770.

Brocklebank, D., Ram, F. and Wright, J. (2000) Differences in inhaler technique between inhaler devices: a systematic review. *Thorax*, **55**, A63–A63.

Brodie, T. and Adalat, S. (2006) Unilateral fixed dilated pupil in a well child. *Archives of Disease in Childhood*, **91**(12), 961.

Brodka-Pfeiffer, K., Langguth, P., Gra, P. and Häusler, H. (2003) Influence of mechanical activation on the physical stability of salbutamol sulphate. *European Journal of Pharmaceutics and Biopharmaceutics*, **56**(3), 393–400.

Broeders, M.E.A.C., Molema, J., Hop, W.C.J. and Folgering, H.T.M. (2003) Inhalation profiles in asthmatics and COPD patients: reproducibility and effect of instruction. *Journal of Aerosol Medicine*, **16**(2), 131–141.

Burgel, P.R. (2011) The role of small airways in obstructive airway diseases. *European Respiratory Review: An Official Journal of the European Respiratory Society*, **20**(119), 23–33.

Burnell, P.K.P., Small, T., Doig, S. *et al.* (2001) Ex-vivo product performance of Diskus (TM) and Turbuhaler (TM) inhalers using inhalation profiles from patients with severe chronic obstructive pulmonary disease. *Respiratory Medicine*, **95**(5), 324–330.

Camenisch, G., Alsenz, J., van de Waterbeemd, H. and Folkers, G. (1998) Estimation of permeability by passive diffusion through Caco-2 cell monolayers using the drugs' lipophilicity and molecular weight. *European Journal of Pharmaceutical Sciences*, **6**(4), 325–333.

Carvalho, T.C., Peters, J.I. and Williams, R.O. (2011) Influence of particle size on regional lung deposition – what evidence is there? *International Journal of Pharmaceutics*, **406**(1–2), 1–10.

Cassidy, J.P., Amin, N., Marino, M. *et al.* (2011) Insulin lung deposition and clearance following technosphere(A (R)) insulin inhalation powder administration. *Pharmaceutical Research*, **28**(9), 2157–2164.

Cegla, U.H.U. (2004) Pressure and inspiratory flow characteristics of dry powder inhalers. *Respiratory Medicine*, **98**(Suppl. 1), S22–S28.

Ceriana, P., Navalesi, P., Rampulla, C. *et al.* (2003) Use of bronchodilators during non-invasive mechanical ventilation. *Monaldi Archives for Chest Disease*, **59**(2), 123–127.

Chawla, A., Taylor, K.M.G., Newton, J.M. and Johnson, M.C.R. (1994) Production of spray dried salbutamol sulphate for use in dry powder aerosol formulation. *International Journal of Pharmaceutics*, **108**(3), 233–240.

Chieng, N., Zujovic, Z., Bowmaker, G. *et al.* (2006) Effect of milling conditions on the solid-state conversion of ranitidine hydrochloride form 1. *International Journal of Pharmaceutics*, **327**(1–2), 36–44.

Chikhalia, V., Forbes, R.T., Storey, R.A. and Ticehurst, M. (2006) The effect of crystal morphology and mill type on milling induced crystal disorder. *European Journal of Pharmaceutical Sciences*, **27**(1), 19–26.

Chow, A.H.L., Tong, H.H.Y., Chattopadhyay, P. and Shekunov, B.Y. (2007) Particle engineering for pulmonary drug delivery. *Pharmaceutical Research*, **24**(3), 411–437.

Clark, A.R., Chambers, C.B., Muir, D. *et al.* (2007) The effect of biphasic inhalation profiles on the deposition and clearance of coarse (6.5 μm) bolus aerosols. *Journal of Aerosol Medicine: Deposition Clearance and Effects in the Lung*, **20**(1), 75–82.

Clark, D.J. and Lipworth, B.J. (1996) Effect of multiple actuations, delayed inhalation and antistatic treatment on the lung bioavailability of salbutamol via a spacer device. *Thorax*, **51**(10), 981–984.

Cooper, C.B. (2006) The connection between chronic obstructive pulmonary disease symptoms and hyperinflation and its impact on exercise and function. *American Journal of Medicine*, **119**, S21–S31.

Cooper, A., Potter, T. and Luker, T. (2010) Prediction of efficacious inhalation lung doses via the use of in silico lung retention quantitative structure-activity relationship models and in vitro potency screens. *Drug Metabolism and Disposition*, **38**(12), 2218–2225.

Corrigan, C.J., Levy, M.L., Dekhuijzen, P.R. *et al.* (2009) The ADMIT series – issues in inhalation therapy. 3) Mild persistent asthma: the case for inhaled corticosteroid therapy. *Primary Care Respiratory Journal: Journal of the General Practice Airways Group*, **18**(3), 148–158.

Crampton, M., Kinnersley, R. and Ayres, J. (2004) Sub-micrometer particle production by pressurized metered dose inhalers. *Journal of Aerosol Medicine*, **17**(1), 33–42.

Crompton, G. (2006) A brief history of inhaled asthma therapy over the last fifty years. *Primary Care Respiratory Journal: Journal of the General Practice Airways Group*, **15**(6), 326–331.

Crosland, B.M., Johnson, M.R. and Matida, E.A. (2009) Characterization of the spray velocities from a pressurized metered-dose inhaler. *Journal of Aerosol Medicine and Pulmonary Drug Delivery*, **22**(2), 1–13.

D'Addio, S.M. and Prud'homme, R.K. (2011) Controlling drug nanoparticle formation by rapid precipitation. *Advanced Drug Delivery Reviews*, **63**(6), 417–426.

Dalby, R. and Suman, J. (2003) Inhalation therapy: technological milestones in asthma treatment. *Advanced Drug Delivery Reviews*, **55**(7), 779–791.

Davies, J.C., Alton, E.W. and Bush, A. (2007) Cystic fibrosis. *British Medical Journal*, **335**(7632), 1255–1259.

Davies, M., Brindley, A., Chen, X.Y. *et al.* (2005) Characterization of drug particle surface energetics and Young's modulus by atomic force microscopy and inverse gas chromatography. *Pharmaceutical Research*, **22**(7), 1158–1166.

Davies, J.C., Davies, G., Gill, D. *et al.* (2011) Safety and expression of a single dose of lipid-mediated cftr gene therapy to the upper and lower airways of patients with CF. *Pediatric Pulmonology*, **281**, 8755–6863.

Denker, B.M. and Nigam, S.K. (1998) Molecular structure and assembly of the tight junction. *American Journal of Physiology: Renal Physiology*, **274**(1), F1–F9.

Deshpande, D., Blanchard, J., Srinivasan, S. *et al.* (2002) Aerosolization of lipoplexes using AERx (R) pulmonary delivery System. *AAPS PharmSci*, **4**(3), 271–278.

Dhand, R. and Sohal, H. (2008) Pulmonary drug delivery system for inhalation therapy in mechanically ventilated patients. *Expert Review of Medical Devices*, **5**(1), 9–18.

Dickhoff, B.H.J., de Boer, A.H., Lambregts, D. and Frijlink, H.W. (2003) The effect of carrier surface and bulk properties on drug particle detachment from crystalline lactose carrier particles during inhalation, as function of carrier payload and mixing time. *European Journal of Pharmaceutics and Biopharmaceutics*, **56**(2), 291–302.

Donnelly, L.E. and Rogers, D.F. (2008) Novel targets and drugs in inflammatory lung disease. *Current Opinion in Pharmacology*, **8**(3), 219–221.

Dubus, J.C. and Dolovich, M. (2000) Emitted doses of salbutamol pressurized metered-dose inhaler from five different plastic spacer devices. *Fundamental and Clinical Pharmacology*, **14**(3), 219–224.

Duddu, S.P., Sisk, S.A., Walter, Y.H. *et al.* (2002) Improved lung delivery from a passive dry powder inhaler using an Engineered PulmoSphere powder. *Pharmaceutical Research*, **19**, 689–695.

Dunbar, C.A. (1997) Atomization mechanisms of the pressurized metered dose inhaler. *Particulate Science and Technology*, **15**(3–4), 253–271.

Dunbar, C.A. and Miller, J.F. (1997) Application of laser diagnostic techniques to characterize the spray issued from a pMDI. *International Journal of Fluid Mechanics Research*, **24**(4–6), 568–577.

Dunbar, C.A., Watkins, A.P. and Miller, J.F. (1997a) An experimental investigation of the spray issued from a pMDI using laser diagnostic techniques. *Journal of Aerosol Medicine: Deposition Clearance and Effects in the Lung*, **10**(4), 351–368.

Dunbar, C.A., Watkins, A.P. and Miller, J.F. (1997b) Theoretical investigation of the spray from a pressurized metered-dose inhaler. *Atomization and Sprays*, **7**(4), 417–436.

Edwards, D.A. (2002) Delivery of biological agents by aerosols. *AIChE Journal*, **48**(1), 2–6.

Edwards, D.A. and Dunbar, C. (2002) Bioengineering of therapeutic aerosols. *Annual Review of Biomedical Engineering*, **4**, 93–107.

Elamin, A.A., Ahlneck, C., Alderborn, G. and Nystrom, C. (1994) Increased metastable solubility of milled griseofulvin, depending on the formation of a disordered surface structure. *International Journal of Pharmaceutics*, **111**(2), 159–170.

Eschenbacher, W.L., Boushey, H.A. and Sheppard, D. (1984) Alteration in osmolarity of inhaled aerosols cause bronchoconstriction and cough, but absence of a permeant anion causes cough alone. *American Review of Respiratory Disease*, **129**(2), 211–215.

Ferron, G.A. (1994) Aerosol properties and lung deposition. *The European Respiratory Journal*, **7**(8), 1392–1394.

Flume, P.A., O'Sullivan, B.P., Robinson, K.A. *et al.* (2007) Cystic fibrosis pulmonary guidelines – Chronic medications for maintenance of lung health. *American Journal of Respiratory and Critical Care Medicine*, **176**, 957–969.

Forbes, B. (2000) Human airway epithelial cell lines for in vitro drug transport and metabolism studies. *Pharmaceutical Science and Technology Today*, **3**(1), 18–27.

Forbes, B. and Ehrhardt, C. (2005) Human respiratory epithelial cell culture for drug delivery applications. *European Journal of Pharmaceutics and Biopharmaceutics*, **60**(2), 193–205.

Foster, N., Mammucari, R., Dehghani, F. *et al.* (2003) Processing pharmaceutical compounds using dense gas technology. *Industrial and Engineering Chemistry Research*, **42**(25), 6476–6493.

Frijlink, H.W. and deBoer, A.H. (2004) Dry powder inhalers for pulmonary delivery. *Expert Opinion Drug Carrier Systems*, **1**(1), 67–86.

Fukunaka, T., Sawaguchi, K., Golman, B. and Shinohara, K. (2005) Effect of particle shape of active pharmaceutical ingredients prepared by fluidized-red jet-milling on cohesiveness. *Journal of Pharmaceutical Sciences*, **94**(5), 1004–1012.

Gaisford, S., Dennison, M., Tawfik, M. and Jones, M.D. (2010) Following mechanical activation of salbutamol sulphate during ball-milling with isothermal calorimetry. *International Journal of Pharmaceutics*, **393**(1–2), 75–79.

Gartlehner, G., Hansen, R.A., Carson, S.S. and Lohr, K.N. (2006) Efficacy and safety of inhaled corticosteroids in patients with COPD: a systematic review and meta-analysis of health outcomes. *Annals of Family Medicine*, **4**(3), 253–262.

Geller, D.E., Weers, J. and Heuerding, S. (2011) Development of an inhaled dry-powder formulation of tobramycin using PulmoSphere (TM) technology. *Journal of Aerosol Medicine and Pulmonary Drug Delivery*, **24**(4), 175–182.

Ghazanfari, T., Elhissi, A.M.A., Ding, Z. and Taylor, K.M.G. (2007) The influence of fluid physicochemical properties on vibrating-mesh nebulization. *International Journal of Pharmaceutics*, **339**(1–2), 103–111.

GINA (Global Initiative for Asthma) (2011) Global Strategy for Asthma Management and Prevention, available from http://www.ginasthma.org/GINA-Report,-Global-Strategy-for-Asthma-Management-and-Prevention (last accessed August 14, 2013).

Gonda, I. (1985) Development of a systematic theory of suspension inhalation aerosols. I. A framework to study the effects of aggregation on the aerodynamic behaviour of drug particles. *International Journal of Pharmaceutics*, **27**(1), 99–116.

Gonda, I. (1996) Inhalation therapy with recombinant human deoxyribonuclease I. *Advanced Drug Delivery Reviews*, **19**(1), 1–110.

Griesenbach, U. and Alton, E.W.F.W. (2009) Gene transfer to the lung: lessons learned from more than 2 decades of CF gene therapy. *Advanced Drug Delivery Reviews*, **61**(2), 128–139.

Griesenbach, U., McLachlan, G., Collie, D. *et al.* (2011) Toxicology studies in support of the UK cf gene therapy consortium's multi-dose clinical trial. *Pediatric Pulmonology*, **46**, 298–298.

Hak-Kim, C. and Gonda, I. (1988) Development of a systematic theory of suspension inhalation aerosols. II. Aggregates of monodisperse particles nebulized in polydisperse droplets. *International Journal of Pharmaceutics*, **41**(1–2), 147–157.

Hall, G.L., Annese, T., Looi, K. and Devadason, S.G. (2011) Usage of spacers in respiratory laboratories and the delivered salbutamol dose of spacers available in Australia and New Zealand. *Respirology*, **16**(4), 639–644.

Hanania, N.A. and Donohue, J.F. (2007) Pharmacologic interventions in chronic obstructive pulmonary disease: bronchodilators. *Proceedings of the American Thoracic Society*, **4**(7), 526–534.

Hess, D.R. (2008) Aerosol delivery devices in the treatment of asthma. *Respiratory Care*, **53**(6).

Heyder, J. (2004) Deposition of inhaled particles in the human respiratory tract and consequences for regional targeting in respiratory drug delivery. *Proceedings of the American Thoracic Society*, **1**(4), 315–320.

Heyder, J., Gebhart, J., Rudolf, G. *et al.* (1986) Deposition of particles in the human respiratory tract in the size range 0.005-15 [mu]m. *Journal of Aerosol Science*, **17**(5), 811–825.

Hiller, F.C. (1991) Health implications of hygroscopic particle growth in the human respiratory-tract. *Journal of Aerosol Medicine: Deposition Clearance and Effects in the Lung*, **4**(1), 1–23.

Hindle, M. and Byron, P.R. (1995) Dose emissions from marketed dry powder inhalers. *International Journal of Pharmaceutics*, **116**(2), 169–177.

Hirst, P.H., Bacon, R.E., Pitcairn, G.R. *et al.* (2001) A comparison of the lung deposition of budesonide from Easyhaler®, Turbuhaler® and pMDI plus spacer in asthmatic patients. *Respiratory Medicine*, **95**(9), 720–727.

Hochrainer, D., Holz, H., Kreher, C. *et al.* (2005) Comparison of the aerosol velocity and spray duration of Respimat® Soft Mist™ Inhaler and pressurized metered dose inhalers. *Journal of Aerosol Medicine: Deposition Clearance and Effects in the Lung*, **18**(3), 273–282.

Hogg, J.C. (2004) Pathophysiology of airflow limitation in chronic obstructive pulmonary disease. *Lancet*, **364**(9435), 709–721.

Hong, J.N., Hindle, M. and Byron, P.R. (2002) Control of particle size by coagulation of novel condensation aerosols in reservoir chambers. *Journal of Aerosol Medicine: Deposition Clearance and Effects in the Lung*, **15**(4), 359–368.

Hoye, J.A., Gupta, A. and Myrdal, P.B. (2008) Solubility of solid solutes in HFA-134a with a correlation to physico-chemical properties. *Journal of Pharmaceutical Sciences*, **97**(1), 198–208.

Irngartinger, M., Camuglia, V., Damm, M. *et al.* (2004) Pulmonary delivery of therapeutic peptides via dry powder inhalation: effects of micronisation and manufacturing. *European Journal of Pharmaceutics and Biopharmaceutics*, **58**(1), 7–14.

Islam, N., Stewart, P., Larson, I. and Hartley, P. (2004a) Effect of carrier size on the dispersion of salmeterol xinafoate from interactive mixtures. *Journal of Pharmaceutical Sciences*, **93**(4), 1030–1038.

Islam, N., Stewart, P., Larson, I. and Hartley, P. (2004b) Lactose surface modification by decantation: are drug-fine lactose ratios the key to better dispersion of salmeterol xinafoate from lactose-interactive mixtures? *Pharmaceutical Research*, **21**(3), 492–499.

Islam, N., Stewart, P., Larson, I. and Hartley, P. (2005) Surface roughness contribution to the adhesion force distribution of salmeterol xinafoate on lactose carriers by atomic force microscopy. *Journal of Pharmaceutical Sciences*, **94**(7), 1500–1511.

Jeffers, D.E. (2005) Relative magnitudes of the effects of electrostatic image and thermophoretic forces on particles in the respiratory tract. *Radiation Protection Dosimetry*, **113**(2), 189–194.

Jeffers, D. (2007) Modelling and analyses do not support the hypothesis that charging by power-line corona increases lung deposition of airborne particles. *Radiation Protection Dosimetry*, **123**(2), 257–261.

Jones, M.D. and Price, R. (2006) The influence of fine excipient particles on the performance of carrier-based dry powder inhalation formulations. *Pharmaceutical Research*, **23**(8), 1665–1674.

Joshi, V., Dwivedi, S. and Ward, G.H. (2002) Increase in the specific surface area of budesonide during storage postmicronization. *Pharmaceutical Research*, **19**(1), 7–12.

Kamin, W.E.S., Genz, T., Roeder, S. *et al.* (2002) Mass output and particle size distribution of glucocorticosteroids emitted from different inhalation devices depending on various inspiratory parameters. *Journal of Aerosol Medicine: Deposition Clearance and Effects in the Lung*, **15**(1), 65–73.

Keller, M. (1999) Innovations and perspectives of metered dose inhalers in pulmonary drug delivery. *International Journal of Pharmaceutics*, **186**(1), 81–90.

Kleinstreuer, C., Zhang, Z. and Donohue, J.F. (2008) Targeted drug-aerosol delivery in the human respiratory system. *Annual Review of Biomedical Engineering*, **10**, 195–220.

Kunka, R., Andrews, S., Pimazzoni, M. *et al.* (2000) Dose proportionality of fluticasone propionate from hydrofluoroalkane pressurized metered dose inhalers (pMDIs) and comparability with chlorofluorocarbon pMDIs. *Respiratory Medicine*, **94**, S10–S16.

Kwok, P.C.L. and Chan, H.-K. (2009) Electrostatics of pharmaceutical inhalation aerosols. *Journal of Pharmacy and Pharmacology*, **61**(12), 1587–1599.

Larhrib, H., Martin, G.P., Marriott, C. and Prime, D. (2003a) The influence of carrier and drug morphology on drug delivery from dry powder formulations. *International Journal of Pharmaceutics*, **257**(1–2), 283–296.

Larhrib, H., Martin, G.P., Prime, D. and Marriott, C. (2003b) Characterisation and deposition studies of engineered lactose crystals with potential for use as a carrier for aerosolised salbutamol sulfate from dry powder inhalers. *European Journal of Pharmaceutical Sciences*, **19**(4), 211–221.

Leach, C.L. (1998) Improved delivery of inhaled steroids to the large and small airways. *Respiratory Medicine*, **92**, 3–8.

Leach, C.L., Davidson, P.J., Hasselquist, B.E. and Boudreau, R.J. (2005) Influence of particle size and patient dosing technique on lung deposition of HFA-beclomethasone from a metered dose inhaler. *Journal of Aerosol Medicine: Deposition Clearance and Effects in the Lung*, **18**(4), 379–385.

Le Brun, P.P.H., de Boer, A.H., Gjaltema, D. *et al.* (1999) Inhalation of tobramycin in cystic fibrosis – Part 1: the choice of a nebulizer. *International Journal of Pharmaceutics*, **189**(2), 205–214.

Le Brun, P.P.H., de Boer, A.H., Heijerman, H.G.M. and Frijlink, H.W. (2000) A review of the technical aspects of drug nebulization. *Pharmacy World and Science*, **22**(3), 75–81.

Leversha, A.M., Campanella, S.G., Aickin, R.P. and Asher, M.I. (2000) Costs and effectiveness of spacer versus nebulizer in young children with moderate and severe acute asthma. *Journal of Pediatrics*, **136**(4), 497–502.

Li, L.L., Mathias, N.R., Heran, C.L. *et al.* (2006) Carbopol-mediated paracellular transport enhancement in Calu-3 cell layers. *Journal of Pharmaceutical Sciences*, **95**(2), 326–335.

Li, H.Y. and Seville, P.C. (2010) Novel pMDI formulations for pulmonary delivery of proteins. *International Journal of Pharmaceutical*, **385**, 73–78.

Longest, P.W., McLeskey, J.T. Jr., and Hindle, M. (2010) Characterization of nanoaerosol size change during enhanced condensational growth. *Aerosol Science and Technology*, **44**(6), 473–483.

Lu, Q., Yang, J., Liu, Z. *et al.* and Nebulized Antibiotics Study Group (2011) Nebulized ceftazidime and amikacin in ventilator-associated pneumonia caused by pseudomonas aeruginosa. *American Journal of Respiratory and Critical Care Medicine*, **184**(1), 106–115.

Lucas, P., Anderson, K. and Staniforth, J.N. (1998) Protein deposition from dry powder inhalers: fine particle multiplets as performance modifiers. *Pharmaceutical Research*, **15**(4), 562–569.

Manford, F., Tronde, A., Jeppsson, A.B. *et al.* (2005) Drug permeability in 16HBE14o-airway cell layers correlates with absorption from the isolated perfused rat lung. *European Journal of Pharmaceutical Sciences*, **26**(5), 414–420.

Martonen, T.B. (1993) Mathematical-model for the selective deposition of inhaled pharmaceuticals. *Journal of Pharmaceutical Sciences*, **82**(12), 1191–1199.

Martonen, T.B. and Katz, I. (1993) Deposition patterns of polydisperse aerosols within human lungs. *Journal of Aerosol Medicine: Deposition Clearance and Effects in the Lung*, **6**(4), 251–274.

Mastrandrea, L.D. and Quattrin, T. (2006) Clinical evaluation of inhaled insulin. *Advanced Drug Delivery Reviews*, **58**(9–10), 1061–1075.

Mathias, N.R., Timoszyk, J., Stetsko, P.I. *et al.* (2002) Permeability characteristics of Calu-3 human bronchial epithelial cells: in vitro-in vivo correlation to predict lung absorption in rats. *Journal of Drug Targeting*, **10**(1), 31–40.

Mathias, N.R., Yamashita, F. and Lee, V.H.L. (1996) Respiratory epithelial cell culture models for evaluation of ion and drug transport. *Advanced Drug Delivery Reviews*, **22**(1–2), 215–249.

Matsukawa, Y., Lee, V.H., Crandall, E.D. and Kim, K.J. (1997) Size-dependent dextran transport across rat alveolar epithelial cell monolayers. *Journal of Pharmaceutical Sciences*, **86**(3), 305–309.

Mehendale, H.M., Angevine, L.S. and Ohmiya, Y. (1981) The isolated perfused lung – a critical evaluation. *Toxicology*, **21**(1), 1–36.

Michael, Y., Chowdhry, B.Z., Ashurst, I.C. *et al.* (2000) The physico-chemical properties of salmeterol and fluticasone propionate in different solvent environments. *International Journal of Pharmaceutics*, **200**(2), 279–288.

Midoux, N., Hosek, P., Pailleres, L. and Authelin, J.R. (1999) Micronization of pharmaceutical substances in a spiral jet mill. *Powder Technology*, **104**(2), 113–120.

Mitchell, J., Bauer, R., Lyapustina, S. *et al.* (2011) Non-impactor-based methods for sizing of aerosols emitted from orally inhaled and nasal drug products (OINDPs). *AAPS PharmSciTech*, **12**(3), 965–988.

Mitchell, J.P., Coppolo, D.P. and Nagel, M.W. (2007) Electrostatics and inhaled medications: influence on delivery via pressurized metered-dose inhalers and add-on devices. *Respiratory Care*, **52**(3), 283–300.

Mitchell, J.P., Nagel, M.W., Nichols, S. and Nerbrink, O. (2006) Laser diffractometry as a technique for the rapid assessment of aerosol particle size from inhalers. *Journal of Aerosol Medicine: Deposition Clearance and Effects in the Lung*, **19**(4), 409–433.

Mitic, L.L., van Itallie, C.M. and Anderson, J.M. (2000) Molecular physiology and pathophysiology of tight junctions – I. Tight junction structure and function: lessons from mutant animals and proteins. *American Journal of Physiology: Gastrointestinal and Liver Physiology*, **279**(2), G250–G254.

Mitsakou, C., Helmis, C. and Housiadas, C. (2005) Eulerian modelling of lung deposition with sectional representation of aerosol dynamics. *Journal of Aerosol Science*, **36**(1), 75–94.

Murnane, D., Martin, G.P. and Marriott, C. (2008a) Developing an environmentally benign process for the production of microparticles: amphiphilic crystallization. *European Journal of Pharmaceutics and Biopharmaceutics*, **69**(1), 72–82.

Murnane, D., Martin, G.P. and Marriott, C. (2008b) Investigations into the formulation of metered dose inhalers of salmeterol xinafoate and fluticasone propionate microcrystals. *Pharmaceutical Research*, **25**(10), 2283–2291.

Nair, A., Clearie, K., Menzies, D. *et al.* (2009) A novel breath-actuated integrated vortex spacer device increases relative lung bioavailability of fluticasone/salmeterol in combination. *Pulmonary Pharmacology and Therapeutics*, **22**(4), 305–310.

National Collaborating Centre for Chronic Conditions (2004) Managing stable COPD. *Thorax*, **59**(Suppl. 1), 139–130.

Newman, S.P., Brown, J., Steed, K.P. *et al.* (1998) Lung deposition of fenoterol and flunisolide delivered using a novel device for inhaled medicines – comparison of RESPIMAT with conventional metered-dose inhalers with and without spacer devices. *Chest*, **113**(4), 957–963.

Newman, S.P. and Chan, H.-K. (2008) In vitro/in vivo comparisons in pulmonary drug delivery. *Journal of Aerosol Medicine and Pulmonary Drug Delivery*, **21**(1), 77–84.

Newman, S. and Gee-Turner, A. (2005) The Omron MicroAir vibrating mesh technology nebuliser, a 21st century approach to inhalation therapy. *Journal of Therapeutic Research*, **5**(4), 29–33.

Newman, S.P. and Newhouse, M.T. (1996) Effect of add-on devices for aerosol drug delivery: deposition studies and clinical aspects. *Journal of Aerosol Medicine: Deposition Clearance and Effects in the Lung*, **9**(1), 55–70.

Nikander, K., Denyer, J., Smith, N. and Wollmer, P. (2001) Breathing patterns and aerosol delivery: impact of regular human patterns, and sine and square waveforms on rate of delivery. *Journal of Aerosol Medicine: Deposition Clearance and Effects in the Lung*, **14**(3), 327–333.

Nikander, K., Turpeinen, M. and Wollmer, P. (1999) The conventional ultrasonic nebulizer proved inefficient in nebulizing a suspension. *Journal of Aerosol Medicine: Deposition Clearance and Effects in the Lung*, **12**(2), 47–53.

Noakes, T. (2002) Medical aerosol propellants. *Journal of Fluorine Chemistry*, **118**(1–2), 35–45.

O'Callaghan, C., Lynch, J., Cant, M. and Robertson, C. (1993) Improvement in sodium cromoglycate delivery from a spacer device by use of an antistatic lining, immediate inhalation, and avoiding multiple actuations of drug. *Thorax*, **48**(6), 603–606.

O'Donnell, D.E., Lam, M. and Webb, K.A. (1999) Spirometric correlates of improvement in exercise performance after anticholinergic therapy in chronic obstructive pulmonary disease. *American Journal of Respiratory and Critical Care Medicine*, **160**(2), 542–549.

O'Leary, M., Balachandran, W., Rogueda, P. and Chambers, F. (2008) The bipolar nature of charge resident on supposedly unipolar aerosols. *Journal of Physics: Conference Series*, **1422**, 012022.

O'Shaughnessy, P.T. and Raabe, O.G. (2003) A comparison of cascade impactor data reduction methods. *Aerosol Science and Technology*, **37**(2), 187–200.

Oversteegen, L. (2008) Inhalation medicines: product differentiation by device. *Innovations in Pharmaceutical Technology*, **26**, 62–65.

Patterson, J.E., James, M.B., Forster, A.H. *et al.* (2005) The influence of thermal and mechanical preparative techniques on the amorphous state of four poorly soluble compounds. *Journal of Pharmaceutical Sciences*, **94**(9), 1998–2012.

Patton, J.S. (1996) Mechanisms of macromolecule absorption by the lungs. *Advanced Drug Delivery Reviews*, **19**(1), 3–36.

Patton, J.S., Bukar, J.G. and Eldon, M.A. (2004) Clinical pharmacokinetics and pharmacodynamics of inhaled insulin. *Clin Pharmacokinet*, **43**(12), 781–801.

Patton, J.S., Fishburn, C.S. and Weers, J.G. (2004) The lungs as a portal of entry for systemic drug delivery. *Proceedings of the American Thoracic Society*, **1**(4), 338–344.

Pauwels, R., Newman, S. and Borgstrom, L. (1997) Airway deposition and airway effects of antiasthma drugs delivered from metered-dose inhalers. *European Respiratory Journal*, **10**(9), 2127–2138.

Peguin, R.O.P.S. and da Rocha, S.R.P. (2008) Solvent-solute interactions in hydrofluoroalkane propellants. *Journal of Physical Chemistry B*, **112**(27), 8084–8094.

Perkins, M.C., Bunker, M., James, J. *et al.* (2009) Towards the understanding and prediction of material changes during micronisation using atomic force microscopy. *European Journal of Pharmaceutical Sciences*, **38**(1), 1–8.

Peyron, I.D., Britto, I.L., Benissan, L.B. and Tardieu, B.Z. (2005) Development and performance of a new hydrofluoroalkane (HFA 134a)-based metered dose inhaler (MDI) of salmeterol. *Respiratory Medicine*, **99**, S20–S30.

Pierart, F., Wildhaber, J.H., Vrancken, I. *et al.* (1999) Washing plastic spacers in household detergent reduces electrostatic charge and greatly improves delivery. *European Respiratory Journal*, **13**(3), 673–678.

Price, R. and Young, P.M. (2004) Visualization of the crystallization of lactose from the amorphous state. *Journal of Pharmaceutical Sciences*, **93**(1), 155–164.

Price, R., Young, P.M., Edge, S. and Staniforth, J.N. (2002) The influence of relative humidity on particulate interactions in carrier-based dry powder inhaler formulations. *International Journal of Pharmaceutics*, **246**(1–2), 47–59.

Pritchard, J.N. (2001) The influence of lung deposition on clinical response. *Journal of Aerosol Medicine*, **14**(Suppl. 1), S19–S26.

Pryor, J.A. (1999) Physiotherapy for airway clearance in adults. *European Respiratory Journal*, **14**(6), 1418–1424.

Rave, K., Bott, S., Heinemann, L. *et al.* (2005) Time-action profile of inhaled insulin in comparison with subcutaneously injected insulin lispro and regular human insulin. *Diabetes Care*, **28**(5), 1077–1082.

Rehman, M., Shekunov, B.Y., York, P. *et al.* (2004) Optimisation of powders for pulmonary delivery using supercritical fluid technology. *European Journal of Pharmaceutical Sciences*, **22**(1), 1–17.

Ressmeyer, A.R., Larsson, A.K., Vollmer, E. *et al.* (2006) Characterisation of guinea pig precision-cut lung slices: comparison with human tissues. *European Respiratory Journal*, **28**(3), 603–611.

Reverchon, E. (1999) Supercritical antisolvent precipitation of micro- and nano-particles. *The Journal of Supercritical Fluids*, **15**(1), 1–21.

Ross, D.L. and Gabrio, B.J. (1999) Advances in metered dose inhaler technology with the development of a chlorofluorocarbon-free drug delivery system. *Journal of Aerosol Medicine: Deposition Clearance and Effects in the Lung*, **12**(3), 151–160.

Rostami, A.A. (2009) Computational modeling of aerosol deposition in respiratory tract: a review. *Inhalation Toxicology*, **21**(4), 262–290.

Rowe, S.M., Miller, S. and Sorscher, E.J. (2005) Cystic fibrosis. *New England Journal of Medicine*, **352**(19), 1992–2001.

Ruecroft, G., Hipkiss, D., Ly, T. *et al.* (2005) Sonocrystallization: the use of ultrasound for improved industrial crystallization. *Organic Process Research and Development*, **9**(6), 923–932.

Sakagami, M. (2006) In vivo, in vitro and ex vivo models to assess pulmonary absorption and disposition of inhaled therapeutics for systemic delivery. *Advanced Drug Delivery Reviews*, **58**(9–10), 1030–1060.

Sakagami, M., Sakon, K., Kinoshita, W. and Makino, Y. (2001) Enhanced pulmonary absorption following aerosol administration of mucoadhesive powder microspheres. *Journal of Controlled Release*, **77**(1–2), 117–129.

Saleem, I., Smyth, H. and Telko, M. (2008) Prediction of dry powder inhaler formulation performance from surface energetics and blending dynamics. *Drug Development and Industrial Pharmacy*, **34**(9), 1002–1010.

Sanders, M. (2007) Inhalation therapy: an historical review. *Primary Care Respiratory Journal*, **16**(2), 71–81.

Sangwan, S., Condos, R. and Smaldone, G.C. (2003) Lung deposition and respirable mass during wet nebulization. *Journal of Aerosol Medicine: Deposition Clearance and Effects in the Lung*, **16**(4), 379–386.

Satir, P. and Sleigh, M.A. (1990) The physiology of cilia and mucociliary interactions. *Annual Review of Physiology*, **52**, 137–155.

Scheuch, G., Kohlhaeufl, M.J., Brand, P. and Siekmeler, R. (2006) Clinical perspectives on pulmonary systemic and macromolecular delivery. *Advanced Drug Delivery Reviews*, **58**(9–10), 996–1008.

Scheuch, G., Kohlhaeufl, M., Moeller, W. *et al.* (2008) Particle clearance from the airways of subjects with bronchial hyperresponsiveness and with chronic obstructive pulmonary disease. *Experimental Lung Research*, **34**(9), 531–549.

Schiavone, H., Palakodaty, S., Clark, A. *et al.* (2004) Evaluation of SCF-engineered particle-based lactose blends in passive dry powder inhalers. *International Journal of Pharmaceutics*, **281**(1–2), 55–66.

Schubert, H. (1981) Principles of agglomeration. *International Chemical Engineering*, **21**(3), 363–377.

Schultz, A., Le Souef, T.J., Venter, A. *et al.* (2010) Aerosol inhalation from spacers and valved holding chambers requires few tidal breaths for children. *Pediatrics*, **126**(6), E1493–E1498.

Schwarze, P.E., Ovrevik, J., Lag, M. *et al.* (2006) Particulate matter properties and health effects: consistency of epidemiological and toxicological studies. *Human & Experimental Toxicology*, **25**(10), 559–579.

Shekunov, B.Y. and York, P. (2000) Crystallization processes in pharmaceutical technology and drug delivery design. *Journal of Crystal Growth*, **211**(1–4), 122–136.

Smaldone, G.C. (2002) Smart nebulizers. *Respiratory Care*, **47**(12), 1434–1441.

Smaldone, G.C. (2006) Advances in aerosols: adult respiratory disease. *Journal of Aerosol Medicine: Deposition Clearance and Effects in the Lung*, **19**(1), 36–46.

Smyth, H.D.C. (2003) The influence of formulation variables on the performance of alternative propellant-driven metered dose inhalers. *Advanced Drug Delivery Reviews*, **55**(7), 807–828.

Smyth, H., Hickey, A.J., Brace, G. *et al.* (2006) Spray pattern analysis for metered dose inhalers I: orifice size, particle size, and droplet motion correlations. *Drug Development and Industrial Pharmacy*, **32**(9), 1033–1041.

Sporty, J.L., Horalkova, L. and Ehrhardt, C. (2008) In vitro cell culture models for the assessment of pulmonary drug disposition. *Expert Opinion on Drug Metabolism and Toxicology*, **4**(4), 333–345.

Srichana, T., Martin, G.P. and Marriott, C. (1998) Dry powder inhalers: the influence of device resistance and powder formulation on drug and lactose deposition in vitro. *European Journal of Pharmaceutical Sciences*, **7**(1), 73–80.

Stahlhofen, W. (1980) Experimental-determination of the regional deposition of aerosol-particles in the human respiratory-tract. *American Industrial Hygiene Association Journal*, **41**(6), 385–398.

Stein, S.W. and Myrdal, P.B. (2004) Theoretical and experimental analysis of formulation and device parameters affecting solution MD1 size distributions. *Journal of Pharmaceutical Sciences*, **93**(8), 2158–2175.

Stein, S.W. and Myrdal, P.B. (2006) The relative influence of atomization and evaporation on metered dose inhaler drug delivery efficiency. *Aerosol Science and Technology*, **40**(5), 335–347.

Stressmann, F.A., Rogers, G.B., Klem, E.R. *et al.* (2011) Analysis of the bacterial communities present in lungs of patients with cystic fibrosis from American and British centers. *Journal of Clinical Microbiology*, **49**(1), 281–291.

Sturton, G., Persson, C. and Barnes, P.J. (2008) Small airways: an important but neglected target in the treatment of obstructive airway diseases. *Trends in Pharmacological Sciences*, **29**(7), 340–345.

Swift, D., Asgharian, B. and Kimble, J.S. (2007) Use of mathematical aerosol deposition models in predicting the distribution of inhaled therapeutic aerosols, in *Inhalation*

Aerosols: Physical and Biological Basis for Therapy, 2nd edn (ed A.J. Hickey), Informa Healthcare, Hoboken, NJ.

Taki, M., Marriott, C., Zeng, X.-M. and Martin, G.P. (2011) The production of 'aerodynamically equivalent' drug and excipient inhalable powders using a novel fractionation technique. *European Journal of Pharmaceutics and Biopharmaceutics*, **77**(2), 283–296.

Tang, P., Chan, H.K. and Raper, J.A. (2004) Prediction of aerodynamic diameter of particles with rough surfaces. *Powder Technology*, **147**(1–3), 64–78.

Tarara, T.E., Hartman, M.S., Gill, H. *et al.* (2004) Characterization of suspension-based metered dose inhaler formulations composed of spray-dried budesonide microcrystals dispersed in HFA-134a. *Pharmaceutical Research*, **21**(9), 1607–1614.

Tarsin, W.Y., Pearson, S.B., Assi, K.H. and Chrystyn, H. (2006) Emitted dose estimates from Seretide® Diskus® and Symbicort® Turbuhaler® following inhalation by severe asthmatics. *International Journal of Pharmaceutics*, **316**(1–2), 131–137.

Technology and Economic Assessment Panel (2009) Montreal Protocol on Substances that Deplete the Ozone Layer Progress Report, United Nations Environment Programme, Nairobi.

Teng, S.L., Wang, P., Zhu, L.J. *et al.* (2009) Experimental and numerical analysis of a lab-scale fluid energy mill. *Powder Technology*, **195**(1), 31–39.

Thiering, R., Dehghani, F. and Foster, N.R. (2001) Current issues relating to anti-solvent micronisation techniques and their extension to industrial scales. *The Journal of Supercritical Fluids*, **21**(2), 159–177.

Thorsson, L., Kenyon, C., Newman, S.P. and Borgstrom, L. (1998) Lung deposition of budesonide in asthmatics: a comparison of different formulations. *International Journal of Pharmaceutics*, **168**(1), 119–127.

Tong, H.H.Y., Shekunov, O.Y., York, P. and Chow, A.H.L. (2005) Surface characterization of saimeterol xinafoate powders by inverse gas chromatography at finite coverage. *Journal of Pharmaceutical Sciences*, **94**(3), 695–700.

Tronde, A., Norden, B., Marchner, H. *et al.* (2003) Pulmonary absorption rate and bioavailability of drugs in vivo in rats: structure-absorption relationships and physicochemical profiling of inhaled drugs. *Journal of Pharmaceutical Sciences*, **92**(6), 1216–1233.

Velaga, S.P., Berger, R. and Carlfors, J. (2002) Supercritical fluids crystallization of budesonide and flunisolide. *Pharmaceutical Research*, **19**(10), 1564–1571.

Verbanck, S., Vervaet, C., Schuermans, D. and Vincken, W. (2004) Aerosol profile extracted from spacers as a determinant of actual dose. *Pharmaceutical Research*, **21**(12), 2213–2218.

Versteeg, H.K., Hargrave, G.K. and Kirby, M. (2006) Internal flow and near-orifice spray visualisations of a model pharmaceutical pressurised Metered Dose Inhaler. *Journal of Physics: Conference Series*, Second International Conference on Optical and Laser Diagnostics, **45**, 207–213.

Vervaet, C. and Byron, P.R. (1999) Drug-surfactant-propellant interactions in HFA-formulations. *International Journal of Pharmaceutics*, **186**(1), 13–30.

Viegi, G., Annesi, I. and Matteelli, G. (2003) Epidemiology of asthma. *European Respiratory Monograph*, **8**(23), 1–25.

Ward, G.H. and Schultz, R.K. (1995) Process-induced crystallinity changes in albuterol sulfate and its effect on powder physical stability. *Pharmaceutical Research*, **12**(5), 773–779.

Weibel, E.R. (1963) *Morphometry of the Human Lung*, Springer-Verlag, Berlin.

Weiler, C., Wolkenhauer, M., Trunk, M. and Langguth, P. (2010) New model describing the total dispersion of dry powder agglomerates. *Powder Technology*, **203**(2), 248–253.

White, S., Bennett, D.B., Cheu, S. *et al.* (2005) EXUBERA: pharmaceutical development of a novel product for pulmonary delivery of insulin. *Diabetes Technology and Therapeutics*, **7**(6), 896–906.

Widdicombe, J. (1997) Airway and alveolar permeability and surface liquid thickness: theory. *Journal of Applied Physiology*, **82**(1), 3–12.

Williams, R.O. and Liu, J. (1998) Influence of formulation additives on the vapor pressure of hydrofluoroalkane propellants. *International Journal of Pharmaceutics*, **166**(1), 99–103.

Wong, W., Fletcher, D.F., Traini, D. *et al.* (2010) Particle aerosolisation and break-up in dry powder inhalers 1: evaluation and modelling of venturi effects for agglomerated systems. *Pharmaceutical Research*, **27**(7), 1367–1376.

World Health Organization (2011) The top 10 causes of death, http://www.who.int/mediacentre/factsheets/fs310/en/index2.html (last accessed 15 August 2013).

Worth Longest, P., Hindle, M. and Das Choudhuri, S. (2009) Effects of generation time on spray aerosol transport and deposition in models of the mouth-throat geometry. *Journal of Aerosol Medicine and Pulmonary Drug Delivery*, **22**(2), 99–112.

York, P., Ticehurst, M.D., Osborn, J.C. *et al.* (1998) Characterisation of the surface energetics of milled dl-propranolol hydrochloride using inverse gas chromatography and molecular modelling. *International Journal of Pharmaceutics*, **174**(1–2), 179–186.

Young, P.M., Edge, S., Traini, D. *et al.* (2005) The influence of dose on the performance of dry powder inhalation systems. *International Journal of Pharmaceutics*, **296**(1–2), 26–33.

Yu, Z.S., Johnston, K.P. and Williams, R.O. (2006) Spray freezing into liquid versus spray-freeze drying: influence of atomization on protein aggregation and biological activity. *European Journal of Pharmaceutical Sciences*, **27**(1), 9–18.

Zanen, P., Go, L.T. and Lammers, J.W.J. (1998) The efficacy of a low-dose, monodisperse parasympathicolytic aerosol compared with a standard aerosol from a metered-dose inhaler. *European Journal of Clinical Pharmacology*, **54**(1), 27–30.

Zeng, X.M., Martin, G.P. and Marriott, C. (2001) *Particulate Interactions in Dry Powder Formulations for Inhalation*, Taylor & Franics, London.

Zeng, X.M., O'Leary, D., Phelan, M. *et al.* (2002) Delivery of salbutamol and of budesonide from a novel multi-dose inhaler Airmax (TM). *Respiratory Medicine*, **96**(6), 404–411.

Zhao, Q.Q. and Schurr, G. (2002) Effect of motive gases on fine grinding in a fluid energy mill. *Powder Technology*, **122**(2–3), 129–135.

Zhou, Q.T., Qu, L., Larson, I. *et al.* (2010) Improving aerosolization of drug powders by reducing powder intrinsic cohesion via a mechanical dry coating approach. *International Journal of Pharmaceutics*, **394**(1–2), 50–59.

Zhou, S.-F., Wang, L.-L., Di, Y.M. *et al.* (2008) Substrates and inhibitors of human multidrug resistance associated proteins and the implications in drug development. *Current Medicinal Chemistry*, **15**(20), 1981–2039.

11

Bioaerosols and Hospital Infections

Ka man Lai[1]*, Zaheer Ahmad Nasir*[2]*, and Jonathon Taylor*[2]

[1]*Department of Biology, Hong Kong Baptist University, China*

[2]*Healthy Infrastructure Research Centre, Department of Civil, Environmental & Geomatic Engineering, University College London, UK*

11.1 The Importance of Bioaerosols and Infections

Increasing threats from new and emerging diseases such as severe acute respiratory syndrome (SARS), flu pandemics, rising numbers of hospital-acquired infections and drug-resistant pathogens like extensively drug-resistant tuberculosis (XDR-TB) have again raised the concern that air plays a key role in disease transmission. World Health Organization (WHO) experts have predicted that '*… at a minimum, between 2 and 7.4 million people might die in the next pandemic*'. (WHO, 2005). The international impact of such diseases was clearly shown in the SARS outbreak in 2002 and 2003. Within a few weeks, SARS had spread from Hong Kong to 26 countries and caused over 8000 cases and 700 deaths worldwide (WHO, 2004). In one study, it was reported that 40 health workers had contracted SARS over a 6-week period in a community hospital; the infection incidence was highest in healthcare assistants (8%), followed by physicians (5%) and nurses (4%) (Ho, Sung and Chan-Yeung, 2003). Hospital environments are particularly vulnerable to infections because people carrying infectious agents and those most at risk of infection come together there. Given the prodigious advances that have been in medical technology and therapies, more patients are becoming immunocompromised during their course of treatment and thus more susceptible to infection (Husain *et al*., 2003; WHO, 2012). The use of antibiotics in the past few decades has also contributed to the new generation of drug-resistant pathogens, turning traditionally easy-to-treat infections into new hospital-acquired infection threats (WHO, 2012). Moreover, modern technology

Aerosol Science: Technology and Applications, First Edition. Edited by Ian Colbeck and Mihalis Lazaridis.

and medical treatment have the potential to provide new habitats and opportunities for microorganisms to propagate, spread and infect their victims (CDC, 2003). The quality of healthcare, hospital design and facility management and hospital hygiene can also influence the risk of infection transmission in hospitals (Ducel, Fabry and Nicolle, 2002). The worldwide perspective on the burden of endemic healthcare-associated infection (HCAI) was systematically reviewed in a WHO report (WHO, 2011). In brief, the hospital-wide prevalence of HCAI (not entirely contributed by bioaerosols) ranged between 5.7 and 19.1% during the reported period, with the pooled HCAI prevalence in mixed patient populations 7.6% in high-income countries. Intensive care units (ICUs) and operating theatres represent the most high-risk environments, with the HCAI prevalence measure between 4.4 and 88.9% there, while patients using invasive devices and ventilators are a high-risk population. The burdens and costs of HCAI include prolonged hospital stay, long-term disability, increased resistance of microorganisms to antimicrobials, a massive additional financial burden for health systems, high costs for patients and their families and excess deaths. According to the same report, HCAIs caused 16 million extra days of hospital stay, 37 000 attributable deaths and annual direct financial losses of approximately €7 billion in Europe. In the United States, HCAIs contributed to approximately 99 000 deaths and caused annual losses of approximately US$6.5 billion in recent years.

Bioaerosols are an important source of infections; they are particularly relevant in the twenty-first century in the wake of various health challenges from new and emerging diseases and within hospital environments. Devising appropriate solutions to reduce the risk of infection transmission can save lives and reduce the overall economic and social burdens.

11.2 Bioaerosol-Related Infections in Hospitals

Table 11.1 lists some examples of bioaerosol-related infections in hospitals. Microorganisms capable of causing infections are known as pathogens. Pathogens can originate either in the environment or from a host (e.g. animals and humans). Those that can use the airborne route for transmission by infecting the respiratory system through inhalation are called airborne pathogens. Although some writers refer pathogens that use an airborne pathway to cause an infection not facilitated through the respiratory route as 'airborne diseases', in this chapter we use the term 'bioaerosol-related infections' to cover all airborne pathways and/or routes of disease transmission. Age-old diseases such as tuberculosis (TB) and emerging ones such as Legionnaires' disease, first diagnosed in the 1970s, are examples of bioaerosol-related infections. Pathogens that have originated from environmental sources, such as *Legionella* spp. and *Aspergillus* spp., may have a diverse range of habitats in the environment, such as soils, water reservoirs and building materials (CDC, 2003). One of the main environmental sources for hospital infections arises from the construction, demolition, renovation and repair of healthcare facilities. During these activities, pathogens become airborne whenever their habitat is disturbed. In particular, the spores of mould species, which have evolved to use airborne pathways for their dispersal, can become airborne with only a slight change in environmental conditions and air movement.

For human-to-human transmission, pathogens contained in respiratory droplets are emitted in these droplets when people talk, sing, cough, sneeze or merely breathe. Respiratory droplets vary in size. Once in the atmosphere, they rapidly evaporate to form droplet nuclei.

Table 11.1 *Examples of bioaerosol-related infections in hospitals.*

Source	Pathogens/diseases	References
People, e.g. patients, healthcare workers and visitors	*Mycobacterium tuberculosis* (TB), varicella-zoster virus (VZV), measles, influenza, rhinoviruses, adenoviruses, respiratory syncytial virus (RSV), methicillin-resistant *Staphylococcus aureus* (MRSA), *Clostridium difficile*, vancomycin-resistant *Enterococcus faecium* (VRE), norovirus	CDC (2003, 2004) and MacCannell *et al.* (2011)
Outdoor environments, e.g. natural and engineered environments outside hospitals (exact sources are normally unknown)	*Aspergillus* spp., *Candida* spp., *Hyalohyphomycetes*, *Phaeohyphomycetes*, *Zygomycetes*	Husain *et al.* (2003) and CDC (2004)
Animals within (e.g. service animals) and around (e.g. pigeons) the hospital	*Coxiella burnetti* (Q fever), tularemia, yersiniosis, blastomycosis, *Cryptococcus neoformans*, *Cryptococcus gattii*	Weber (1979) and CDC (2003)
Infections related to the construction, demolition, renovation and repair of healthcare facilities, as well as poorly maintained infrastructure, such as old and abandoned filters and ventilation systems	*Aspergillus* spp., *Mucoraceae*, *Penicillium* spp.	CDC (2003, 2004)
Water reservoir-related infections, e.g. water in engineering and building systems and medical devices and equipment; hospital-acquired pneumonia (HAP); ventilator-associated pneumonia (VAP)	*Legionella* spp., nontuberculous mycobacteria (NTM), *Pseudomonas aeruginosa*, *Acinetobacter* spp., *Enterobacter* spp.	CDC (2003, 2004) and WHO (2011)
Surgical-site infections (SSIs)	*Staphylococcus aureus*, *Staphylococcus epidermidis*, group A beta-haemolytic *Streptococci*	CDC (2003)
Aerial disseminated infections, e.g. from fomites, environmental surfaces, laundry and bedding, cleaning, clinical waste and treatment	Aerosolisation of pathogens from contaminated materials and settled particles; all pathogens in this table with human origins can be re-suspended	CDC (2003)

Nuclei that have an aerodynamic size less than 5 μm will become airborne and can linger in the air for a long time. Infections that occur through these droplet nuclei, such as TB, are referred to as having 'airborne transmission', whereas infections that occur as a result of the droplets are known as 'droplet-borne transmissions'. Influenza is widely believed to be transmitted predominantly via the droplet-borne route, although airborne transmission can also occur (CDC, 2004; WHO, 2009a). The general assumption is that droplet-borne transmission has an infection range of about 1 m, due to the sedimentation of droplets larger than 5 μm within this range. This 1 m distance is used in infection control of droplet-borne diseases (CDC, 2004; WHO, 2009a).

Droplets can also be produced through various mechanisms from water reservoirs present in the hospital environment, such as nebulisers, water baths, hydrotherapy, sinks, faucet aerators, showers, humidifiers, ventilators and flushing toilets, all of which break water into droplets (CDC, 2004). The biological composition of these droplets is associated with the microbial content in the water reservoirs. Whether a pathogen can transmit through the airborne or the droplet-borne route will depend on its survival ability in the air. For example, *Legionella* spp. can travel for long periods of time and over long distances, while close contact is generally required for the transmission of other Gram-negative pathogens such as *Pseudomonas aeruginosa* (CDC, 2004).

In addition to airborne/droplet-borne transmission from environmental and human sources, many hospital activities can generate vast amounts of infectious bioaerosols from fomites (CDC, 2003). These pathogens do not infect the respiratory system directly through the air; instead, they infect susceptible people through contamination of surfaces and subsequently of the hands, or else infect via the direct deposition of bioaerosols on wounds, as in the case of surgical-site infections (SSIs). Hospital cleaning, processing and transportation of medical waste, laundry and bed-making are common sources of bioaerosols within the hospital environment. The air plays a role in dispersing these bioaerosols and controlling their environmental pathways after they are emitted from their sources.

In this chapter, we explore bioaerosol-related infections and discuss how aerosol science and technology have contributed to hospital infection control. We assume that the reader has some knowledge in the area of aerosol science. We first briefly review bioaerosol properties and their deposition in the human respiratory system and then explain the three components (pathogens, hosts and pathways) in the chain of infection within the hospital environment, the different phases of infection control, from hospital design to demolition, and the leading disciplines involved in each phase. One of the main contributions of aerosol science and technology to infection control is an understanding of hospital aerobiology, starting with the source, aerosolisation, stabilisation, transport, deposition, physical and biological properties and decay of bioaerosols. This knowledge has opened up numerous opportunities for infection control among different disciplines in hospital design, construction and operation. The presence of vulnerable patients poses a complicated challenge to the conducting of bioaerosol experiments in hospitals, where access is difficult and sometimes restricted. Laboratory bioaerosol experiments have therefore helped researchers to further their understanding of the physicochemical and biological properties of bioaerosols, particularly under differing climatic conditions, and to research, develop and test air-cleaning devices and other technologies used in infection control and aerosol research. In conjunction with laboratory experiments, numerical modelling tools contribute to the prediction and simulation of bioaerosol dynamics and particle dispersal, and have been shown to be useful in hospital design and the optimisation of the effectiveness of infection-control measures. Various free-standing air-cleaning devices and built-in air-cleaning systems have been developed on the basis of different air-cleaning technologies, such as filtration, dilution and deactivation. Some of these technologies are now standard infection-control measures, while others are still under development.

11.3 Bioaerosol Properties and Deposition in Human Respiratory Systems

Several items in the literature have reviewed bioaerosol properties and aerosol deposition in relation to the physiology of the human respiratory system (ICRP, 1994; Hinds, 1999; Sosnowski, 2011). In short, the respiratory system acts as a filter to capture aerosols through impaction, interception, diffusion and settling mechanisms. The aerodynamic size of an aerosol is one of the main factors that determines its site and efficiency of deposition in the respiratory system. Other factors include the shape, hygroscopicity and surface properties of the aerosol, the respiratory tract geometry and the breathing pattern. In general, bioaerosols smaller than 5 μm can travel to the lung, and those larger than 5 μm are mostly filtered by the nostrils. As in other filters, the lowest particle size (diameter) filtration efficiency in the respiratory system is between 0.1 and 1.0 μm, where none of these mechanisms is dominant (Sosnowski, 2011). Bioaerosol properties are influenced by the nature of the microbes, their source and aerosolisation mechanism and climatic conditions. Droplet aerosols discharged from aqueous sources such as water reservoirs and respiratory systems are subject to evaporation and subsequent formation into droplet nuclei. The rate of evaporation and the ultimate size of these droplets are determined by relative humidity (RH) and atmospheric temperature (Hinds, 1999).

11.4 Chain of Infection and Infection Control in Hospitals

Infection requires three basic components in order to form a complete chain: *pathogens* (from infected hosts or contaminated environments, at adequate dose and with sufficient virulence); *pathways* (appropriate to the mode of transmission); and susceptible *hosts* (with the correct portal of entry into the individual) (Figure 11.1). Hospital environments can (and do) provide and impact on all of these components in order to cause infection.

The aim of infection control is to break this chain by eliminating any of the components or finding a barrier to block their passage. For example, if TB patients are isolated from the general population, the link between pathogens and susceptible individuals is severed. Increasing the supply of fresh air by opening windows can effectively dilute and remove TB aerosols and consequently reduce the level of TB exposure, even if TB patients and susceptible individuals are present in the same environment. Ideally, infection control should start from the early design and planning phase of a hospital and follow through until the end of its service. The type and number of patients, the nature of their diseases and treatments, the corresponding infection-control strategy and the infection-control features of the hospital should be decided and integrated into the initial design plan. Importantly, when change

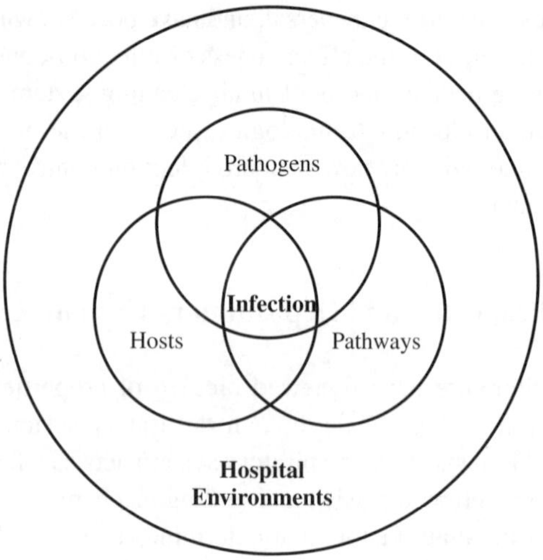

Figure 11.1 *Chain of infection in hospital environments.*

is needed over the course of the hospital's lifetime, a new design, construction, environmental and infection control plan should be assessed. Hospitals represent one of the most complex building types, due to the presence of diverse but interdependent functional units, such as wards, laboratories, isolation rooms, offices and public areas. These units can all harbour pathogens and allow them to pass through and cross-contaminate people and environments via movement of individuals, air, water, waste and other contaminated objects. They can directly or indirectly contribute to infection as an individual cause or as a series of linked events. In recent years, the concept of 'built-in' infection control in the hospital environment has been strongly advocated (Wilson and Ridgway, 2006; Ulrich, 2006). In the United Kingdom, the Department of Health is the agent responsible for the provision of guidelines on building and managing hospitals (via the Health Technical Memorandum, HTM; Department of Health, 2007a, b). In the US, the Centers for Disease Control and Prevention (CDC) produce various guidelines for infection control. After a hospital is built, facilities are commissioned, and from this point onwards, the facility management team takes charge of the maintenance and servicing of the hospital and its environment. However, with different governance and management structures in every hospital, the scope of facility management and service provision can vary widely. In the UK, there is heated debate over whether the private finance initiative (PFI) scheme in building and servicing hospitals compromises infection control as a result of poor building work, poor facilities management and high operational and maintenance costs (Lister, 2003). Facility management ensures functionality of the built environment but does not secure infection control. For example, as previously mentioned, opening windows can increase fresh air supply for TB aerosol dilution. However, a functioning window that can be opened and closed does not guarantee that adequate air dilution will be achieved. The infection-control team, comprising infection-control nurses, consultant microbiologists and clinical scientists, is responsible for day-to-day infection surveillance and control; its work is heavily directed by infection incidents and is grounded in microbiology and epidemiological approaches. From an environmental point of view, the CDC and the Healthcare Infection Control Practices Advisory Committee (HICPAC) recommend infection-control guidelines in seven areas: Air, Water, Environmental Services, Environmental Sampling, Laundry and Bedding, Animals in Health-Care Facilities and Regulated Medical Waste (CDC, 2003). Table 11.2 illustrates the leading disciplines across different hospital building and operation phases – from hospital construction to demolition – and shows the environmental infection-control components. In discussions surrounding the application of aerosol science and technology in infection control, the aim of this table is to elaborate a wide scope of aerosol science and technology applications by different disciplines and in different phases of the hospital lifetime.

11.5 Application of Aerosol Science and Technology in Infection Control

11.5.1 Understanding Hospital Aerobiology and Infection Control

Various techniques used in the sampling of aerosols and bioaerosols contribute to an understanding of their sources, aerosolisation mechanisms, environmental pathways and transmission processes, and ultimately to the development of tools and strategies for infection control. Pasquarella, Pitzurra and Savino (2000) review the advantages and disadvantages of different bioaerosol sampling techniques in hospital applications, broadly divided into active and passive sampling. Active sampling includes impactors, filtration samplers, impingers, centrifugal samplers, electrostatic precipitation samplers and thermal precipitation samplers. These all collect a known volume of air, meaning that the level of contamination is expressed as a measurement of the number of microbial agent (e.g. colony forming units (CFUs), DNA, ATP, enzymes and plaque forming units) per volume unit of air. Passive sampling requires no mechanical pumps and relies on the settling of aerosols on to collection devices. Passive sampling with settling plates is routinely used in hospitals to estimate the fall-out of airborne particles from the air, and thus the level of surface biocontamination. In monitoring the total aerosol levels (viable or nonviable, biological or nonbiological), various particle counters can be used to indicate the emission and contamination of aerosols in the environment. Passive sampling is recommended by the CDC as a tool for monitoring airborne particle control in construction areas within the hospital environment (CDC, 2003).

Figure 11.2 depicts the aerobiology dynamics from different exemplar sources and succinctly describes how human factors, built environments and environmental parameters affect bioaerosol transport, decay, dispersal and transmission. This kind of information is the foundation for the development and implementation by various disciplines, as shown in Table 11.2, of specific infection-control measures. Moreover, by understanding the processes and factors that impact on hospital aerobiology, specific indicators can be identified and applied to the monitoring of indoor air environments, such as operating theatres. In the UK, HTM 03-01 recommends that bioaerosol (bacteria and fungi) levels not exceed 10 CFU/m^3 of air when the theatre is empty and that they not exceed 180 CFU/m^3 averaged over any 5 minute period, unless there is a high level of personnel and activity in the room (Department of Health, 2007b). Dharan and Pittet (2002) report that, with the high-efficiency particulate-arresting (HEPA) filters operating in the theatre ventilation, there is a tendency to apply particle counting as an industrial standard for cleanroom technology

Table 11.2 *Leading disciplines in environmental infection control across hospital building and operation phases.*

Building and operation phases/ environmental components	Planning and design	Construction	Maintenance	Operation and surveillance	Infection-control response and mitigation	Renovation and demolition
Air: ventilation, air conditioning, air distribution, airflow direction, air quality Water: water supply, water use, water storage and maintenance, water quality Surfaces: building materials, high-touch areas, cleaning effectiveness, porosity, biological and moisture absorption and retention properties Environmental services: cleaning, hygiene, environmental management, laundry and bedding, clinical waste (treatment and disposal)	Architects, planners and designers, engineers	Construction contractors, engineers	Facilities managers	Infection-control team, healthcare workers, hospital epidemiologists	Infection-control team, healthcare workers	Architects, construction contractors, engineers

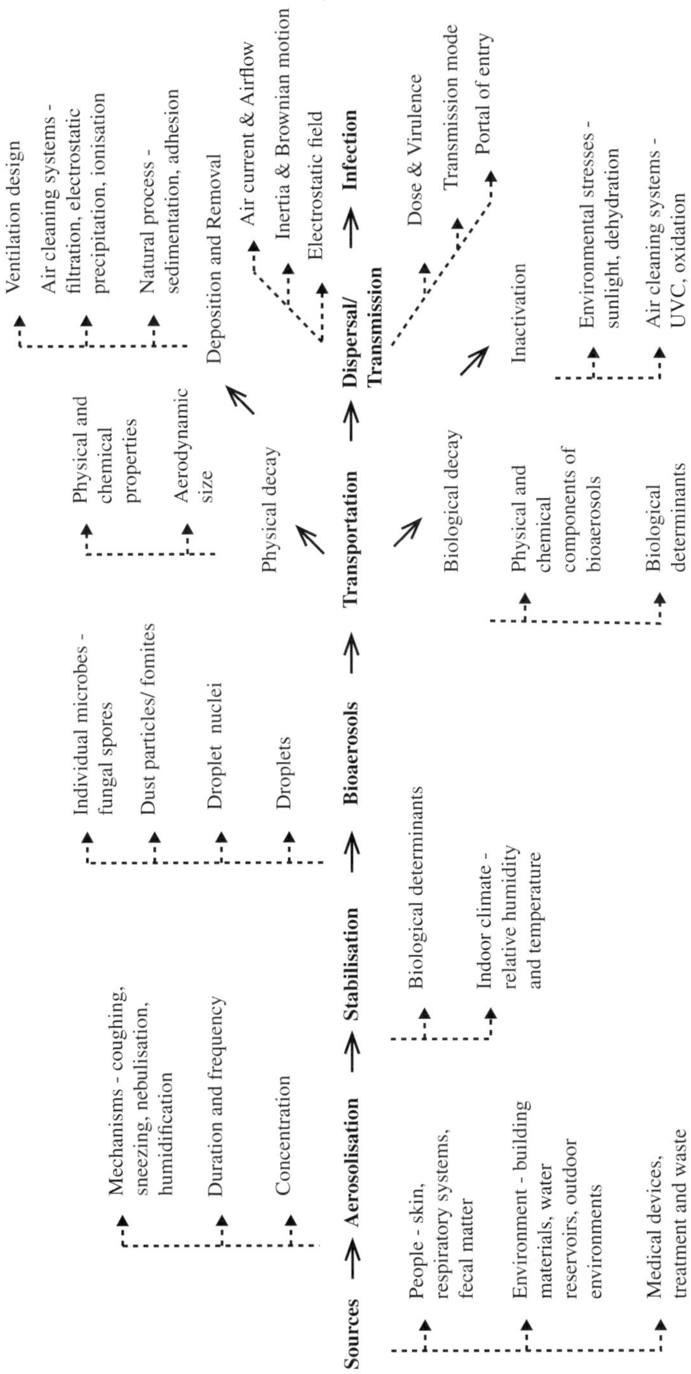

Figure 11.2 *Aerobiology dynamics in hospital environments.*

in hospitals. However, Landrin, Bissery and Kac (2005) find no correlation between particle counts and bacterial counts in their study of operating theatres, and thus particle counting cannot replace microbiology sampling. Given the understanding that infection and infection control are processes and that various factors can alter and magnify their impacts collectively in real time, Pankhurst *et al.* (2011) expand on the uses of particle counting, from a spot-check tool to a means of assessing particle profile (particle size and number) dispersal over time and space under various operational scenarios. Their approach sheds light on the aerosol pathway and illustrates how different activities and building and equipment settings involving the hospital environment, people and aerosol interaction can affect the dilution, removal, flow direction and distribution of aerosols.

Through understanding hospital aerobiology, a better aerosol and bioaerosol sampling strategy can be developed by which to capture the critical times and locations of sampling and infection control. Future development of an aerosol-based monitoring and assessment framework will allow infection-control teams to trace a source, predict and prevent current and future transmission and develop a site-specific infection-control strategy. Several optical detection devices based on fluorescence and/or elastic scattering have been developed in order to distinguish biological from nonbiological aerosols, and these will further advance our present knowledge and technological applications (Huang *et al.*, 2008; Sivaprakasam *et al.*, 2011).

11.5.2 Bioaerosol Experiments and Models

The development of laboratory facilities that allow research into the physicochemical (e.g. evaporation rate, aerodynamic size and surface charges) and biological (e.g. infectivity, culturability and survivability) properties of bioaerosols and the development of air-cleaning technology have allowed significant advances in basic aerosol science and its applications in infection control. The Hendersen apparatus, which channels bioaerosols to an animal chamber, and Goldberg rotating drums, which hold bioaerosols in the air for long periods of time, were among the earliest equipment designed to test the infectivity and survivability of bioaerosols, respectively (Henderson, 1952; Goldberg *et al.*, 1958; Goldberg, 1971). The fundamental research relating to the infective dose–response and survival time of bioaerosols at various temperatures and RH levels was established based on this early work. Tang (2009) have reviewed the effect of environmental parameters (mainly temperature and RH) on the survival of airborne infectious agents (viruses, bacteria and fungi). In general, viral aerosols survive longer at low temperature (below 10 °C) than at high temperature (above 30 °C), while bacterial aerosols decay faster above about 24 °C. In terms of the effect of RH, lipid-enveloped viruses (e.g. influenza, SARS, respiratory syncytial virus, measles, rubella, chickenpox and para-influenza viruses) survive longer at lower RH (20–30%) than do non-lipid-enveloped viruses (e.g. adenoviruses and rhinoviruses), which survive longer at RH between 70 and 90%. These properties may contribute to the seasonality effect of various respiratory diseases; SARS, which peaks during winter time, is linked to the susceptibility of the virus to high temperatures and RH after the winter season (WHO, 2004). RH has a less clear effect on different groups of bacteria. However, Gram-positive bacteria generally survive longer in the air than do Gram-negative bacteria, due to their cell-wall structure and to the fact that some Gram-positive bacteria can form spores, such as

Clostridium difficile, which make them persistent and extremely difficult to clean in the hospital environment. Fungal spores have similar environmental properties to bacterial spores.

Wind tunnels are another laboratory facility that has been used widely for the research, development and testing of aerosol and bioaerosol sampling devices, such as the development of IOM (Institute of Occupational Medicine) personal samplers (Mark *et al.*, 1990). A room-sized test chamber has also been used in various studies to mimic hospital environments (isolation rooms are highly susceptible to the spread of airborne pathogens) and to test the effectiveness of air-cleaning technology (e.g. upper-room ultraviolet germicidal irradiation (UVGI); Miller *et al.*, 2002). These laboratory facilities are the result of knowledge and skills developed in the field of aerosol science in the past several decades; facilities include the design of safe laboratory environments for bioaerosol research, development of aerosolisation methods, sampling of bioaerosols, analysis of physical and chemical bioaerosol properties and measurement of aerosol movement and distribution.

Recently, from the accumulated knowledge of the physicochemical and biological properties of bioaerosols under different climate conditions, mathematical models have been developed to describe these properties and to apply this knowledge to predict bioaerosol dispersal and behaviour in different environments. Using modelling techniques, Parienta *et al.* (2011) have reported a theoretical analysis of the motion and evaporation of exhaled respiratory droplets.

11.5.3 Numerical Analysis of Particle Dispersion in Hospitals

The modelling of infection transmission is a complex process, involving such factors as pathogen characteristics, host susceptibility and the media through which infection occurs. Airflow models such as multizone models and computational fluid dynamics (CFD) have been used to simulate and investigate airflow and contaminant dispersion in healthcare facilities in order to support hospital design and operation. These tools can be used to design ventilation systems and spatial layout, assist in disease outbreak investigation and optimise the effectiveness of infection-control measures.

Multizone models can be used to simulate the movement of air and bioaerosols from room to room, by considering each room as a separate node in the model and the potential airflow paths as linkages. Such models are limited, however, in that they do not consider local airflow within rooms, such as an air disturbance caused by the repositioning of furniture. They are nevertheless computationally simple to solve and have demonstrated good estimates of the performance of entire buildings (Asfour and Gadi, 2007). Multizone models have been used to predict the movement of viruses around hospitals (Lim, Cho and Kim, 2011) and in isolation wards (Yik and Powell, 2003) and to examine the efficiency or air-cleaning devices in a hospital environment (Nicas and Miller, 1999).

CFD is a more complex method of calculating airflow, where the volumes of rooms are discretised into specific cells and physical models are used to compute the movement of air within an individual room according to the boundary conditions. Various CFD models are commercially available and are used widely in industry and research to simulate airborne particle dispersal and deposition; they are also widely applied in various aspects of infection control. Liu and Zhai (2007) have reviewed and assessed the criteria used to identify appropriate CFD models (lazy particle models, isothermal particle models and free dropping particle models) for a simulation of different types of aerosol. According to their findings, the Stoke number and the evaporation effectiveness number are critical justification factors in determining appropriate models.

Operating theatres are one of highest infection-risk environments in hospitals. CFD models have been applied to investigate ventilation design and the effect of human activities, operating procedures and interior layout on airborne particle dispersal and deposition. Memarzadeh and Manning (2002) have used CFD models to evaluate the effects of diffuser type and location, ventilation flow rate, supply temperature and exhaust location on the risk of SSIs and to determine the efficiency of various ventilation systems in removing particles. Wu, Zheng and Chu (2011) have found that the maximum contaminant removal rate in operating theatres is obtained when the air diffuser is installed directly above the surgical bed and the air exhaust vent is situated on the wall, close to the ground. Rui, Guangbei and Jihong (2008) have carried out numerical simulations to calculate the particle dispersal in two newly built ISO class 5 and class 6 operating rooms. Chow and Yang (2005) and Chow, Lin and Bai (2006) have observed that variations in air supply velocity and medical lamp positions can seriously affect the movements of infectious particles. According to Zoon *et al.* (2010), different lamp shapes result in different flow patterns, but in their study this has no significant influence on infection risk. With regard to the influence of people movement on contaminant transport, a study by Brohus, Balling and Jeppesen (2006) shows a significant risk of contaminant transport from less clean zones to ultra-clean zones during orthopaedic operations under laminar flow conditions.

In the ward environment, Mazumdar *et al.* (2010) have studied the effects of movement by visitors and caretakers, changing of patients' bed sheets and the swinging of entrance doors on contaminant concentration in an inpatient ward with displacement ventilation. Their results show that moving objects can cause a 10–90 second oscillation of the contaminant concentration distribution and that the variation in contaminant concentration is greatly influenced by the proximity of moving objects to the contaminant source. Zhao *et al.* (2009) have simulated the number of particles emitted from a nurse's body that enter into the breathing zone of a patient, and found that the highest probability of some particles reaching the patient's body occurs when the nurse leans over the patient rather than standing nearby. Ip *et al.* (2007) simulated the probability of the spread of infectious aerosols from patients with respiratory infections using oxygen masks; they recommended that an area within at least 0.4 m of a patient using an oxygen mask should be considered a potential nosocomial hazard zone. There are many other examples of the use of CFD methods in improving infection control in hospitals, and it is likely that this application will expand in the future.

11.5.4 Air-Cleaning Technologies

Environmental exposure to bioaerosols can be reduced through a combination of air-cleaning technologies, such as filtration, ventilation dilution, airflow direction and deactivation (Figure 11.2). These technologies can be built into the building infrastructure or used as free-standing units. It is critical in hospital environments to maintain a high quality of clean air, free of pathogens, in order to limit the spread of infection, which makes indoor air cleaners an important consideration in building design and operation. Several books and guidelines review various technologies related to air cleaning in indoor environments and hospitals (Department of Health, 2007a, b; Kowalski, 2012). In this section, we discuss filtration, air dilution and direction, UVGI, negative-air ionisation and passive systems in order to illustrate a wide range of aerosol science and technology approaches and to emphasise the potential issues that need to be considered.

11.5.4.1 Filtration

Filtration systems physically remove contaminants from indoor air and can be installed in the heating ventilation and air conditioning (HVAC) system or in a portable filtration unit. In hospitals, HVAC filtration systems usually consist of a combined prefilter and filter unit, with the prefilter removing large particles upstream to the heating/cooling coil and the main filter. Main filters in hospital settings should have a minimum efficiency of 90% for particles between 1 and 5 μm, but in more critical areas with immunocompromised patients, a higher standard filter should be used. In the UK, the HTM 03-01 guide lists the appropriate grades for filters in specific settings (Department of Health, 2007a). In the US, filtration requirements for different types of hospital room have been proposed by ASHRAE and AIA (ASHRAE, 1999; AIA, 2001). Filters may also have an antimicrobial coating, in order to minimise the ability of pathogens to survive on them. Any decision to upgrade the HVAC filtration system using higher-efficiency filters needs to consider the possible increased loss of pressure through the system, which can lead to reduced airflow and poorer ventilation overall. Furthermore, leakage within the HVAC system may mean that the increased filtration provided by more efficient filters will not result in higher-quality indoor air. Portable air filtration units are also used in hospitals for filtration of indoor air, but their benefits are not as well established as those of fixed filtration units. Portable HEPA filter units have proven to be effective in removing *Aspergillus* from hospital environments (Abdul Salam *et al.*, 2010).

11.5.4.2 Air Dilution and Direction

Ventilation dilution in indoor air is an effective way of reducing contaminant levels. In the UK, recommended ventilation rates vary according to hospital settings, as shown in the HTM 03-01 guidelines (Department of Health, 2007a, b). Operating theatres have the highest ventilation requirement of above 25 air changes per hour (ACH), while ventilation rates in general wards are recommended to be set to 6 ACH. In some resource-limited countries, mechanical ventilation may not be available; the WHO has provided a guideline for the design of a natural ventilation system for infection control in these settings (WHO, 2009b). Natural ventilation and hybrid or mixed-mode ventilation have also been recommended in wider areas, in order to cut the energy demand of running a mechanical system (WHO, 2009b).

Among all the procedures for tackling airborne pathogens, the most effective way of fully eradicating contaminants from the indoor environment is through source control. Local exhaust ventilation systems can be installed around patient areas in order to rapidly remove contaminated air. Local exhaust ventilation methods include the installation of tents and booths around an infectious patient; these systems have been employed in hospital environments as a method of infection control at source (Leung and Chan, 2006). The source must be known in order for this method to be effective, however.

Adjustment of pressure differentials so that air and contaminants are directed downwards from high- to low-pressure zones is a common engineering design in infection-control settings. *Positive-pressure* rooms ensure that air flows from them (e.g. from the rooms of patients who are susceptible to infection) to the surrounding environment, whereas *negative-pressure* rooms channel air from the surrounding environment to a patient's room, in order to avoid dispersal of infectious agents from them to the surrounding area (e.g. from the rooms of TB patients). In operating theatres, air flows governed by pressure differentials move from ultra-clean zones to clean zones, to corridors and finally to disposal zones (Department of Health, 2007a, b). When establishing pressure differentials, care needs to be taken to avoid cracks or imperfections in the build, which might allow air to balance between zones. Furthermore, the opening of doors can lead to a short-term breakdown in pressure differentials, leading to the spread of infection (Tang *et al.*, 2005).

11.5.4.3 Deactivation

Ultraviolet Germicidal Irradiation (UVGI). UVGI systems (254 nm) are used to disinfect indoor air in portable units, upper-room systems and HVAC systems. Their efficacy depends on the UV susceptibility of the bioaerosols (affected by RH, temperature, coating and the group of microorganisms) and the UV dose, a factor of irradiation intensity and time influenced by the air-mixing levels to which the bioaerosols are exposed (Miller *et al.*, 2002). UV radiation can be harmful, so the cleaning unit should be arranged such that no person is mistakenly or inadvertently exposed directly to the light. Units are typically installed in the ducts of air-handling units ('in-duct') or on ceilings or walls, so as to direct light upwards and not towards anyone in the room ('upper-room') (First *et al.*, 1999a, b). While it may not be sufficient to provide adequate air cleaning on its own, UVGI can be used as a supplemental air disinfectant when combined with indoor filtration units, as their impacts are additive. The CDC recommends UVGI as a supplemental engineering control in hospitals (CDC, 1994).

The technologies discussed here have been studied widely and have been extensively proven to work, both experimentally and epidemiologically. However, negative-air ionisation and passive systems coated with antimicrobial materials, nanoparticles and photocatalytic agents have been less studied in the hospital environment. Clinical and epidemiological data on these technologies are somewhat limited, and their modes of action are not yet fully understood. They are discussed in this chapter in order to highlight their complexity and to illustrate the wide contexts in which aerosol science may be applied to infection control.

Negative-Air Ionisation. In theory, negative-air ionisers generate negative-air ions that collide with bioaerosols and charge the bioaerosol surface. The charged bioaerosols tend to aggregate and fall from the air as a result of their weight, which is greater than that of the original bioaerosols. Kerr *et al.* (2006) found that using negative-air ionisers reduced infections by *Acinetobacter* but not by methicillin-resistant *Staphylococcus aureus* (MRSA) in an ICU. These data would therefore seem to support the idea that the airborne route of transmission is significant in the epidemiology of nosocomial *Acinetobacter* infections. However, a later study shows that negative-air ionisers can also charge the surface of plastic medical equipment and may contribute to the reduction of surface contamination by bacteria, due to bacterial repulsion from the surface (Shepherd *et al.*, 2010). Escombe *et al.* (2009) reported that negative-air ionisation prevents most airborne TB transmission detectable by guinea pig sampling. The charged bioaerosols have nevertheless raised the question of whether they will increase their deposition in the lungs (Jeffers, 2007). We observe that this technology is not included in any CDC or HTM guidelines. It is possible that technologies introduced into the environment can have unexpected outcomes, which could either help to clean the air or else potentially provoke unwanted side effects. This is one of the major challenges in testing the effects and the effectiveness of air-cleaning devices, since various human and environmental factors can interfere with the test results.

Passive Systems. Passive systems have been developed to help purify indoor air by deactivating contaminants that come in contact with building surfaces. Coatings such as antimicrobial materials, nanoparticles and photocatalytic oxidation (PCO) can be applied to wall surfaces in order to perform passive decontamination in reaction with contaminants or indoor air. PCO has been utilised on wall tiles in surgical theatres in Japan (Fujishima, Hashimoto and Watanabe, 1999), where bacterial counts on the tiles decreased to negligible levels in 1 hour, and indoor air counts significantly decreased. Passive systems technology shows a passive pathway in deactivating bioaerosols, which is a completely different approach from that of air cleaning and requires turbulent air mixing so as to increase the contact frequency between bioaerosols and surfaces.

11.6 Conclusion

We have shown in this chapter that aerosol science and technology have advanced our understanding and control of bioaerosols in the hospital environment. Various disciplines, factors and technological opportunities can impact on the attempt to reduce infection risks. Advances in aerosol science and technology will ensure that new approaches and applications emerge in the future. The application of new technologies and practices in hospitals must be underpinned by robust laboratory and clinical evidence, and must also address patient benefits and cost concerns. Procurement, operational, maintenance and ethical issues have to be evaluated in the initial stages. Furthermore, in the assessment of new technologies for clinical use, applications will need to abide by industrial and health-sector guidelines. The effectiveness of different air-cleaning systems/devices depends not only on the specific technology used but also on various product designs and environmental factors, such as the source of bioaerosol emission, air mixing, air-flow pattern, humidity, exposure time, occupants' activities, volume of enclosed space and the types and operation of HVAC system. Technology and product development and validation must consider all of these requirements and influencing factors in achieving the ultimate aim of protecting public health.

References

Abdul Salam, Z.H., Karlin, R.B., Ling, M.L. and Yang, K.S. (2010) The impact of portable high-efficiency particulate air filters on the incidence of invasive aspergillosis in a large acute tertiary-care hospital. *American Journal of Infection Control*, **38**(4), e1–e7.

AIA (2001) *Guidelines for Design and Construction of Hospitals and Health Care Facilities*, The American Institute of Architects Press, Washington, DC.

Asfour, O.S. and Gadi, M.B. (2007) A comparison between CFD and Network models for predicting wind-driven ventilation in buildings. *Building and Environment*, **42**(12), 4079–4085.

ASHRAE (1999) Health care facilities, in *ASHRAE Handbook: HVAC Applications, Fundamentals*, American Society of Heating, Refrigerating, and Air-Conditioning Engineers, Atlanta.

Brohus, H., Balling, K.D. and Jeppesen, D. (2006) Influence of movements on contaminant transport in an operating room. *Indoor Air*, **16**(5), 356–372.

CDC (1994) Guidelines for preventing the transmission of Mycobacterium tuberculosis in health care facilities. *MMWR. Morbidity and Mortality Weekly Report*, **43**, 1–17.

CDC (2003) Guideline for environmental infection control in health-care facilities. *MMWR. Morbidity and Mortality Weekly Report*, **52**(RR10), 1–42.

CDC (2004) Guidelines for preventing health-care-associated pneumonia, Department of Health and Human Services. *MMWR. Morbidity and Mortality Weekly Report*, **53**(RR03), 1–36.

Chow, T., Lin, Z. and Bai, W. (2006) The integrated effect of medical lamp position and diffuser discharge velocity on ultra-clean ventilation performance in an operating theatre. *Indoor and Built Environment*, **15**(4), 315–331.

Chow, T. and Yang, X. (2005) Ventilation performance in the operating theatre against airborne infection: numerical study on an ultra-clean system. *Journal of Hospital Infection*, **59**(2), 138–147.

Department of Health (2007a) *Health Technical Memorandum HTM 03-01: Specialised Ventilation for Healthcare Premises – Part A: Design and Validation*, TSO, Norwich.

Department of Health (2007b) *Health Technical Memorandum 03-01: Specialised Ventilation for Healthcare Premises – Part B: Operational Management and Performance Verification*, TSO, Norwich.

Dharan, S. and Pittet, D. (2002) Environmental controls in operating theatres. *Journal of Hospital Infection*, **51**(2), 79–84.

Ducel, G., Fabry, J. and Nicolle, L. (2002) *Prevention of Hospital-Acquired Infections – Practical Guide*, WHO Press, Geneva.

Escombe, A.R., Moore, D.A., Gilman, R.H. *et al.* (2009) Upper-room ultraviolet light and negative air ionization to prevent tuberculosis transmission. *PLoS Medicine*, **6**(3), e1000043.

First, M.W., Nardell, E.A., Chaisson, W. and Riley, R. (1999a) Guidelines for the application of upper room ultraviolet germicidal irradiation for preventing transmission of airborne contagion- Part I: basic principles. *ASHRAE Transactions*, **105**, 869–876.

First, M.W., Nardell, E.A., Chaisson, W. and Riley, R. (1999b) Guidelines for the application of upper room ultraviolet germicidal irradiation for preventing transmission of airborne contagion- Part II: design and operation guidance. *ASHRAE Transactions*, **105**, 877–887.

Fujishima, A., Hashimoto, K. and Watanabe, T. (1999) *TiO_2 Photocatalysis: Fundamentals and Applications*, BKC, Tokyo.

Goldberg, L.J. (1971) Naval biomedical research laboratory, programmed environment, aerosol facility. *Applied Microbiology*, **21**(2), 244–252.

Goldberg, L., Watkins, H., Boerke, E. and Chatigny, M. (1958) The use of a rotating drum for the study of aerosols over extended periods of time. *American Journal of Hygiene*, **68**(1), 85–93.

Henderson, D.W. (1952) An apparatus for the study of airborne infection. *Journal of Hygiene*, **50**(1), 53–68.

Hinds, W.C. (1999) *Aerosol Technology: Properties, Behavior, and Measurement of Airborne Particles: Properties, Behavior and Measurement of Airborne Particles*, Wiley-Interscience, New York.

Ho, A.S., Sung, J.J. and Chan-Yeung, M. (2003) An outbreak of severe acute respiratory syndrome among hospital workers in a community hospital in Hong Kong. *Annals of Internal Medicine*, **139**(7), 564–567.

Huang, H.C., Pan, Y.L., Steven, C.H. *et al.* (2008) Real-time measurement of dual-wavelength laser-induced fluorescence spectra of individual aerosol particles. *Optic Express*, **16**(21), 16523–16528.

Husain, S., Alexander, B.D., Munoz, P. *et al.* (2003) Opportunistic mycelial fungal infections in organ transplant recipients: emerging importance of non-Aspergillus mycelial fungi. *Clinical Infectious Diseases*, **37**(2), 221–229.

ICRP (1994) Human respiratory tract model for radiological protection. A report of a Task Group of the International Commission on Radiological Protection. *Annals of the ICRP*, **24**(1–3), 1–482.

Ip, M., Tang, J.W., Hui, D.S. *et al.* (2007) Airflow and droplet spreading around oxygen masks: a simulation model for infection control research. *American Journal of Infection Control*, **35**(10), 684–689.

Jeffers, D. (2007) Modelling and analyses do not support the hypothesis that charging by power-line corona increases lung deposition of airborne particles. *Radiation Protection Dosimetry*, **123**(2), 257–261.

Kerr, K.G., Beggs, C.B., Dean, S.G. *et al.* (2006) Air ionisation and colonisation/infection with methicillin-resistant Staphylococcus aureus and acinetobacter species in an intensive care unit. *Intensive Care Medicine*, **32**(2), 315–317.

Kowalski, W.J. (2012) *Hospital Airborne Infection Control*, CRC Press, Florida.

Landrin, A., Bissery, A. and Kac, G. (2005) Monitoring air sampling in operating theatres: can particle counting replace microbiological sampling? *Journal of Hospital Infection*, **61**(1), 27–29.

Leung, M. and Chan, A.H.S. (2006) Control and management of hospital indoor air quality. *Medical Science Monitor*, **12**(3), 17–23.

Lim, T., Cho, J. and Kim, B.S. (2011) Predictions and measurements of the stack effect on indoor airborne virus transmission in a high-rise hospital building. *Building and Environment*, **46**(12), 2413–2424.

Lister, J. (2003) The PFI Experience, Voices from the Front Line, A PFI report for UNISON.

Liu, X. and Zhai, Z. (2007) Identification of appropriate CFD models for simulating aerosol particle and droplet indoor transport. *Indoor and Built Environment*, **16**(4), 322–330.

MacCannell, T., Umscheid, C. A., Agarwal, R. K. *et al.*: the Healthcare Infection Control Practices Advisory Committee (HICPAC) (2011) Guidelines for the Prevention and Control of Norovirus Gastroenteritis Outbreak in Health Care Setting, CDC, Atlanta, GA (Online), http://www.cdc.gov/hicpac/norovirus/norovirus-references.html (last accessed 8 August 2013).

Mark, D., Vincent, J. H., Lynch, G. *et al.*, (1990) The Development of a Static Sampler for the Measurement of Inspirable Aerosol in the Ambient Atmosphere with Special Reference to PAHs. Research Report TM/90/06, IOM, pp. 1–59.

Mazumdar, S., Yin, Y., Guity, A. *et al.* (2010) Impact of moving objects on contaminant concentration distributions in an inpatient ward with displacement ventilation. *HVAC and R Research*, **16**(5), 545–556.

Memarzadeh, F. and Manning, A.P. (2002) Comparison of operating room ventilation systems in the protection of the surgical site. *ASHRAE Transactions*, **108**(2), 3–15.

Miller, S.L., Hernandez, M., Fennelly, K. *et al.*, (2002) Efficacy of Ultraviolet Irradiation in Controlling the Spread of Tuberculosis, Report NTIS PB2003-103816 NIOSH, October, pp. 1–80.

Nicas, M. and Miller, S.L. (1999) A multi-zone model evaluation of the efficacy of upper-room air ultraviolet germicidal irradiation. *Applied Occupational and Environmental Hygiene*, **14**(5), 317–328.

Pankhurst, L., Taylor, J., Cloutman-Green, E.A. *et al.* (2011) Can clean-room particle counters be used as an infection control tool in hospital operating theatres? *Indoor and Built Environment*, **21**, 381–391. doi: 10.1177/1420326X11409467

Parienta, D., Morawska, L., Johnson, G.R. *et al.* (2011) Theoretical analysis of the motion and evaporation of exhaled respiratory droplets of mixed composition. *Journal of Aerosol Science*, **42**(1), 1–10.

Pasquarella, C., Pitzurra, O. and Savino, A. (2000) The index of microbial air contamination. *Journal of Hospital Infection*, **46**(4), 241–256.

Rui, Z., Guangbei, T. and Jihong, L. (2008) Study on biological contaminant control strategies under different ventilation models in hospital operating room. *Building and Environment*, **43**(5), 793–803.

Shepherd, S.J., Beggs, C.B., Smith, C.F. *et al.* (2010) Effect of negative air ions on the potential for bacterial contamination of plastic medical equipment. *BMC Infectious Diseases*, **10**, 92.

Sivaprakasam, V., Lin, H.B., Huston, A.L. and Eversole, J.D. (2011) Spectral characterization of biological aerosol particles using two-wavelength excited laser-induced fluorescence and elastic scattering measurements. *Optic Express*, **19**(7), 6191–6208.

Sosnowski, T.R. (2011) Importance of airway geometry and respiratory parameters variability for particle deposition in the human respiratory tract. *Journal of Thoracic Disease*, **3**(3), 153–155.

Tang, J.W. (2009) The effect of environmental parameters on the survival of airborne infectious agents. *Journal of the Royal Society Interface*, **6**(6), S737–S746.

Tang, J.W., Eames, I., Li, Y. *et al.* (2005) Door-opening motion can potentially lead to a transient breakdown in negative-pressure isolation conditions: the importance of vorticity and buoyancy airflows. *Journal of Hospital Infection*, **61**(4), 283–286.

Ulrich, R.S. (2006) Evidence-based health-care architecture. *Lancet*, **368**, S38–S39.

Weber, W. (1979) Pigeon associated people diseases. Indianapolis, Indiana: Bird Control Seminars Proceedings, Paper 21 (Online), http://digitalcommons.unl.edu/icwdmbird control/21 (last accessed 8 August 2013).

WHO (2004) *SARS Risk Assessment and Preparedness Framework*, WHO Press, Geneva.

WHO (2005) *WHO Checklist for Influenza Pandemic Preparedness Plan*, WHO Press, Geneva.

WHO (2009a) Influenza (Seasonal). Fact Sheet N 211, WHO, Geneva (Online), http://www .who.int/mediacentre/factsheets/fs211/en/index.html (last accessed 8 August 2013).

WHO (2009b) *Natural Ventilation for Infection Control in Health Care Settings*, WHO Press, Geneva.

WHO (2011) *Report on the Burden of Endemic Health Care-Associated Infection Worldwide, Clean Care is Safer Care, A Systematic Review of the Literature*, WHO Press.

WHO (2012) *The Evolving Threat of Antimicrobial Resistance: Options for Action*, WHO Press, Geneva.

Wilson, A.P.R. and Ridgway, G.L. (2006) Reducing hospital-acquired infection by design: the new University College London Hospital. *Journal of Hospital Infection*, **62**(3), 264–269.

Wu, Y., Zheng, Y. and Chu, K. (2011) Study on contaminant removal rate and ventilation efficiency under different HVAC vent design of operating room. *Advanced Science Letters*, **4**(6), 2361–2368.

Yik, F.W.H. and Powell, G. (2003) Review of isolation ward ventilation design and evaluation by simulation. *Indoor and Built Environment*, **12**(1–2), 73–79.

Zhao, B., Yang, C., Chen, C. *et al.* (2009) How many airborne particles emitted from a nurse will reach the breathing zone/body surface of the patient in ISO Class-5 single-bed hospital protective environments? – a numerical analysis. *Aerosol Science and Technology*, **43**(10), 990–1005.

Zoon, W., van der Heijden, M., Loomans, M. and Hensen, J. (2010) On the applicability of the laminar flow index when selecting surgical lighting. *Building and Environment*, **45**(9), 1976–1983.

12

Nanostructured Material Synthesis in the Gas Phase

Peter V. Pikhitsa and Mansoo Choi
Division of WCU Multiscale Mechanical Design, School of Mechanical and Aerospace Engineering, Seoul National University, South Korea

12.1 Introduction

Nanostructured materials made of assembled nanoparticles can manifest various properties unattainable in usual homogeneous bulk materials. Electron energy levels, plasmonic states, magnetic states, and so on can be very specific to individual nanoparticles. When nanoparticles are assembled into a material, they may constitute artificial bulk materials with interesting properties. First among these are transport and transmission properties, whether heat transfer, electromagnetic waves, spin waves, or electric charge transportation. The transport properties can be very sensitive to the environment, due to systems of interfaces and grain boundaries, especially for nanoparticles arranged in low-dimensional chain-like structures. Other properties include thermodynamical properties: the capacity for charges, as in Li batteries; superparamagnetic properties; the effects of radiation transformation, as in a solar cell; catalytic activity; and so on.

Two technical questions then arise: how to create nanoparticles with the desired properties and how to arrange them into useful structures. A vast number of methods of obtaining nanostructured materials have been developed, and sometimes the two tasks can be solved in a one-step synthesis process, during which the nanoparticles are arranged into such structures. Individual nanoparticles can be generated by virtually any high-energy chemical or physical process, including wet chemistry, solid–solid interaction, heat, plasma, electron beam, or ion irradiation, laser ablation, and flame. Usually, various charging processes

Aerosol Science: Technology and Applications, First Edition. Edited by Ian Colbeck and Mihalis Lazaridis.

and fast termination of cluster growth due to fast quenching prevent the coalescence and ripening of nanoparticles into the micrometer size range.

In this chapter, we will consider a few methods of nanostructure synthesis, some of which utilize peculiar physical effects, as well as the useful properties that synthesized materials can manifest.

12.2 Aerosol-Based Synthesis

One of the most popular methods of synthesizing nanostructured materials is the aerosol-based synthesis of nanoparticles. A recent review can be found in Schmidt-Ott (2011). Aerosol methods have many advantages: they are cleaner than wet chemistry; they are readily and rapidly controlled by gas flow rates; and they are easily accessible for diagnostics *in situ*. The stream of nanoparticles can be manipulated with and guided by electric, magnetic, or other fields to arrange/assemble the nanoparticles into useful structures. Additionally, the gases can be turned into plasma, either within flames or by glow and spark discharges. The transparency of the gas flow allows the reaction zone to be accessed by laser beams. Fast quenching or direct condensation into the solid phase allows precise coating. The characteristic milliseconds duration provides the fastest rate of nanoparticle generation, which is useful for deposition and mass production.

Let us consider some methods and interesting effects, which, while sometimes being outside the mainstream of research, may highlight the potential of aerosol-based synthesis for nanostructured materials. Unlike Schmidt-Ott (2011), who concentrates only on the synthesis of individual nanoparticles, we will address here some assembly methods and emphasize physical effects found in the assembled nanostructured materials.

12.3 Flame Synthesis

Nanoparticles are often generated in flames; the soot formed by carbon nanoparticles of varying morphology and degree of agglomeration is a well-known example. Another is the generation of metal oxides. Certain properties of metal-oxide nanoparticles, such as superparamagnetism and catalytic activity, depend crucially on size reduction. Among the preparation methods for size-controlled nanoparticles, a gas-phase route using a flame has certain advantages, one of which is that it can continuously produce very pure metal-oxide nanoparticles (Pratsinis, 1998). Diffusion flames, unlike premixed flames, provide an opportunity to precisely control the growth environment using gas velocities and burner configuration. Precursors such as metal chlorides and metal acetylacetonates driven into a flame by carrier gases are oxidized to generate metal-oxide nanoparticles. This normally leads to a gradual and well-controllable growth in the nanoparticle size as the distance from the burner edge z increases. However, even in this simple configuration, iron III oxide nanoparticles formed by a gas-phase route from an iron-containing precursor iron acetylacetonate within an oxyhydrogen flame, which initially grew to a size of 20 nm, were found to abruptly transform into 3 nm nanoparticles as the distance from the burner edge increased (Yang *et al.*, 2003). It was suggested that the mechanism for nanoparticle fragmentation might be a thermal instability caused by a thermally-induced phase transformation from a metastable phase to a stable phase of iron oxide (Yang *et al.*, 2003). If that were the case, however, the temperature of the oxyhydrogen flame would have played a decisive role and the phenomenon would have been material-specific, as such a phase transformation works only for iron oxide. Yet, after further experiments looking at the flame synthesis of tin oxide (SnO_2) nanoparticles from tin tetrachloride injected into an oxyhydrogen flame cone revealed the same fragmentation phenomenon, the explanation that the phase transformation is a driving mechanism was put into question (Pikhitsa *et al.*, 2007). In addition, it turned out that hydrogen flow, not the flame temperature, was the main source of control over the fragmentation of nanoparticles of both oxides (Figure 12.1).

It is natural to suspect that hydrogen is responsible for the nanoparticle fragmentation. Indeed, removal of oxygen from the surface as a result of a reduction in hydrogen is known to lead to oxide surface corrugation and carving (e.g. for titania), and even fragmentation (known for blue tungsten oxide). The reason for the surface instability is the surface diffusion of excess metal atoms, which tend to come together and terminate the surface of

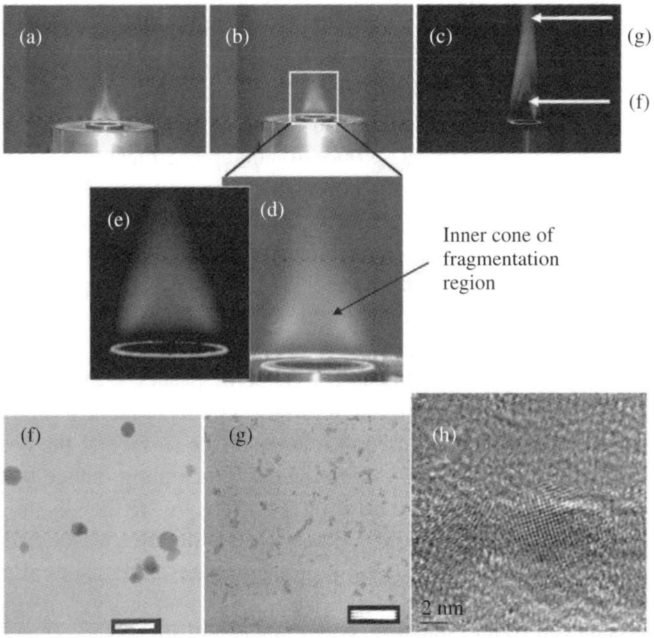

Figure 12.1 *Photographic images of flame cones and transmission electron microscopy (TEM) images for iron oxide nanoparticles. (a) Appearance of the flame cone in the case of no fragmentation at a hydrogen : oxygen ratio of 3 : 4 at 1345 °C. (b) Appearance of the flame cone in the case of fragmentation at a hydrogen : oxygen ratio of 4 : 5 at 1650 °C. (c) Appearance of the flame cone in the case of fragmentation at a hydrogen : oxygen ratio of 14 at 1100 °C (considerably reduced). (d,e) Magnified part of (c) taken with and without a flash light, respectively. The fragmentation region is clearly distinguished. (f) A TEM image of iron-oxide nanoparticles collected inside the cone in (c) from the region at z = 20 mm. (g) Outside the cone region at z = 100 mm (sampling regions are shown by arrows in (c), scale bar is 50 nm). (h) High-resolution TEM image of fragmented particles from (g). Reprinted with permission from Pikhitsa* et al. *(2007). Copyright © 2007, American Institute of Physics. See plate section for colour version.*

a nano-oxide. Carving as a result of oxygen removal is not possible in places where metal atoms are abundant, and due to the uneven distribution of metal and oxygen-rich regions, a surface instability develops following mechanical strain. The tiny nanoparticles then start peeling off the surface and eventually the whole primary large nanoparticle ends up with a plethora of them. This example (Pikhitsa *et al.*, 2007) demonstrates that even in a simple flame configuration, one may face unexpected and material nonspecific effects, which can be utilized for nanoparticle size control beyond the effect of standard flame synthesis (Vemury, Pratsinis, and Kibbey, 1997; Teleki *et al.*, 2009).

More sophisticated flame synthesis involves a heterogeneous flame structure, leading to coating of the nanoparticles to preserve and protect their properties from the reactive environment (Teleki *et al.*, 2009). Synthesis of composite nanoparticles in coating mode attracts much attention, since composite nanoparticles may not only exhibit unique properties that would not be expected from the individual constituents of the composites, but might also be necessitated in order to ensure the stability of particles.

In particular, coated composite nanoparticles have shown enhanced properties compared to noncoated particles. For example, more efficient photoluminescence (PL) and stability were reported from silica-coated CdSe quantum dots (Selvan, Tan, and Ying, 2005), and silica-coated maghemite particles (Teleki *et al.*, 2009) were shown to be more easily dispersed in organic and aqueous solutions (Santra *et al.*, 2001). Layer-by-layer assembly of a silica-coated Fe_3O_4 nanoparticle was proposed, and coated magnetic particles showed a reduction in cooperative magnetization switching between particles due to the barrier effect of the silica layer (Aliev *et al.*, 1999). Specific superparamagnetism of the tiniest Fe_3O_4 coated nanoparticles totally preserves the magnetism of naked iron-oxide nanoparticles. A simple *in situ* flame-coating method has been developed by designing a new type of coflow diffusion flame burner with a sliding unit (Figure 12.2).

The sliding unit has been shown to be very effective in finding the position where the precursor for the coating layer should meet with the core particles. $SiCl_4$ was injected into the coating unit and $TiCl_4$ was injected into the central nozzle as a host precursor. In order to find a proper position for the coating unit relative to the surface of the burner, particles were collected at $h_p = 160$ mm using a thermophoretic sampling device by changing the sliding coating unit at different vertical positions (h_r). They were then examined through a transmission electron microscope (TEM). When the coating unit was placed at low positions ($h_r = 5$ and 45 mm) (Figure 12.3a,b), large aggregates and spherical particles were observed together and energy-dispersed spectroscopy (EDS) analysis revealed that they were SiO_2 aggregates and TiO_2 spheres. At the lower positions of $h_r = 5$ and 45 mm, diffused $SiCl_4$ vapor was likely to form small aggregates by chemical reaction due to high flame temperatures at these low flame heights. By sliding the coating unit vertically to a further, higher position at $h_r = 80$ mm, it was possible to prevent the formation of independent SiO_2 particles, since flame temperatures decrease in the downstream region. As shown in Figure 12.3c, silica-coated titania nanoparticles were prepared. A magnified TEM image confirmed that the surface of the spherical TiO_2 nanoparticle was completely coated with a thin layer of SiO_2 of about 2.4 nm thickness (Figure 12.3d).

Suppression of silica aggregate formation where $h_r = 80$ mm was double-checked in a TEM from the powder collected on the water-cooled quartz tube. An additional experiment where only $SiCl_4$ was injected into the coating unit at $h_r = 80$ mm, without the core particle precursor of $TiCl_4$, was also carried out in order to confirm the suppression of silica

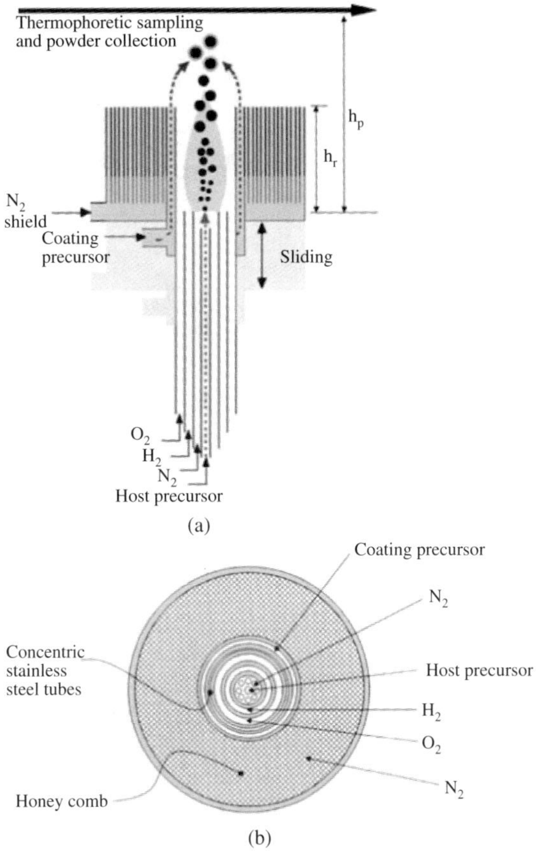

Figure 12.2 *Schematic illustration of the modified H_2/O_2 diffusion flame burner with a sliding-type coating unit. (a) Side view: h_r and h_p denote the distance between the burner surface and the coating unit surface and the distance of particle sampling from the burner surface, respectively. (b) Top view. Reproduced with permission from Sheen* et al. *(2009). Copyright © 2009, Springer Science+Business Media B.V.*

aggregates. SiO_2-coated TiO_2 nanoparticles were prepared first, and the coating of particle surfaces was examined by both direct observation of the particles through a TEM and zeta-potential measurements (Figure 12.4). The TEM results showed an absence of particles, indicating that the $SiCl_4$ vapor precursor was ventilated without particle formation in the case of $h_r = 80$ mm. The suppression of aggregate formation indicated that the coating mechanism in Sheen *et al.* (2009) could be a result of cluster scavenging and chemical vapor deposition on the surface of host particles, rather than the formation of silica aggregates and subsequent collision of the aggregates with host particles. Mean core sizes varied from 28 to 109 nm and mean the coating thickness was about 2.4 nm for silica-coated titania particles. Simply by changing chemical precursors, it was demonstrated that SiO_2-coated SnO_2, SnO_2-coated TiO_2 and SiO_2–SnO_2-coated TiO_2 nanoparticles could also be synthesized.

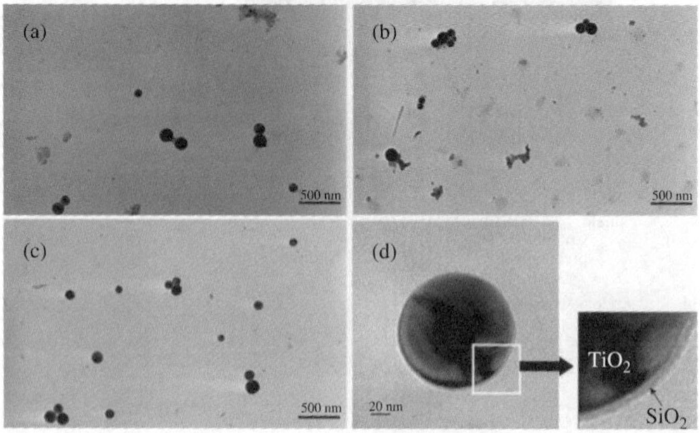

Figure 12.3 TEM images of nanoparticles collected at different sliding unit positions: (a) $h_r = 5$ mm, $h_p = 160$ mm; (b) $h_r = 45$ mm, $h_p = 160$ mm; (c) $h_r = 80$ mm, $h_p = 160$ mm. (d) Magnified TEM image of a SiO_2-coated TiO_2 nanoparticle, $h_r = 80$ mm, $h_p = 160$ mm. Reproduced with permission from Sheen et al. (2009). Copyright © 2009, Springer Science+Business Media B.V.

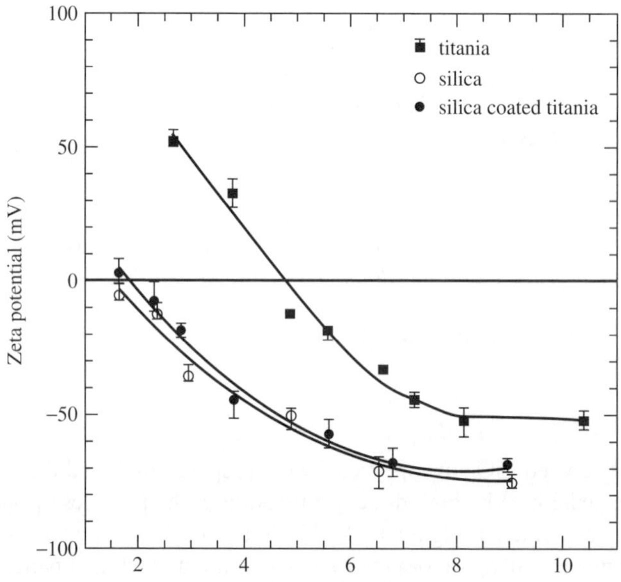

Figure 12.4 Zeta-potential values of SiO_2, TiO_2, and SiO_2-coated TiO_2 nanoparticles; $h_r = 80$ mm, $h_p = 160$ mm. Reproduced with permission from Sheen et al. (2009). Copyright © 2009, Springer Science+Business Media B.V.

An example of a nanostructured carbon material generated from a 'perpetual coating' process is the aerosol-based growth of carbon nanotubes from gaseous hydrocarbons, where the first generated iron metal nanoparticles catalyze the growth of single-walled carbon nanotubes from their surfaces. Each iron nanoparticle stays in the head of the growing nanotube and the nanotube makes a growing long tail (Nasibulin *et al.*, 2007). The nanoparticle radius dictates the radius of the nanotube. However, the dynamics of the steady-state synthesis brings some peculiar inhomogeneities into the growing process of the nanotube. A novel hybrid material that combines fullerenes and the single-walled nanotubes (SWNTs) into a single structure in which the fullerenes are covalently bonded to the outer surface of the SWNTs has been discovered (Nasibulin *et al.*, 2007). These fullerene-functionalized SWNTs, called 'NanoBuds', were selectively synthesized in two different one-step continuous methods, during which fullerenes were formed on iron-catalyst particles together with SWNTs during CO disproportionation. The field-emission characteristics of NanoBuds suggest that they may possess advantageous properties compared with SWNTs or fullerenes alone, or in their nonbonded configurations. NanoBuds have been synthesized in two different continuous aerosol (floating catalyst) reactors by using particles grown *in situ* via ferrocene vapor decomposition and premade iron-catalyst particles produced by a hot-wire generator (HWG) (Figure 12.5).

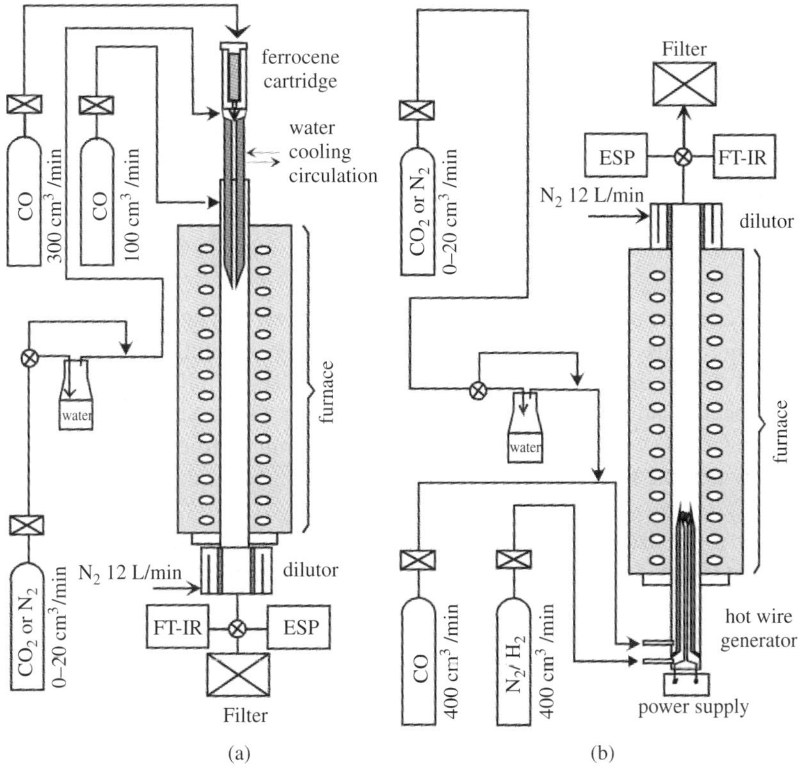

Figure 12.5 *Schematics of the experimental set-ups: (a) ferrocene reactor and (b) HWG method. Adapted with permission from Nasibulin* et al. *(2007).*

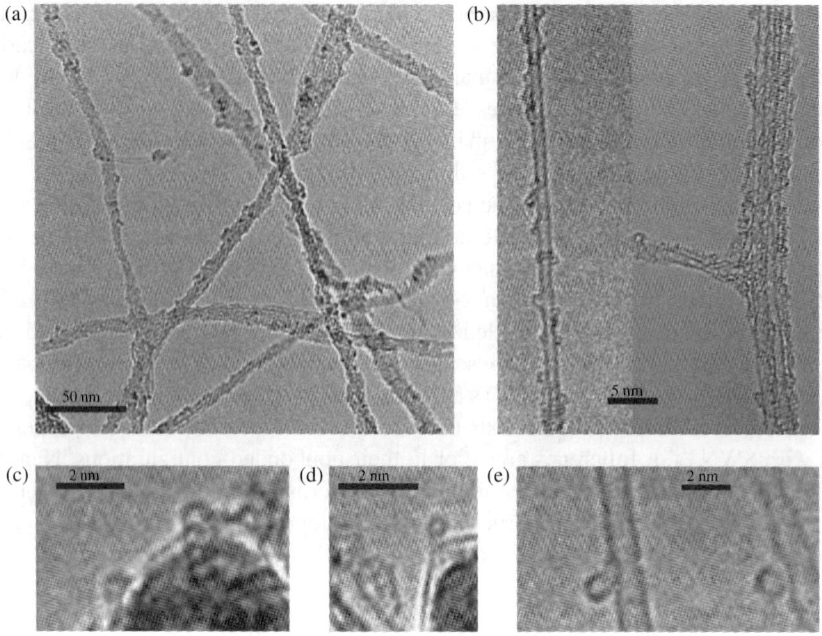

Figure 12.6 *TEM observation of NanoBud structures. (a) Low-magnification TEM image of a sample showing SWNTs seemingly covered by amorphous carbon. (b) Intermediate-magnification TEM images of the sample, revealing the presence of spherical structures on the surfaces of the SWNTs. (c–e) High-resolution TEM images showing fullerenes located on an iron-catalyst particle (c) and on a carbon layer covering a catalyst particle (d), and NanoBuds in which a fullerene is combined with an SWNT (e, left) and attached to the surface of an SWNT (e, right). Adapted with permission from Nasibulin* et al. *(2007).*

TEM images of the product at low magnifications suggest that most synthesized SWNTs have an 'amorphous coating' (Figure 12.6a). However, careful investigations (Figure 12.6b–e) reveal that much of the coating actually consists of fullerenes. Their spherical nature has been confirmed by tilting samples within a TEM. Statistical measurements of the spherical cages on the surfaces of SWNTs determined from high-resolution TEM images show that the majority of the measured sizes can be attributed to the presence of C_{42} and C_{60}.

Interestingly, the size distribution of fullerenes suggests the presence of C_{20} fullerenes, the smallest possible dodecahedron. The key parameter required for NanoBud synthesis in both types of reactor was found to be the presence of trace concentrations of H_2O vapor and CO_2, thereby demonstrating the generality of the effect of these etching agents on catalytic fullerene/SWNT coformation. In an attempt to control the degree of SWNT coverage with fullerenes (i.e. the degree of SWNT functionalization), the reactor temperature was varied, as were the concentrations of water vapor and carbon dioxide. It was found that these parameters noticeably affected the amount of fullerenes on the surfaces of the SWNTs. The introduction of H_2O and CO_2 into the ferrocene reactor revealed that the optimum reagent concentrations were between 45 and 245 ppm for water and between 2000 and

6000 ppm for CO_2, and the highest fullerene density achieved was more than one fullerene per nanometer, with fullerenes arranged in a continuous stream along an SWNT.

12.4 Flame and Laser Synthesis

Using additional external energy sources, such as electrical discharge through the flame (Vemury, Pratsinis, and Kibbey, 1997) and heating of the growing nanoparticles *in situ* with a laser, also leads to nanoparticle modifications. A powerful infrared CO_2 laser can be used on flames in order to control the morphology, size, and crystallinity of nanoparticles. One example is the morphology and size transformation of SiO_2 nanoparticles (Lee and Choi, 2000). The coalescence or sintering characteristic time depends strongly on temperature and follows an Arrhenius expression; that is, exponential decay of the sintering characteristic time as particle temperature increases while the collision characteristic time for the free molecular regime decays according to $1/T^{1/2}$, where T is the nanoparticle temperature. Therefore, the sintering characteristic time can be controlled nearly independently of the collision time through rapid heating of aggregate nanoparticles. An attempt was made by Lee and Choi (2000) to control the characteristic sintering time using CO_2 laser irradiation on aggregate nanoparticles generated in a flame, as shown in Figure 12.7. This altered the morphology of the nanoparticles, from chain-like aggregates to more spherical forms. Aggregates irradiated by a high-power continuous-wave (CW) CO_2 laser beam can be heated to high temperature and then sintered to become more sphere-like nanoparticles, due to the enhanced sintering rate.

Since spherical nanoparticles have much smaller collision cross-sections than do volume-equivalent aggregates (Pratsinis, 1998), much slower growth can be expected. As a result, the size of the transformed spherical nanoparticles resulting from CO_2 laser irradiation should become smaller than that of the aggregates which originally existed in the flame. In this way, the synthesis of smaller, spherical nanoparticles can be achieved using laser irradiation on aggregates formed in a flame. A cooled substrate can be placed just above the laser irradiation point to deposit the transformed nanoparticles.

A new approach to the control of the size and morphology of flame-generated nanoparticles using laser irradiation was thus proposed (Lee and Choi, 2000), and the effects of laser irradiation in the flame synthesis of nanoparticles were confirmed by measurements of light scattering and the observation of TEM photos of locally captured nanoparticles. Depending on the irradiation height of a CO_2 laser beam in a flame, significantly different mechanisms can be found. At a low irradiation height, the particle-generation effect becomes dominant, due to the gas absorption of laser power. This results in an increase in Ar-ion laser light-scattering intensity and is confirmed from TEM observation. At an intermediate height, the effect of the sintering of aggregates is observed, resulting in smaller spherical particles as the laser power increases. A higher carrier gas flow rate can even result in a change from nonspherical nanoparticles to smaller, spherical nanoparticles as the laser power increases. Control of the sintering characteristic time using CO_2 laser irradiation in a flame seems to be a promising way of producing smaller, nanosized, and spherical nanoparticles, even for a high carrier gas flow rate (Lee and Choi, 2000). The radial distributions of scattering intensity and morphological change have also been studied by Lee and Choi (2000).

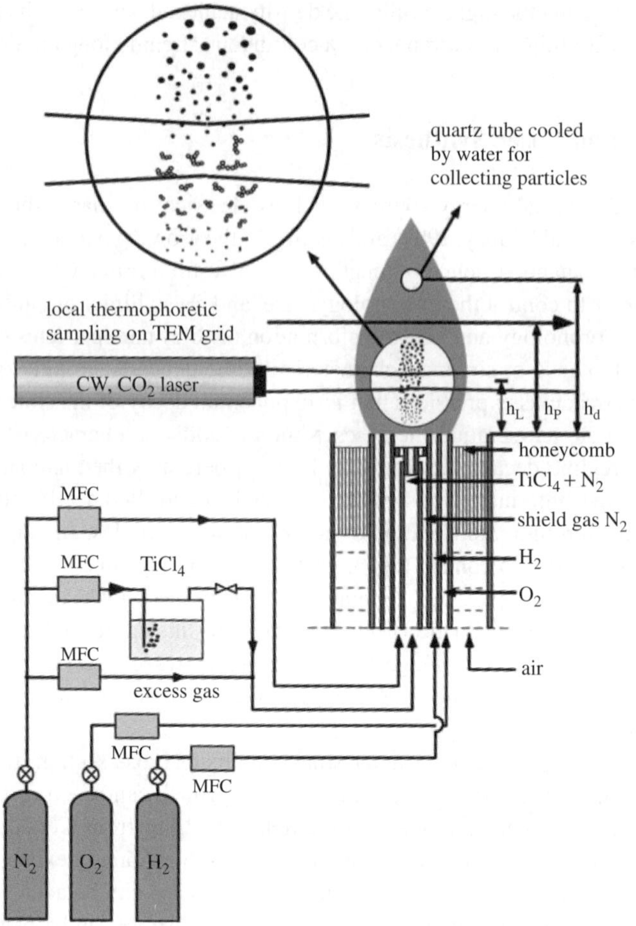

Figure 12.7 *Schematic of coalescence-enhanced synthesis for smaller and unagglomerated nanoparticles with controlled crystallinity at high concentrations: OD (outer diameter of annular space, mm); ID (inner diameter, mm). Reprinted from Lee and Choi (2000). Copyright © 2000, with permission from Elsevier.*

The size, morphology, and crystalline phase of a nanoparticle determine the properties of a nanostructured material. Therefore, mastery of the control of properties ultimately requires control of the size, morphology, and phase of nanoparticles. Highly pure nanoparticles can be produced from various aerosol methods; however, agglomeration is considered almost unavoidable when nanoparticles are generated at the high concentrations necessary for practical application. Efforts to control agglomeration have had only limited success. Lee and Choi (2002) and Lee, Yang, and Choi (2001) have reported that enhancement of the coalescence of nanoparticles using laser-beam irradiation on aggregates formed in flames might be a solution for this problem and successfully controls the size, morphology, and crystalline phase of high-concentration nanoparticles of silica and titania. This principle was demonstrated by not only synthesizing smaller and unagglomerated nanoparticles but also generating them in high concentrations. In addition, it was shown that the method was capable of controlling even the crystalline phase of titania nanoparticles. Surprisingly, stable rutile titania particles have been transformed into metastable anatase, and the weight percent of each phase could be controlled. Studies (Lee and Choi, 2002; Lee, Yang, and Choi, 2001) using CO_2 laser irradiation have revealed that the laser irradiation can alter the phase from rutile to anatase. For example, 16.6% rutile wt% in particles collected at 65 mm without CO_2 laser irradiation becomes almost 0% when the CO_2 laser beam is irradiated at 25 mm with laser power P = 326 W (see Figure 12.8). It should be emphasized that the original rutile content at 25 mm without laser irradiation was 10.3%. This indicates that the rutile particles at 25 mm could be transformed into anatase by the laser irradiation. This result can be explained by the possible occurrence of melting due to irradiation and subsequent recrystallization after passing the irradiation zone (beam diameter: 3 mm). It is necessary to allow a sufficiently long time for atomic rearrangement, leading to a more compact and stable rutile phase, but the cooling time in the system is very short: on the order of 1 ms. Therefore, the initial solid phase that can be recrystallized from melted titania might prefer a metastable anatase phase. Furthermore, the melting process should reduce the necking area of the aggregates, correspondingly decreasing the surface free energy, which has been considered a driving force in transforming anatase to rutile. Therefore, a more stabilized anatase phase can be produced, and this may be one of the reasons why anatase particles recrystallized at the low irradiation position of $h_L = 11$ mm maintain their phase even after reaching the collecting position $h_p = 65$ mm.

An even more sophisticated example is the composite nanoparticles that have been extensively studied as a way of improving the characteristics of single-component nanomaterials or of generating different properties from their own constituents (Vollath and Szabo, 2004). In particular, composite SiO_2/TiO_2 nanoparticles have attracted much attention due to their

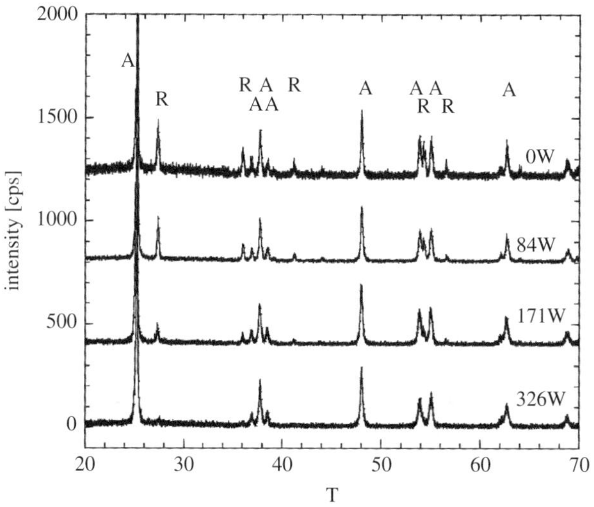

Figure 12.8 *Variations in X-ray diffraction profiles of TiO_2 particles for different CO_2 laser powers; $h_L = 25$ mm, $h_p = 65$ mm. Reprinted from Lee and Choi (2002). Copyright © 2002, with permission from Elsevier.*

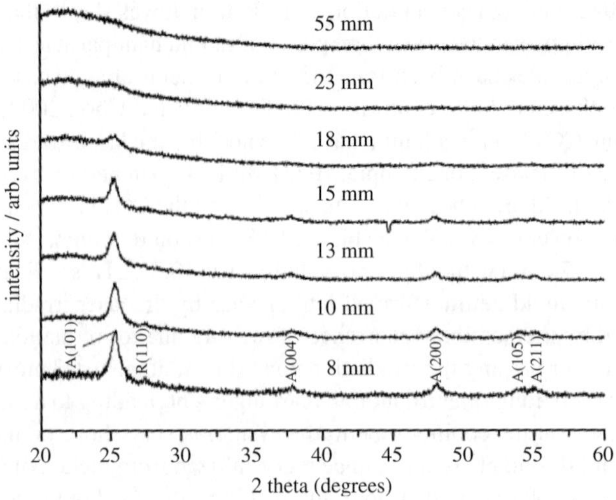

Figure 12.9 *X-ray diffraction patterns of collected composite powder with increasing height from the burner surface, without laser irradiation. Reprinted from Sheen* et al. *(2012). Copyright © 2012, with permission from Elsevier.*

novel catalytic, mechanical, and optical characteristics (Stark and Pratsinis, 2002). Several preparation procedures have been studied, including flame hydrolysis, impregnation, precipitation, and sol–gel methods (Dutoit, Schneider, and Baiker, 1995). Again, among these preparation methods, flame aerosol synthesis has the advantage that it can produce materials of high purity in a single-step continuous process. A CW CO_2 laser is irradiated on SiO_2/TiO_2 composite nanoparticles generating within a flame, in order to control *in situ* the crystalline phase. When the nanoparticles have been created and are growing along the flame then, depending on the composition, the anatase phase may dissolve into an amorphous mixed phase with the silica. However, under laser irradiation the anatase phase survives much better. The most interesting and puzzling result is that not only is the laser irradiation capable of protecting the already existing anatase phase from dissolution, but it can *recover* the crystallinity from the amorphous phase even after the anatase phase has completely dissolved at some height over the burner head (see Figures 12.9 and 12.10).

Applied above that height, not only is the laser irradiation shown to transform originally amorphous composite nanoparticles into crystalline ones, but it also controls the crystallite size of TiO_2 embedded in SiO_2 nanoparticles by altering the irradiating laser power (Figure 12.10).

X-ray diffraction analysis and TEM analysis have confirmed that the size of TiO_2 anatase crystallites in silica nanoparticle increases with increasing irradiation laser power (Figure 12.11). It has been shown that the corresponding absorption edge of SiO_2/TiO_2 composite nanoparticles can be controlled by changing the crystallite size of TiO_2 inside SiO_2 nanoparticles.

12.5 Laser-Induced Synthesis

Usually, the laser-induced synthesis of nanoparticles deals with the laser ablation or evaporation of a solid target. However, some gas-phase chemical reactions can be initiated

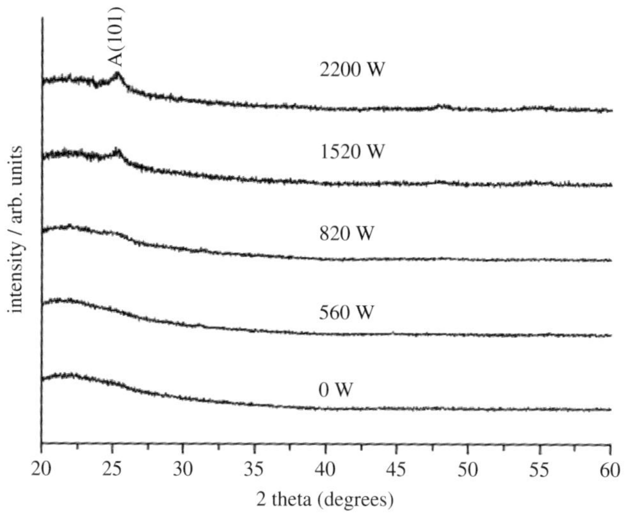

Figure 12.10 *X-ray diffraction pattern evolution for particles obtained in a flame with laser irradiation for different laser powers; $h_P = 55\,mm$, $h_L = 28\,mm$. Reprinted from Sheen* et al. *(2012). Copyright © 2012, with permission from Elsevier.*

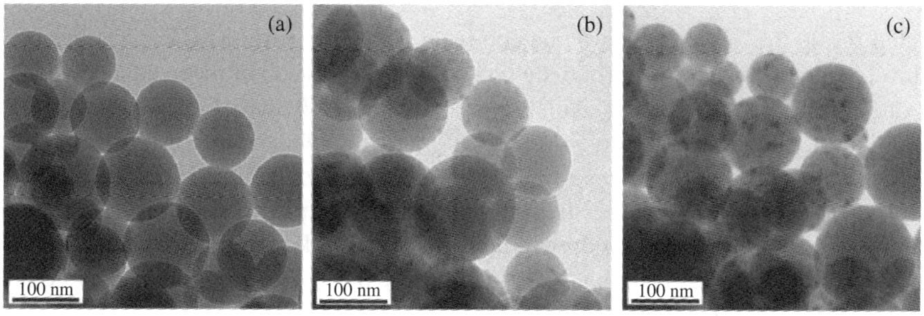

Figure 12.11 *High-resolution TEM images of SiO_2/TiO_2 composite nanoparticles obtained with an increasing power of irradiating CO_2 laser; $h_P = 55\,mm$, $h_L = 28\,mm$: (a) 0 W, (b) 820 W, (c) 2200 W. Recovery of the anatase phase from the amorphous one can clearly be seen. Reprinted from Sheen* et al. *(2012). Copyright © 2012, with permission from Elsevier.*

and supported exclusively by the laser irradiation (also called 'laser photolysis' or 'laser pyrolysis'; Schmidt-Ott, 2011) without any assistance from oxygen or any other oxidizers, like the well-known dissociation of ethylene C_2H_4 molecules under irradiation by the CW CO_2 infrared laser beam with wavelength 10.56 μm, which is efficiently absorbed by the molecules being resonant to the frequency of the wag mode of the CH_2 group at $949\,cm^{-1}$. The multiphoton absorption leads to the molecular dissociation, so that carbon and hydrogen are released at sufficiently high laser power density. Carbon and hydrocarbon clusters produced by the dissociation can further condense into carbon nanoparticles, mostly carbon soot when the laser power is not very high. Additionally, polycyclic aromatic hydrocarbons (PAHs) are generated. Moreover, some types of carbon nanoparticle cannot be generated at all without the laser assistance. Indeed, experiments into the irradiation of acetylene

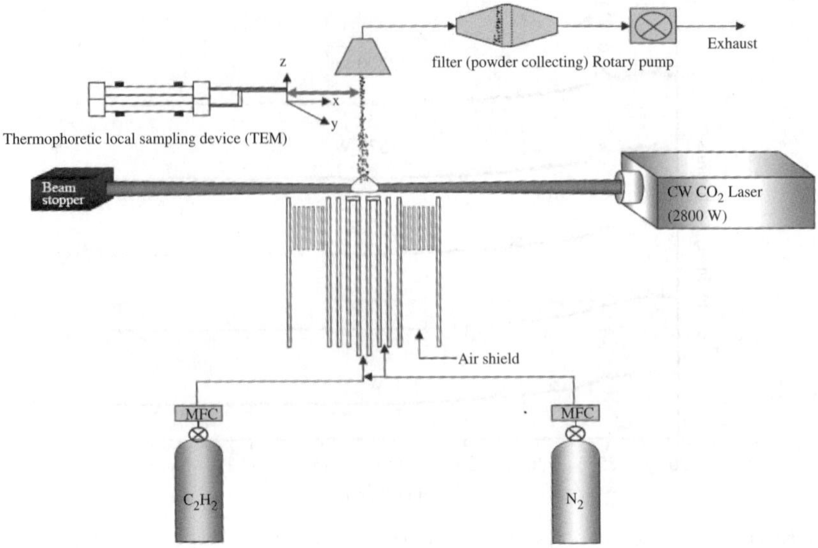

Figure 12.12 The experimental set-up. The acetylene flow interacts with the laser beam. The acetylene flow is protected from oxygen in the air by a nitrogen gas shield. Reprinted with permission from Pikhitsa et al. *(2005). Copyright 2005 by the American Physical Society.*

injected inside the oxyhydrogen flame cone and irradiated with the CW CO_2 laser led to the production of polyhedral carbon-shell nanoparticles instead of the usual acetylene soot that one would expect from the oxidation of acetylene in flames (Choi *et al.*, 2004). The generation of carbon shells with a developed graphene layered structure was a puzzle in itself in Choi *et al.* (2004), because the role of the oxyhydrogen flame was not understood. In fact, the sp2 graphene layered structure helped the nanoparticles survive the high temperature while passing the oxyhydrogen flame feather, unlike PAH or soot particles, which could form even without the laser action. Virtually no particles were produced without laser irradiation when the oxyhydrogen flame was on.

Later it was established (Pikhitsa *et al.*, 2005) that the reactions leading to the production of carbon shells could occur without any oxyhydrogen supporting flame and even when the acetylene was protected from oxygen, despite the fact that acetylene does not absorb the 10.56 μm laser irradiation (Figure 12.12).

The acetylene gas temperature in this reaction zone flow was very low: below 500 °C. This hints that athermal processes were responsible for the generation of carbon shells in Figure 12.13. But the reaction of acetylene dissociation can still be self-supported through steady-state reaction intermediates that do absorb the laser irradiation. The spectroscopy of the much diluted (so-called 'blue') acetylene or ethylene flames irradiated with the CO_2 laser beam proves that there is a substantial interaction between the laser beam and the intermediates, and that this interaction is especially pronounced for acetylene, which contains less hydrogen and thus allows more C_2 dimers (which are the precursor for carbon shells) to be generated without hydrogen passivation (Figure 12.14).

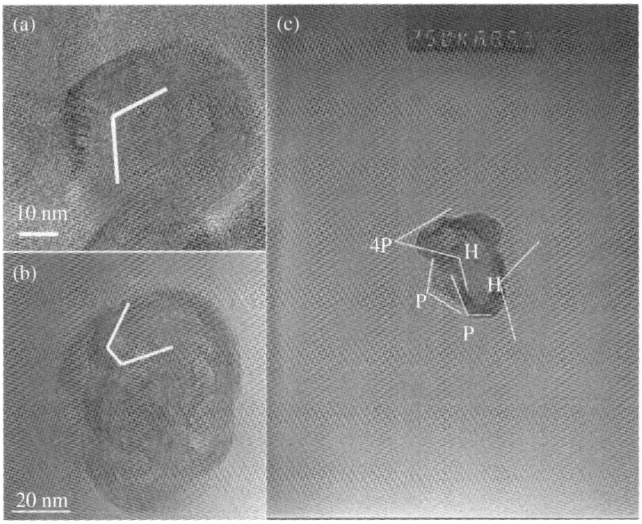

Figure 12.13 *The geometry of various type shell-shaped carbon nanoparticles, seen in TEM images. (a,b) The apex angle, 112.88°, for a five-member ring corner is marked with white lines. (c) The pentagon and heptagon defect-induced angles are marked with the letters P and H, respectively. Reprinted with permission from Pikhitsa* et al. *(2005). Copyright © 2005 by the American Physical Society.*

To start the laser-assisted reaction, the intermediates should be ignited only using another source. This initial ignition can begin with an ordinary acetylene diffusion flame under the laser-beam irradiation and then, after the flame is extinguished by the inert gas shield (nitrogen or argon) that blocks atmospheric oxygen, the reaction can continue without any oxygen or any flame, and only the CW CO_2 laser supporting the self-sustained reaction of acetylene dissociation to produce carbon nanoparticles, with some interesting results. Again, pure acetylene, which does not absorb the laser beam frequency, is very efficient in the C_2H_2/laser synthesis of carbon polyhedral shell-shaped nanoparticles through the laser-absorbing intermediates. The peculiar electronic structure of these nanoparticles, namely their polyhedral corner defect states (Figure 12.13), provides specific optical properties. For example, they show a famous 220 nm extinction hump in the absorption curve (Pikhitsa *et al.*, 2005) found in interstellar dust. It is important that the laser-induced synthesis directly from N5-grade acetylene produces no impurities in the carbon nanoparticles, which makes them interesting to use as templates and as a well-organized material when the nanoparticles are assembled. Two examples of the capabilities of nanostructured materials made of polyhedral carbon shells are given in the following paragraphs: cold electron field emission and H_2 and CO gas sensing at room temperature.

The electronic defects in these nanoparticles explain the high enhancement factor for their cold electron field emission. Indeed, cold electron field emission from carbon particles was tested against carbon nanotubes (Choi *et al.*, 2004; Altman, Pikhitsa, and Choi (2004a, b) and it was found that the former performed even better than the latter, which

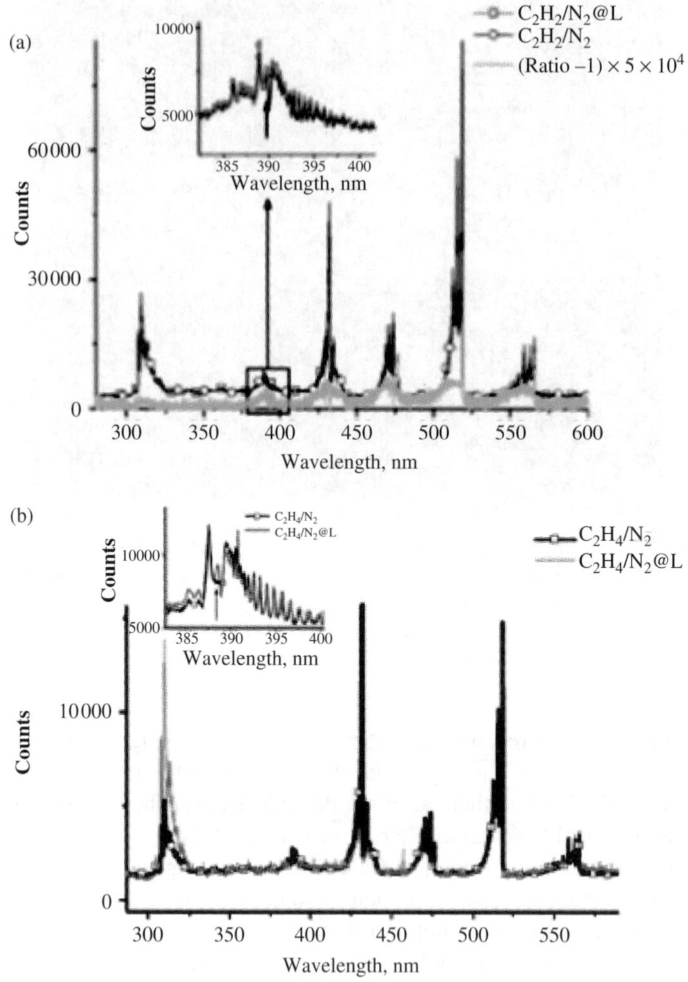

Figure 12.14 The spectra of nitrogen-diluted acetylene blue flame, the flame under laser irradiation, and the ratio (laser on/laser off) of the spectra. (a) The line profile demonstrates the increase in numbers of C_2 dimers in the flame due to the laser irradiation. (b) On contrast, a nitrogen-diluted ethylene flame demonstrates only OH-band enhancement at 310 nm, due to there being more hydrogen in the molecules. Insets show the CN/CH violet system and the (0-0) band head of the CN(B-X) violet system (arrow). Adapted with permission from Pikhitsa, Kim, and Choi (2011).

were assumed to have a huge electric field enhancement factor due to their high aspect ratio. The puzzle of why carbon-shell nanoparticles showed nearly the same performance despite not having this aspect ratio was solved by Altman, Pikhitsa, and Choi (2004a, b). It turned out that the structure of carbon nanoparticles dictates the existence of electron states that allow two-electron processes during field emission. The manifestation of this peculiar mechanism is the characteristic 'knee' in the Fowler–Nordheim plot (Figure 12.15). There being one electron trapped near the surface of the carbon nanoparticle prompts another

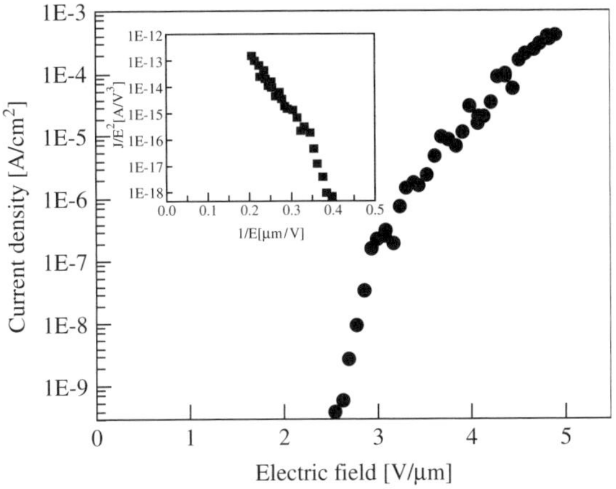

Figure 12.15 *Typical cold electron field emission from shell-shaped carbon nanoparticles. The turn-on voltage is 2.5 V/μm. Inset shows the Fowler–Nordheim plot, with a characteristic 'knee'. Adapted with permission from Choi* et al. *(2004).*

electron to leave the material for vacuum (Altman, Pikhitsa, and Choi, 2004a, b). The electric field of the first electron greatly enhances the local electric field and therefore creates a huge enhancement factor, which had previously been ascribed to the huge aspect ratio of the nanotubes. It then became clear that the caps on the round tips of nanotubes actually perform in the same way as nanoparticles. This example demonstrates the usefulness of properly nanostructured material.

The other example of a great performance by carbon nanostructured material is the room-temperature H_2 and CO gas sensing by carbon nanoparticle chains (Kim *et al.*, 2011). Defects, combined with the high conductivity of the individual nanoparticles, can produce high sensitivity to the environment, which can be utilized in sensors. The electron transport through a chain is restricted by the charge transport between nanoparticles, and this feature, together with the chemical sensitivity of the resistance to hydrogen molecules even at room temperature, can be used. It is quite understood that for hydrogen gas, sensing of any temperature elevation is not welcome, because of the danger of ignition when oxygen is available. Yet there was no success in sensing CO and H_2 at room temperature without the use of any catalysts or hybrids. However, it was found (Kim *et al.*, 2011) that it was possible to use a crystalline shell-shaped carbon nanoparticle (SCNP)-based gas sensor to detect CO and H_2 at room temperature. A SCNP-based gas sensor reversibly detects reducing gas molecules such as CO and H_2 at room temperature, both in air and in an inert atmosphere. Those SCNPs were functionalized by nitric acid treatment, which was critical in showing the gas-sensing property, and were arranged into chains or films. The SCNP sensor could recover even in an inert atmosphere, which suggests a different mechanism is responsible for sensing H_2 molecules at room temperature than that applied in conventional catalyst-based gas sensors. The SCNPs were synthesized through CW CO_2 laser irradiation of pure acetylene flow (Pikhitsa *et al.*, 2005). Pure acetylene was fed into the burner and an acetylene flame was formed by ignition. An infrared CO_2 laser (Bystronic, BTL2800) irradiated the acetylene flame and nitrogen gas was used as a shield gas to block the reaction from the ambient air. Following laser irradiation, crystalline carbon nanoparticles could be produced without any catalysts. Carbon soot particles from acetylene flame were prepared as well. The carbon particles generated were collected by a filtering system connected to a vacuum pump. The filter was placed above the burner to collect the generated particles directly from the particle stream. SCNPs were patterned into sensor devices by the ion-induced focusing method (Kim *et al.*, 2006). The ion-induced focusing approach (Kim *et al.*, 2006) requires a well-dispersed charged aerosol form of SCNPs. In order to obtain a well-dispersed solution of hydrophobic SCNPs, chemical treatment with nitric and sulfuric acid was conducted at 110 °C. Acid treatment can make the surfaces of SCNPs hydrophilic by attaching functional groups such as carboxylic and hydroxyl groups; it can separate SCNPs and thus maintain the SCNPs to be well dispersed in the solvent. Au (electrode, 50 nm)/Ti (adhesion layer, 10 nm) was formed on the SiO_2 (thermal oxidation). A chemically functionalized SCNP-based gas sensor works for low concentrations of CO and H_2 at room temperature, even without the Pd or Pt catalysts commonly used to split H_2 molecules into reactive H atoms, while metal-oxide gas sensors and bare carbon-nanotube-based gas sensors used to sense CO and H_2 molecules can operate only at elevated temperatures. The splitting of H_2 molecules may occur on numerous graphene platelet edges, as suggested by Browning *et al.* (2002). Additional dissociative adsorption can occur on the extremely curved parts of the outer shells of acid-treated SCNPs, like the predicted H_2 dissociation on the small diameter of SWCNTs (Tada, Furuya, and Watanabe, 2001). It was suggested by Kim *et al.* (2006) that the mechanism of the most efficient dissociation could be the one that utilizes both graphene edges and their functionalization with oxygen groups. Indeed, the hydrogen molecule may dissociate on the edges of the oxidized/functionalized graphene sheets (the OH and C=O edge groups may be relevant) in the following reversible scheme. First, $OH + H_2 + M \rightarrow H + H_2O + M$ produces H atoms (M denotes the platelets). After dissociation, the migration of H atoms can resemble the spillover process when H atoms migrate over CO, only instead of Pd or Pt, the acid-treated carbon platelet edges now play their catalytic role in the dissociation (Psofogiannakis and Froudakis, 2009). Second, the reverse 'association' process takes place when $H_2O + M \rightarrow O + H_2 + M$ (see Figure 4 in Zhu *et al.*, 2002). Here the bound water molecule plays the role of a mere intermediate and therefore the recovery process does not require ambient oxygen and can occur in an inert N_2 atmosphere, as was observed (see Figure 12.16).

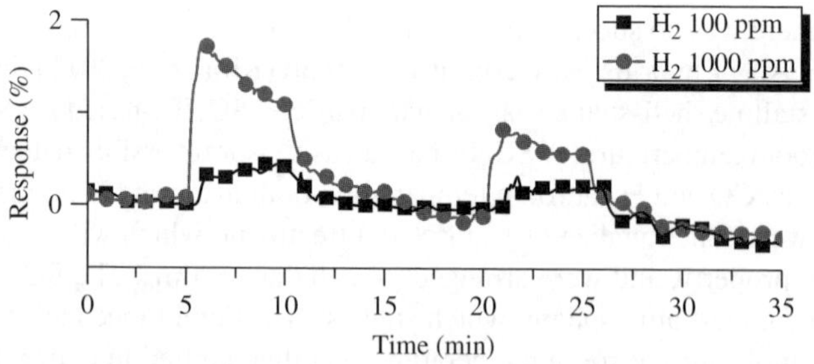

Figure 12.16 *Sensing performance of an acid-treated soot-particle sensor at room temperature. The response to hydrogen was obtained in nitrogen atmosphere in order to check the recovery mechanism. From Kim* et al. *(2011). © IOP Publishing. Reproduced by permission of IOP Publishing. All rights reserved. See plate section for colour version.*

A pristine SCNP-based gas sensor was also examined in Kim *et al.* (2011), to elucidate the role of functional groups formed on the surfaces of functionalized SCNP. A pristine SCNP gas sensor showed no response to reducing gases at room temperature but a significant response at elevated temperatures, indicating a different sensing mechanism from a chemically functionalized SCNP sensor.

12.6 Metal-Powder Combustion

A metal combustion method is one way of preparing nanoparticles, mostly metal nanooxides. Several studies have reported on the optimization of metal combustion processes for use as a solid propellant for rockets, a refractory material, a catalyst, and so on. The combustion of solid precursors can take place with the type of self-propagating combustion well known with Mg metal. A conventional metal combustion method requires the reactant to be incorporated with a cylindrical pellet and ignition process. In this process, the reactant in the pellet, ignited by an external source such as a tungsten coil, a laser, and a microwave, either locally or uniformly initiates an exothermic reaction and maintains self-burning, generating oxide particles. Yet a steady-state generation of nanoparticles would be desirable. Such a flame metal combustion method for continuously synthesizing metal-oxide nanoparticles was realized by Yang *et al.* (2010). Micro-sized metal powder precursors were injected into a hydrogen–oxygen flame through a newly designed feeder system (Figure 12.17) in order to be completely converted into metal-oxide nanoparticles with good crystallinity. There is no limit to the selection of precursors, including metal, metal chloride, and metal organic powders, for the synthesis of nanoparticles. Various oxides, such as magnesia (Figure 12.18), iron oxides, and zinc oxide, have been demonstrated to be successfully generated using this method. The sizes of oxide nanoparticles are controlled by varying flame temperatures and residence times.

One of the more impressive aspects of metal combustion is the possibility of controlling doping, at least at low doping levels; it allows one to discover and assign an interesting pair of zero-phonon lines (ZPLs) in Mn^{2+}-doped MgO nanoparticles (Altman *et al.*, 2003). The emission from nanoscale MgO is of particular interest, related to the recent finding of the growth of carbon nanotubes on MgO nanocubes and the prediction of a new type of nanoscale optoelectronic circuitry based on MgO nanocrystals (Vajtai *et al.*, 2002). Zero-phonon sharp lines have been found in macroscopic PL from MgO nanocrystals and MgO films produced by the burning out of Mg particles in air (Altman *et al.*, 2003). These lines are attributed to the emission of different ions (Mn^{2+} and Cr^{3+}) in the MgO matrix. Mn^{2+} ZPLs have never been seen in such a system as bulk $MgO{:}Mn^{2+}$, despite intensive study, which makes Altman *et al.*'s (2003) finding a fundamental one. The sharpness of the observed lines in comparison with those from a quantum dot array is of interest for optical nanoelectronics.

The nanocrystalline MgO can be easily synthesized by burning out Mg particles in air. The particles collected from MgO smoke are perfect cubes, with an average size of about 40 nm. By deposition of the product of Mg combustion on to a substrate, MgO films are also produced. Figure 12.19 shows TEM images of nanocrystalline MgO nanocrystals and film. X-ray diffraction analysis demonstrates a high crystallinity of MgO synthesized. In order

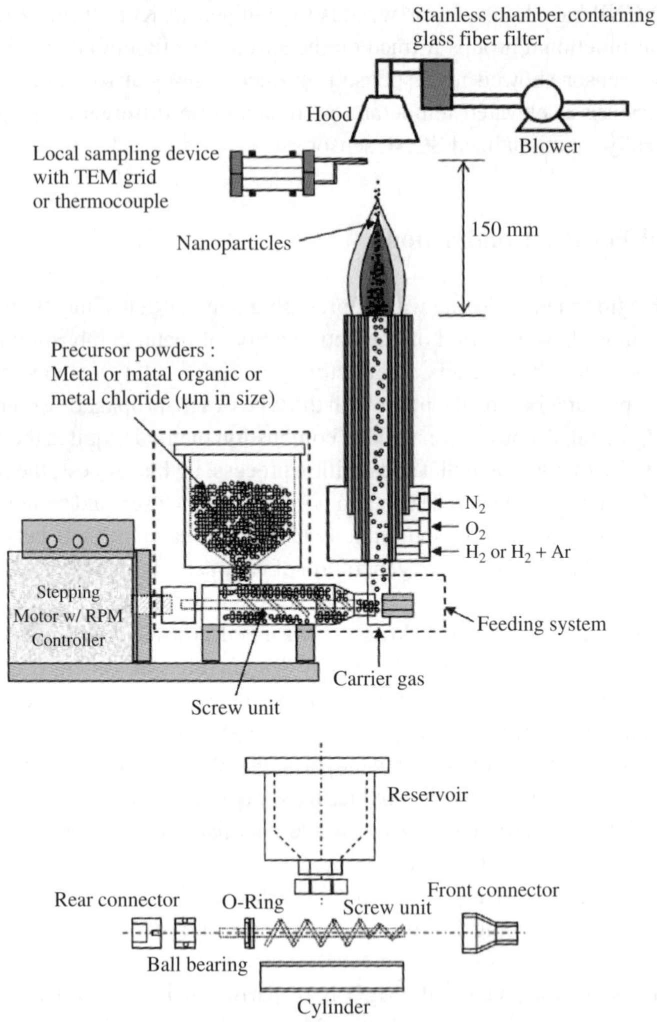

Figure 12.17 *Experimental set-up of the flame metal combustion method and details of the newly devised feeding system. Micro-sized metals such as Mg (dia.: 50–100 μm, oval, 98% purity), Zn (dia.:10–40 μm, sphere, 90% purity), Fe (dia.: 3–13 μm, oval, 90% purity), metal organics (Fe ($C_5H_7O_2)_3$), and metal chloride ($PtCl_2$) precursor powders between 50 and 100 μm are fed through the center nozzle of the co-axial burner by the continuous feeding system and the carrier gas. Solid precursors stocked in the reservoir tank are supplied by a screw-type rod inside a cylinder. The feeding rate of precursors is controlled by the RPM (revolution per minute) controller connected to a stepping motor, which rotates 0.08° per step. These precursor powders are carried by nitrogen gas and injected into a hydrogen–oxygen flame. Hydrogen and oxygen are injected through the two concentric annuli next to the center nozzle to create a hydrogen–oxygen diffusion flame. Compressed dry air is fed through the outer tube, equipped with a honeycomb to stabilize the flame. The flame temperature is controlled by the $[H_2]/[O_2]$ ratio and measured using a B-type thermocouple. The measured flame temperatures are corrected for radiative heat loss. Metal oxide nanoparticles are collected using a localized thermophoretic sampling device. The morphology of metal-oxide nanoparticles is observed using a transmission electron microscope (TEM, model JEM-2000, JEOL). Reprinted from Yang* et al. *(2010). Copright © 2010, Elsevier.*

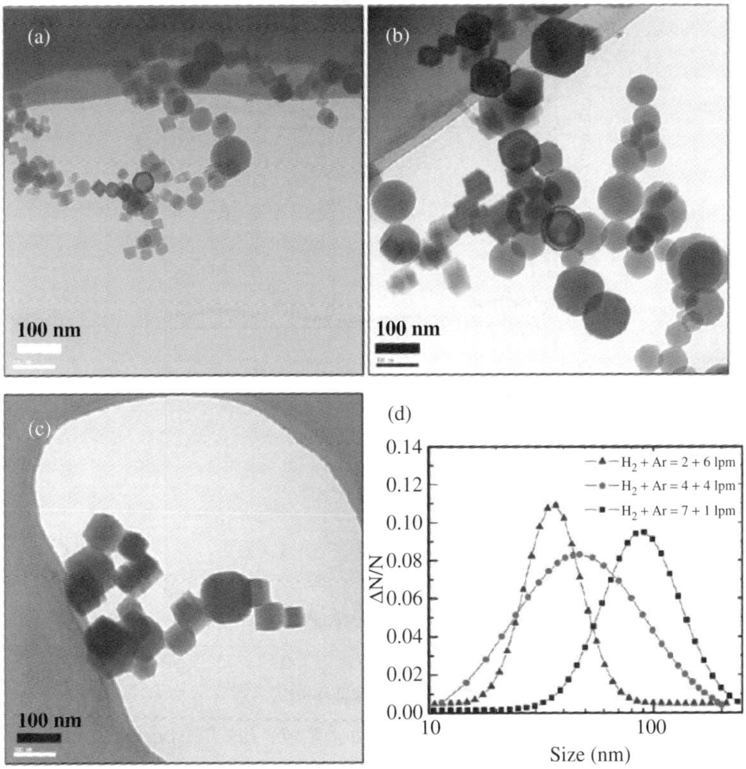

Figure 12.18 *TEM images of magnesium oxide nanoparticles prepared at different hydrogen flow rates. (a) 2 lpm hydrogen gas. (b) 4 lpm hydrogen gas. (c) 7 lpm hydrogen gas. (d) Particle size distributions for all three cases. Reprinted from Yang* et al. *(2010). Copright © 2010, Elsevier. See plate section for colour version.*

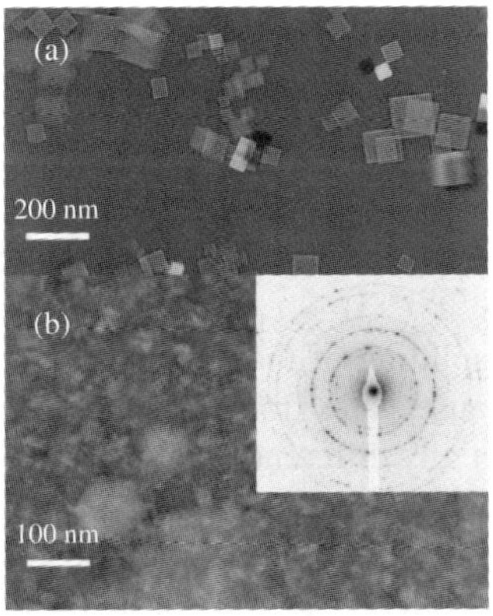

Figure 12.19 *TEM images of nanocrystalline MgO. (a) MgO nanocrystals. (b and inset) MgO film. The electron diffraction pattern in (b) confirms the high crystallinity of the film. Reprinted with permission from Altman* et al. *(2003). Copyright © 2003, Rights managed by AIP Publishing LLC.*

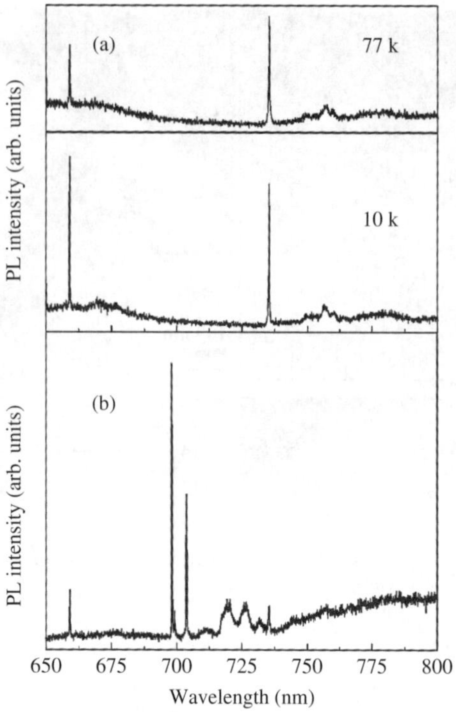

Figure 12.20 *CW PL spectra from MgO of nanoscale. (a) PL spectra from an MgO film doped with Mn excited by an He-Cd laser at different temperatures. (b) PL spectrum from nanocrystalline MgO doped with Mn and Cr excited by an Ar-ion laser at 10 K. Reprinted with permission from Altman* et al. *(2003). Copyright © 2003, Rights managed by AIP Publishing LLC.*

to dope MgO with transition-metal-impurity Mn and Cr, mixtures of pure Mg particles and Mn- or Cr-containing salts are burnt.

The ZPLs (Figure 12.20a) found for Mn^{2+} transitions can be assigned using the Tanabe–Sugano diagram for the $3d^5$ system. The low-energy 735 nm ZPL undoubtedly comes from the lowest possible transition in the system, namely the transition from the lowest excited $^4T_1(^4G)$ state to the ground $^6A_1(^6S)$ state. It can be easily shown that the energy of the next excited state, $^4T_2(^4G)$, cannot match the 659 nm ZPL, while the 735 nm ZPL is already assigned to the $^4T_1(^4G)$–$^6A_1(^6S)$ transition. At the same time, a transition from a state that arises from splitting of the ^2I term on to the ground $^6A_1(^6S)$ state can produce the 659 nm ZPL. Because states from different electron terms are responsible for two observed ZPLs, two essentially different Huang–Rhys parameters may correspond to the transitions. The absence of phonon replicas for the 659 nm ZPL (which is likely weakly coupled with phonons) and the existence of them for the 735 nm ZPL (see Figure 12.20a) support this idea. The aforementioned idea of utilizing the ^2I term in order to match the observation is the only way to explain such an unforeseen fact as the simultaneous appearance of two ZPLs originating from the same transition ion in the matrix. The typical CW PL spectrum measured at 10 K for the Cr-doped sample (which obviously contains some amount of Mn impurity) is shown in Figure 12.20b. Note that ZPLs originating from Mn^{2+} transitions in a cubic crystalline field were previously seen only in the PL spectra of narrow-band materials. The thermal quenching and the PL yield observed in the present work on ZPLs are similar to those for a PL when the emitting states are excited by close electron states. With such a wide-gap substance as MgO, these states should lie deep inside the forbidden band, and therefore cannot exist in ordinary MgO crystals. The energy transfer between the environment and excited states of Mn^{2+} ion thus does not occur, making the Mn^{2+} ion excitation impossible. This explains why, in usual MgO crystals, Mn^{2+}-originating ZPLs are not observed. Note that if nanoparticles are generated by vapor condensation during combustion, a high concentration of low-coordination defects (corners, terraces, etc.) on the particle surface and of defects in the substance will occur. Just the presence of these defects, which is a distinctive feature of particles synthesized by combustion, leads to the existence of states deep inside the forbidden band and to the possibility of exciting an Mn^{2+} ion in MgO matrix, and therefore to the appearance of the Mn-originating ZPLs in PL. We can then claim that nanoscale oxides generated by metal combustion form a distinctive class of materials and, of particular interest, allow the excitation of perfect ZPLs.

12.7 Spark Discharge

Spark discharge at atmosphere is a simple and reliable way of generating nanometer-sized particles in various gas-phase synthesis methods (Biskos *et al.*, 2008). It can generate charged aerosols using a simple set-up, which can be exploited to construct nanostructures via a controlled electric field (Krinke *et al.*, 2002). Rod-to-rod-type spark discharge generators (RR-SDGs) have been widely studied and the effects of their parameters have been well defined (Tabrizi *et al.*, 2009). A point-to-grid configuration has been studied in order to demonstrate that aerosol production by corona in clean gas at atmospheric pressure is related to the development of plasma filaments that reach the low-field electrode grid in streamer and spark-discharge regimes with positive-point discharge, while aerosol generation occurs only in a spark regime with negative-point discharge (Borra *et al.*, 1998). Although an SDG is known to produce nanometer-sized particles, charged aerosols of less than 10 nm diameter at high concentrations tend to get agglomerated due to the post-discharge electrostatic agglomeration of bipolar nanoparticles resulting from post-discharge diffusion charging in bipolar ion clouds, which has been studied by measuring post-spark ion density for the RR-SDG. Prevention of the agglomeration and generation of a larger amount of smaller-sized charged aerosols from this facile method could further strengthen its advantages and widen its range of applications. The agglomeration of particles in an SDG might be lessened by controlling such operational parameters as spark frequency, spark energy, and the flow rate of a carrier gas over about 101 pm (Tabrizi *et al.*, 2009). In such systems as furnace reactors and flame reactors, ions have been utilized to prevent the agglomeration of particles (Nakaso *et al.*, 2003).

A pin-to-plate-type SDG (Han *et al.*, 2012) has been studied and compared with a conventional RR-SDG. The pin-to-plate-type generator has asymmetric electrodes, comprising a pin with a sharp tip as positive electrode and a grounded plate with a narrow exit hole at the

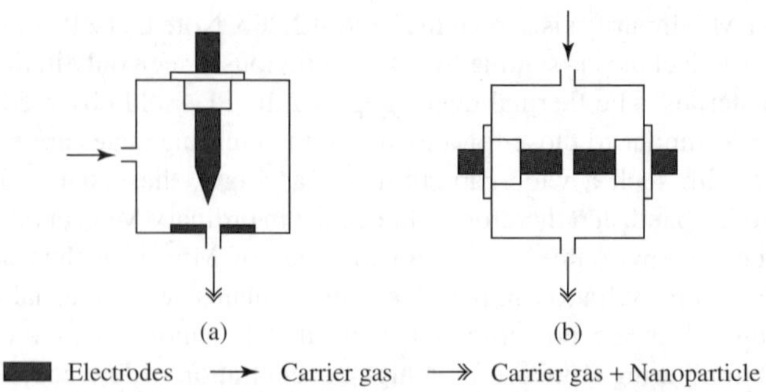

Figure 12.21 *Two different configurations of spark discharge generator. (a) Pin-to-plate-type electrode configuration. (b) Rod-to-rod-type electrode configuration. Reprinted with permission from Han* et al. *(2012). Copyright © 2012, Elsevier.*

center (Figure 12.21a). In comparison to the conventional RR-SDG (Figure 12.21b), the pin-to-plate-type generator produces much smaller, unagglomerated charged nanoparticles with a narrower size distribution and nearly equal charge distribution at higher concentrations (Figure 12.22). In-Sn alloy, silver, and copper nanoparticles have been tested using these two different configurations, and their size, morphology, and charge state analyzed with respect to spark parameters and flow pattern in each configuration. The generation of much smaller, unagglomerated nanoparticles with a narrow size distribution via the pin-to-plate-type generator can be mainly attributed to a much faster transport of as-produced particles than is achieved by the rod-to-rod type.

12.8 Assembling Useful Nanostructures

After nanoparticles have been generated, the question is how to assemble them in order to make use of their interaction and eventually produce well-controlled nanostructured materials. The use of such nanoparticles as building blocks in fabricating quantum, photonic, or biological nanodevices requires a robust particle patterning. Various approaches have been reported, utilizing different physical forces, such as microfluidic flow field, magnetic field, aerodynamic inertial force, DNA scaffold, directed printing assembly, and nanoxerography. Recently, several techniques for enhancing the resolution of particle patterns have been reported, including ion-induced parallel focusing of charged aerosols, a p-n junction-based electrostatic method, and electrostatic funneling using surface chemical treatment in liquid phase, which enables particle patterns smaller than the original patterns to be generated, due to their focusing ability.

There is a most precise way of focusing generated nanoparticles, through spark discharge into prescribed places on a substrate. This method utilizes nanoscopic electric field lenses that naturally appear in any patterned structure if it is properly charged (Kim *et al.*, 2006). The development of nanodevices that exploit the unique properties of nanoparticles will require high-speed methods for patterning surfaces with nanoparticles over large areas and with high resolution. Moreover, the technique will need to work with both conducting and nonconducting surfaces. The ion-induced parallel-focusing approach (Kim *et al.*, 2006)

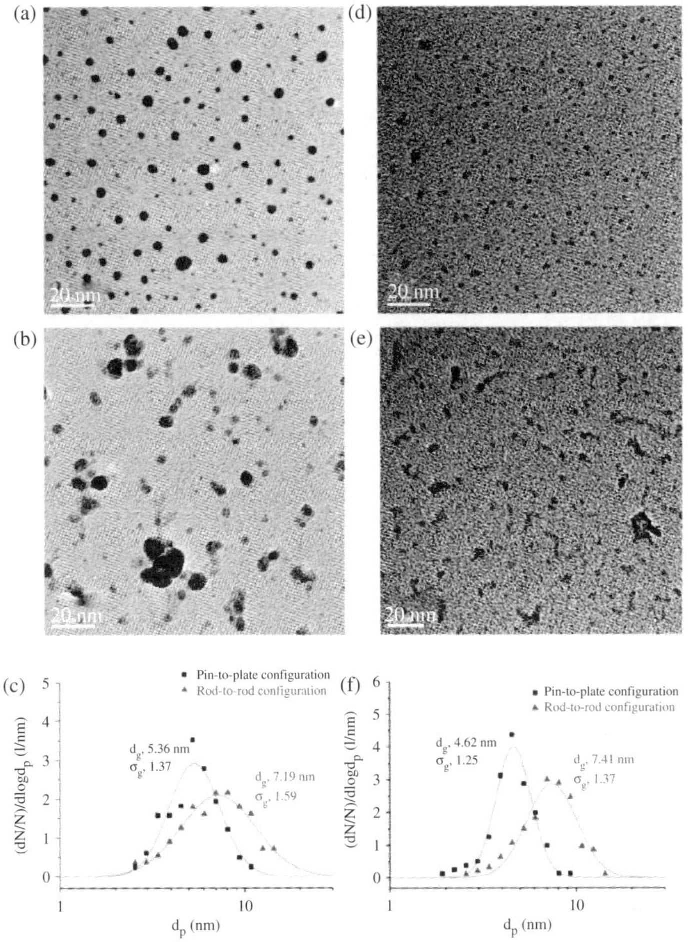

Figure 12.22 *Silver and copper nanoparticles generated by spark discharge with a gap distance of 2.5 mm, argon gas flow rate of 3.5 lpm, and applied positive potential of 5 kV. (a) TEM image of silver particles for pin-to-plate-type SDG. (b) TEM image of silver particles for RR-SDG. (c) Size distributions of silver particles. (d) TEM image of copper particles for pin-to–plate-type SDG. (e) TEM image of copper particles for RR-SDG. (f) Size distributions of copper particles. Reprinted with permission from Han* et al. *(2012). Copyright © 2012, Elsevier. See plate section for colour version.*

satisfies all requirements. Charged monodisperse aerosol nanoparticles are deposited on to a surface patterned with a photoresist (PR), while ions of the same polarity are introduced into the deposition chamber in the presence of an applied electric field. The ions accumulate on the PR, modifying the applied field to produce nanoscopic electrostatic lenses that focus the nanoparticles on to the exposed parts of the surface. It has been demonstrated that the technique can produce high-resolution patterns at high speed on both conducting (p-type silicon) and nonconducting (silica) surfaces. Moreover, the feature sizes in the nanoparticle patterns are significantly smaller than those in the original PR pattern (Figure 12.23).

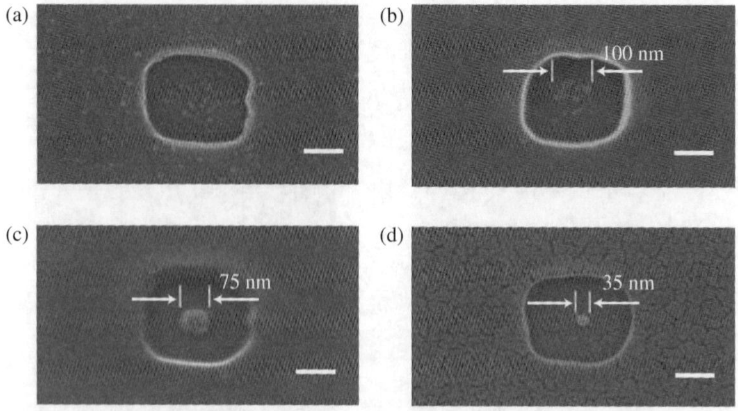

Figure 12.23 *Control of focusing with an increase of N_2 ion flow rate on a substrate with 230 nm-wide and 135 nm-thick PR patterns. Scale bars are 100 nm. (a) Particle deposition with no ion injection. (b–d) Particle deposition with ion injection: (b) 3 l/min (ion concentration, $N_i = 3.31 \times 10^5\ cm^{-3}$), 100 nm particle array consisting of 10 nm Ag particles; (c) 4 l/min ($N_i = 4.73 \times 10^5\ cm^{-3}$), 75 nm array of 10 nm Ag particles; (d) 6 l/min ($Ni = 6.13 \times 10^5\ cm^{-3}$), 35 nm array of 10 nm Ag particles. Adapted with permission from Kim* et al. *(2011).*

The successful approach (Kim *et al.*, 2006) led to the development of a nanoxerography utilizing alternating bipolar surface charge patterns (Lim *et al.*, 2012). The bipolar charge patterns are formed by first depositing positive ions uniformly on a dielectric surface and then transferring negative charges via the conformal contact of a conductive stamp under a given potential. Unlike a conventional nanoxerography that utilizes unipolar charge patterns, the concept (Lim *et al.*, 2012) generates convex-shaped equipotential planes with a large curvature that can act as nanoscopic electrostatic lenses (Figure 12.24).

Through these lenses, positively charged aerosol nanoparticles are focused into the center region of the negative surface charge pattern, leading to a significant reduction of the particle deposition width. Furthermore, the positive surface charge region generates a repelling field that can prevent the deposition of noise particles. The focusing capability can be controlled by changing the ion flow rates. Numerical simulations of the trajectory and deposition of charged particles have been carried out to support the experimental results. The repulsive component is a shape-forming one for the deposit (Kim *et al.*, 2006; Lee *et al.*, 2011) and can be manipulated with the positive charge density. For any given negative (attractive for positive nanoparticles) charge density and for given sizes of the initial xerography pattern, one can always find some critical positive charge density below which there is deposition of a certain size and above which there is no deposition at all (all the nanoparticles are repelled). By manipulating charge density below its critical value, one can focus the size of the nanoxerography feature as illustrated in Figure 12.24, from the initial pattern size down to a one-nanoparticle size resolution.

It is possible to build arrays of 3D nanoparticle structures by injecting both ions and charged aerosols under a given electric field into a chamber containing a PR prepatterned silicon substrate. Ion deposition on to a PR surface generates a nanoscopic electrostatic

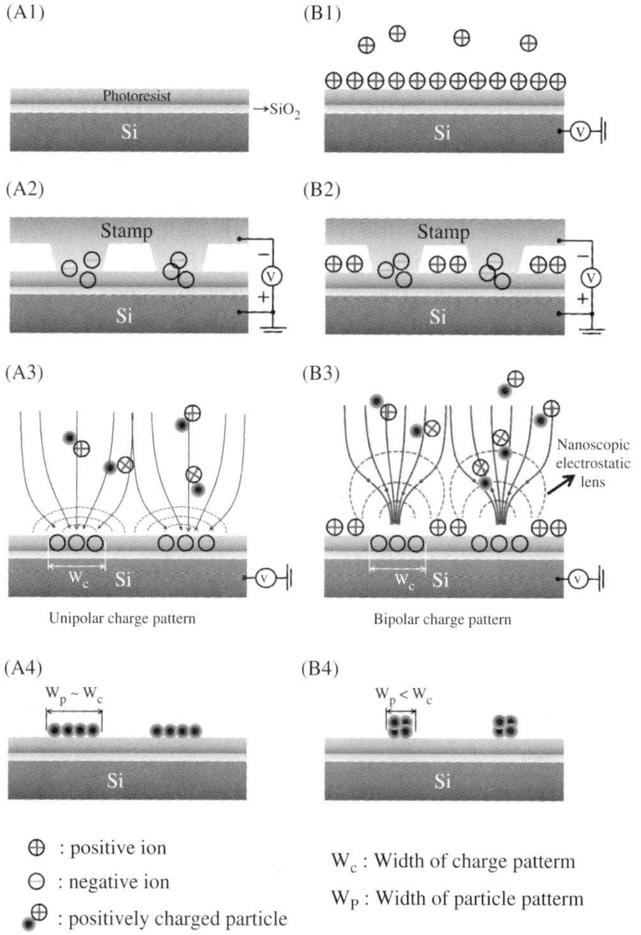

Figure 12.24 *Schematic diagram of conventional and contemporary nanoxerography. (A1) A nonconductive polymer-coated substrate is prepared. (A2) A conductive stamp comes in contact with a PR surface and negative charges are transferred on to the PR. (A3,A4) Due to the Coulombic force, nanoparticles are attracted to the negative surface charge pattern area. (B1) Positive ions are deposited on the PR surface. (B2) A prepatterned metal-coated stamp is contacted with the PR surface and transfers negative charges on to the PR, which replace the positive ions on the contacting area. (B3) Positively charged particles are focused toward the center of the negatively charged surface via a nanoscopic electrostatic lens, induced by alternating bipolar surface charges. (B4) Nanoparticle pattern sizes are much reduced compared to the original negative charge patterns. Reprinted with permission from Lim* et al. *(2012). Copyright © 2012, American Institute of Physics. See plate section for colour version.*

lens, which focuses charged nanoparticles into the center of the open space within each PR pattern in a parallel fashion; this is the principle of ion-induced focusing (Kim *et al.*, 2006) used for the early growth stage. Ions accumulated on the PR create the electric field that reflects incoming nanoparticles from the PR surface and thus prevents their deposition on to it. This repelling electric field forms a kind of electrical scaffold into which charged nanoparticles are not allowed to penetrate. The effect of the electrical scaffold plays a critical role in determining 3D nanoparticle structures. In addition, the enhanced field near the surface of the growing nanoparticle structure, which is defined here as the antenna effect, plays a role in constructing a 3D nanoparticle structure (Figure 12.25).

The spark discharge configuration (see Figure 12.26) is different from a conventional rod-to-rod-type discharge chamber, and this design (Han *et al.*, 2012) is critical to producing the smaller-sized nanoparticles necessary to the construction of stable 3D nanostructures.

It is notable that the shape of a 3D nanoparticle structure is completely different from its root shape; for example, the four-leaf-clover shape shown in Figure 12.27a is different from the root shape of a square. While ion-induced focusing cannot explain this, the electrical scaffold effect caused by repelling electric field from a PR surface plays an important

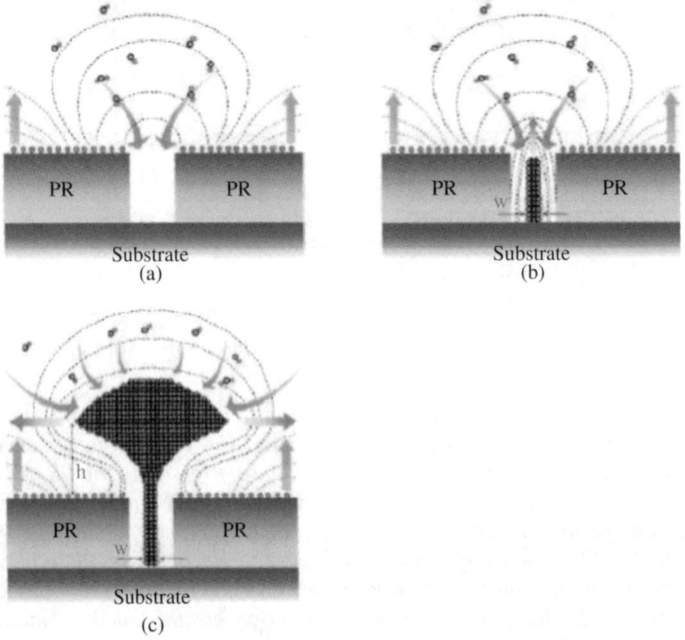

Figure 12.25 *Scheme for the 3D assembly of nanoparticles. (a) Ion-induced electrodynamic focusing guides charged nanoparticles into the center region of the opened Si substrate. (b) Vertical growth of nanostructure within the PR pattern is caused by ion-induced focusing and the antenna effect. (c) Further growth in the lateral direction results from an enhanced antenna effect near the side of the 3D nanoparticle structure. The repelling field is shown on the PR surface. Reprinted with permission from Lee* et al. *(2011). Copyright © 2011, American Chemical Society. See plate section for colour version.*

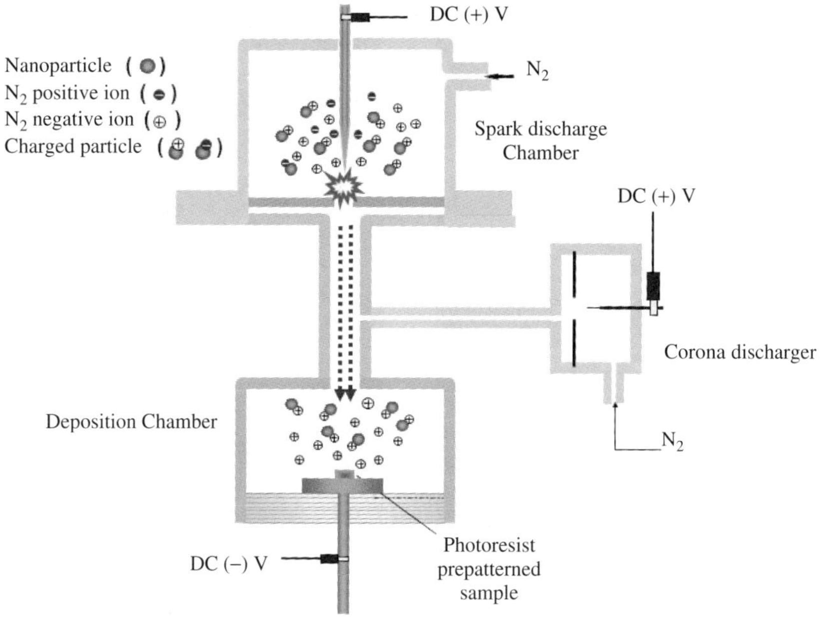

Figure 12.26 *Experimental set-up for the generation and deposition of charged nanoparticles by spark discharge. Charged nanoparticles and ions with mostly positive charge are generated when a positive voltage is applied to the pin electrode and the plane-type electrode is grounded. These are fed into the deposition chamber under an N2 flow rate of 4 l/min. 4 kV negative potential is applied to the substrate in order to attract ions and charged nanoparticles toward it. Therefore, only positively charged ions and nanoparticles move toward the substrate, and a small portion of ions and particles with negative charge cannot be deposited on to it. A corona discharger provides N2 positive ions, which are accumulated on the PR surface prior to deposition. The ions accumulate on the PR surface along the electric field line and form ion-induced electrostatic lens. Charged nanoparticles are convergently guided by the lens and placed at the desired position upon the substrate. As the deposition time increases, various 3D nanoparticle structures, such as hemispheres, pillars, mushrooms, and four-leaf clovers, are obtained from the originally 500 nm-wide square PR pattern (Figure 12.27a) and the 3D nanoparticle structures, such as 3D crosses and flowers, are obtained from originally cross-shaped PR patterns with one side of 500 nm (Figure 12.27b). As observed early at 4 minutes, nanoparticles are convergently guided only within the center region of the PR pattern, due to the ion-induced focusing mentioned earlier (see Figure 12.27a). Reprinted with permission from Lee* et al. *(2011). Copyright © 2011, American Chemical Society. See plate section for colour version.*

role in determining the overall shape of a 3D nanoparticle structure. Note that pure Laplacian growth forbids the closing of the gap between the structural elements, yet the 'petals' at 90 minutes in Figure 12.27a nearly contact each other due to the deposition of those nanoparticles, which move in a ballistic regime, thus deviating from the Laplacian field lines. One can see the conditions of the deposition for which the 3D network will eventually be fully connected.

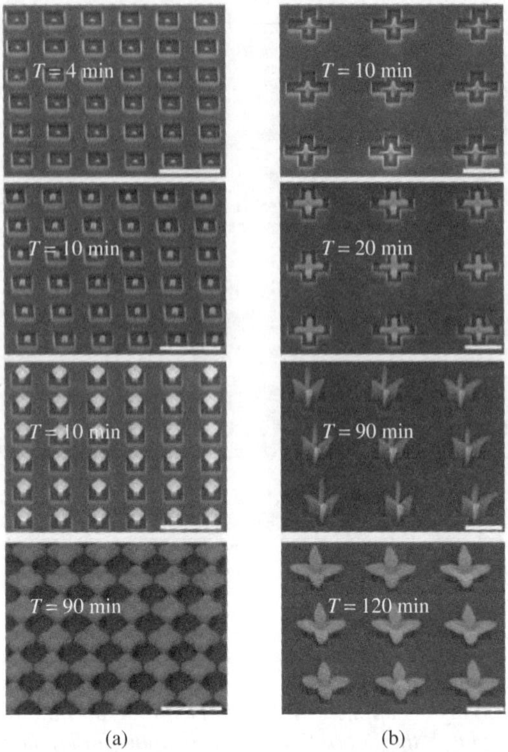

Figure 12.27 *Time-dependent growth of 3D nanoparticle structure arrays. The combination of an ion-induced scaffold and the antenna effect realizes a selective formation of 3D nanoparticle structure arrays on the PR pattern. 3D nanoparticle structure arrays composed of copper nanoparticles are formed within (a) PR square patterns with one side of 500 nm for deposition times of 4, 10, 40, and 90 minutes and (b) PR cross patterns with one side of 500 nm for 10, 20, 90, and 120 minutes. All scale bars are 1.5 μm. Reprinted with permission from Lee* et al. *(2011). Copyright © 2011, American Chemical Society.*

The fact the flower nanostructure is conductive demonstrates several plasmonic effects resulting from the resonance collective electron waves induced by the external electromagnetic field of light. One of them is the well-known surface enhanced Raman scattering (SERS), where the resonance plasmons enhance the local electric fields acting on molecules lying on the surface and thus make possible the molecule diagnostics for an extremely low molecular concentration.

In order to enable Raman measurements (Lee *et al.*, 2011), 3D patterned SERS substrates were immersed in a 10^{-3} M ethanol solution of thiophenol (Aldrich) for 3 hours and then dried for several minutes. Raman spectra were obtained by a LabRam HR model confocal Raman microscope with an LN_2-cooled charge coupled device (CCD) multichannel detector (Jobin-Yvon, France). An argon ion laser with wavelength 514 nm mean spot size of approximately 700 nm was used as the excitation source. The laser power radiated to the

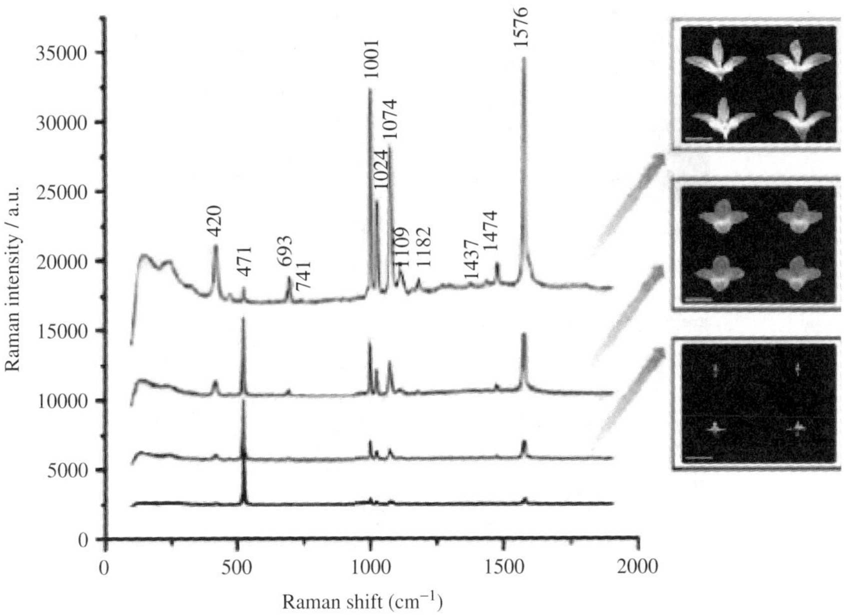

Figure 12.28 Raman spectra of 10^{-3} M ethanolic solution of thiophenol on composite nanoparticle structure samples composed of copper and silver nanoparticles, compared with the Raman spectra from a 2D flat pattern of Ag nanoparticles deposited on the Si substrate. Composite 3D nanoparticle structures were prepared by accumulating silver nanoparticles on copper nanoparticle structures made with different deposition times (top box: 20 minutes for copper and 30 minutes for silver; middle box: 60 minutes for copper and 60 minutes for silver; bottom box: 90 minutes for copper and 90 minutes for silver). The bottom curve shows the case for 2D flat patterns. Reprinted with permission from Lee et al. (2011). Copyright © 2011, American Chemical Society.

sample was about 0.02 mW and the acquisition time was 100 seconds. Figure 12.28 shows the 17-times enhancement of the signal intensity and the improvement in the Raman signal quality, which allow one to discern hidden spectral features.

The other plasmonic effect is the redistribution of the spectral intensity of the white light scattered by the plasmonic structure. Such a scattering is responsible for the colors of butterfly wings. It can be measured in the configuration of dark-field microscopy, where only the scattered light spectrum is detected, without any background white light (Lee *et al.*, 2011). In one experiment, dark-field micrographs were obtained with a true-color CCD camera (DP72, Olympus) directly aligned to a conventional reflected-light microscope (BX51, Olympus) that used a 100 W halogen lamp as a white-light illumination source. A 100× dark-field objective lens was used for dark-field imaging, in combination with a ring-shaped mirror, which separated the illumination-beam path from the detection-beam path. In dark-field imaging mode, only scattered light was collected through the objective lens, which was sent to the CCD camera for imaging.

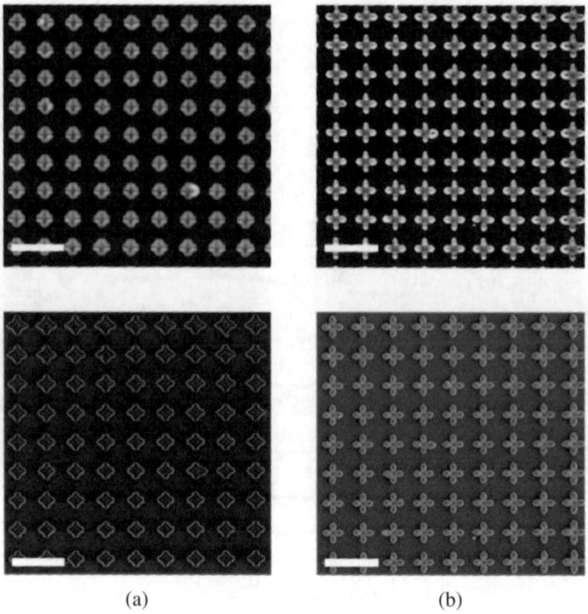

Figure 12.29 Dark-field scattering image (top) and scanning electron microscopy (SEM) image (bottom) of (a) silver-nanoparticle-covered and (b) gold-film-covered 3D copper nanoparticle structure arrays. The scale bar is 5 μm. Reprinted with permission from Lee et al. (2011). Copyright © 2011, American Chemical Society. See plate section for colour version.

The colors of flower petals originate exclusively in the plasmonic modes generated in this nanostructured metal (Figure 12.29). The strong interaction between constituent nanoparticles produces a virtually homogeneous metal structure, with its own characteristic plasmon modes, which reflect the fourfold symmetry of the flower and provide its colorful beauty.

12.9 Conclusions

Aerosol synthesis has proven to be an effective method for a number of nanostructured materials. Primary nanoparticles are produced in a very dispersed phase and can be guided into any useful structure, whether an ordinary deposition on a substrate or some sophisticated patterning and assembly. The intrinsic properties of nanoparticles can be peculiar in themselves and differ from those properties shown by the same materials in bulk phases. Additional heterogeneity, such as doping, coating, and composing/assembly, creates further potential for manipulating these properties. Due to the innumerable contacts between the nanoparticles in the material, such a nanostructured condensed material can have additional peculiarities such as high sensitivity to the environment and a specific response to electron transport and electric and magnetic fields.

References

Aliev, F.G., Correa-Duarte, M.A., Mamedov, A. *et al.* (1999) Layer-bylayer assembly of core-shell magnetite nanoparticles: effect of silica coating on interparticle interactions and magnetic properties. *Advanced Materials*, **11**, 1006–1010.

Altman, I.S., Pikhitsa, P.V. and Choi, M. (2004a) Electron field emission from nanocarbons: a two-process model. *Applied Physics Letters*, **84**(7), 1126–1128.

Altman, I.S., Pikhitsa, P.V. and Choi, M. (2004b) Two-process model of electron field emission from nanocarbons: temperature effect. *Journal of Applied Physics*, **96**(6), 3491–3493.

Altman, I.S., Pikhitsa, P.V., Jeong, J.I. *et al.* (2003) Line spectra from doped nano-oxide: a design for nanooptics. *Applied Physics Letters*, **83**(18), 3689–3691.

Biskos, G., Vons, V., Yurteri, C. and Schmidt-Ott, A. (2008) Generation and sizing of particles for aerosol-based nanotechnology. *KONA Powder Particle Journal*, **26**, 13–35.

Borra, J., Goldman, A., Goldman, M. and Boulaud, D. (1998) Electrical discharge regimes and aerosol production in point-to-plane DC high-pressure cold plasmas: aerosol production by electrical discharges. *Journal of Aerosol Science*, **29**, 661–674.

Browning, D.J., Gerrard, M.L., Lakeman, J.B. *et al.* (2002) Studies into the storage of hydrogen in carbon nanofibers: proposal of a possible reaction mechanism. *Nano Letters*, **2**, 201–205.

Choi, M., Altman, I.S., Kim, Y. *et al.* (2004) Formation of shell-shaped carbon nanoparticles above a critical laser power in irradiated Acetylene. *Advanced Materials*, **16**, 1721–1725.

Dutoit, D.C.M., Schneider, M. and Baiker, A. (1995) Titania–silica mixed oxides: I. Influence of sol–gel and drying conditions on structural properties. *Journal of Catalysis*, **153**, 165–176.

Han, K., Kim, W., Yu, J. *et al.* (2012) A study of pin-to-plate type spark discharge generator for producing unagglomerated nanoaerosols. *Journal of Aerosol Science*, **52**, 80–88.

Kim, H., Kim, J., Yang, H. *et al.* (2006) Parallel patterning of nanoparticles via electrodynamic focusing of charged aerosols. *Nature Nanotechnology*, **1**, 117–121.

Kim, D., Pikhitsa, P.V., Yang, H. and Choi, M. (2011) Room temperature CO and H_2 sensing with carbon nanoparticles. *Nanotechnology*, **22**, 485501.

Krinke, T.J., Deppert, K., Magnusson, M.H. *et al.* (2002) Microscopic aspects of the deposition of nanoparticles from the gas phase. *Journal of Aerosol Science*, **33**, 1341–1359.

Lee, D. and Choi, M. (2000) Control of size and morphology of nano particles using CO_2 laser during flame synthesis. *Journal of Aerosol Science*, **31**(10), 1145–1163.

Lee, D. and Choi, M. (2002) Coalescence enhanced synthesis of nanoparticles to control size, morphology and crystalline phase at high concentrations. *Journal of Aerosol Science*, **33**, 1–16.

Lee, D., Yang, S. and Choi, M. (2001) Controlled formation of nanoparticles utilizing laser irradiation in a flame and their characteristics. *Applied Physics Letters*, **79**(15), 2459–2461.

Lee, H., You, S., Pikhitsa, P.V. *et al.* (2011) Three-dimensional assembly of nanoparticles from charged aerosols. *Nano Letters*, **11**, 119–124.

Lim, K., Lee, J., Lee, H. *et al.* (2012) Nanoxerography utilizing bipolar charge patterns. *Applied Physics Letters*, **101**, 203106.

Nakaso, K., Han, B., Ahn, K. *et al.* (2003) Synthesis of non-agglomerated nanoparticles by an electrospray assisted chemical vapor deposition (ES-CVD) method. *Journal of Aerosol Science*, **34**, 869–881.

Nasibulin, A.G., Pikhitsa, P.V., Jiang, H. *et al.* (2007) A novel hybrid carbon material. *Nature Nanotechnology*, **2**, 156–161.

Pikhitsa, P.V., Kim, Y.J., Kim, M.W. *et al.* (2005) Optical manifestations of the condensation of topological defects in continuous layer multi shell carbon nanoparticles. *Physical Review B*, **71**, 073402; Pikhitsa, P.V., Kim, D. and Choi, M. CO_2 laser-driven reactions in pure acetylene flow, arXiv: 109.3331.

Pikhitsa, P.V., Kim, D. and Choi, M. (2011) Nitrogen/argon diluted acetylene and ethylene blue flames under infrared CO_2 laser irradiation. *AIP Advances*, **1**, 032142.

Pikhitsa, P.V., Yang, S., Kim, J. *et al.* (2007) Fast fragmentation of metal oxide nanoparticles via reduction in oxyhydrogen flame. *Applied Physics Letters*, **90**, 163106.

Pratsinis, S.E. (1998) Flame aerosol synthesis of ceramic powders. *Progress in Energy and Combustion Science*, **24**, 197–219.

Psofogiannakis, G.M. and Froudakis, G.E. (2009) DFT study of hydrogen storage by spillover on graphite with oxygen surface groups. *Journal of the American Chemical Society*, **131**, 15133–15135.

Santra, S., Tapec, R., Theodoropoulou, N. *et al.* (2001) Synthesis and characterization of silica coated iron oxide nanoparticles in microemulsion: the effect of nonionic surfactants. *Langmuir*, **17**, 2900–2906.

Schmidt-Ott, A. (2011) Aerosol methods for nanoparticle synthesis and characterization, in *Handbook for Nanophysics, Principles and Methods* (ed K.D. Sattler), Chapter 30, CRC Press, New York, pp. 1–19.

Selvan, S.T., Tan, T.T. and Ying, J.Y. (2005) Robust, non-cytotoxic, silica-coated CdSe quantum dots with efficient photoluminescence. *Advanced Materials*, **17**, 1620–1625.

Sheen, S., Yang, S., Jun, K. and Choi, M. (2009) One-step flame method for the synthesis of coated composite nanoparticles. *Journal of Nanoparticle Research*, **11**, 1767–1775.

Sheen, S., Yang, S., Jun, K. *et al.* (2012) Crystallinity control of flame generated composite nanoparticles by laser irradiation. *Powder Technology*, **229**, 246–252.

Stark, W.J. and Pratsinis, S.E. (2002) Aerosol flame reactors for manufacture of nanoparticles. *Powder Technology*, **126**, 103–108.

Tabrizi, N., Ullmann, M., Vons, V. *et al.* (2009) Generation of nanoparticles by spark discharge. *Journal of Nanoparticle Research*, **11**, 315–332.

Tada, K., Furuya, S. and Watanabe, K. (2001) *Ab initio* study of hydrogen adsorption to single-walled carbon nanotubes. *Physical Review B*, **63**, 155405.

Teleki, A., Suter, M., Kidambi, P.R. *et al.* (2009) Hermetically coated superparamagnetic Fe_2O_3 particles with SiO_2 nanofilms. *Chemistry of Materials*, **21**, 2094–2100.

Vajtai, R., Wei, B.Q., Zhang, Z.J. *et al.* (2002) Building carbon nanotubes and their smart architectures. *Smart Materials and Structures*, **11**, 691–698.

Vemury, S., Pratsinis, S.E. and Kibbey, L. (1997) Electrically controlled flame synthesis of nanophase TiO_2, SiO_2, and SnO_2 powders. *Journal of Materials Research*, **12**(4), 1031–1042.

Vollath, D. and Szabo, D.V. (2004) Synthesis and properties of nanocomposites. *Advanced Engineering Materials*, **6**, 117–127.

Yang, S., Jang, Y., Kim, C. *et al.* (2010) A flame metal combustion method for production of nanoparticles. *Powder Technology*, **197**, 170–176.

Yang, S., Yi, J., Son, S. *et al.* (2003) Fragmentation of Fe_2O_3 nanoparticles driven by a phase transition in a flame and their magnetic properties. *Applied Physics Letters*, **83**, 4842–4844.

Zhu, Z.H., Finnerty, J., Lu, G.Q. *et al.* (2002) Molecular orbital theory calculations of the H_2O-carbon reaction. *Energy Fuel*, **16**, 847–854.

13

The Safety of Emerging Inorganic and Carbon Nanomaterials

L. Reijnders
IBED, University of Amsterdam, The Netherlands

13.1 Introduction

Nanomaterials or nanoparticles are often defined as having a diameter < 100 nm in at least one dimension (Fubini, Ghiazza, and Fenoglio, 2010; Oberdörster, 2010; Cushen *et al.*, 2011; Maynard, Warheit, and Philbert, 2011). Such nanomaterials may have a natural origin, as exemplified by nanoparticulate mercuric sulfides (Zhang *et al.*, 2012). They may be present in conventional materials, such as carbon black and the paint additives calcium carbonate and talc, and in emissions from fuel combustion (Kunzli, 2011; Reijnders, 2012, van Broekhuizen *et al.*, 2012). They may also be generated from conventional materials under ambient conditions, as exemplified by the generation of Cu and Ag nanoparticles from silver and copper objects exposed to liquid water or humidity (Glover, Miller, and Hutchison, 2011). Nanoparticles may furthermore originate in welding (Dasch and D'Arcy, 2008). Wear, tear, and processing of conventional materials can also lead to nanoparticle release. For instance, operating electromotors can release Cu nanoparticles and diamond processing can give rise to the generation of carbon nanoparticles (Scymczak, Menzela, and Kecka, 2007; Beniwal and Shivgotra, 2009).

Inorganic and carbonaceous nanoparticles are also increasingly being engineered for a variety of applications. These are the emerging nanomaterials, of which persistent engineered inorganic and carbon nanoparticles will be considered here. In this context, 'persistent' refers to nanoparticles that are poorly soluble or insoluble in water. Examples of emerging persistent engineered inorganic and carbon nanomaterials are given in Box 13.1.

Aerosol Science: Technology and Applications, First Edition. Edited by Ian Colbeck and Mihalis Lazaridis.

Box 13.1 Examples of emerging persistent inorganic and carbon nanomaterials currently engineered for practical application (Wynne *et al.*, 2008; Peters *et al.*, 2009; Ferreira, Filho, and Alves, 2010; Won, Nersisyan, and Won, 2010; Cushen *et al.*, 2011; Reijnders, 2012)

Elements:

Ag	Gd
Au	La
C (nanotubes, fullerenes, graphene nanoplatelets, nanoparticulate carbon black)	Mn
Fe	Pb
	Pt and related metals of the platinum group

Oxides, hydroxides, and nitrides:

Al_2O_3	Sc_2O_3
CeO_2	Sb_2O_5
CuO	SiO_2
CoO	V_2O_5
Fe_2O_3, Fe_3O_4	TiO_2
MnO	TiN
$Ni(OH)_2$	ZnO

Other:

Clays (e.g., montmorillonite)	Sc_3NC_{80}
CdSe	Silicate nanoplatelets
CdTe	SnS_2
Lithium titanate	WC
SiC	ZnS

For actual applications, the persistent engineered inorganic and carbon nanoparticles mentioned in Box 13.1 may undergo surface modification, including functionalization, capping, and coating (e.g., Zhang, Su, and Mao, 2006; Jeong 2009; Ma *et al.*, 2010; Chafer-Pericas, Maquiera, and Puchades, 2011; Thio *et al.*, 2011; Piccapietra, Sigg, and Behra, 2012; Rivet *et al.*, 2012).

Engineered nanoparticles can be released into the working and wider environment during the production and industrial handling and application of such nanoparticles (Kuhlbusch *et al.*, 2011; van Broekhuizen and Reijnders, 2011). Some engineered nanomaterials have dispersive applications in consumer products, including spray cans, and in medicines and fuels (Reijnders, 2008; Quadros and Marr, 2011; Reijnders, 2012). Use of such products can lead to human exposure to emerging nanoparticles. Applications in fuels, consumer products, and medicines can also lead to releases into the environment (e.g., Mueller and Nowack, 2008; Gottschalk *et al.*, 2010; Quadros and Marr, 2011). Nanoparticles may furthermore be released from products on which they are fixed or in which they are embedded (e.g., Benn and Westerhoff, 2008; Mueller and Nowack, 2008; Wohlleben *et al.*, 2011). Examples thereof are the release of TiO_2 from polymer matrixes linked to polymer degradation (Wohlleben *et al.*, 2011) and the release of Ag nanoparticles from textile fabrics (Benn and Westerhoff, 2008). In the case of fires, the release of small particles from fire-retarded nanocomposite polyamide 6 and polypropylene may be increased if compared with virgin polymers (Rhodes, Smith, and Stec, 2011).

Environmental release of emerging nanoparticles can lead to the exposure of humans and other organisms. This raises the question of safety. The safety of persistent inorganic and carbon nanomaterials has been studied to a limited extent. In part this follows from the long and widely held assumption that hazard (potential harm) and risk (chance of harm occuring) are independent of particle size (Maynard *et al.*, 2011). This assumption is now known to be incorrect for many persistent engineered inorganic and carbon nanoparticles (Oberdörster, 2010; Elsaesser and Howard, 2011; Maynard *et al.*, 2011).

Available studies regarding the safety of emerging nanomaterials have mainly focused on persistent engineered nanoparticles that lack surface treatments such as functionalization or coating. They have considered nanomaterial risks and hazards linked to human toxicity and ecotoxicity. There may be other hazards and risks linked to the use of engineered nanomaterials; widespread application of Ag nanoparticles might lead to the emergence of increased Ag resistance in hazardous microbes, which in turn might make it harder to combat human infections (Silver, 2003; Silver, Phung, and Silver, 2006). Also, large discharges of Ag, CuO, and ZnO nanoparticles might negatively affect wastewater treatment, leading to increased levels of hazardous substances in surface waters (Luna-delRisco, Orupold, and Dubourguier, 2011; Sheng and Liu, 2011).

A relatively large amount of safety-related research has looked at the relation between inhalation exposure to persistent inorganic and carbon nanoparticles and human health (Oberdörster, 2010; Maynard *et al.*, 2011; Song *et al.*, 2011; Yokel and MacPhail, 2011). However, data on hazard and exposure to these airborne nanoparticles are still fragmentary, which implies that risk estimates are highly uncertain (Oberdörster, 2010; Kuhlbusch *et al.*, 2011; Maynard *et al.*, 2011, Paur *et al.*, 2011; Stebounova *et al.*, 2011; Yokel and MacPhail, 2011). Moreover, in air there may be interactions between primary nanoparticles and substances and other particles, which could give rise to the formation of assemblages or secondary particles (aggregates and agglomerates) with hazardous properties quite different to those of the original nanoparticles (Kreyling *et al.*, 2009; Lankoff *et al.*, 2012).

Studies regarding the human health hazards and risks of other types of human exposure, such as ingestion, have been more fragmentary than those regarding inhalation (Reijnders, 2012). Ecotoxicity studies regarding the engineered nanoparticles mentioned in Box 13.1 have been more limited than the study of human health-related hazards and risks (Bhatt and Tripathi, 2011; Menard, Drobne, and Jemec, 2011; Turco *et al.*, 2011; Petersen *et al.*, 2011). Moreover, interactions in the environment can give rise to a wide variety of assemblages and conversion products with a different ecotoxicity to the original nanoparticles (Auffan *et al.*, 2010; Chinnapongse, MacCuspie, and Hackley, 2011; Menard *et al.*, 2011; Turco *et al.*, 2011; van Hoecke *et al.*, 2011; Mudunkotuwa, Pettibone, and Grassian, 2012; Nowack *et al.*, 2012).

In this chapter, the studies regarding the relation between inhalation exposure to engineered persistent inorganic and carbon nanoparticles and human health will first be summarized. Measures conducive to safety in view of hazard and potential exposure to such airborne nanomaterials, especially in the workplace, will then be examined. Inhalation of nanoparticles may be followed by clearance of inhaled nanoparticles from the lungs, and this can subsequently lead to ingestion (Sturm, 2007; Sturm and Hofmann, 2009; Beamish, Osornio-Vargas, and Wine, 2011). Thus, studies regarding the safety of nanoparticle ingestion will also be briefly reviewed. Finally, ecotoxicity studies relevant to emerging persistent inorganic and carbon nanomaterials will be considered.

13.2 Human Health and Inhaled Persistent Engineered Inorganic and Carbon Nanomaterials

Inhaled primary persistent inorganic and carbon nanoparticles can penetrate deeply into the lungs and be deposited there (Oberdörster, 2010; Maynard *et al.*, 2011). Clearance of such particles from the lungs occurs but may be partial (Oberdörster, 2010; Maynard *et al.*, 2011). Deposited persistent primary inorganic and carbon nanomaterials can be hazardous (Fubini *et al.*, 2010; Oberdörster 2010; Maynard *et al.*, 2011; Paur *et al.*, 2011; Yokel and MacPhail, 2011), their effects on the respiratory system including inflammation, granulomas, fibrosis, and cancer (Peters *et al.*, 2009; Oberdörster, 2010; Maynard *et al.*, 2011; Yokel and MacPhail, 2011). Inhaled primary nanoparticles may be translocated from the respiratory system depending on such factors as size, shape, and surface charge (Choi *et al.*, 2010; Fubini *et al.*, 2010; Oberdörster, 2010; Aschberger *et al.*, 2011; Kim *et al.*, 2011; Maynard *et al.*, 2011, Yokel and MacPhail, 2011), for example to the cardiovascular system (Kreyling *et al.*, 2002, 2009; Choi *et al.*, 2010). This can lead to the deposition of nanoparticles in the organs, including the cardiovascular system, liver, kidney, spleen, stomach, brain, and testis. Deposition of nanomaterials in these organs can cause inflammation, changes in immune responses (immunomodulation), and, in case of the brain, changes in neurotransmitter levels (Fubini *et al.*, 2010; Oberdörster, 2010; Hubbs *et al.*, 2011; Maynard *et al.*, 2011; Yokel and MacPhail, 2011). Translocation of nanoparticles across the placenta and generation of reactive oxygen species due to nanoparticle-induced signaling across the placental barrier have also been reported (Braydich-Stolle *et al.*, 2010; Hougaard *et al.*, 2010; Sood *et al.*, 2011; Zhang *et al.*, 2011a). Reactive oxygen species in turn may trigger developmental toxicity to the fetus, damage to fetal DNA, and pregnancy complications (Braydich-Stolle *et al.*, 2010; Hougaard *et al.*, 2010; Sood *et al.*, 2011; Zhang *et al.*, 2011a).

Furthermore, deposition of nanomaterials in the lungs may lead to increased levels of stressors such as cytokines and oxidized biomolecules in the bloodstream and to changes in the activity of the autonomous nervous system (Gomez-Mejiba *et al.*, 2009; Horie and Fujita, 2011; Paur *et al.*, 2011). The latter can initiate cardiac arrhythmias, while the former may trigger cardiovascular diseases and chronic inflammation in organs and affect fetal development (Gomez-Mejiba *et al.*, 2009; Jackson *et al.*, 2011). Furthermore, nanomaterial-induced inflammation of the lungs can lead to platelet–leukocyte aggregates, which have been linked to the progression of arteriosclerosis (Tabuchi and Kuebler, 2008).

Translocation is also possible from the nasal area to the brain, via the olfactory nerve and bulb (Maynard *et al.*, 2011). This can lead to changes in neurotransmitter levels and to enhanced inflammation (Oberdörster, 2010; Hubbs *et al.*, 2011).

Persistent inorganic and carbon nanomaterials may differ widely in their inhalation hazard. For example, C_{60} fullerenes are much less hazardous than long, rigid carbon nanotubes (Borm and Castranova, 2009). A number of determinants of the hazards and/or risks of persistent engineered inorganic and carbon nanoparticles are emerging from currently available research.

A first important determinant of hazard is size. Size is important for such characteristics as penetration of the lungs, uptake by cells, intracellular interactions, and excretion (Kreyling *et al.*, 2002; Semmler-Behnke *et al.*, 2008; Pelley, Daar, and Saner, 2009; Choi *et al.*, 2010; Park *et al.*, 2011). For instance, zwitterionic cysteine-coated nanodots have been found to be rapidly excreted at sizes $< \sim 5.5$ nm, whereas the excretion of much larger nanodots is relatively poor (Pelley *et al.*, 2009). The cytotoxicity of 1.4 nm Au particles has been found to be sixfold higher than that of 0.8, 1.2, and 1.8 nm Au nanoparticles (Aillon *et al.*, 2009). Primary nanoparticle size also impacts surface area (Ramachandran *et al.*, 2011; Klaine *et al.*, 2012), which is another determinant of hazard.

Persistent inorganic and carbon nanoparticles can form assemblages in air and in the lungs. This can lead to changes in human health hazards, when compared with primary particles (e.g., Kreyling *et al.*, 2009; Reijnders, 2012). There are indications, however, that hazards may be conserved to the extent that the surface area of the original nanoparticles remains available for reactivity (Rabolli *et al.*, 2011).

Surface characteristics are important determinants of hazard. Aside from surface area, these characteristics include surface reactivity (e.g., the generation of reactive oxygen species and the release of toxic substances), surface charge, and hydrophilicity (Duffin *et al.*, 2007; Limbach *et al.*, 2007; Monteiller *et al.*, 2007; Ayres *et al.*, 2008; Clift *et al.*, 2008; Hussain *et al.*, 2009; Choi *et al.*, 2010; Marambio-Jones and Hoek, 2010; Napierska *et al.*, 2010; Chen *et al.*, 2008; Kayat *et al.*, 2011; Koike and Kobayashi, 2011; Rim *et al.*, 2011; Soenen *et al.*, 2011; Zhu *et al.*, 2011; Rivet *et al.*, 2012). In a study of TiO_2 and carbon black nanoparticles, for example, Monteiller *et al.* (2007) reported that toxicity to epithelial cells varied by surface area. Choi *et al.* (2010) found that noncationic nanoparticles < 34 nm were rapidly translocated from the lungs, whereas cationic nanoparticles of a similar size were not. This may imply that noncationic nanoparticles are more hazardous than cationic nanoparticles to organs other than the respiratory system. The release of Ag from Ag nanoparticles, of Cd from CdSe nanodots, and of Fe, Cu, Ni, and Zn ions from metal-oxide nanoparticles are determinants of nanoparticle hazard (Pelley *et al.*, 2009; Marambio-Jones and Hoek, 2010; Horie and Fujita, 2011; Zhu *et al.*, 2011). The abundance and distribution of silanols at the surface on SiO_2 nanoparticles have been suggested to be major determinants of nanosilica hazard (Napierska *et al.*, 2010). Functionalization by polydimethylamine strongly increases the cytotoxic effect of iron-oxide nanoparticles to cortical neurons (Rivet *et al.*, 2012).

Furthermore, shape, structure, and rigidity may be relevant to the hazards of persistent inorganic and carbon primary nanoparticles (Sayes, Reed, and Warheit, 2007; Borm and Castranova, 2009; Hamilton *et al.*, 2009; Longmire *et al.*, 2011; Napierska *et al.*, 2010; Oberdörster, 2010; Maynard *et al.*, 2011; Reijnders, 2012). There is some evidence that, *ceteris paribus*, rigid nanomaterials may be more hazardous than flexible nanomaterials

(Reijnders, 2012): crystalline nanosilica appears to be more hazardous than amorphous SiO_2 nanoparticles (Napierska *et al.*, 2010) and, as mentioned before, long carbon nanotubes appear to be much more hazardous than fullerenes (C_{60}) (Borm and Castranova, 2009). Frustrated phagocytosis has been suggested to be a major cause of the difference between long carbon nanotubes and fullerenes (Borm and Castranova, 2009): fullerenes are far more easily removed from the lungs by macrophages than are long carbon nanotubes (Borm and Castranova, 2009) and long carbon nanotubes tend to be more cytotoxic than fullerenes (Jia *et al.*, 2005). Frustrated phagocytosis has also been suggested as a reason for the relatively strong inflammogenic effect in lungs and in the pleural space of graphene nanoplatelets, compared with nanoparticulate carbon black (Schinwald *et al.*, 2012).

Finally, in determining risk, the number of nanoparticles to which one is exposed makes a difference (Borm and Castranova, 2009; Oberdörster, 2010; Maynard *et al.*, 2011).

There are differences in individual responses to inhaled nanoparticles. For example, there are differences in response to the inhalation of amorphous silica nanoparticles, suggesting that the hazard of such nanoparticles may be relatively large for old people, children, and diabetics and relatively low for healthy people of medium adult age (Chen *et al.*, 2008; Schuepp and Sly, 2012).

A variety of molecular mechanisms are thought to be involved in the inhalation hazard of persistent inorganic and carbon nanomaterials. These include the release of toxic substances and the generation of reactive oxygen species, such as superoxides and hydroxyl radicals (Mahtab *et al.*, 2007; Oberdörster, 2010; Elsaesser and Howard, 2011; Liu *et al.*, 2011; Maynard *et al.*, 2011; Paur *et al.*, 2011; Shukla *et al.*, 2012). Interactions of nanoparticles with protein and DNA negatively affect the functionality of these biomolecules, and in the case of nanoclay, physical damage to biological structures may also contribute to nanoparticle hazard (Bhattacharya and Mukherjee, 2008; Horie *et al.*, 2009; Myllynen, 2009; An and Jin, 2011; Gagner *et al.*, 2011; Lordan, Kennedy, and Higginbotham, 2011). The interaction of nanoparticles with biomolecules that leads to the formation of agglomerates may influence the biological effects, too (Horie and Fujita, 2011). These molecular mechanisms are also thought to be relevant to other types of human exposure (Reijnders, 2012) and to ecotoxicity (e.g., Ivask *et al.*, 2012; Yang *et al.*, 2012).

A tool that might be useful in achieving safety regarding airborne exposure to emerging nanomaterials is exposure standards. These are now emerging for workplaces, though not yet for non-workplace exposures (Kunzli, 2011; van Broekhuizen and Reijnders, 2011). There are some cases in which the available toxicity data have been considered adequate for deriving health-based exposure limits as workplace standards (NIOSH, 2010, 2011). In the USA, NIOSH has proposed a health-based exposure limit for TiO_2 nanoparticles in workplace air on the basis of available toxicity data; specifically, data linking tumors to exposure. This proposed standard is 0.3 mg m^{-3} as a time-weighted average for up to 10 hours per day across a 40 hours working week. This standard is a factor 8 stricter than the US standard for fine TiO_2 particles (NIOSH, 2011). NIOSH has also published a draft standard for a health-based exposure limit regarding carbon nanotubes and carbon nanofibers based on available toxicity data. In this case, hazard would justify an 8 hour time-weighted average exposure limit of between 0.2 and 2.0 μg m^{-3} air, but, due to a higher upper limit of detection, 7.0 μg m^{-3} has been proposed as a draft exposure limit (NIOSH, 2010).

Another approach to standard setting has been proposed by Pauluhn (2011). This is a generic mass-based approach based on the assumption that the particle displacement volume corresponding with the prevention of an overload of nanoparticles, as determined in studies with rats, should be the basis of a standard setting. One problem with Pauluhn's proposal is the uncertainty over the correctness of extrapolating the overload established in rats to human health hazard (Maynard *et al.*, 2011).

In many cases, available toxicity data are insufficient for the establishment of exposure standards. This has led to strategies invoking the precautionary principle, which allows for government interventions in the absence of conclusive scientific proof. As a result thereof, precautionary standards, provisional nano-reference values, and benchmark levels have been proposed (van Broekhuizen and Reijnders, 2011; van Broekhuizen *et al.*, 2012). For example, provisional nano-reference values for exposure to (bio)persistent nanoparticles in the workplace have been accepted by representatives of trade unions and employers' organizations in the Netherlands (van Broekhuizen *et al.*, 2012). For long, rigid nanofibers and nanotubes, the 8 hour time-weighted provisional nano-reference value is 0.01 particles cm^{-3} and for granular nanomaterials it is 20 000 or 40 000 particles cm^{-3}, dependent on density (van Broekhuizen *et al.*, 2012). These provisional nano-reference values are linked to three previously mentioned determinants of risk and/or hazard: number, shape, and rigidity. According to the representatives of the Dutch trade unions and employers' organizations, when the provisional nano-reference values are exceeded, the the nanoparticle emissions' source(s) should be identified and measures should be taken to reduce them.

A third approach to standard setting has been proposed by the British Standards Institute (BSI) (2007). This approach is based on scaling from current health-based standards for large particles. For instance, for insoluble nanomaterials, the BSI suggests a benchmark value of $0.066 \times$ the current workplace exposure limit for large particles (macroparticles). For nanoparticles to which larger particles are classified as carcinogenic, mutagenic, asthmagenic, or a reproductive toxin, the proposed standard is $0.1 \times$ the current workplace exposure limit for macroparticles.

Still another approach has been suggested by Hesterberg *et al.* (2010), who reviewed studies in which human volunteers were exposed to diesel exhaust particles. They reported a no-adverse-effect level between about 30 000 and 50 000 particles cm^{-3} and suggested this might be used as a reference value for estimating the potential human health hazards of nanoparticles. This approach would lead to standards roughly similar to the Dutch provisional nano-reference values for granular nanomaterials.

Attempts to meet such workplace standards, or more generally to reduce the risk of airborne exposure, currently tend to focus on elimination, substitution, and engineering controls as close as possible to the sources of hazardous substances (Ayotte, 2011; Reijnders, 2012). Options for substitution, elimination, and engineering controls are given in Table 13.1.

13.3 Human Health Hazards and Risks Linked to the Ingestion of Persistent Inorganic Nanomaterials

The study of hazards and risks linked to the ingestion of nanoparticles has so far been very limited. Depending on the size and character of a particle's surface, translocation of persistent inorganic nanomaterials from the intestines to the cardiovascular system is a possibility (Florence, 2005; Powell *et al.*, 2010; Cushen *et al.*, 2011). Relatively more is known about

Table 13.1 *Options for reducing human health risk in the face of exposure to airborne inorganic and carbon nanoparticles (Pelley* et al.*, 2009; Woskie, 2010; Ayotte, 2011; Fleury* et al.*, 2011; Yokel and MacPhail, 2011; Reijnders, 2012).*

Risk-reduction option	Example(s)
Elimination/substitution	
Replacement by less hazardous substances	Substitution of Cr and Ni by Fe in carbon-nanotube production; elimination of metal catalysts in carbon-nanotube production
Suppression of major determinants of hazard; reduction of 'intrinsic' nanoparticle hazard	Suppression of the generation of reactive oxygen species by coating or doping; reduction of the presence of metal contamination in carbon nanotubes; increased solubility of fullerenes
Engineering controls to reduce the release of nanoparticles	
Increase of yields of nanoparticles in conformity with product specifications, to reduce waste-handling hazards	Can be applied to many nanoparticle production processes
Well-contained *in situ* production of nanoparticles on/in material used for coating or embedding; use of master batches with high concentrations of nanoparticles instead of nanoparticle powders	Production of self-cleaning glass; production of polymer-nanoparticle nanocomposites
Well-contained transport of nanoparticles	Can be applied to many nanomaterial production processes
Design to reduce nanoparticle release	Catalytic converters in motorcars
Immobilization of nanoparticles on substrates instead of free nanoparticles	Application of nano TiO_2 for the degradation of hazardous substances
Wet chemistry in nanoparticle production	Generation of Al_2O_3 nanoparticles
Use of glove boxes and hoods	Laboratory-scale work with nanoparticles
Use of local ventilation and uninterruptible high-efficiency particulate air (HEPA) filters	Near vessels that mix nanoparticulate materials and near vapor-deposition reactors during clean-out; filtering of exhausts containing nanoparticles

the effects of ingesting Ag nanoparticles (colloidal silver), which have a longstanding use as a medicine (Panyala, Pena-Mendez, and Havel, 2008; Zhang *et al*., 2011b) and may be applied in spray cans (Quadros and Marr, 2011). Ingestion of Ag nanoparticles can lead to Ag deposits in organs, the preferential organ for deposition being dependent on particle size and surface characteristics (Lankveld *et al*., 2010; Loeschner *et al*., 2011). Large exposures to ingested colloidal silver are linked to dysfunction of the central nervous system, liver, and kidneys and to severe immunological responses (Panyala *et al*., 2008; Zhang *et al*., 2011b). Large exposures to nanoparticulate Ag may also give rise to silver deposits in the mucous membranes and under the skin (argyria) (Panyala *et al*., 2008; Zhang *et al*., 2011b).

There is some evidence for intestinal inflammation and immunomodulation by ingested titania nanomaterials (Reijnders, 2008). In ingestion by mice, large exposures to amorphous nanosilica have been associated with negative impacts on liver function (So, Jang, and Han, 2008).

13.4 Ecotoxicity of Persistent Inorganic and Carbon Nanomaterials

Studies relevant to the ecotoxicity of persistent inorganic and carbon nanoparticles are extremely fragmentary when compared with what would allow for confident estimates of hazard and risk (Pelley *et al.*, 2009; Menard *et al.*, 2011; Turco *et al.*, 2011; Petersen *et al.*, 2011). Whereas measurements of actual occupational exposure to airborne engineered nanoparticles have been published (cf. Kuhlbusch *et al.*, 2011; van Broekhuizen *et al.*, 2012), there are almost no empirical data directly relevant to the presence and fate of engineered nanoparticles in aquatic and soil ecosystems, in which airborne nanoparticles may end up (Farré, Sanchis, and Barcelo, 2011). Likewise, very little is known about the environmental behavior of engineered nanomaterials, their conversion products, or the assemblages which they may form (Hristozov *et al.*, 2012; Nowack *et al.*, 2012).

Some work modeling the future presence of persistent inorganic and carbon nanoparticles in the environment has been carried out, but the outcomes are highly uncertain, as studies of the actual fate of released nanoparticles that might be used to check the assumptions made are very limited (e.g., Turco *et al.*, 2011; Jensen-Eckelman *et al.*, 2012; Ramanan *et al.*, 2012). The variety of assemblages that can be formed in aquatic and soil ecosystems is moreover presumably much larger than the variety of agglomerates containing engineered nanoparticles in the air (Darlington *et al.*, 2009; Ottofuelling, von der Kammer, and Hofman, 2011; van Hoecke *et al.*, 2011) and the chances of the surfaces of nanoparticles being different from the original surfaces (e.g., due to aging and the degradation of their coatings; Auffan *et al.*, 2010; Mudunkotuwa *et al.*, 2012) will be much increased compared with normal human workplace exposure (e.g., Pereira *et al.*, 2011). Empirical studies regarding ecotoxicity as they are conducted in practice tend moreover to provide less information about the physicochemical characteristics of the nanoparticles tested than do studies relevant to airborne human exposure (Menard *et al.*, 2011).

Most ecotoxicological research has been done on TiO_2, Ag, and some types of carbon nanoparticle (nanotubes and fullerenes). Available studies regarding TiO_2 nanoparticles have examined the exposure of microorganisms, algae, higher plants, freshwater fish, aquatic invertebrates, and terrestrial invertebrates (Du *et al.*, 2011; Menard *et al.*, 2011). Modeling studies concerning the consequences of widespread application of TiO_2 nanoparticles suggests that concentrations in treated wastewater may exhibit ecotoxicity (Mueller and Nowack, 2008; Gottschalk *et al.*, 2010). As noted before, however, the outcomes of modeling studies are highly uncertain. Available ecotoxicity studies suggest that particle size and particle surface area may be determinants of ecotoxicity (Menard *et al.*, 2011).

Studies regarding the ecotoxicity of Ag nanoparticles have highlighted the release of Ag ions and the formation from the latter of nanoparticulate Ag compounds (Liu and Hurt, 2010). A relatively high mobility and stability of capped and carbonate-coated Ag nanoparticles in surface waters has also been reported (Thio *et al.*, 2011; Piccapietra *et al.*, 2012). It has been noted that widespread application of Ag nanoparticles might negatively affect the functioning of wastewater treatment facilities, which rely on microbial conversions (Luna-delRisco *et al.*, 2011; Sheng and Liu, 2011). This may in turn, *ceteris paribus*, lead to increased water pollution by organic compounds. Whether wastewater treatment facilities would actually be negatively impacted by the use of Ag nanoparticles is highly uncertain, not in the least because sensitivity varies greatly between the different bacteria involved in wastewater treatment and because bacteria in biofilms tend to be less vulnerable to Ag than are bacteria in pure cultures (Sheng and Liu, 2011). A modeling study concerning future applications of Ag nanoparticles suggests that treated wastewater may exhibit ecotoxicity to Ag nanoparticles, negatively affecting prokaryotes, fish, and invertebrates (Gottschalk *et al.*, 2010; Fabrega *et al.*, 2011), but again the correspondence of this outcome with the real world is highly uncertain.

Available studies concerning carbon nanomaterials do not yet seem to allow for conclusions to be made about their actual or expected impact on ecosystems (Petersen *et al.*, 2011; Petersen and Henry, 2011; Turco *et al.*, 2011).

13.5 Conclusion

Persistent inorganic and carbon nanoparticles are increasingly being engineered for practical application but can be hazardous to humans. A relatively great deal is known about the human health hazards of inhaled nanoparticles, which may give rise to respiratory disease and to negative effects in other organs, including the cardiovascular system. Determinants of inhaled nanoparticle risk and/or hazard are size, surface characteristics, shape, rigidity, structure, and the formation of assemblages. A major molecular mechanism underlying the inhalation hazard of nanoparticles is the generation of reactive oxygen species, but other mechanisms such as the release of toxic substances and interactions with proteins and DNA may also contribute. Human health hazards might be linked to the ingestion of persistent inorganic and carbon nanoparticles after their clearance from the lungs. Hazards and risks to ecosystems are highly uncertain. Options for reducing the human hazard linked to the inhalation of engineered nanomaterials include the elimination and substitution of hazardous nanoparticles and the use of engineering controls.

References

Aillon, K.L., Xie, Y., El-Gendy, N. *et al.* (2009) Effects of nanomaterial physicochemical properties on in vitro toxicity. *Advanced Drug Delivery Reviews*, **61**, 457–466.

An, H. and Jin, B. (2011) DNA exposure to buckminsterfullerene (C60): toward DNA stability, reactivity and replication. *Environmental Science and Technology*, **45**, 6608–6616.

Aschberger, K., Micheletti, C., Sokull-Klüttgen, B. and Christensen, F.M. (2011) Analysis of currently available data for characterizing the risk of engineered nanoparticles to the environment and human health – lessons learnt from four case studies. *Environment International*, **37**, 1143–1156.

Auffan, M., Pedeutor, M., Rose, J. *et al.* (2010) Structural degradation at the surface of TiO_2-based nanomaterial used in cosmetics. *Environmental Science and Technology*, **44**, 2689–2694.

Ayotte, P.R. (2011) Are classical process safety concepts relevant to nanotechnology applications? *Journal of Physics Conference Series*, **304**, 012071.

Ayres, J.G., Borm, P., Cassee, F.R. *et al.* (2008) Evaluating the toxicity of airborne particulate matter and nanoparticles by measuring oxidative stress potential. *Inhalation Toxicology*, **20**, 75–99.

Beamish, L.A., Osornio-Vargas, A.R. and Wine, E. (2011) Air pollution: an environmental factor contributing to intestinal disease. *Journal of Crohn's & Colitis*, **5**, 279–286.

Beniwal, R. and Shivgotra, V.K. (2009) An elementary framework for judging the cardiovascular toxicity of carbon soot: experiences from an occupation health survey of diamond industry workers. *Cardiovascular Toxicology*, **9**, 194–200.

Benn, T.M. and Westerhoff, P. (2008) Nanoparticle silver released into water from commercially available sock fabrics. *Environmental Science and Technology*, **42**, 4133–4139.

Bhatt, I. and Tripathi, B.N. (2011) Interaction of engineered nanoparticles with various components in the environment and possible strategies for their risk assessment. *Chemosphere*, **82**, 308–317.

Bhattacharya, R. and Mukherjee, P. (2008) Biological properties of 'naked' metal nanoparticles. *Advanced Drug Delivery Reviews*, **60**, 1289–1306.

Borm, P. and Castranova, V. (2009) Toxicology of nanomaterials: permanent interactive learning. *Particle and Fibre Toxicology*, **6**, 28.

Braydich-Stolle, L.K., Lucas, B., Schrand, A. *et al.* (2010) Silver nanoparticles disrupt GDNF/Fyn kinase signaling in spermatogonial stem cells. *Toxicological Sciences*, **116**, 577–589.

British Standards Institute (2007) PD 6699-2-2007. *Nanotechnologies – Part 2: Guide to Safe Handling and Disposal of Manufactured Nanomaterials*, BSI, London.

van Broekhuizen, P. and Reijnders, L. (2011) Building blocks for a precautionary approach to the use of nanomaterials. *Risk Analysis*, **31**, 1646–1657.

van Broekhuizen, P., van Broekhuizen, F., Cornelissen, R. and Reijnders, L. (2012) Workplace exposure to nanoparticles and the application of provisional nanoreference values in times of uncertain risks. *Journal of Nanoparticle Research*, **14**, 1–25.

Chafer-Pericas, C., Maquiera, A. and Puchades, R. (2011) Functionalized inorganic nanoparticles used as labels in solid-phase immunoassays. *Trends in Analytical Chemistry*. doi: 10.1016/j.trac.2011.07.011.

Chen, Z., Meng, H., Xing, G. *et al.* (2008) Age-related differences in pulmonary and cardiovascular responses to SiO_2 nanoparticle inhalation: nanotoxicity has susceptible population. *Environmental Science and Technology*, **42**, 8985–8992.

Chinnapongse, S.L., MacCuspie, R.I. and Hackley, V.A. (2011) Persistence of singly dispersed silver nanoparticles in natural freshwaters, synthetic seawater, and simulated estuarine waters. *Science of the Total Environment*, **409**, 2442–2450.

Choi, H.S., Liu, W., Liu, F. *et al.* (2010) Design conditions for tumor targeted nanoparticles. *Nature Nanotechnology*, **5**, 42–47.

Clift, M.J.D., Rothen-Rutishauser, B., Brown, D.M. *et al.* (2008) The impact of different nanoparticle surface chemistry on uptake and toxicity in a murine macrophage cell line. *Toxicology and Applied Pharmacology*, **232**, 418–427.

Cushen, M., Kerry, J., Morris, M. *et al.* (2011) Nanotechnologies in the food industry-recent developments, risks and regulation. *Trends in Food Science and Technology*. doi: 10.1016/j.tifs.2011.10.006

Darlington, T.K., Neigh, A.M., Spencer, M.T. *et al.* (2009) Nanoparticle characteristics affecting environmental fate and transport through soil. *Environmental Toxicology and Chemistry*, **28**, 1191–1199.

Dasch, J. and D'Arcy, J. (2008) Physical and chemical characterization of airborne particles from welding operations in automotive plants. *Journal of Occupational Health and Hygiene*, **5**, 444–454.

Du, W., Sun, Y., Ji, R. *et al.* (2011) TiO_2 and ZnO nanoparticles negatively affect wheat growth and soil enzyme activities in agricultural soil. *Journal of Environmental Monitoring*, **12**, 822–828.

Duffin, R., Tran, L., Brown, D. *et al.* (2007) Proinflammogenic effects of low toxicity and metal nanoparticles in vivo and in vitro: highlighting the role of particle surface area and surface reactivity. *Inhalation Toxicology*, **19**, 849–856.

Elsaesser, A. and Howard, C.V. (2011) Toxicology of nanoparticles. *Advanced Drug Delivery Reviews*. doi: 10.1016/j.addr.2011.09.001.

Fabrega, J., Luoma, S.N., Tyler, C.R. *et al.* (2011) Silver nanoparticles: behavior and effects in the aquatic environment. *Environment International*, **37**, 517–531.

Farré, M., Sanchis, J. and Barcelo, D. (2011) Analysis and assessment of the occurrence, the fate and the behavior of nanomaterials in the environment. *Trends in Analytical Chemistry*, **30**, 517–527.

Ferreira, O.P., Filho, A.G.S. and Alves, O.L. (2010) Recycling dodecylamine intercalated vanadate nanotubes. *Journal of Nanoparticle Research*, **12**, 367–372.

Fleury, D., Bomfim, J.A.S., Vignes, A. *et al.* (2011) Identification of main exposure scenarios in the production of CNT-polymer nanocomposites by melt-moulding process. *Journal of Cleaner Production*. doi: 101.1016/j.jclepro.2011.11.009.

Florence, A.T. (2005) Nanoparticle uptake by the oral route: fulfilling its potential? *Drug Discovery Today: Technologies*, **2**, 75–81.

Fubini, B., Ghiazza, M. and Fenoglio, I. (2010) Physico-chemical features of engineered nanoparticles relevant to their toxicity. *Nanotoxicology*, **4**, 347–363.

Gagner, J.E., Lopez, M.D., Dordick, J.S. and Siegel, R.W. (2011) Effect of gold nanoparticle morphology on adsorbed protein structure and function. *Biomaterials*, **32**, 7241–7252.

Glover, R.D., Miller, J.M. and Hutchison, J.E. (2011) Generation of metal nanoparticles from silver and copper objects: nanoparticle dynamics on surfaces and potential sources of nanoparticles in the environment. *ACSNANO*, **11**, 8950–8957.

Gomez-Mejiba, S.E., Zhai, Z., Akram, H. *et al.* (2009) Inhalation of environmental stressors and chronic inflammation: autoimmunity and neurodegradation. *Mutation Research*, **674**, 62–72.

Gottschalk, F., Sonderer, T., Scholz, R.W. and Nowack, B. (2010) Possibilities and limitations of modeling environmental exposure to engineered nanomaterials by probabilistic material flow analysis. *Environmental Toxicology and Chemistry*, **29**, 1036–1048.

Hamilton, R.F., Wu, N., Porter, D. *et al.* (2009) Particle length-dependent titanium dioxide nanomaterials toxicity and bioactivity. *Particle and Fibre Toxicology*, **6**, 35.

Hesterberg, T.W., Long, C.M., Lapin, C.A. *et al.* (2010) Diesel exhaust particles (DEP) and nanoparticle exposures: what do DEP human clinical studies tell us about potential human health hazards of nanoparticles. *Inhalation Toxicology*, **23**, 679–694.

van Hoecke, K., de Schampelaere, K.A.C., van der Meeren, P. *et al.* (2011) Aggregation and ecotoxicity of CeO_2 nanoparticles in synthetic and natural waters with variable pH, organic matter concentration and ionic strength. *Environmental Pollution*, **159**, 970–976.

Horie, M. and Fujita, K. (2011) Toxicity of metal oxide nanoparticles. *Advances in Molecular Toxicology*, **4**, 145–178.

Horie, M., Nishio, K., Fujita, K. *et al.* (2009) Protein adsorption of ultrafine metal oxide and its influence on cytotoxicity towards cultured cells. *Chemical Research in Toxicology*, **22**, 543–553.

Hougaard, K.S., Jackson, P., Jensen, K.A. *et al.* (2010) Effects of prenatal exposure to surface coated nanosized titanium dioxide (UV-Titan). A study in mice. *Particle and Fibre Toxicology*, **7**, 16.

Hristozov, D.R., Gottardo, S., Critto, A. and Marcomini, A. (2012) Risk assessment of engineered nanomaterials: a review of available data and approaches from a regulatory perspective. *Nanotoxicology*. doi: 10.3109/17435390.2011. 626534

Hubbs, A.F., Mercer, R.R., Benkovic, S.A. *et al.* (2011) Nanotoxicology – a pathologist's perspective. *Toxicologic Pathology*, **39**, 301–324.

Hussain, S., Boland, S., Baeza-Squiban, A. *et al.* (2009) Oxidative stress and proinflammatory effects of carbon black and titanium dioxide nanoparticles: role of surface area and internalized amount. *Toxicology*, **260**, 142–149.

Ivask, A., Suarez, E., Patel, T. *et al.* (2012) Genome-wide bacterial toxicity screening uncovers mechanisms of toxicity of cationic polystyrene nanomaterials. *Environmental Science and Technology*. doi: 10.1021/es203087m.

Jackson, P., Hougaard, K.S., Vogel, U. *et al.* (2011) Exposure of pregnant mice to carbon black by intracheal installation: toxicogenomic effects in dams and offspring. *Mutation Research*. doi: 10.1016/j.mrgentox.2011.09.018

Jensen-Eckelman, M., Mauter, M.S., Isaacs, J. and Elimelech, M. (2012) New perspectives on nanomaterial aquatic exotoxicity: production impacts exceed direct exposure impacts for carbon nanotubes. *Environmental Science and Technology*. doi: 10.1021/es203409a

Jeong, G. (2009) Surface functionalization of single-walled carbon nanotubes using metal nanoparticles. *Transactions of Nonferrous Metals Society China*, **19**, 1009–1012.

Jia, G., Wang, H., Yan, L. *et al.* (2005) Cytotoxicity of carbon nanomaterials: single-wall nanotube, multi-wall nanotube, and fullerene. *Environmental Science and Technology*, **39**, 1378–1383.

Kayat, J., Gajbhiye, V., Tekade, R.K. and Kumar, N. (2011) Pulmonary toxicity of carbon nanotubes: a systematic report. *Nanomedicine, Nanotechnology, Biology and Medicine*, **7**, 40–49.

Kim, J.S., Song, K.S., Lee, J.K. *et al.* (2011) Toxicogenomic comparison of multiwall carbon nanotubes (MWCNTs) and asbestos. *Archives of Toxicology*. doi: 10.1007/s00204-011-0770-6

Klaine, S.J., Koelmans, A.A., Horne, N. *et al.* (2012) Paradigms to assess the environmental impact of manufactured nanomaterials. *Environmental Toxicology and Chemistry*, **31**, 3–14.

Koike, E. and Kobayashi, T. (2011) Chemical and biological oxidative effects of carbon black nanoparticles. *Chemosphere*, **65**, 946–951.

Kreyling, W.G., Semmler-Behnke, M., Erbe, F. *et al.* (2002) Translocation of ultrafine insoluble iridium particles from lung epithelium to extrapulmonary organs is size dependent but very low. *Journal of Toxicology and Environmental Health, Part A*, **65**, 1513–1520.

Kreyling, W.G., Sammler-Behnke, M., Seitz, J. *et al.* (2009) Size dependence of the translocation of inhaled iridium and carbon nanoparticle aggregates from the lungs of rats to the blood and secondary target organs. *Inhalation Toxicology*, **21**, 55–60.

Kuhlbusch, T.A.J., Asbach, C., Fissan, H. *et al.* (2011) Nanoparticle exposure at nanotechnology workplaces: a review. *Particle and Fibre Toxicology*, **8**, 22.

Kunzli, N. (2011) From bench to policies: ready for a nanoparticle air quality standard? *European Heart Journal*, **32**, 2613–2615.

Lankoff, A., Sandberg, W.J., Wegierek-Ciuk, A. *et al.* (2012) The effect of agglomeration state of silver and titanium dioxide nanoparticles on cellular response of HepG2, A549 an THP-1 cells. *Toxicology Letters*, **208**, 197–213.

Lankveld, D.P.K., Oomen, A.G., Krystek, P. *et al.* (2010) The kinetics of tissue distribution of silver nanoparticles of different sizes. *Biomaterials*, **31**, 8350–8361.

Limbach, L., Wick, P., Manser, O.P. *et al.* (2007) Exposure of engineered nanoparticles to human lung epithelial cells: influence of chemical composition and catalytic activity on oxidative stress. *Environmental Science and Technology*, **41**, 4158–4163.

Liu, J. and Hurt, R.H. (2010) Ion release kinetics and particle persistence in aqueous nanosilver colloids. *Environmental Science and Technology*, **44**, 2169–2175.

Liu, X., Sen, S., Liu, J. *et al.* (2011) Antioxidant deactivation on graphenic nanocarbon surfaces. *Small*, **7**, 2775–2785.

Loeschner, K., Hadrup, N., Qvortrup, K. *et al.* (2011) Distribution of silver in rats following 28 days of repeated oral exposure to silver nanoparticles or silver acetate. *Particle and Fibre Toxicology*, **8**, 18.

Longmire, M.R., Ogawa, M., Choyke, P.L. and Kobayashi, H. (2011) Biologically optimized nano-sized molecules and particle; more than just size. *Bioconjugate Chemistry*, **22**, 993–1000.

Lordan, S., Kennedy, J.E. and Higginbotham, C.L. (2011) Cytotoxic effects induced by unmodified and organically modified nanoclays in the human hepatic HepG2 cell line. *Journal of Applied Toxicology*, **31**, 27–35.

Luna-delRisco, M., Orupold, K. and Dubourguier, H. (2011) Particle size effect of CuO and ZnO on biogas and methane production during anaerobic digestion. *Journal of Hazardous Materials*, **189**, 603–608.

Ma, P., Siddiqui, N.A., Marom, G. and Kim, J. (2010) Dispersion and functionalization of carbon nanotubes for polymer-based nanocomposites: a review. *Composites: Part A*, **41**, 1345–1367.

Mahtab, R., Sealey, S.M., Hunyadi, S.E. *et al.* (2007) Influence of the nature of quantum dot surface cations on interactions with DNA. *Journal of Inorganic Biochemistry*, **101**, 559–564.

Marambio-Jones, C. and Hoek, E.M.V. (2010) A review of the antibacterial effects of silver nanomaterials and potential implications for human health and the environment. *Journal of Nanoparticle Research*, **12**, 1531–1551.

Maynard, A.D., Warheit, D.B. and Philbert, M.A. (2011) The new toxicology of sophisticated materials: nanotoxicology and beyond. *Toxicological Sciences*, **120**, S109–S129.

Menard, A., Drobne, D. and Jemec, A. (2011) Ecotoxicity of nanosized TiO_2, review of in vivo data. *Environmental Pollution*, **159**, 577–584.

Monteiller, C., Tran, L., MacNee, W. *et al.* (2007) The pro-inflammatory effects of low-toxicity low solubility particles, nanoparticles and fine particles, on epithelial cells in vitro: the role of surface area. *Occupational and Environmental Medicine*, **64**, 609–615.

Mudunkotuwa, I., Pettibone, J.M. and Grassian, V.H. (2012) Environmental implications of nanoparticle aging in the processing and fate of copper-based nanomaterials. *Environmental Science and Technology*. doi: 101021/es203851d.

Mueller, N.C. and Nowack, B. (2008) Exposure modeling of engineered nanoparticles in the environment. *Environmental Science and Technology*, **42**, 4447–4453.

Myllynen, F. (2009) Damaging DNA from a distance. *Nature Nanotechnology*, **4**, 795–796.

Napierska, D., Thomassen, L.C.J., Lison, D. *et al.* (2010) The nanosilica hazard: another variable entity. *Particle and Fibre Toxicology*, **7**, 39.

NIOSH (2010) Occupational exposure to carbon nanotubes and nanofibers. *Current Intelligence Bulletin*, **161-A**, 1–149.

NIOSH (2011) Occupational exposure to titanium dioxide. *Current Intelligence Bulletin*, **63**, 1–119.

Nowack, B., Ranville, J.F., Diamond, S. *et al.* (2012) Potential scenarios for nanomaterial release and subsequent alteration in the environment. *Environmental Toxicology and Chemistry*, **31**, 50–59.

Oberdörster, G. (2010) Safety assessment for nanotechnology and nanomedicine: concepts of nanotoxicology. *Journal of Internal Medicine*, **267**, 89–105.

Ottofuelling, S., von der Kammer, F. and Hofman, T. (2011) Commercial titanium dioxide nanoparticles in both natural an synthetic water: comprehensive multidimensional testing and prediction of aggregation behavior. *Environmental Science and Technology*, **45**, 10045–10052.

Panyala, N.R., Pena-Mendez, E.M. and Havel, J. (2008) Silver or silver nanoparticles: a hazardous threat to the environment and human health? *Journal of Applied Biomedicine*, **6**, 117–129.

Park, M.V.D.Z., Neigh, A.M., Vermeulen, J.P. *et al.* (2011) The effect of particle size on the cytotoxicity, inflammation, developmental toxicity and genotoxicity of silver nanoparticles. *Biomaterials*, **32**, 9810–9817.

Pauluhn, J. (2011) Poorly soluble particulates: searching for a unifying denominator of nanoparticles and fine particles for DNEL estimation. *Toxicology*, **279**, 176–188.

Paur, H., Cassee, F.G., Teeguarden, J. *et al.* (2011) In-vitro cell exposure studies for the assessment of nanoparticle toxicity in the lung – a dialog between aerosol science and biology. *Journal of Aerosol Science*, **42**, 668–692.

Pelley, J.L., Daar, A.S. and Saner, M.A. (2009) State of academic knowledge on toxicity and biological fate of quantum dots. *Toxicological Sciences*, **112**, 276–296.

Pereira, R., Rocha-Santos, J.A.P., Antunes, F.E. *et al.* (2011) Screening evaluation of ecotoxicity and genotoxicity of soils contaminated with organic and inorganic nanoparticles: the role of aging. *Journal of Hazardous Materials*, **194**, 345–354.

Peters, T.M., Elzey, S., Johnson, R. *et al.* (2009) Airborne monitoring to distinguish engineered nanomaterials from incidental particles for environmental health and safety. *Journal of Occupational and Environmental Hygiene*, **6**, 73–81.

Petersen, E.J. and Henry, T.B. (2011) Methodological considerations for testing the ecotoxicity of carbon nanotubes and fullerenes. *Environmental Toxicology and Chemistry*. doi: 10.1002/etc.710.

Petersen, E.J., Zhang, L., Mattison, N.T. *et al.* (2011) Potential release pathways, environmental fate and ecological risks of carbon nanotubes. *Environmental Science and Technology*, **45**, 9837–9856.

Piccapietra, F., Sigg, L. and Behra, R. (2012) Colloidal stability of carbonate coated silver nanoparticles in synthetic and natural freshwater. *Environmental Science and Technology*, **46**, 818–825.

Powell, J.J., Faria, N., Thomas-McKay, E. and Pele, L.C. (2010) Origin and fate of dietary nanoparticles and microparticles in the gastrointestinal tract. *Journal of Autoimmunology*, **34**, J226–J233.

Quadros, M.E. and Marr, L.C. (2011) Silver nanoparticles and total aerosols emitted by nanotechnology related consumer spray products. *Environmental Science and Technology*, **45**, 10713–10719.

Rabolli, V., Thomassen, L.C.J., Uwambayinema, F. *et al.* (2011) The cytotoxic activity of amorphous silica nanoparticles, is mainly influenced by surface area and not by aggregation. *Toxicology Letters*, **206**, 197–203.

Ramachandran, G., Ostaat, M., Evans, D.E. *et al.* (2011) A strategy for assessing workplace exposures to nanomaterials. *Journal of Occupational Health and Environmental Hygiene*, **8**, 673–685.

Ramanan, B., Holmes, W.M., Sloan, W.T. and Phoenix, V.R. (2012) Investigation of nanoparticle transport inside coarse grained geological media using magnetic resonance imaging. *Environmental Science and Technology*, **46**, 360–366.

Reijnders, L. (2008) Hazard reduction in nanotechnology. *Journal of Industrial Ecology*, **12**, 297–306.

Reijnders, L. (2012) Human health hazards of persistent inorganic and carbon nanoparticles. *Journal of Materials Science*, **47**, 5061–5073.

Rhodes, J., Smith, C. and Stec, A.A. (2011) Characterisation of soot particulates from fire retarded and nanocomposite materials and their toxicological impact. *Polymer Degradation and Stability*, **96**, 277–284.

Rim, K., Jim, S., Han, J. *et al.* (2011) Effects of carbon black to inflammation and oxidative DNA damages in mouse macrophages. *Molecular and Cellular Toxicology*, **7**, 415–423.

Rivet, C.J., Yuan, Y., Borca-Tasciuc, D. and Gilbert, R.J. (2012) Altering iron oxide nanoparticle surface properties induce cortical neuron cytotoxicity. *Chemical Research in Toxicology*. doi: 10.1021/tx200369s

Sayes, C.M., Reed, K.L. and Warheit, D.B. (2007) Assessing toxicity of fine and nanoparticles: comparing in vitro measurements to in vivo pulmonary toxicity profiles. *Toxicological Sciences*, **97**, 163–180.

Schinwald, A., Murphy, F.A., Jones, A. *et al.* (2012) Graphene based nanoplatelets: a new risk to the respiratory system as a consequence of their unusual aerodynamic properties. *ACSNANO*. doi: 10.1021/nn204229f

Schuepp, K. and Sly, P.D. (2012) The developing respiratory tract and its specific needs in regard to ultrafine particulate matter exposure. *Pediatric Respiratory Reviews*. doi: 10.1016/j.prrv.2011.08.002.

Scymczak, W., Menzela, N. and Kecka, L. (2007) Emission of ultrafine copper particles by universal motors controlled by phase angle modulation. *Journal of Aerosol Science*, **38**, 520–531.

Semmler-Behnke, M., Kreyling, W.G., Lipka, J. *et al.* (2008) Biodistribution of 1.4 and 18 nm gold nanoparticles in rats. *Small*, **4**, 2108–2111.

Sheng, Z. and Liu, Y. (2011) Effects of silver nanoparticles on wastewater biofilms. *Water Research*, **45**, 6039–6050.

Shukla, R.K., Kumar, A., Gurbani, D. *et al.* (2012) TiO_2 nanoparticles induce oxidative DNA damage and apoptosis in human liver cells. *Nanotoxicology*. doi: 10.3109/17435390.2011.629747.

Silver, S. (2003) Bacterial silver resistance: molecular biology and misuses of silver compounds. *FEMS Microbiology Research*, **27**, 341–353.

Silver, S., Phung, L.T. and Silver, G. (2006) Silver as biocides in burn and wound dressings and bacterial resistance to silver compounds. *Journal of Industrial Microbiology and Biotechnology*, **33**, 627–634.

So, S.J., Jang, I.S. and Han, C.S. (2008) Effect of micro/nano silica particle feeding on mice. *Journal of Nanoscience and Nanotechnology*, **8**, 5367–5371.

Soenen, S.J., Rivera-Gil, P., Montenegro, J. *et al.* (2011) Cellular toxicity of inorganic nanoparticles: common aspects and guidelines for improved nanotoxicity evaluation. *Nano Today*, **6**, 446–465.

Song, Y., Li, X., Wang, L. *et al.* (2011) Nanomaterials in humans: identification, characteristics and potential damage. *Toxicologic Pathology*, **39**, 841–849.

Sood, A., Salih, S., Roh, D. *et al.* (2011) Signaling of DNA damage and cytokines across cell barriers exposed to nanoparticles depends on barrier thickness. *Nature Nanotechnology*, **6**, 824–833.

Stebounova, L.V., Morgan, H., Grassian, V.H. and Brenner, S. (2011) Health and safety implications of occupational exposure to engineered nanomaterials. *WIREs Nanomedicine and Nanobiotechnology*. doi: 10.1002/wnan.174.

Sturm, R. (2007) A computer model for the clearance of insoluble particles from the tracheobronchial tree of the human lung. *Computers in Biology and Medicine*, **37**, 680–690.

Sturm, R. and Hofmann, W. (2009) A theoretical approach to the deposition and clearance of fibers with variable size in the human respiratory tract. *Journal of Hazardous Materials*, **170**, 210–218.

Tabuchi, A. and Kuebler, W.M. (2008) Endothelium-platelet interactions in inflammatory lung disease. *Vascular Pharmacology*, **49**, 141–150.

Thio, B.J.R., Montes, M., El Mohsen, M.A. *et al.* (2011) Mobility of capped silver nanoparticles under environmentally relevant conditions. *Environmental Science and Technology*. doi: 10.1012/es203596w.

Turco, R.F., Bischoff, M., Tong, Z.H. and Nies, L. (2011) Environmental implications of nanomaterials: are we studying the right thing? *Current Opinion in Biotechnology*, **22**, 527–532.

Wohlleben, W., Brill, S., Meier, M.W. *et al.* (2011) On the lifecycle or nanocomposites: comparing released fragments and their in vivo hazards from three release mechanisms and four nanocomposites. *Small*, **7**, 2384–2395.

Won, H.I., Nersisyan, H.H. and Won, C.W. (2010) Combustion synthesis of nano-sized tungsten carbide powder and effects of sodium halides. *Journal of Nanoparticle Research*, **12**, 493–500.

Woskie, S. (2010) Workplace practices for engineered nanomaterial manufacturers. *Nanomedicine and Nanobiotechnology*, **2**, 685–692.

Wynne, J.H., Buckley, J.L., Coumbe, C.E. *et al.* (2008) Reducing hazardous material and environmental impact through recycling of scandium nanomaterial waste. *Journal of Environmental Science and Health Part A*, **43**, 357–360.

Yang, X., Gondikas, A., Marinakos, S.M. *et al.* (2012) The mechanism of silver nanoparticle toxicity is dependent on dissolved silver and surface coating in Caenorhabdatis elegans. *Environmental Science and Technology*. doi: 10.1021/es2024171t.

Yokel, R.A. and MacPhail, R.C. (2011) Engineered nanomaterials: exposures, hazards and risk prevention. *Journal of Occupational Medicine and Toxicology*, **6**, 7.

Zhang, R., Bai, Y., Zhang, B. *et al.* (2011a) The potential health risk of titania nanoparticles. *Journal of Hazardous Materials*. doi: 10.1016/j.jhazmat. 2011.11.022.

Zhang, B., Luo, Y. and Wang, Q. (2011b) Development of silver/α-lactalbumin nanocomposites: a new approach to reduce silver toxicity. *International Journal of Antimicrobial Agents*, **38**, 502–509.

Zhang, T., Kim, B., Levard, C. *et al.* (2012) Methylation of mercury by bacteria exposed to dissolved nanoparticulate, and microparticulate mercuric sulfides. *Environmental Science and Technology*. doi: 10.1021/es203181m

Zhang, M., Su, L. and Mao, L. (2006) Surfactant functionalization of carbon nanotubes (CNTs) for layer-by-layer assembling of CNT multi-layer films and fabrication of gold nanoparticle/CNT nanohybrid. *Carbon*, **44**, 276–283.

Zhu, M., Wang, B., Wang, Y. *et al.* (2011) Endothelial dysfunction and inflammation induced by iron oxide nanoparticle exposure: risk factors for early atherosclerosis. *Toxicology Letters*, **201**, 162–171.

14

Environmental Health in Built Environments

Zaheer Ahmad Nasir
Healthy Infrastructure Research Centre, Department of Civil, Environmental & Geomatic Engineering, University College London, UK

14.1 Environmental Hazards and Built Environments

Environmental health in built environments is central to public health. The existence of a wide range of contaminants of both indoor and outdoor origin, the great source strength per area and the high proportion of time spent in various indoor environments mean that exposure can have a significant impact on human well-being. The amount of time spent in different built environments (residential, occupational, transport and recreational) varies across the globe. However, people around the world typically spend a higher proportion of their time inside than outside.

Broadly, 'built environment' refers to 'human-made space in which people live, work, and recreate on a day-to-day basis' (Roof and Oleru, 2008). It is a material, spatial and cultural product of human labour that combines physical elements and energy in forms necessary to life, work and play (Doleman and Brooks, 2011). In the twenty-first century, humans are exposed to a range of physical, chemical, biological and ergonomic hazards in various built environments. Worldwide, there is considerable variation in the degree of exposure to these hazards, due to noticeable differences in their types and strengths; they are closely linked to socioeconomic developments. Environmental, social, economic, political, technological and climatic changes are constantly altering principles and practices in the design, construction, operation and management of built environments, leading to shifts in

Aerosol Science: Technology and Applications, First Edition. Edited by Ian Colbeck and Mihalis Lazaridis.

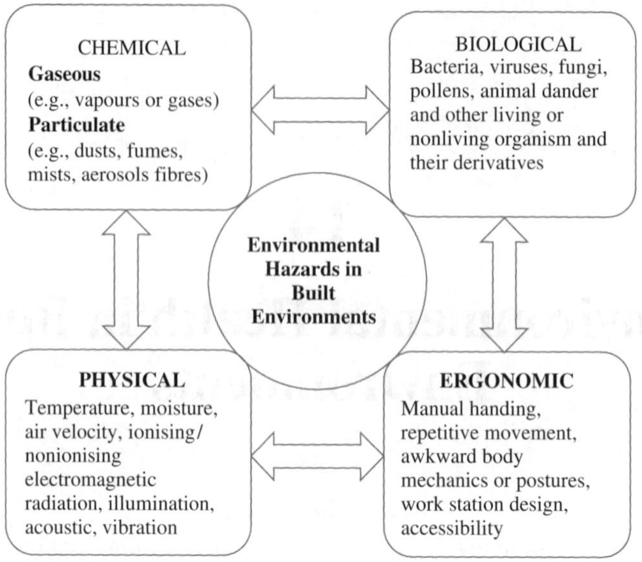

Figure 14.1 *Environmental hazards influencing the well-being of the occupants of various built environments.*

exposure patterns and having the potential to create new exposure pathways. Figure 14.1 depicts some environmental hazards that can influence human well-being.

There is good body of knowledge on the routes of exposure to these hazards (inhalation, dermal contact and/or absorption, ingestion), their levels in different built environments and their possible association with ill-health effects. However, what is less understood is the role of diverse human, social, economic, political and environmental factors in the design, construction, use and management of different built environments and how complexity and connectivity among these might influence exposure pathways to these hazards. The degree of risk to public health in different built environments depends on the nature and strength of the harmful effects of a hazard, the social and physical vulnerability of the population, the susceptibility of the persons exposed and the concentration and duration of exposure. Hence, numerous factors can greatly influence the magnitude and duration of exposure.

Most of the current exposure models use either outdoor concentrations of contaminants or individual contaminants in different built environments to predict human exposure. However, in order to estimate the total risk of exposure, there is a dire need to calculate the cumulative exposure to all of the environmental hazards across built environments. This approach will assist greatly in gaining a holistic understanding of environmental health issues in different built environments and in designing the mitigation and prevention strategies that will help reduce the total population exposure to environmental hazards.

These hazards have a variety of sources, including ambient air pollutants, building materials, furniture and furnishing (especially wet or damp furnishing), occupants and their personal care, household cleaning, maintenance and leisure activities, combustion sources (oil, gas, kerosene, coal and wood appliances, tobacco products), photocopying/printing, underground transport, pesticides, organic chemicals, building technologies (air-handling systems, water distribution systems), indoor plants/animals, outdoor plants/animals, insect infestation, agricultural and livestock production, waste management, manufacturing and packing, electrical equipment, poor ergonomics and accessibility. A range of design, construction, operation and management factors can greatly influence the risk of exposure to environmental hazards in built environments. It is notable that while the design and construction of the built environment are intrinsic, its use and management are largely dependent on the inhabitants' behaviour, which can be influenced by a wide range of factors. Figure 14.2 reflects various factors influencing the design, construction, use and management of the built environment that might implicate exposure to different environmental hazards.

It is also worthwhile to mention that built environments undergo change both spatially and temporally and that the design, construction, use and management of built environments and of different built forms in a geographical region are the result of complex political, socioeconomic and environmental factors. These factors constantly alter the parameters influencing built environments, leading to a wide variation in the magnitude of exposure to environmental hazards across the globe. This is evidenced by the substantial differences in the environmental burden of disease among low-, middle- and high-income countries.

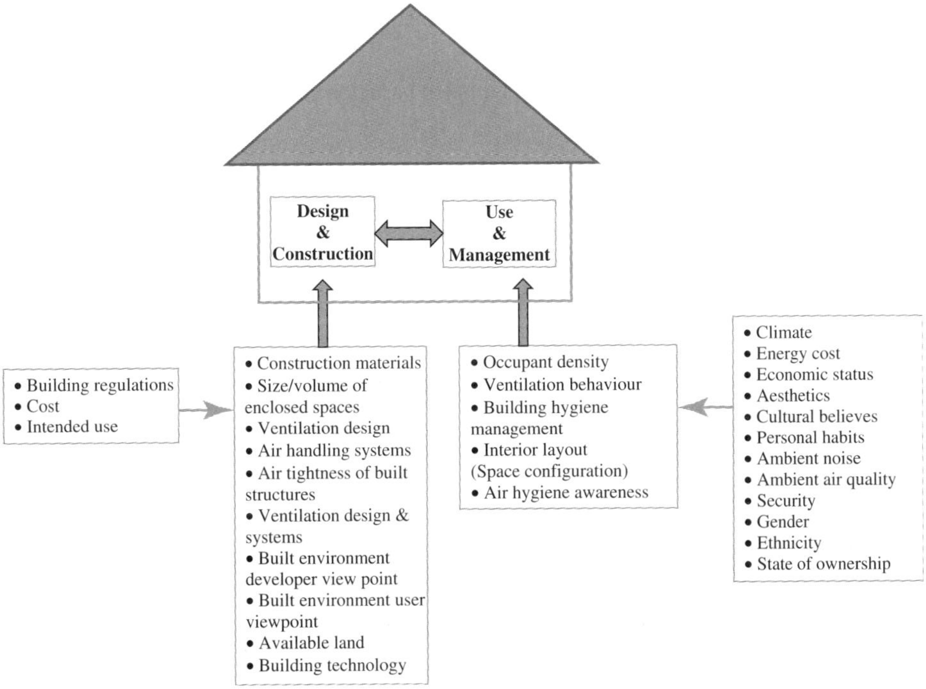

Figure 14.2 *Factors influencing the design, construction, use and management of various built environments.*

14.2 Particulate Contaminants

Particulate matter (small solid and liquid particles of various sizes, shapes and chemical properties suspended in the air) is a major air pollutant. It can be of both anthropogenic and natural origin and is classified according to its size, generation processes and ability to enter the human respiratory system (Morawska and Salthammer, 2003). In brief, particles can be categorized according to their size based on: (i) their observed modal distribution, (ii) the 50% cut-off diameter or (iii) dosimetric variables related to human exposure. In category (i), several subcategories can be observed:

Nucleation mode: particles with diameter <10 nm.
Aitken mode: particles with diameter 10 nm <d <100 nm.
Accumulation mode: particles with diameter 0.1 μm <d <1 μm.
Ultrafine particles: particles in the Aitkin and nucleation modes.
Fine particles: includes the nucleation, Aitkin and accumulation modes.
Coarse particles: particles with diameter between 2 and 10 μm. Here the most common division is PM_{10} and $PM_{2.5}$, and this has been widely used in ambient air-quality standards.

Classification according to the site of deposition in the lungs includes:

Inhalable fraction: mass fraction of total airborne particles that is inhaled through the nose and mouth.
Thoracic fraction: mass fraction of inhaled particles that penetrates beyond the larynx.
Extrathoracic fraction: mass fraction of total inhalable particles that fails to penetrate beyond the larynx.
Respirable fraction: mass fraction of total inhaled particles that infiltrates up to the alveolar region.

Particles in the air are a complex mixture generated by a wide range of sources and exhibit a significant variation in their shape and size distribution (see Chapter 2). In addition, once airborne, they are subject to multiple processes, interactions and reactions, leading to changes in their chemical composition and size. This is due to a variety of physical and chemical processes, such as nucleation (new particle formation), condensation, evaporation, coagulation, deposition (both wet and dry), activation due to water and other gaseous species and aqueous-phase reactions (Finlayson-Pitts and Pitts, 2000; Seinfeld and Pandis, 2006). Fine particles are generated mainly by combustion processes, gas-to-particle conversion and photochemical processes, while coarse particles result from mechanical processes. Different terminology has been used to refer to their sizes, generation processes and composition characteristics (see Chapter 1).

The fate of the particles in the air is largely determined by their physical properties, particularly their size. Aerosol particle size spans over several orders of magnitude, ranging from almost macroscopic dimensions down to near molecular sizes (Hinds, 2005; Seinfeld and Pandis, 2006). Particle size distribution is a crucial parameter that determines the dynamics of aerosols in the atmosphere and their transport, deposition and residence time (Colbeck and Lazaridis, 2010).

With reference to built environments, particles present in enclosed environments can have both indoor and outdoor sources. These include but are not limited to: human occupants (i.e. skin, hair, respiratory aerosols, etc.), plants, pets, household cleaning and maintenance activities, building material and furnishings, combustion appliances, personal-care products and tobacco products (Morawska and Salthammer, 2003). There is a considerable variation in the contribution of different sources to the concentration of indoor particles, and the relative importance of a source depends upon its emission strength and toxicity. Some sources release particles more or less continuously, while others are intermittent. In most cases, the particles generated by different sources or even from the same source can differ in their concentration according to a number of factors. According to Morawska (2004), these factors are:

1. type, nature and number of sources
2. source use characteristics
3. building characteristics
4. outdoor concentration of pollutants
5. infiltration or ventilation rates
6. air mixing
7. removal rates by surfaces, chemical transformation or radioactive decay
8. existence and effectiveness of air contaminant removal systems
9. meteorological parameters.

All these factors vary to a great extent in different built environments and are greatly influenced by elements of design, construction, use and management. There is thus a substantial variation in the concentration of particles and in their subsequent exposure in various indoor environments.

14.2.1 Transport and Behaviour of Particles in Built Environments

Upon emission from a source, particles can undergo a range of physical and chemical process, which can also transform their concentration (see Chapters 2 and 7). These processes include: sedimentation, deposition on surfaces, coagulation and changes by evaporation or condensation. The degree of transformation and the residence time of the particles are largely dependent on the source type. For example, emission from combustion sources can undergo much more rapid changes than can mechanically formed dust (Morawska, 2004). The following are the most important processes affecting indoor particle concentrations after emission: infiltration, penetration of outdoor particles indoors, particle deposition, resuspension of particles, ventilation, phase change, mixing and coagulation. A number of studies have focussed on measuring and modelling indoor particle dynamics. In an extensive review, Chen and Zhao (2011) found a large variation in infiltration (0.3–0.82 for $PM_{2.5}$ and 0.17–0.52 for PM_{10}) and penetration factors (0.6–1.0 for particles >0.05 and <2 μm) in real buildings. These are size-dependent and are influenced by many other factors (air exchange rate, indoor/outdoor pressure difference, wind direction and speed, the geometry of cracks in building envelopes). Similarly, many studies have quantified deposition and resuspension rates, finding that deposition of particles on indoor surfaces is size-dependent and that many mechanisms can contribute to this process (Lai, 2002; Wallace, Emmerich and Howard-Reed, 2004; He, Morawska and Gilbert, 2005; Chen, Yu and Lai, 2006; Hussein *et al.*, 2006, 2009; Hamdani *et al.*, 2008). Advection and turbulent diffusion are generally strong enough to rapidly transport particles from core air to boundary layers, and thus their deposition on various surfaces. For ultrafine particles, Brownian diffusion is an important deposition mechanism, while gravitational settling amplified by impaction is responsible for coarse-particle deposition. On the other hand, accumulation-mode particles deposit the least effectively (Nazaroff, 2004). Among other factors affecting the deposition are increased air flow, surface area (Thatcher *et al.*, 2002) and surface charges (Lai, 2006).

With reference to resuspension, many household activities and airflows in ventilation ducts can result to particle resuspension. A number of studies have demonstrated that a variety of factors (particle size, surface material and roughness, air velocity, particle composition, air flow characteristics, relative humidity) influence the introduction of particles previously deposited on various indoor surfaces into the air (Wu, Davidson and Russell, 1992; Thatcher and Layton, 1995; Ibrahim, Dunn and Brach, 2003, 2004; Ferro, Kopperud and Hildemann, 2004; Gomes, Freihaut and Bahnfleth, 2007; Qian and Ferro, 2008; Mukai, Siegel and Novoselac, 2009; Shaughnessy and Vu, 2012; Goldasteh, Ahmadi and Ferro, 2012a,2012b; Boor, Siegel and Novoselac, 2013; Kassab *et al.*, 2013). In addition, apart from affecting the levels of indoor particles, resuspension may provide an exposure pathway for allergens and for semivolatile species.

Ventilation, the process of exchanging indoor air with that outdoors, is an important factor in quantifying the emission strength of indoor sources. The three main modes are mechanical ventilation, natural ventilation and infiltration or leakage flow. He *et al.* (2005) have shown that ventilation rate is an important factor affecting deposition rates for particles in the size range 0.08–1.0 μm. Ventilation rate varies significantly with the type and location of a building, the climate and the lifestyle of the inhabitants (Morawska, 2004). Mechanical ventilation systems are commonly equipped with fibrous filters to limit the introduction of outdoor air pollutants indoors. The efficiency of a filtration system depends on the filter properties and the aerodynamic size of the particles (Jamriska, Morawska and Ensor, 2003).

Additionally, indoor particles can undergo phase-change processes involving vapours and gases that lead to particle generation and hence affect their number and mass concentration (Morawska, 2004). A number of phase-change processes have been reported in previous studies, such as growth associated with changing humidity conditions (Dua and Hopke, 1996), phase division of semivolatile organic compounds found between the gas phase and sorbed on to indoor airborne particles (Weschler and Shields, 2003), formation of secondary organic particulate matter (Fan *et al.*, 2003; Sarwar *et al.*, 2003) and dissociation of outdoor particles into their gaseous constituents (Lunden *et al.*, 2003). Coagulation does not affect the mass concentration but can change the size distribution of particles, so it can be important when concentrations are high. However, Morawska (2004) has argued that the significance of coagulation in affecting indoor concentration is lower than that of other process.

It is clear from this that many factors play major roles in determining the fate of indoor particles and it is thus difficult to study individual processes. However, it is important to highlight that most of these processes will be greatly affected by various elements of the design, construction, use and management of built environments. In an extensive review of personal exposure to indoor particulate matter and risk assessment, Morawska *et al.* (2013) suggested that in developed countries, 10–30% of the total burden of disease from particulate-matter exposure was due to that generated indoors.

14.3 Gas Contaminants

The gaseous class encompasses contaminants that exist as atoms or free molecules in air and can be divided into two important subclasses: gases and vapours. Table 14.1 shows the main gaseous contaminants of particular significance in built environments, along with examples and brief descriptions of their occurrence and use.

It is evident that many elements of the construction, use and management of built environments possess the potential to contaminate the air with a range of gaseous contaminants. A great number of studies have been carried out to investigate the levels of these contaminants, particularly volatile organic compounds and inorganic gases (CO, NO_X, SO_2, O_3, NH_3), in various built environments across the globe. Volatile organic compounds and carbonyl compounds are currently attracting the greatest attention, especially in high-income countries, due to their emission from building materials, furnishings (carpets, composite-wood furniture), interior decoration (wall coverings, paints, adhesives) and household cleaning products. A recent review by Sarigiannis *et al.* (2011) concerning organic compounds (benzene, toluene, xylenes, styrene, acetaldehyde, formaldehyde, naphthalene, limonene, α-pinene and ammonia) has reported that there is a great variation in their levels within and among European countries due to differences in their sources and emission strengths.

14.3.1 Biological Hazards

Biological agents are ubiquitous and can originate from almost any natural or artificial surface, and each source can give rise to an entirely unique assemblage of bioaerosols (see Chapter 16). Wind action on soil, agitation of open water, raindrop impaction, animal farming facilities, compositing, harvesting, HVAC (heating, ventilation and air conditioning) systems, industrial processes, food processing units and wastewater/sewage treatment plants are some examples of major outdoor sources. Indoor bioaerosols are mostly of outdoor origin (Burge, 1990; Levetin *et al.*, 1995), but building materials, carpets, pets, plants and HVAC systems can contribute to their levels, particularly in moisture-damage spaces, which can provide the substrate/conditions for microbes to grow and multiply (Lehtonen, Reponen and Nevalainen, 1993; DeKoster and Thorne, 1995; Ren *et al.*, 2001). Microorganisms can get indoors through HVAC systems, doors, windows, cracks in the walls and the potable drinking-water system. They can also be brought in on people's shoes and clothes (Pasanen *et al.*, 1989). Once in the indoor environment, a range of abiotic factors (water, humidity, temperature, nutrients, oxygen and light) determines their growth on indoor surfaces. Indeed, built environments are complex ecosystems in which there is a complicated relationship between humans, microorganisms and physical structures. With the growing use of molecular approaches, a number of studies have been carried out to elucidate the microbiology of built environments (Kelley and Gilbert, 2013; Robertson *et al.*, 2013). It has been found that building attributes (sources of ventilation air, airflow rates, relative humidity and temperature) influence the diversity and composition of indoor bacterial communities (Kembel *et al.*, 2012).

A large body of knowledge is available concerning the levels of bioaerosols in different built environments, but very often the focus is on their allergenic and toxic potential, with less attention being paid to airborne infection. A range of factors can enhance the probability of disease transmission in built environments (see Chapter 11): enhanced exposure of

Table 14.1 Brief overview of the main chemical families of gaseous air contaminants and their occurrence and use (Adapted from ASHRAE, 2009).

Family	Example	Occurrence and use
Inorganic contaminants		
Oxidants	Ozone, nitrogen dioxide	Both members are corrosive and act as respiratory irritants
Reducing agents	Carbon monoxide	Carbon monoxide is a toxic fuel-combustion product
Acid gases	Carbon dioxide, hydrogen chloride, hydrogen fluoride, hydrogen sulfide, nitric acid, sulfur dioxide, sulfuric acid	Carbon dioxide and hydrogen sulfide are only weakly acidic. Hydrogen sulfide is the main agent in sewer gas. Other members are corrosive and respiratory irritants. Some are important outdoor contaminants
Nitrogen compounds	Ammonia, hydrazine, nitrous oxide	Ammonia is used in cleaning products. It is a strong irritant. Hydrazine is used as an anticorrosion agent. Nitrous oxide is used as an aesthetic
Organic contaminants		
Chlorinated hydrocarbons	Carbon tetrachloride, chloroform, dichloromethane, 1,1,1-trichloroethane, trichloroethylene, tetrachloroethylene, p-dichlorobenzene	Dichlorobenzene is a solid used as an air freshener. The others shown are liquids and are effective nonpolar solvents. Some are used as degreasers or in the dry-cleaning industry
Alcohols	Methanol, ethanol, 2-propanol (isopropanol), 3-methyl 1-butanol, ethylene glycol, 2-butoxyethanol, phenol	Alcohols are strongly polar. Some are used as solvents in water-based products. Phenol is used as a disinfectant. 3-Methyl 1-butanol is emitted by some moulds
Aldehydes	Formaldehyde, acetaldehyde, acrolein, benzaldehyde	Formaldehyde, acetaldehyde and acrolein have unpleasant odours and are strong irritants
Ketones	2-Propanone (acetone), 2-butanone (MEK), methyl isobutyl ketone (MIBK), 2-hexanone	Ketones are medium-polarity chemicals. Acetone and 2-hexanone are emitted by some moulds
Esters	Ethyl acetate, vinyl acetate, butyl acetate	Esters are medium-polarity chemicals. Some have pleasant odours and are added as fragrances to consumer products

Table 14.1 *(continued)*

Family	Example	Occurrence and use
Aromatic hydrocarbons	Benzene, toluene, p-xylene, styrene, 1,2,4 trimethyl benzene, naphthalene, benz-α-pyrene	Benzene, toluene and xylene are widely used as solvents and in manufacturing, and are ubiquitous in indoor air. Naphthalene is used as a moth repellent
Terpenes	a-Pinene, limonene	A variety of terpenes are emitted by wood. The two listed here have pleasant odours and are used as fragrances in cleaners, perfumes and so on
Heterocyclics	Ethylene oxide, tetrahydrofuran, 3-methyl furan, 1, 4-dioxane, pyridine, nicotine	Most heterocyclics are of medium polarity. Ethylene oxide is used as a disinfectant. Tetrahydrofuran and pyridine are used as solvents. Nicotine is a component of tobacco smoke

individuals to infectious people in a small, enclosed spaces, inadequate ventilation, recirculation of contaminated air, increased duration of exposure and susceptibility of the exposed person. Over the years, it has been established that poor housing structures and conditions (inadequate ventilation, limited space and overcrowding) enhance the risk of airborne transmission of diseases. For example, tuberculosis (TB) has long been associated with crowded and poorly ventilated environments (CCDA, 2007), as there is an increased possibility of inhaling droplets expelled by infectious individuals (Wanyeki *et al.*, 2006). Additionally, ventilation and air movement in built environments have been strongly associated with airborne transmission of infectious diseases (Li *et al.*, 2007).

Upon the discovery of antibiotics over half a century ago, far-reaching predictions were made of the end of the infectious disease era. However, over the past 2 decades it has seemed that we are losing the battle against them. Environmental, social, economic, technological and climatic changes are constantly altering disease transmission cycles, which is leading to shifts in the distribution, prevalence and severity of existing and newly emerging infectious diseases. Those spread via airborne transmission have the potential to cause global pandemics; they are a serious threat to twenty-first-century humanity. Figure 14.3 shows airborne biological threats from diverse anthropogenic activities, settings and natural and man-initiated phenomena.

In health-care facilities, the airborne transmission of measles virus, chickenpox virus (*Varicella*), rhinoviruses, adenoviruses, *Staphylococcus aureus*, *Mycobacterium tuberculosis* and *Aspergillus* is a global problem (CDC, 2003). Aside from human sources, certain medical procedures (e.g. intubation, bronchoscopy), building maintenance activities (e.g. construction, demolition, renovation and repair) and daily cleaning procedures (e.g. laundry, bedding, clinical waste) enhance the risk of health-care-associated infections (HCAIs). Globally, HCAIs are responsible for significant mortality and financial burden

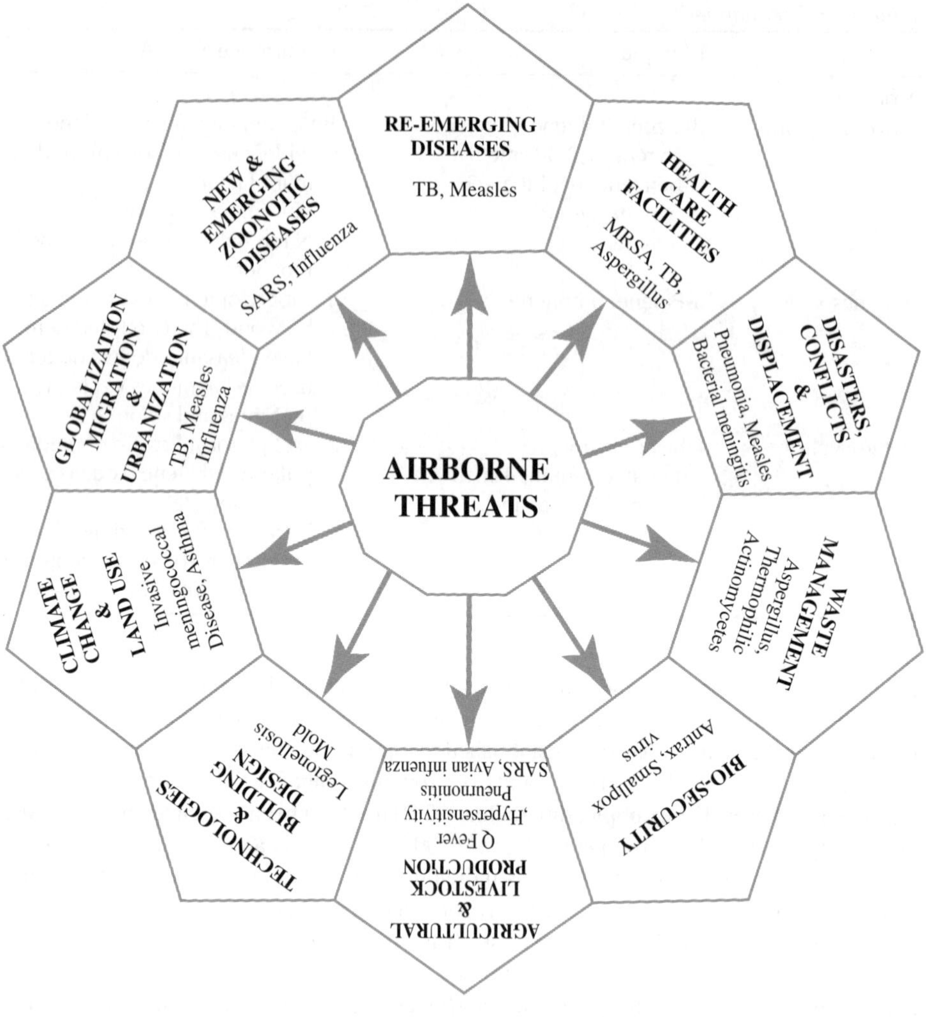

Figure 14.3 *Current and future threats from airborne disease transmission.*

in health-care systems. According to the World Health Organization (WHO), HCAI prevalence in developed and developing countries was 7.6 and 10.1%, respectively, in 2011. In Europe, HCAIs are responsible for 16 million extra days of hospital stay, 37 000 attributable deaths and annual financial losses of up to €7 billion (WHO, 2011).

Similarly, waste-management processes are associated with increased exposure to airborne microorganisms and their toxins. Various mechanical activities associated with these processes generate significant numbers and types of airborne microorganism, which may pose a potential health hazard to workers and those in their vicinity. However, the degree of human exposure to airborne microorganisms may vary with the type and capacity of these facilities, the activities performed there and meteorological conditions (Swan *et al.*, 2003; Stagg *et al.*, 2010; Korzeniewska, 2011; Dehghani *et al.*, 2012).

Natural disasters and conflicts can also provide a paradise for disease transmission, due to massive population displacement, sudden environmental changes, loss of public utilities, disruption of basic health facilities and people's increased vulnerability. Airborne diseases such as pneumonia, measles and bacterial meningitis have an epidemic potential during droughts, tsunami, tornados, earthquakes and flooding. Overcrowding, poor ventilation and inadequate hygiene in closed shelters, lack of nutrition and limited access to medical facilities all enhance the risk of acute respiratory infections (WHO, 2006; Watson, Gayer and Connolly, 2007; Kouadio *et al.*, 2012).

Climate, seasonality and climate change affect the prevalence and geographical distribution of airborne diseases. For example, the epidemic activity of respiratory syncytial virus (RSV) is related to meteorological conditions. In tropical climatic conditions, RSV infection peaks in summer and early autumn, while in temperate climates, RSV infection is highest in the winter (Yusuf *et al.*, 2007). Similarly, the incidence and geographical distribution of invasive meningococcal disease (IMD) is likely to be affected by climate change (Palmgren, 2009). The recent emergence of *Cryptococcosis gatti* (naturally restricted to tropical and subtropical regions) in the more temperate climates of the Pacific Northwest of the USA has been hypothesised to be linked to climate change (Cooney, 2011). Climate change might also increase the duration of the pollen season and change the spatial distribution of vegetation and regional pollen production (Beggs and Bambrick, 2005; IPCC, 2007), causing people with allergic rhinitis and asthma to be at increased risk of exacerbation. Additionally, more frequent precipitation events, along with urban densification and land use changes, especially in flood plains, have increased the risk and severity of urban flooding, which can have a profound impact on airborne disease transmission. For instance, flooding can trigger extensive growth of biological agents (e.g. mould, bacteria) due to the dampness of surfaces, structures and the air in buildings, leading to enhanced exposure of occupants to aeroallergens, mycotoxins, endotoxins, ß-glucans, volatile organic compounds and infections (WHO, 2009).

Airborne transmission of new and emerging zoonotic disease holds serious epidemic potential. For example, severe acute respiratory syndrome (SARS) and avian flu may become pandemics. The 1918–1919 H1N1 influenza pandemic, which killed an estimated 50–100 million people, clearly shows the severity and consequences of such an event, and the recent spread of H5N1 avian influenza viruses highlights the high risk of pandemic emergence (Morens and Fauci, 2007). Due to the intensification of human-, pathogen- and environment/climate-related factors influencing the emergence of infectious diseases, it is very likely that most of the future infectious disease outbreak caused by either a novel or a recurring agent will be zoonotic (Cascio *et al.*, 2011). Apart from serious harm to human health, a pandemic could result in great social and economic damage.

Along with the new and emerging diseases, airborne transmission of reemerging infectious diseases is also posing an overwhelming threat to humanity. For example, TB is an increasing problem worldwide. In 2009, 1.7 million people around the world died from TB, and there were an estimated 9.4 million new cases (WHO, 2010a). The emerging threat of multidrug-resistant tuberculosis (MDR-TB) and extensively drug-resistant tuberculosis (XDR-TB) is a serious global health concern. In 2008, there were an estimated 390 000–510 000 cases of MDR-TB: 3.6% of all incident TB cases. 5.4% of MDR-TB cases were found to have XDR-TB (WHO, 2010b). Recently cases of totally drug-resistant tuberculosis (TDR-TB) have been reported from Mumbai, India (Loewenberg, 2012).

Globalisation, increased travel, migration and urbanisation may influence the prevalence, geographical distribution and severity of infectious diseases in certain population groups, the reemergence of previously known pathogens and the emergence of new pathogens, due to changes in biological, social and environmental factors associated with these phenomena (WHO, 2004; Ka-Wai Hui, 2006; Alirol *et al.*, 2011). The total number of international migrants has increased from an estimated 178 million in 2000 to 213 million in 2010 (UN DESA, 2009). In 2010, for the first time, more than 50% of the world's population was living in urban areas (United Nations, 2010). Urbanisation is increasingly affecting the epidemiological characteristics of infectious diseases and is having a profound effect on global health. In low-income countries, urbanisation can lead to the development of slums and shanty towns, with poor housing conditions that facilitate the proliferation of airborne diseases. High population density and an increased amount of shared space, heterogeneity in health status and increased mobility in urban areas all enhance exposure to influenza, measles and TB. In fact, densely populated cities have become hubs for the global spread of airborne diseases such as SARS and H1N1 influenza (Alirol *et al.*, 2011).

The continuous shift from traditional agriculture and livestock production to intensive industrial production systems in both developing and developed countries poses a significant threat to global public health, with the risk of not only the increased prevalence of already known infectious diseases but also the emergence of new pathogens. Bacterial concentrations with multidrug resistance have been recovered up to 150 m downwind of a swine-farming installation (Gibbs *et al.*, 2006). Similarly, the emergence of livestock-associated Methicillin-resistant *S. aureus* ST398 (LA-MRSA) and its transmission to humans clearly shows potential for the emergence of novel pathogens (Smith and Pearson, 2011). In agricultural work environments, bioaerosols and their secondary metabolites can be a cause of allergic rhinitis, toxic pneumonitis, hypersensitivity pneumonitis and asthma (Dutkiewicz *et al.*, 2011).

Technologies such as air-handling systems, cooling towers and architectural fountains have the potential to harbour infectious pathogens (e.g., *Legionella*) and aid in their airborne transmission. The increased use of mechanical air-conditioning and wet-cooling systems may lead to an increase in cases of Legionnaires' disease. Building design can also enhance the exposure of occupants to airborne pathogens. Energy-efficient and environmentally sustainable buildings (green buildings) have been of great interest in recent years. However, certain of their components may provide ecological niches for the growth of microorganisms and for pathways of dispersal in enclosed spaces. For example, the use of green roofs and green atriums may proliferate human pathogens associated with soil (Morey, 2010). In addition, the 'build-tight' approach (reduced dilution of indoor air) may increase the likelihood of airborne infection. Current technologies and building design are more focussed on energy conservation than public health protection.

Moreover, biological agents might be secretly prepared, transported and intentionally released by terrorists and rogue states. The airborne transmission of these weaponised pathogens poses a significant risk to the world population. For example, the deliberate dissemination of *Anthrax* through postal services in the USA in 2001 clearly shows the threat (Lane, La Montagne and Fauci, 2001). In addition, accidental release of biological agents from lab facilities is also of great concern and could cause a significant threat to the public. The current research on avian influenza viruses (H5N1) and the recent report of recombinant virus (ferret-transmissible H5 HA) has generated wide concern regarding the development and maintenance of these modified strains (Fouchier, García-Sastre and Kawaoka, 2012; Muller, 2012). We are therefore facing a wide range of airborne threats, which necessitates a new focus on control of airborne biological agents.

A range of design, construction, use and management factors (presented in Figure 14.2) may implicate airborne disease transmission. The SARS outbreak in a housing estate (Amoy Gardens) in Hong Kong in 2003 demonstrated how elements of housing and the environment can contribute to the airborne transmission of infectious diseases (Yu *et al.*, 2004). Similarly, outbreaks of Legionnaires' disease highlight the potential for built environments to act as reservoirs of infectious pathogens and aid in their transmission. There is thus a dire need to rethink the role of built environments in airborne disease transmission. Greater attention must be given to the implications of existing practices in the design, construction, use and management of built environments for airborne disease transmission.

14.3.2 Physical Hazards

Physical hazards in built environments involve thermal conditions (temperature, humidity, air velocity, radiant energy), electromagnetic radiation and visual and acoustic conditions. Physical factors can not only impact directly on occupants but also influence other environmental quality factors. British Standard BS EN ISO 7730 defines thermal comfort as 'that condition of mind which expresses satisfaction with the thermal environment'. It is a person's psychological state of mind and a range of environmental and personal factors that influence the occupant's satisfaction with thermal conditions in different built environments. According to the British Health and Safety Executive (HSE), air temperature, radiant temperature, air velocity, humidity, clothing insulation and metabolic rates all contribute to a worker's thermal comfort. The predicted mean vote (PMV) and percentage people dissatisfied (PPD) index for general thermal comfort and the draught rate (DR) for local thermal discomfort have been widely used to measure thermal comfort. Details of the methods used to predict general thermal sensation and thermal dissatisfaction and the criteria requirements can be found in BS EN ISO 7730, BS EN ISO 10551 (subjective method) and ANSI/ASHRAE Standard 55-2010. Recently, Dear *et al.* (2013) reviewed thermal comfort research over the last 20 years and discussed the major trends, progress and paradigm shifts in this domain.

Design criteria for environmental factors affecting the thermal comfort of people in built environments are available (e.g. CIBSE Guide A – Environmental Design), but selection of suitable design conditions is still very difficult due to the subjective nature of comfort perception. The main goal is to keep most people thermally comfortable most of the time. According to the HSE, a reasonable limit for the minimum number of people who should be thermally comfortable in a given environment is 80% of occupants.

Over the last century, considerable changes have been made in the electromagnetic fields within built environments. An escalating increase in the amount of electrical equipment and the extent of wireless infrastructure has raised concerns over electromagnetic pollution. The health effects of electromagnetic fields and emission standards have been discussed in detail by Clements-Croome (2012). Electromagnetic compatibility is now a key design consideration (Lock, 2012) and recently a number of papers have looked at the implications of changing the electromagnetic nature of built environments for human well-being (Jamieson *et al.*, 2010, 2011; Xing, 2012). Light can have biological, behavioural and psychological effects on people (e.g. circadian rhythm, body temperature, mood/behaviour, activity levels). There is a good body of literature available on the effects of light on human well-being and there is a growing trend to maximise the use of daylight in buildings (CIE, 2004; Webb, 2006; Boyce, 2010; Todorovic and Kim, 2012; Veitch and Galasiu, 2012). Sound and vibration can also affect well-being: excessive noise can cause annoyance, interference with the intelligibility of speech and even hearing damage; similarly, vibration can have both physical and biological effects, depending on its magnitude and frequency. Codes of practice and guidelines for noise and vibration control in different built environments have been proposed by various organisations (e.g. British standards, ASHRAE).

14.3.3 Ergonomic Hazards

According to the International Ergonomic Association, ergonomics (or 'human factors') is 'the scientific discipline concerned with the understanding of the interactions among humans and other elements of a system, and the profession that applies theoretical principles, data and methods to design in order to optimize human well being and overall system performance'. Ergonomic design of the various elements in built environments (e.g. work station layout) assists people in interacting with the environment comfortably and efficiently, leading to increased productivity and a healthy working experience. Good ergonomics allows the environment to fit the physical activities of the occupants, rather than forcing the occupants to fit their activities to their environment. On the other hand, poor ergonomic design can result not only in low productivity but also in many visual and musculoskeletal disorders/cumulative trauma disorders/repetitive strain injuries. In addition, poor space ergonomics may result in alterations being made during building occupancy, which could have a significant impact on indoor environmental health. For example, the location and number of work stations, cabinets, printers and other equipment can disturb air flows, hygrothermal regulation and ventilation. During refurbishment or alteration, the need for effective ventilation may be overlooked. Often these alterations are focussed on efficiency of space usage or aesthetics and they are likely to be made without changes to the original ventilation design. Environmental health issues often arise due to poor space layout and high occupant densities, especially in work-related built environments. It is very likely that a space with low occupant density based on floor space will have high 'net occupant density' due to the concentration of work stations in a small space. Studies have shown an improvement in productivity and comfort and a decrease in musculoskeletal complaints following from good ergonomics design characteristics (Dainoff, Fraser and Taylor, 1982; Smith and Bayeh, 2003; Hemphälä and Eklund, 2012).

Exposure to environmental health hazards in built environments is a complex process. The proximal environmental determinants for enhanced exposure may include inadequate ventilation, ineffective air mixing, poor ergonomics and inadequate facility management. The distal determinants are much more complex and are often overlooked. Built environments undergo change both spatially and temporally, and their design and management in a geographical region are the result of complex political, socioeconomic and environmental factors (Figure 14.4). Political factors include policies and regulation

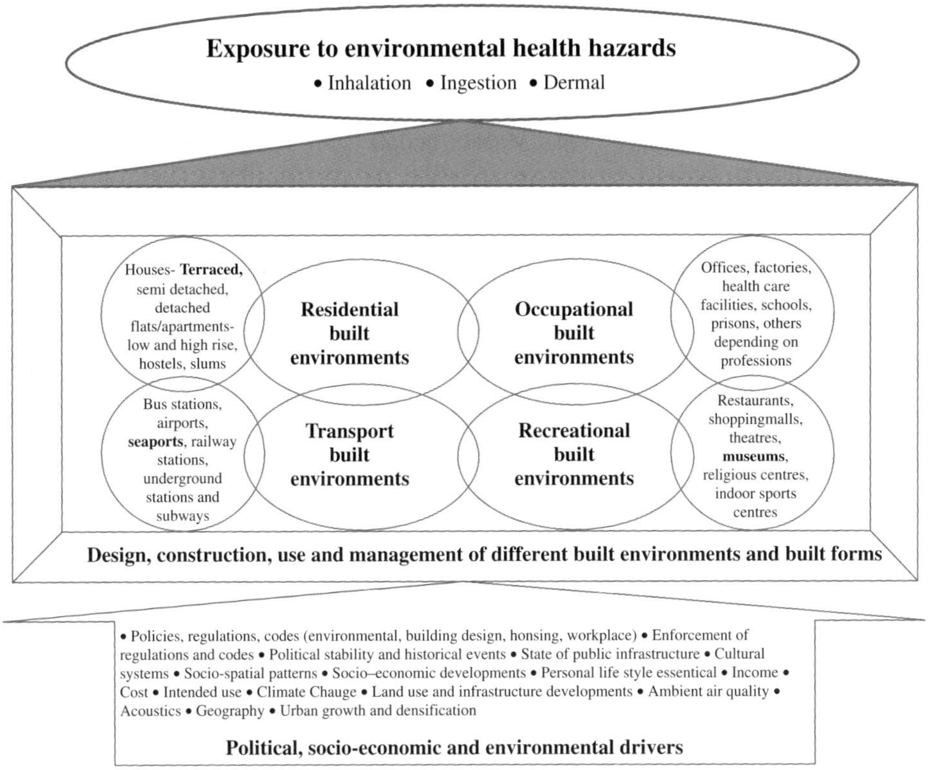

Figure 14.4 *Exposure to environmental health hazards in various built environments.*

concerning design and construction, as well as their enforcement. Socioeconomic factors include personal lifestyle essentials and cultural systems. Environmental factors include land use, urban growth and densification. The design and implementation of effective interventions/strategies to reduce the risk of exposure to different environmental hazards in built environments can thus only be achieved by following a holistic system approach that takes into account the ways in which political, social, economic and natural drivers affect the design, construction, use and management of different built environments.

14.3.4 Ventilation and Environmental Hazards

Ventilation is one of the most crucial factors in reducing the risk of exposure to a range of air contaminants in built environments. Ventilation design over the past century has seen a shift in its role from infection control to mere thermal comfort. During the nineteenth and early twentieth centuries, when TB and other infectious diseases were common, high ventilation rates were recommended to minimise the risk of infection. With the decline of epidemic diseases, due to improvements in sanitation, hygiene and public health, the rationale for ventilation began to shift from contagion control to the creation of comfortable conditions and odour removal (Maston and Sherman, 2004). The energy crisis in the late 1970s led to further reductions in ventilation rates through the sealing of buildings, in order to reduce infiltration and increase energy efficiency. However, later in that decade the prevalence of sick building syndrome was attributed to inadequate ventilation resulting from this (Mendell and Fine, 1994). Modern buildings are designed to be energy efficient rather than to enhance environmental health and prevent illness, but over the past few decades increases in infectious diseases and respiratory illnesses have highlighted the importance of ventilation. Hobday (2010) has stated that the concept that housing should support and promote the health and well-being of its occupants has been surpassed by energy efficiency concerns and that over the years, interaction between housing and medical officials has declined. In a review of the role of ventilation in the airborne transmission of infectious agents, Li *et al.* (2007) concluded that ventilation and air movement are strongly associated with the spread of disease in built environments. In addition, they asserted that most epidemiological studies have used inadequate methods to study ventilation and air flow movement, reflecting the lack of interaction among epidemiologists, microbiologists and building engineers. Data on ventilation rates are limited, but the available information highlights that the use and management of a ventilation system greatly impact on its overall performance. An increased focus on energy conservation has led to the introduction of energy-efficient designs and the building of highly insulated, airtight structures with greater use of mechanical-ventilation-with-heat-recovery (MVHR) systems. A recent review of ventilation in European dwellings has concluded that despite the growing recognition of ventilation as an important component of healthy dwelling, in practice ventilation is poor and a large proportion of dwellings are underventilated (lower than $0.5\ h^{-1}$) (Dimitroulopoulou, 2011). Furthermore, in various built environments, especially occupational, transport and other recreational built environments, the operation, maintenance and management of facilities (e.g. HVAC systems, water supply systems) by management can have a considerable impact on the overall environmental health. Moreover, the degree of interaction between facilities management and the organisations using the built environment, and their knowledge of cumulative exposure to various environmental hazards, has a vital role to play in ensuring appropriate environmental health. The lack of knowledge and awareness of various environmental hazards among the occupants of built environments, the personnel responsible for health and safety at work and facilities management can lead to enhanced exposure to various environmental hazards. For instance, incorrect operation and poor maintenance of ventilation and air-conditioning systems during a building's occupancy can lead to inadequate ventilation rates and poor air distribution, which in turn enhances the risk of exposure to a range of air contaminants. Furthermore, it is worth highlighting that very often the need for ventilation is governed by the psychological and physiological needs of the occupants and that spaces with hybrid ventilation design (mechanical and natural ventilation) can lead to uncontrolled ventilation and air mixing depending on the locations of the opened windows and doors and how long they remain open. Natural ventilation can provide higher air exchange rates at a lower cost and for lower maintenance than can mechanical ventilation. However, it is important to take into account the quality of the outdoor air. A large number of studies have shown an association between ambient air pollution and morbidity. What is required is not only a healthy design for new buildings but also the healthy use of existing built environments, in order to lessen the exposure to different environmental hazards.

14.3.5 Energy-Efficient Built Environments, Climate Change and Environmental Health

Due to their cumulative impact on natural resource depletion and degradation, waste generation and accumulation, ecosystems and climate change, governments around the world are putting forward new polices, laws and standards to reduce the environmental impacts of built environments. Consequently, depending on the implementation of these laws, stakeholders face new, complex and rapidly changing challenges, leading to significant changes in the design and construction of built environments. In order to reduce carbon dioxide emissions, energy efficiency has emerged as the guiding paradigm for the creation of a new kind of built environment. However, an increased focus on energy efficiency had raised the concern that various factors involved in the design, construction and operation of energy-efficient built environments might increase the vulnerability of their occupants to environmental hazards, especially airborne pollutants. In recent years, governments across the developed world have focussed their attentions on housing, due to its considerable share in energy use (e.g., codes for sustainable homes include 'PassivHaus', 'Smart Energy Home', 'R-2000 homes', 'indoor airPLUS'). The UK government intends that all new homes in England will be zero-carbon by 2016. This approach/policy entails greater airtightness of the building envelope in order to improve energy efficiency, and concerns have been raised by builder and owners over issues surrounding poor indoor air quality (Davis and Harvey, 2008). Homes built to level 4 of the code for sustainable homes are expected to use MVHR in order to comply with energy-use and indoor-air-quality requirements. Crump *et al.* (2009) carried out an extensive review of indoor air quality in highly energy-efficient homes, of ventilation performance in dwellings construction and of ventilation provision in highly energy-efficient homes and concluded that there is a dearth of knowledge in this area and that more research is needed.

In recent years, the long-term challenges of climate change and its potential effects on indoor air quality and public health have been addressed in numerous publications. A report by the Institute of Medicine (IOM) concluded that the extensive body of literature highlights that poor indoor environmental quality is creating health problems but that there is inadequate evidence to associate climate-change-induced alterations in the indoor environment with any specific adverse health outcomes. However, climate change has the potential to worsen existing indoor environmental problems and to introduce new ones (IOM, 2011). Based on the IOM report, Spengler (2012) and Nazaroff (2013) have summarised and discussed the consequences of climate change for indoor air quality. A report by Crump (2011) on the health impacts arising from changes in the indoor environment in the UK also considered changes caused by the mitigation and adoption of actions as a result of climate change. Furthermore, a recent report by the UK Health Protection Agency on the health effects of climate change (HPA, 2012) concluded that climate change has the potential to modify risks in the indoor built environments by exacerbating the existing health risks associated with indoor air pollutants.

Florence Nightingale (1820–1910) recognised that 'the connection between health and the dwellings of the population is one of the most important that exists'. This statement is still true today. It is evident that aerosol science is central to improving environmental health in the built environment. It draws on numerous branches of science, including ventilation engineering, aerobiology, chemistry, nanotechnology, occupational exposure and more.

References

Alirol, E., Getaz, L., Stoll, B. *et al.* (2011) Urbanisation and infectious diseases in a globalised world. *The Lancet Infectious Diseases*, **11**, 131–141.

ASHRAE (2009) Air contaminants, in *ASHRAE Handbook – Fundamentals (I-P Edition)*, American Society of Heating, Refrigerating and Air-Conditioning Engineers, Inc., pp. 11-1–11-22.

Beggs, P.J. and Bambrick, H.J. (2005) Is the global rise of asthma an early impact of anthropogenic climate change? *Environmental Health Perspectives*, **113**(8), 915–919.

Boor, B.E., Siegel, J.A. and Novoselac, A. (2013) Wind tunnel study on aerodynamic particle resuspension from monolayer and multilayer deposits on linoleum flooring and galvanized sheet metal. *Aerosol Science and Technology*. doi: 10.1080/02786826.2013.794929.

Boyce, P.R. (2010) Review: the impact of light in buildings on human health. *Indoor and Built environment*, **19**, 8–20.

Burge, H. (1990) Bioaerosols: prevalence and health effects in the indoor environment. *Journal of Allergy and Clinical Immunology*, **86**, 687–701.

Cascio, A., Bosilkovski, M., Rodriguez-Morales, A.J. and Pappas, G. (2011) The socioecology of zoonotic infections. *Clinical Microbiology and Infection* 17, 336–342.

CCDA (2007) Housing Conditions that Serve as Risk Factors for Tuberculosis Infection and Disease. *Canada Communicable Disease Report*, **33**(ACS 9).

CDC (Centers for Disease Control and Prevention) (2003) Guidelines for Environmental Infection Control in Health-care Facilities: Recommendations of CDC and the Healthcare Infection Control Practices Advisory Committee (HICPAC), U.S. Department of Health and Human Services, Centers for Disease Control and Prevention (CDC), Atlanta, GA.

Chen, F., Yu, S.C.M. and Lai, A.C.K. (2006) Modeling particle distribution and deposition in indoor environments with a new drift-flux model. *Atmospheric Environment*, **40**, 357–367.

Chen, C. and Zhao, B. (2011) Review of relationship between indoor and outdoor particles: I/O ratio, infiltration factor and penetration factor. *Atmospheric Environment*, **45**(2), 275–288.

CIE (2004) Ocular Lighting Effects on Human Physiology and Behaviour, Commission Internationale de l'Eclairage Publication, p. 158.

Clements-Croome, D. (ed) (2012) *Electromagnetic Environments and Health in Building*, Routledge.

Colbeck, I. and Lazaridis, M. (2010) Aerosols and environmental pollution. *Naturwissenschaften*, **97**, 117–131.

Cooney, C.M. (2011) Climate change and infectious disease: is the future here? *Environmental Health Perspectives*, **119**, a394.

Crump, D. (2011) *Climate Change – Health Impacts Due to Changes in the Indoor Environment; Research Needs*, Institute of Environment and Health, Cranfield University.

Crump, D., Dengel, A. and Swainson, M. (2009) Indoor Air Quality in Highly Energy Efficient Homes: A Review, NHBC Foundation Report NF18, National House Building Council, Milton Keynes.

Dainoff, M.J., Fraser, L. and Taylor, B.J. (1982) Visual, musculoskeletal, and performance differences between good and poor VDT workstations. *Proceedings of the Human Factors and Ergonomics Society Annual Meeting*, **26**(2), 144. Sage Publications.

Davis, I. and Harvey, V. (2008) Zero Carbon: What does it Mean to Homeowners and Housebuilders? NHBC Foundation Report NF9, April 2008.

Dear, R., Akimoto, T., Arens, E. *et al.* (2013) Progress in thermal comfort research over the last twenty years. *Indoor Air*. doi: 10.1111/ina.12046.

Dehghani, R., Asadi, M.A., Charkhloo, E. *et al.* (2012) Identification of fungal communities in producing compost by windrow method. *Journal of Environmental Protection*, **3**, 61–67.

DeKoster, J.A. and Thorne, P.S. (1995) Bioaerosol concentrations in noncompliant, complaint and intervention homes in the Midwest. *American Industrial Hygiene Association Journal*, **56**, 576–580.

Dimitroulopoulou, C. (2011) Ventilation in European dwellings: a review. *Building and Environment*, **47**, 109–125.

Doleman, R. and Brooks, D.J. (2011) A strategy to articulate the facility management knowledge categories within the built environment, in *Proceedings of the 4th Australian Security and Intelligence Conference* (eds D.J. Brooks and C. Valli), Secau –Security Research Centre, School of Computer and Security Science, Edith Cowan University, Perth, Western Australia, pp. 58–67.

Dua, S.K. and Hopke, P.K. (1996) Hygroscopic growth of assorted indoor aerosols. *Aerosol Science and Technology*, **24**, 151–160.

Dutkiewicz, J., Cisak, E., Sroka, J. *et al.* (2011) Biological agents as occupational hazards-selected issues. *Annals of Agricultural and Environmental Medicine: AAEM*, **18**, 286.

Fan, Z.H., Lioy, P., Weschler, C. *et al.* (2003) Ozone-initiated reactions with mixtures of volatile organic compounds under simulated indoor conditions. *Environmental Science and Technology*, **37**, 1811–1821.

Ferro, A.R., Kopperud, R.J. and Hildemann, L.M. (2004) Source strengths for indoor human activities that resuspend particulate matter. *Environmental Science and Technology*, **38**, 1759–1764.

Finlayson-Pitts, B.J. and Pitts, J.N. (2000) *Chemistry of the Upper and Lower Atmosphere*, Academic Press.

Fouchier, R.A.M., García-Sastre, A. and Kawaoka, Y. (2012) Pause on avian flu transmission studies. *Nature*. doi: 10.1038/481443a.

Gibbs, S.G., Green, C.F., Tarwater, P.M. *et al.* (2006) Isolation of antibiotic-resistant bacteria from the air plume downwind of a swine confined or concentrated animal feeding operation. *Environmental Health Perspectives*, **114**, 1032.

Goldasteh, I., Ahmadi, G. and Ferro, A. (2012a) A model for removal of compact, rough, irregularly shaped particles from surfaces in turbulent flows. *The Journal of Adhesion*, **88**, 766–786.

Goldasteh, I., Ahmadi, G. and Ferro, A.R. (2012b) Wind tunnel study and numerical simulation of dust particle resuspension from indoor surfaces in turbulent flows. *Journal of Adhesion Science and Technology*. doi: 10.1080/01694243.2012.747729.

Gomes, C., Freihaut, J. and Bahnfleth, W. (2007) Resuspension of allergen containing particles under mechanical and aerodynamic disturbances from human walking. *Atmospheric Environment*, **41**, 5257–5270.

Hamdani, S.E., Limam, K., Abadie, M.O. and Bendou, A. (2008) Deposition of fine particles on building internal surfaces. *Atmospheric Environment*, **42**, 8893–8901.

He, C., Morawska, L. and Gilbert, D. (2005) Particle deposition rates in residential houses. *Atmospheric Environment*, **39**, 3891–3899.

Health Protection Agency (2012) *Health Effects of Climate Change in the UK 2012: Current Evidence, Recommendations and Research Gaps*, UK Health Protection Agency, London.

Hemphälä, H. and Eklund, J. (2012) A visual ergonomics intervention in mail sorting facilities: effects on eyes, muscles and productivity. *Applied Ergonomics*, **43**, 217–229.

Hinds, W.C. (2005) Aerosol properties, in *Aerosols Handbook, Measurement, Dosimetry and Health Effects* (eds Ruzer L.S. and Harley N.H.), CRC Press, Boca Raton, FL, pp. 19–33.

Hobday, R. (2010) *Designing Houses for Health – A Review*, Commissioned by the VELUX Company Ltd.

Hussein, T., Glytsos, T., Ondracek, J. *et al.* (2006) Particle size characterization and emission rates during indoor activities in a house. *Atmospheric Environment*, **40**, 4285–4307.

Hussein, T., Hruska, A., Dohanyosova, P. *et al.* (2009) Deposition rate on smooth surfaces and coagulation of aerosol particles inside a test chamber. *Atmospheric Environment*, **43**, 905–914.

Ibrahim, A.H., Dunn, P.F. and Brach, R.M. (2003) Microparticle detachment from surfaces exposed to turbulent air flow: controlled experiments and modeling. *Journal of Aerosol Science*, **34**, 765–782.

Ibrahim, A.H., Dunn, P.F. and Brach, R.M. (2004) Experiments and validation of a model for microparticle detachment from a surface by turbulent air flow. *Journal of Aerosol Science*, **35**, 805–821.

IOM (2011) *Climate Change, The Indoor Environment, and Health*, The National Academies Press, Washington, DC.

IPCC (Intergovernmental Panel on Climate Change) (2007) Climate Change 2007: Impacts, Adaptation and Vulnerability, IPCC, Geneva.

Jamieson, I.A., Holdstock, P., ApSimon, H.M. and Bell, J.N.B. (2010) Building health: the need for electromagnetic hygiene? *IOP Conference Series: Earth and Environmental Science*, **10**(1), 012007.

Jamieson, I.A., Jamieson, S.S., ApSimon, H.M. and Bell, J.N.B. (2011) Grounding and human health – a review. *Journal of Physics: Conference Series*, **301**(1), 012024.

Jamriska, M., Morawska, L. and Ensor, D.S. (2003) Control strategies for sub-micrometer particles indoors: model study of air filtration and ventilation. *Indoor Air*, **13**, 96–105.

Kassab, A.S., Ugaz, V.M., King, M.D. and Hassan, Y.A. (2013) High resolution study of micrometer particle detachment on different surfaces. *Aerosol Science and Technology*, **47**, 351–360.

Ka-Wai Hui, E. (2006) Reasons for the increase in emerging and re-emerging viral infectious diseases. *Microbes and Infection*, **8**, 905–916.

Kelley, S.T. and Gilbert, J.A. (2013) Studying the microbiology of the indoor environment. *Genome Biology*, **14**, 1–9.

Kembel, S.W., Jones, E., Kline, J. *et al.* (2012) Architectural design influences the diversity and structure of the built environment microbiome. *The ISME Journal*, **6**, 1469–1479.

Korzeniewska, E. (2011) Emission of bacteria and fungi in the air from wastewater treatment plants-a review. *Frontiers in Bioscience*, **3**, 393.
Kouadio, I.K., Aljunid, S., Kamigaki, T. *et al.* (2012) Infectious diseases following natural disasters: prevention and control measures. *Expert Review of Anti-infective Therapy*, **10**, 95–104.
Lai, A.C.K. (2002) Particle deposition indoors: a review. *Indoor Air*, **12**, 211–214.
Lai, A.C.K. (2006) Investigation of electrostatic forces on particle deposition in a test chamber. *Indoor and Built Environment*, **15**, 179–186.
Lane, H.C., La Montagne, J. and Fauci, A.S. (2001) Bioterrorism: a clear and present danger. *Nature Medicine*, **7**, 1271–1273.
Lehtonen, M., Reponen, T. and Nevalainen, A. (1993) Everyday activities and variation of fungal spore concentrations in indoor air. *International Biodeterioration and Biodegradation*, **31**, 25–39.
Levetin, E., Shaugnessy, R., Fisher, E. *et al.* (1995) Indoor air quality in schools: exposure to fungal allergens. *Aerobiologia*, **11**, 27–34.
Li, Y., Leung, G.M., Tang, J.W. *et al.* (2007) Role of ventilation in airborne transmission of infectious agents in the built environment – a multidisciplinary systematic review. *Indoor Air*, **17**, 2–18.
Lock, K.S. (2012) EMC design in the built environment, in *Electromagnetic Compatibility (APEMC), Asia-Pacific Symposium on 2012*, IEEE, pp. 21–24.
Loewenberg, S. (2012) India reports cases of totally drug-resistant tuberculosis. *The Lancet*, **379**(9812), 205.
Lunden, M.M., Revzan, K.L., Fischer, M.L. *et al.* (2003) The transformation of outdoor ammonium nitrate aerosols in the indoor environment. *Atmospheric Environment*, **37**, 5633–5644.
Maston, N.E. and Sherman, M.H. (2004) Why we ventilate our houses-an historical look, in *ACEEE Summer Study on Energy Efficiency in Buildings*, Vol. **7**, American Council for an Energy Efficient Economy, Washington, DC, Pacific Grove, CA, pp. 241–250.
Mendell, M. and Fine, L. (1994) Building ventilation and symptoms – where do we go from here? *American Journal of Public Health*, **84**, 346.
Morawska, L. (2004) Indoor particles combustion products and fibers, in *The Hand Book of Environmental Chemistry. Part – F Indoor Air Pollution*, Vol. **4** (ed Pluschke P.), Springer-Verlag, Berlin, Heidelberg, pp. 117–147.
Morawska, L., Afshari, A., Bae, G. *et al.* (2013) Indoor aerosols: from personal exposure to risk assessment. *Indoor Air*. doi: 10.1111/ina.12044.
Morawska, L. and Salthammer, T. (2003) Fundamental of indoor particles and and settled dust, in *Indoor Environments* (eds L. Morawska and T. Salthammer), Wiley-VCH Verlag GmbH & Co. KGaA, Weinheim, pp. 3–46.
Morens, D.M. and Fauci, A.S. (2007) The 1918 influenza pandemic: insights for the 21st century. *Journal of Infectious Diseases*, **195**, 1018.
Morey, P.R. (2010) Climate Change and Potential Effects on Microbial Air Quality in the Built Environment, Preliminary Draft 2010, http://www.epa.gov/iaq/pdfs/climate_and_microbial_iaq.pdf (last accessed 9 August 2013).
Mukai, C., Siegel, J.A. and Novoselac, A. (2009) Impact of airflow characteristics on particle resuspension from indoor surfaces. *Aerosol Science and Technology*, **43**, 1–11.

Müller, V. (2012) A plea for caution: huge risks associated with lab-bred flu. *Viruses*, **4**, 276–279.

Nazaroff, W.W. (2004) Indoor particle dynamics. *Indoor Air*, **14**(Suppl. 7), 175–183.

Nazaroff, W.W. (2013) Exploring the consequences of climate change for indoor air quality. *Environmental Research Letters*, **8**, 015022.

Palmgren, H. (2009) Meningococcal disease and climate. *Global Health Action*, **2**. doi: 10.3402/gha.v2i0.2061.

Pasanen, A.L., Kalliokoski, P., Pasanen, P. *et al.* (1989) Fungi carried from farmers work into farm homes. *American Industrial Hygiene Association Journal*, **50**, 631–633.

Qian, J. and Ferro, A. (2008) Resuspension of dust particles in a chamber and associated environmental factors. *Aerosol Science and Technology*, **42**, 566–578.

Ren, P., Jankun, T.M., Belanger, K. *et al.* (2001) The relation between fungal propagules in indoor air and home characteristics. *Allergy*, **56**, 419–424.

Robertson, C.E., Baumgartner, L.K., Harris, J.K. *et al.* (2013) Culture-independent analysis of aerosol microbiology in a metropolitan subway system. *Applied and Environmental Microbiology*. doi: 10.1128/AEM.00331-13.

Roof, K. and Oleru, N. (2008) Public health: Seattle and King county's push for the built environment. *Journal of Environmental Health*, **71**, 24–27.

Sarigiannis, D.A., Karakitsios, S.P., Gotti, A. *et al.* (2011) Exposure to major volatile organic compounds and carbonyls in European indoor environments and associated health risk. *Environment International*, **37**, 743–765.

Sarwar, G., Corsi, R., Allen, D. and Weschler, C. (2003) The significance of secondary organic aerosol formation and growth in buildings: experimental and computational evidence. *Atmospheric Environment*, **37**, 1365–1381.

Seinfeld, J.H. and Pandis, S.N. (2006) *Atmospheric Chemistry and Physics, From Air Pollution to Climate Change*, 2nd edn, John Wiley & Sons, Inc.

Shaughnessy, R. and Vu, H. (2012) Particle loadings and resuspension related to floor coverings in a chamber and in occupied school environments. *Atmospheric Environment*, **55**, 515–524.

Smith, M.J. and Bayeh, A.D. (2003) Do ergonomics improvements increase computer workers' productivity?: an intervention study in a call centre. *Ergonomics*, **46**, 3–18.

Smith, T.C. and Pearson, N. (2011) The emergence of Staphylococcus aureus ST398. *Vector-Borne and Zoonotic Diseases*, **11**, 327–339.

Spengler, J.D. (2012) Climate change, indoor environments, and health. *Indoor Air*, **22**, 89–95.

Stagg, S., Bowry, A., Kelsey, A. and Crook, B. (2010) Bioaerosol Emissions from Waste Composting and the Potential for Workers' Exposure. Health and Safety Executive Research Report 786, Health and Safety Laboratory Harpur Hill, Buxton, p. SK17 9JN.

Swan, J., Gilbert, E., Kelsey, A. and Crook, B. (2003) Occupational and Environmental Exposure to Bioaerosols from Composts and Potential Health Effects– A Critical Review of Published Data. HSE Report RR130, HSE Books, Sudbury.

Thatcher, T.L., Lai, A.C.K., Moreno-Jackson, R. *et al.* (2002) Effects of room furnishings and air speed on particle deposition rates indoors. *Atmospheric Environment*, **36**, 1811–1819.

Thatcher, T.L. and Layton, D.W. (1995) Deposition re-suspension and penetration of particles within residence. *Atmospheric Environment*, **29**, 1487–1497.

Todorovic, M.S. and Kim, J.T. (2012) Beyond the science and art of the healthy buildings daylighting dynamic control's performance prediction and validation. *Energy and Buildings*, **46**, 159–166.

UN DESA (2009) United Nations Department of Economic and Social Affairs Trends in International Migrant Stock: The 2008 Revision, http://esa.un.org/migration/index.asp? (last accessed 9 August 2012).

United Nations (2010) Department of Economic and Social Affairs, Population Division. World Urbanization Prospects: The 2009 Revision. CD-ROM Edition – Data in Digital Form (POP/DB/WUP/Rev.2009), United Nations, New York.

Veitch, J.A. and Galasiu, A.D. (2012) The Physiological and Psychological Effects of Windows, Daylight, and View at Home: Review and Research Agenda. NRC-IRC Research Report RR-325, NRC Institute for Research in Construction, doi: http://dx.doi.org/10.4224/20375039.

Wallace, L.A., Emmerich, S.J. and Howard-Reed, C. (2004) Effect of central fans and induct filters on deposition rates of ultrafine and fine particles in an occupied townhouse. *Atmospheric Environment*, **38**, 405–413.

Wanyeki, I., Olson, S., Brassard, P. *et al.* (2006) Dwellings, crowding, and tuberculosis in Montreal. *Social Science & Medicine*, **63**, 501–511.

Watson, J.T., Gayer, M. and Connolly, M.A. (2007) Epidemics after natural disasters. *Emerging Infectious Diseases*, **13**, 1.

Webb, A.R. (2006) Considerations for lighting in the built environment: non-visual effects of light. *Energy and Buildings*, **38**, 721–727.

Weschler, C.J. and Shields, H.C. (2003) Experiments probing the influence of air exchange rates on secondary organic aerosols derived from indoor chemistry. *Atmospheric Environment*, **37**, 5621–5631.

WHO (2004) Globalization and Infectious Diseases: A Review of the Linkages, Special Topics No. 3, Geneva.

WHO (2006) Communicable Diseases Following Natural Disasters. Risk Assessment and Priority Interventions. Programme on Disease Control in Humanitarian Emergencies Communicable Diseases Cluster, World Health Organization, Geneva.

WHO (2009) *WHO Guidelines for Indoor Air Quality: Dampness and Mould*, WHO Regional Office for Europe, Copenhagen.

WHO (2010a) Global Tuberculosis Control, WHO report 2010, Geneva.

WHO (2010b) Multidrug and Extensively Drug-Resistant TB (M/XDR-TB): 2010 Global Report on Surveillance and Response, Geneva.

WHO (2011) Report on the Burden of Endemic Health Care-associated Infection Worldwide, World Health Organization, Switzerland.

Wu, Y.L., Davidson, C.I. and Russell, A.G. (1992) Controlled wind tunnel experiments for particle bounceoff and resuspension. *Aerosol Science and Technology*, **17**, 245–262.

Xing, J.Y. (2012) Electromagnetic radiation on human health hazards and protective measures in modern society. *Advanced Materials Research*, **518**, 1022–1026.

Yu, I.T.S., Li, Y., Wong, T.W. *et al*. (2004) Evidence of airborne transmission of the severe acute respiratory syndrome virus. *New England Journal of Medicine*, **350**(17), 1731–1739.

Yusuf, S., Piedimonte, G., Auais, A. *et al*. (2007) The relationship of meteorological conditions to the epidemic activity of respiratory syncytial virus. *Epidemiology and Infection*, **135**, 1077–1090.

15

Particle Emissions from Vehicles

Jonathan Symonds
Cambustion, UK

15.1 Introduction

Aerosols from vehicle engines originate from five sources: fuel, fuel additives, inlet air, lubrication oil and the mechanical breakdown of preexisting materials. The latter can also form from other sources in the vehicle (e.g., brake dust). Of those formed in the engine, there are four main types of aerosol: carbonaceous, organic, sulfate and ash. These usually appear in combination. Fuel and oil contribute to all four fractions; fuel additives, air and mechanical breakdown contribute to the ash fraction. A typical aerosol from a heavy-duty diesel engine is 41% carbon, 13% ash, 14% sulfate/water, 25% unburnt oil and 7% unburnt fuel (Kittelson, 1998).

The number-based size spectrum (Figure 15.1) of an engine aerosol normally includes some of at least three distinct, lognormal modes (Kittelson, 1998). Homogenous nucleation of volatile materials in the exhaust (or ash particles) can form the so-called *nucleation mode*, which is usually smaller than 30 nm in size, with a narrow geometric standard deviation (<1.5). After this comes the *accumulation mode*, which is normally between 60 and 200 nm, with a general standard deviation (GSD) between 1.5 and 2.0. This is where the carbonaceous ('soot') agglomerate particles are usually found. Particles larger than this are referred to as the *coarse mode* and consist mostly of material produced outside the engine, such as brake dust and reentrained soot from the walls of the exhaust system. On a particle number basis, the nucleation- and accumulation-mode particles vastly dominate, but when weighted by mass, the coarse mode can contribute around 20% of the total particulate mass.

Aerosol Science: Technology and Applications, First Edition. Edited by Ian Colbeck and Mihalis Lazaridis.

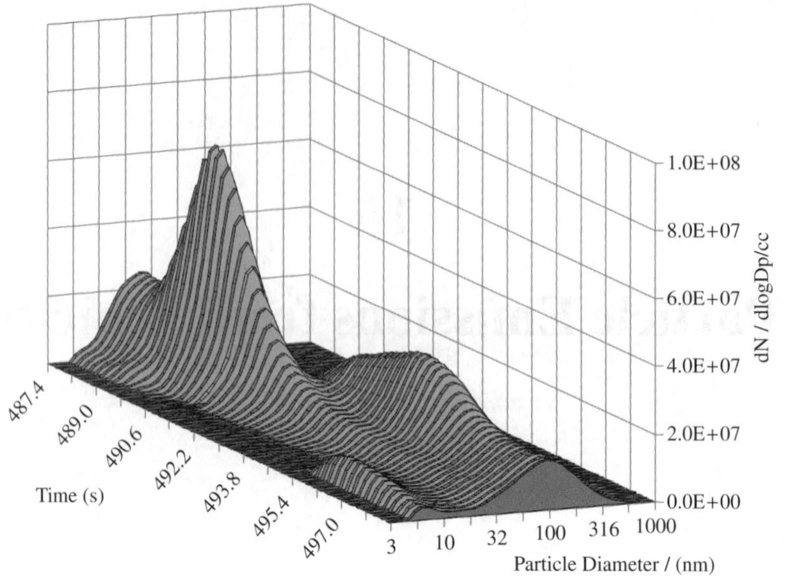

Figure 15.1 *Typical heavy-duty diesel particle size spectrum, showing nucleation and accumulation modes appearing during a transient. Data courtesy of West Virginia University.*

15.2 Engine Concepts and Technologies

15.2.1 Air–Fuel Mixture

As we shall see, one of the most important parameters affecting particle formation during combustion is the air–fuel ratio (AFR; e.g. Heywood, 1988, Section 3.3). This is expressed in terms of the *mass* fraction of air to fuel. If exactly enough air is present to burn all the fuel present then the mixture is called *stoichiometric*. If too much fuel is present compared with the air, it is called a *rich* mixture, whereas if too little fuel is present, it is known as a *lean* mixture. The ratio of the overall AFR to the stoichiometric AFR is represented by λ. Hence, for rich mixtures. $\lambda < 1$, and for lean mixtures, $\lambda > 1$.

The 'ideal' combustion of a stoichiometric or lean mixture of oxygen and a simple hydrocarbon fuel may be represented as:

$$C_xH_y + \lambda\left(x + \tfrac{y}{4}\right)(O_2 + 3.76N_2) \rightarrow xCO_2 = \frac{y}{2}H_2O + (\lambda - 1)\left(x + \frac{y}{4}\right)O_2$$
$$+3.76\lambda\left(x + \frac{y}{4}\right)N_2.$$

If a rich mixture is burnt then the lack of oxygen leads, in the first instance, to the production of carbon monoxide, and then to the production of carbonaceous soot. In reality, all engine combustion processes generate small but finite quantities of CO, NO, NO_2, unburnt fuel (uHC) and particulate matter (PM).

Due to the short time scales involved in combustion compared with the vaporisation of fuel and the mixing of fuel vapour and air, the localised AFR in parts of the cylinder can be quite different from the overall AFR. Depending on the method of introduction of fuel and air into the cylinder, locally rich regions can lead to soot formation (Greeves and Wang, 1982).

If fully vaporised fuel and air are well mixed, the combustion occurs by way of a *premixed flame* (e.g. typical gasoline engine combustion). If, however, the combustion process is dominated by the rate at which fuel and air mix, then combustion will proceed by way of a *diffusion flame* (e.g. typical diesel engine combustion).

15.2.2 Spark-Ignition Engines

A spark-ignition (SI) engine relies upon an electrical discharge to ignite a fuel–air mixture. SI engines are usually fuelled by gasoline, though increasingly liquefied petroleum gas (LPG), compressed natural gas (CNG), ethanol, bioethanol or another biofuel, often mixed with gasoline, is used instead. The fuel and air may be premixed in the intake system of the engine, as in carburetted or port fuel-injected (PFI) engines, or the fuel may mix with the air in-cylinder, after being injected directly into it. SI engines use either a four-stroke (intake, compression, power, exhaust) or two-stroke (intake–compression, power–exhaust) working cycle.

The engine management system in a PFI engine will attempt (under closed-loop control) to keep the mixture near stoichiometric, which for gasoline means an AFR of approximately 14.5. Though significantly better fuel economy is achieved with slightly lean mixtures, the simultaneous oxidation of CO and uHCs and reduction of NO_X requires the use of both a stoichiometric mixture and a 'three-way' catalyst. Closed-loop control of the fuelling to achieve stoichiometry is achieved by the use of oxygen sensors in the exhaust gas. The stoichiometric and well-mixed fuel/air conditions of PFI gasoline engines also mean that the carbonaceous particle emissions from the fuel in these engines are small.

In recent years, gasoline direct injection (GDI) engines have increased in popularity. In these, gasoline is directly injected into the cylinder, rather than into the intake port. This method of fuelling has a number of advantages. In a PFI engine, the flow through the inlet valve consists of air and fuel vapour. If fully vaporised, about 5% by volume of the flow is fuel vapour, which represents a 5% reduction of the air flow into the engine – and a 5% reduction in the maximum power. In addition, when the fuel evaporates in the cylinder, the cylinder charge is cooler than it is when the evaporation takes place in the port. This allows the engine to be run with a higher compression ratio and hence a higher efficiency. Finally, there are benefits associated with better AFR control, since in a PFI engine some of the fuel entering the cylinder is from previous injections, in which it was deposited on the manifold wall.

In a stoichiometric GDI engine, fuel is normally injected immediately after the exhaust valve closes, which maximises the time available for mixing. Some GDI engines operate, for part of the operating envelope, in *stratified mode*, and fuel is injected in the latter stages of compression. In principle, this allows a portion of the cylinder to contain a flammable mixture in the region of the spark plug, surrounded by air. If such a mode were 'perfect', the engine could be run without throttling, as the power output would be modulated by the relative volumes of the region containing the flammable mixture and the air. Sadly, the high turbulence levels in the cylinder lead to significant mixing, and stratified combustion has not been widely adopted.

In general, GDI engines suffer from higher PM emissions that do PFI engines (e.g. Price *et al.*, 2006). This is due to their reduced air/fuel homogeneity and to impingement of fuel on the piston and cylinder surfaces. These higher emissions cause liquid fuel pools to be formed, which eventually burn by diffusion, again causing higher carbonaceous soot emission (Witze and Green, 2005).

GDI engines are becoming much more common due to their increased efficiency and reduced CO_2 emission, particularly in the USA and Japan, where light-duty diesel vehicles offering similar advantages have not been as widely adopted as they have in Europe. However, the increased particulate emissions are a drawback, and the regulation and reduction of these emissions is a current area of intense research.

15.2.3 Compression-Ignition Engines

Compression-ignition (CI) engines are usually referred to as 'diesel engines', after the inventor Rudolf Diesel. Diesel engines have a high thermal efficiency, due to their high compression ratio (of around 15 : 1 to 22 : 1). The compression ratio of gasoline engines is limited by the onset of 'knocking'; that is, autoignition of part of the mixture late in the burning process. In most diesel engines, the fuel is injected directly into a bowl formed in the piston, near the end of the compression stroke. Ignition then occurs spontaneously in the mixture formed by evaporating fuel and air, due to the temperature generated during the compression stroke (diesel fuel is formulated to have a low autoignition temperature compared to gasoline). Although the use of ever-higher injection pressures leads to enhanced fuel–air mixing, rich combustion inevitably occurs in the fuel vapour plume as it mixes with air. Depending on the crank angle and the location within the cylinder, λ values between 0 and infinity exist, unlike in a stoichiometric PFI gasoline engine, where λ is unity, or nearly so, everywhere. Inevitably, significant PM is formed, and though much of it is oxidised later in the combustion process, significant quantities leave through the exhaust valves. The levels of carbonaceous particle emissions from individual diesel engines are also much higher than those from individual GDI engines.

15.2.4 Two-Stroke Engines

A two-stroke engine completes the combustion cycle in one revolution of the crankshaft. Two-stroke engines are commonly used in motorcycles and horticultural equipment when a high power density is required. In their simplest form, an air, fuel and lubrication oil mixture enters the combustion chamber via the crankcase and a port in the cylinder wall linking the under and top sides of the piston. As the piston rises, the port is covered; when it reaches the top, ignition via a spark plug occurs. As the piston descends, a second port in the wall of the cylinder is uncovered and the exhaust gases exit. The particulate emission levels from two-stroke engines can be high, due to the lubrication oil present in the mixture and to short-circuiting of the fuel–air mixture from the intake port directly to the exhaust port; indeed, visible smoke is often produced. Fuel short-circuiting can be reduced by modern direct-injection technology. The particle size spectrum from two-stroke engines can contain some quite large particles (between 500 and 1000 nm), thought to be condensed lubrication oil, in addition to the nucleation and accumulation modes (Hands *et al.*, 2010). Two-stroke scooters are particularly popular in emerging economies, and particulate emissions from these are currently unregulated.

15.2.5 Gas-Turbine Engines

Particles emitted from jet engines are of particular concern, as they are emitted directly into the upper atmosphere and lower troposphere, where they can directly affect climate forcing (Fortuin *et al.*, 1995). The soot particles produced by these engines tend to be smaller than those produced by diesel engines, usually sub-60 nm (Petzold *et al.*, 2011), and primary spherical soot particles are often directly detected in their unagglomerated form (Schmid *et al.*, 2011). In addition to carbonaceous soot, organic matter and hydrated sulfate are also seen, which form nucleation aerosols in the engine wake. Ground-based gas turbines (for power generation, propulsion, etc.) also produce PM. Unlike piston engines, particle filtration on these devices is not practicable.

15.3 Particle Formation

15.3.1 In-Cylinder Formation

The combustion process converts hydrocarbon fuel molecules, each of which has just a few carbon atoms, into soot particles, which contain many thousands of carbon atoms (Figure 15.2). The initial complex and varied collection of reactions that occurs during fuel combustion with limited oxygen is known as pyrolysis. The first stage sees the formation of poly-aromatic hydrocarbon (PAH) molecules from the fuel hydrocarbon molecules (e.g. Eastwood, 2008, Chapter 3). These PAH molecules, initially in the form of vapour, then undergo a nucleation process to form nuclei particles of less than 3 nm in size. These nuclei particles then grow through the addition of carbon until *primary* soot particles emerge at between 20 and 50 nm. Electron microscopy shows that these primary particles are spherical. The primary particles then agglomerate together, forming fractal soot agglomerate particles. During all of these processes, a competing oxidation process removes carbon, and final levels of carbon are as much a function of this removal process as they are of the creative processes (Pipho, Ambs and Kittelson, 1986).

During the initial premixed burn stage, the jet of liquid fuel starts to vaporise at its tip as a result of air entrainment (Dec, 1997). Ignition then takes place within the fuel–air mix at the end of the jet and the pyrolysis processes start the conversion of fuel into PAH. The hot nascent soot first appears in pockets and then spreads to the edge of the plume. The soot particles surrounded by the flame envelope at the edge of the plume grow rapidly compared with those within the plume. The process then enters the mixing controlled-combustion stage, until the end of the injection event. Fuel is partially burnt at the tip of the fuel jet,

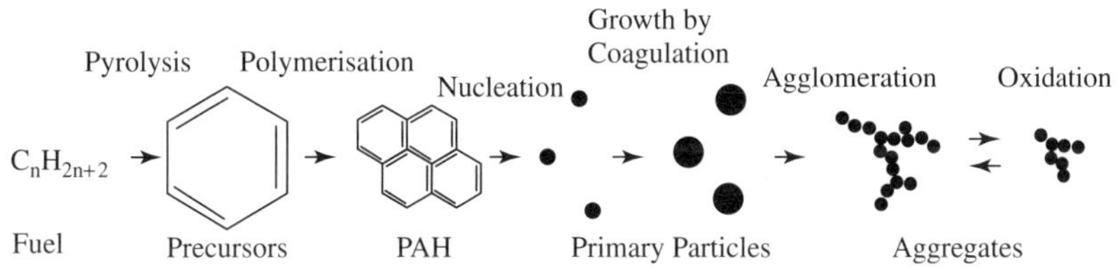

Figure 15.2 *Soot formation.*

and the products of partial combustion proceed to the tip of the plume, where they enter the diffusion flame. The process of soot formation starts in the premixed flame, and the soot precursors grow as they head towards the plume head, where some oxidation then occurs in the diffusion flame. After fuel injection terminates, the plume collapses. The quality of the jet as the fuel pressure drops may become poor, thereby momentarily increasing soot output before the end of combustion.

The ash fraction arises from inorganic molecules present in the fuel, oil or fuel additives, along with a small amount entrained from mechanical wear of the engine. A certain amount of airborne dust may be present, though the air intake filter should limit this. Common constituents of the ash fraction are metal oxides, particularly oxides of calcium, magnesium and zinc, which arise from the lubrication oil, and iron oxides, which arise from corrosion of the engine materials (Jung, Kittelson and Zachariah, 2003). As oil enters the periphery of the combustion chamber via the piston rings or valve stems, it is not subjected to such intense combustion conditions as the fuel, and therefore the aerosol constituents arising from oil may be chemically quite close to the substances in the original oil.

Additives in fuel designed to improve engine operation (e.g. the anti-knock agent ferrocene) or to catalyse diesel particulate filter (DPF) regeneration (e.g. cerium; Section 15.6.3) can form ash particles in the exhaust (but in the latter case, since a DPF is fitted, these ash particles would be removed before release into the environment). The ash particles produced from fuel additives tend to be in the same size region as the 'nucleation' mode; that is, <50 nm (Gidney *et al.*, 2010). Unlike the primary particle precursors of soot agglomerates (which are of a similar size), ash particles do not tend to form larger aggregates. Ash components are of low volatility; therefore, their conversion from the gas to the particulate phase occurs in the heat of the early formation processes, and certainly within the combustion chamber rather than in the exhaust system. Metal fuel additives can in fact have a soot-suppressing effect (Howard and Kausch, 1980). For example, ions formed during chemiionisation of additives in hydrocarbon flames can suppress soot agglomeration by means of mutual electrostatic repulsion.

The organic fraction usually results from unburnt fuel molecules, caused by locally rich regions in the cylinder (e.g., late vaporisation of fuel pools on the cylinder walls; Kato *et al.*, 1997) or an over-mixed charge that is too weak to support combustion. The compounds present in the aerosol phase may be essentially chemically unchanged from their form in the fuel or else may undergo some chemical synthesis in the combustion chamber. Some may be the PAH soot precursor molecules, i.e. molecules that have not evolved as far as soot formation (Fujiwara, Tosaka and Murayama, 1993). Sufficient quantities of unburnt fuel can lead to *white smoke* formation at the tailpipe. These particles, unlike the nanoparticles produced by combustion, are large enough to be optically detectable. Less volatile hydrocarbons from lubrication oil that pass by the piston ring set are another source of organic aerosol; this is particularly true in two-stroke engines, where the intake air and oil are mixed.

In general, soot formation increases with engine load. As noted earlier, in diesel engines a higher load means increased rich combustion. Also, high load leads to higher combustion temperatures, which means that the processes of soot formation are fully completed. At lower loads, the processes may not complete, and the engine aerosol will consist mainly of nucleation-mode materials created from the organic compounds intermediate in the soot formation process. Furthermore, when soot is produced, material that might otherwise form a nucleation mode (including sulfate) is often absorbed on to the soot, thereby suppressing nucleation mode formation (Abdul-Khalek, Kittelson and Brear, 2000). This results in a characteristic (at times mutually exclusive) interplay of nucleation and accumulation modes as a function of load over a legislated drive cycle or real-world driving when examined with a fast-response particle sizer (e.g. Campbell *et al.*, 2006).

Organic and sulfurous vapour particle precursors often do not enter the particulate phase until they reach the exhaust systems or even the air, where condensation nucleation occurs due to cooling and dilution.

15.3.2 Evolution in the Exhaust and Aftertreatment Systems

Both solid and volatile fractions of engine exhaust aerosol undergo further processes in the exhaust system. At high concentrations, the accumulation-mode soot particles undergo agglomeration, increasing in mean size and decreasing in number but maintaining the same overall mass. As the aerosol cools, the soot particles may adsorb volatile materials, forming a coating. The organic fraction can continue to undergo chemical reactions even in the exhaust systems; for example, PAH can continue to react down to 250 °C (Williams, Perez and Griffing, 1985).

The sulfate aerosol fraction consists primarily of sulfuric acid, as a condensation nucleation aerosol (Abdul-Khalek, Kittelson and Brear, 2000). Sulfur exists in both fuel and lubrication oil, although in many markets its content in fuel is significantly reduced. During combustion, the sulfur oxidises to form mostly sulfur dioxide, with some sulfur trioxide. The sulfuric acid forms from the hydrolysis of the SO_3, but as there is little of this present as a result of combustion itself, the biggest source of the acid is from the preconversion of SO_2 to SO_3, which usually occurs in the aftertreatment system:

$$SO_2 + \frac{1}{2}O_2 \rightarrow SO_3 \xrightarrow{H_2O} H_2SO_4$$

Modern diesel vehicles are fitted with a diesel oxidation catalyst (DOC) in the exhaust system in order to ameliorate CO, hydrocarbon and soluble organic fraction (SOF) emissions. However, the DOC is also the main source of conversion of SO_2 to SO_3. The use of low-sulfur diesel mitigates the problem of sulfate formation.

15.3.3 Noncombustion Particle Sources

Sources of coarse-mode particles include: dust from brake linings, tyre wear, road-surface wear, engine wear and rust, dust and scale from the exhaust and catalyst system. Lubrication oil is another source of noncombustion aerosol, as well as of combustion aerosol. The fumes from an engine's crankcase contain fine particulate matter (including 'blow-by', which escapes from the combustion chamber via the piston ring set) and larger oil drops on the scale of many microns, whether the engine is being fired or even just rotated by means of the starter motor (Johnson, Hargrave and Reid, 2011). In the USA and Europe, crankcase fumes must either be eliminated by means of a closed crankcase ventilation system (CCV) or be vented to the exhaust system (in which case the total exhaust is still subject to relevant PM emission limits).

15.3.4 Evolution in the Environment

Nucleation of the organic and sulfate fractions into the aerosol phase usually takes place as the aerosol cools in the exhaust system or upon dilution at the end of the tailpipe. When the exhaust finally reaches the environment at the end of the tailpipe, two effects dominate: the exhaust is rapidly cooled and the partial pressure of chemical constituents in the gas phase is reduced. The cooling effect increases the saturation ratio, driving particle nucleation and growth, and dilution decreases the vapour pressure of the constituents, suppressing nucleation and growth and, indeed, eventually shrinking and evaporating volatile particles (Abdul-Khalek, Kittelson and Brear, 2000). Upon addition of a small amount of dilution, the cooling effect dominates (and hence so do nucleation and growth), but the exhaust temperature quickly reaches that of the air, so adding further dilution beyond this point just serves to reduce the concentration of the gas-phase constituents, leading particle shrinkage and evaporation to dominate.

In the near wake of a moving vehicle, the flow becomes complex, with turbulent mixing dominating (Carpentieri, Kumar and Robins, 2011). On longer length scales, further from the vehicle, the wakes from vehicles interact and the dispersion is dominated by the geometry of the locality. For example, *street canyons*, formed by high buildings on either side of a road, tend to allow levels of particulate to build up. While nucleation-mode particles may continue to mutate, the accumulation-mode particles can remain for up to a month at the concentrations present in the atmosphere, in the absence of rain (Kumar *et al.*, 2010).

15.4 Impact of Vehicle Particle Emissions

15.4.1 Health and Environmental Effects

Acute exposure to diesel-engine aerosol causes irritation (in the eyes, throat and bronchial tubes) and respiratory problems (coughing, exacerbation of asthma) (US EPA, 2002). Chronic effects include likely carcinogenic activity (US EPA, 2002) and an increased risk of heart disease (Miller *et al.*, 2007). The fact that pedestrians in urban environments are placed so close to vehicles and their exhausts, and that the slow speed of urban vehicles creates less dilution in the wake than is found at high speed, means that levels of exposure for pedestrians can easily exceed concentrations that can cause health effects (Buzzard, Clark and Guffey, 2009). Larger particles are deposited in the nose and upper airways, while the smallest nanoparticles can reach the alveolar region and even enter into the bloodstream.

Black carbon, including engine soot, absorbs light and thus acts as a positive radiative forcer and contributes to global warming (Ramanathan and Carmichael, 2008).

15.4.2 Legislation

In order to counter the health and environmental impacts of engine-sourced nanoparticles, most territories have introduced some sort of limit on the levels produced. Limits on particulate mass have been in force in the USA [1] since the 1990s (for diesel vehicles) and in Europe since 2000 (Table 15.1). Limits differ for light- and heavy-duty vehicles, and there

[1] California tends to have its own set of emissions standards, often leading the rest of the USA.

Table 15.1 *European light-duty particulate emission standards.*

Stage	Date (for type approval)	Diesel		Direct injection gasoline	
		PM (mg/km)	*PN* (#/km)	*PM* (mg/km)	*PN* (#/km)
Euro 3	January 2000	50	–	–	–
Euro 4	January 2005	25	–	–	–
Euro 5a	September 2009	5*	–	5*	–
Euro 5b	September 2011	5*	6.0×10^{11}		
Euro 6	September 2014	5*	6.0×10^{11}	5*	6.0×10^{12} (6.0×10^{11} from 2017)

[a] 4.5 mg/km if measured by the PMP procedure.
Source: www.dieselnet.com, last accessed 8 August 2013.

are numerous special categories such as low emission vehicle (LEV) standards. The method used to measure particulate mass is almost universally filter paper; see Section 15.5.2.

Vehicles being tested for emissions are set up for monitoring on a chassis dynamometer (a 'rolling road') and driven on a standard drive cycle. These are prescribed by legislation, and involve accelerations, decelerations, gear changes and steady-state cruising. In the USA, the FTP-75 (Federal Test Procedure) is common; as it was based on a real-world drive, it has an irregular speed profile. In Europe, the NEDC (New European Drive Cycle) is used for light-duty testing. Its speed profile is much more uniform, consisting of three identical 'urban' speed-profile patterns, followed by an 'extra-urban' phase, which includes cruises at up to 120 kph. Emissions are usually expressed in terms of amount of particulate per kilometre or mile. Testing is normally conducted from a cold start – a condition that can lead to increased particulate emission at the start of the test. A number of world-harmonised drive cycles are currently under development for use in future legislation.

As emissions standards have become increasingly tight, the filter paper method has become somewhat difficult to use in practice. The mass collected during a test has become very small and avoiding the effect of artefacts has become difficult, or at least very expensive to avoid. In addition, some concern has been raised by those studying health effects that a mass measurement does not adequately relate to health risks. In response to this concern, the United Nations Economic Commission for Europe (UNECE) commissioned a study into alternative methods of particulate-level measurement for use in future European legislation. The Particle Measurement Programme (PMP) undertook experimental investigations into the efficacy of the current methods and, after suggesting a *solid*-particle-number-based standard (see Section 15.5.3), undertook correlation exercises in order to demonstrate the method's practicality, repeatability, reproducibility and robustness (Andersson *et al.* 2010). The light-duty report found that the particle number emission from DPF-equipped diesel engines was less than 2.0×10^{11} #/km, with a typical repeatability of 30%. Conventional diesel vehicles produced emissions of around $10^{13}-10^{14}$ #/km; that is, 2 orders of magnitude higher than the DPF-equipped vehicles. In additional, particle number emissions from those direct-injection gasoline vehicles tested were around 10^{13} #/km. In 2006, the European Parliament endorsed the suggested solid-particle-number-based measurement system as part of the Euro 5/6 phases of emissions testing legislation in Europe. The limit for diesel light-duty vehicles was set as

6.0×10^{11} #/km, and in 2011 it was proposed that this would also apply to direct-injection gasoline vehicles from 2017, with an interim limit 1 order of magnitude higher than that being introduced from 2014 (Table 15.1).

15.5 Sampling and Measurement Techniques

15.5.1 Sample Handling

The method by which an engine aerosol sample is collected and transported to an analyser is a most important factor in determining the quality of a subsequent measurement. It is usually necessary to dilute and/or cool an engine aerosol sample before measurement, due to limits upon the concentration and temperature set by the instrumentation used. However, dilution and cooling will inevitably change the nature of the aerosol (Lyyränen *et al.*, 2004). As with the real-world dilution that occurs inside and outside a vehicle's exhaust system, the dilution and cooling necessary for measurement will affect the volatile particle fraction. Cooling with little or no dilution will lead to a large degree of nucleation and condensation – material in the gas phase will enter the aerosol phase and then be measured. If sufficient dilution is used, the partial pressure of the volatile material in the gas phase will be reduced, thereby preventing supersaturation and condensation into the particle phase. This process can be further accelerated by using hot dilution (Kawai, Goto and Odaka, 2004). It comes down to a question of what needs to be measured: a representative sample of aerosol, as would be found in the real environment, or (accepting that this depends on so many variables and conditions as to be almost subjective) a measurement of the much less mutable solid fraction.

As an alternative to extra dilution, a thermodenuder can be used to remove volatile species, if desired. This consists of a hot tube, which ensures volatile material is present only in the gas phase, and a region filled with activated charcoal, which then absorbs the volatile material, preventing subsequent renucleation. One advantage of a thermodenuder is that it does not require additional dilution, which enables the use of less sensitive instrumentation. The use of a *catalytic stripper* (Kittelson *et al.*, 2004) is another alternative way of removing volatiles. Both thermodenuders and catalytic strippers exhibit some solid particle losses due to diffusion.

The most common on-line method of sampling from an exhaust pipe is the constant volume sampler (CVS; e.g., Burtscher, 2005). In its simplest form, this is a large-diameter tube, with the exhaust pipe fitted into one end and a fan causing a constant volumetric flow to be drawn at the other. The overall flow is much larger than the maximum exhaust flow and the additional flow drawn causes dilution of the exhaust, on the order of 10 : 1. The level of dilution varies with the changes in exhaust flow that occur across a test cycle, but the key to the CVS system is that it is simple to calculate instantaneous rates of emission, whether for particles or for gas-phase emissions. Regardless of the actual concentrations that emerge from the exhaust, the flux of a species in the CVS system is the concentration as measured in the CVS system (for particles, say, in N/cc) multiplied by the volumetric flow (cc/s, giving N/s). It is then simple to integrate the flux over time to give a total particle number (or mass) and then to divide by the distance travelled in the test to give the legislated quantity in N/km (or mg/km). By contrast, if one were to sample directly from the exhaust, it would be necessary to know the flow in the exhaust pipe. Given the pernicious nature of the species present in the exhaust flow, it is not usually possible to measure this flow directly using a mass-flow meter. It can be inferred from the air intake flow to the engine, with an addition to take account of the fuel burnt, but this is considerably more complex and more prone to error than the use of a CVS system. It is usually desirable to filter the clean air intake to the CVS system, especially if it is to be used to sample a DPF-equipped vehicle with relatively low emissions.

A less bulky and less expensive alternative to a CVS system is a partial-flow dilution tunnel. Such tunnels sample a defined proportion of the exhaust flow, but in order to maintain this, measurement of the exhaust flow and a rapid control system are required.

Given the high temperatures of exhaust gases and the fact that most instruments require a cool sample, it is important to avoid thermophoretic loss, which can occur when a sample is gradually cooled. It is therefore usually preferential to cool as quickly as possible by adding cold air, rather than allowing a sample to gradually cool in a length of pipe. Loss of particles by diffusion can be a significant source of error if sufficiently narrow and long tubing is used. It is also vital to use electrically conductive tubing, in order to avoid loss of particles by electrophoresis. This usually means stainless steel where possible, or special electrically conductive silicone or PTFE tubing where flexibility is required, though those nonmetallic materials can deteriorate and emit particles at even moderately high temperatures.

Particles deposited on the walls of the sampling and transport system may at times re-entrain back into the aerosol phase. After re-entrainment, these particles tend to be large, and their effect can be reduced by using a cyclone before the measurement system in order to remove particles above the size of interest. It has been reported that certain types of tubing can absorb hydrocarbons from the exhaust gases, subsequently releasing them in an unpredictable manner, creating additional nucleation-mode particles (Maricq *et al.*, 1999).

15.5.2 Mass Measurement

Most legislation concerning particle emissions from vehicles is still expressed in terms of particle mass collected on filter paper. After dilution (e.g., with a CVS system), a known mass flow of sample is drawn through one (or more) filter papers (often a backup filter is used in series with the main filter in order to catch any material that passes through the latter). The filters are weighed on a sensitive balance both before and after the test and the total particulate mass is taken as the difference – usually on the order of a few milligrams. As the flows in the CVS system and through the filter paper are known, scaling the filter paper mass to represent the total mass over the test is simply a matter of using the ratio of those flows as a scaling factor.

The species adsorbed on the paper include not only the solid carbonaceous and ash fractions but also any volatile material that has condensed. It is almost impossible to know if the volatile material adsorbed would have been in the aerosol phase under real-world conditions, and the levels present are very dependent upon the dilution and sampling conditions. As vehicles become ever cleaner, and in particular with the widespread introduction of DPFs, the level of mass which has to be measured on the filter paper is approaching the practical detection limit for the technique (Liu *et al.*, 2009). Various techniques have been used to improve detectability using filter paper. For example, United States Environmental Protection Agency (EPA) regulation 40 CFR 1066 specifies measures to deal with buoyancy and electrostatic charge.

Various on-line measurement techniques used in the automotive industry give either a direct or an indirect measurement of particle mass. Opacity (light extinction) has been used for many years as a surrogate for particle mass. These instruments either work directly by shining light through the aerosol or indirectly by capturing the particulate matter on a filter paper and measuring the 'blackness' of the paper. The former technique is cross-sensitive to NO_2, which is usually present in diesel exhaust. The latter technique is more sensitive but is pseudo-continuous. Two other techniques, the tapered element oscillating microbalance (TEOM) and the quartz crystal microbalance (QCM), both acquire particles on a collection device and use the resonant frequency to measure the mass present. Another device used is the photo acoustic soot sensor (PASS), which uses a light source to irradiate the soot and a sensitive microphone to pick up the resulting sound, which is then used to infer particle mass concentration.

Instruments are available that use opposing electrical and centrifugal forces between concentric spinning cylinders and apply a voltage to classify charged aerosol particles by their mass–charge ratio, such as the aerosol particle mass analyser (APM; Ehara, Hagwood and Coakley, 1996) and the centrifugal particle mass analyser (CPMA; Olfert and Collings, 2005). These have been used as primary standards to measure engine particle mass concentrations, and a comparison with filter-paper measurements shows that the artefact resulting from absorbed volatiles on the filter paper can account for 50% of the total filter-paper mass (Park, Kittelson and McMurry, 2003). Thermogravimetric analysis (TGA) is often used to determine the percentage weight of a sample attributed to carbonaceous soot.

It is also possible to calculate particle mass concentration by appropriately weighting the data from a particle-sizing instrument, assuming a density and a *fractal dimension* – see Section 15.5.5.

15.5.3 Solid-Particle-Number Measurement

Mass measurement by filter paper has several notable limitations (Liu *et al.*, 2009):

- As engines become progressively cleaner, it becomes harder to accurately measure the low masses of soot produced.
- Filter paper can be affected by artefacts caused by condensed volatile species that would perhaps 'normally' be in the gas phase rather than the particulate phase. In many cases, the mass of such volatile material can outweigh the soot produced by a modern engine.
- Mass measurement is inherently biased towards the presence of larger particles. There is much evidence that smaller nanoparticles have greater detrimental health effects than larger nanoparticles; thus, a limitation on particle mass may not correlate all that well with improved health outcomes.

In order to address these concerns in the European Union, a *solid-particle-number* measurement method has been introduced, in time for the Euro 6 stage of emissions standards (Andersson *et al.*, 2007; UNECE R83). The actual particle number counting is performed with a condensation particle counter (CPC). These are more than sufficiently sensitive to make this measurement, usually having a concentration range (in single-particle counting mode) of between <1 and 10 000 particles per cubic centimetre. Above this range, CPCs often operate in photometric mode, where an estimate of particle number is made from bulk optical scattering by particles, rather than by counting individual particles; they therefore require empirical calibration. The upper limit of a CPC in count mode is usually several orders of magnitude less than the concentration of particulate from a typical diesel vehicle, and thus some form of dilution is required upstream of the CPC if count mode is to be used (as mandated).

The reason the methodology chooses to measure only solid particles, and thus exclude volatile particles, is that the latter are difficult to measure with any repeatability. Also, little medical research has been conducted to confirm that there are health effects associated with volatile particles – unlike the solid particles, which can persist in the lungs. Nucleation-mode particles can readily be created from gas-phase species and destroyed – processes that are highly sensitive to the exact ambient environmental, sampling and dilution conditions used. In order to make any sort of meaningful comparison of vehicles with standards, or to be able to spot trends caused by engineering changes, it is very important for such measurements to be repeatable.

The key to the removal of volatile particles in the system is the dilution systems upstream of the CPC (Figure 15.3). Two stages of dilution are used. The first heats the aerosol to 150 °C and serves to both evaporate existing volatile particles and reduce the partial pressure of gas-phase species to prevent their formation into particles. This stage is followed by an evaporation tube, where the sample is heated to 300 °C for ~0.2 seconds to evaporate any further semivolatile material. After this tube, the second stage reduces the gas partial pressures further in order to prevent renucleation, and also cools the sample rapidly (to prevent thermophoretic loss) to allow measurement by the CPC.

The type of diluter used for the first stage is often a rotating disc diluter, which consists of a disc with a series of blind- or through-holes parallel to the axis of rotation, into which concentrated aerosol is deposited by a flow from the sample source. As the disc rotates, the holes transfer the sample to a flow of clean air, which then scavenges the particles and dilutes them. The dilution ratio achieved is controlled by the speed of rotation – the lower the speed, the higher the dilution. Alternative systems use careful measurement of the aerosol flow and controlled metered dilution with clean air. It is usually not possible to measure particle-laden flow using conventional mass-flow controllers, so often the pressure drop across an orifice plate is used.

The CPC mandated for use in this type of system is designed, by varying the internal saturator temperature, to have a d_{50} size cut-off of 23 nm. This is to ensure that nucleation-mode material which is not removed by the dilution stages or the evaporation tube is simply not counted by the CPC. However, this is controversial as not all particles <23 nm can be said to be volatile. Solid ash particles can be smaller than this size, and while not volatile, they would be removed by the CPC cut-off point (Gidney, Twigg and Kittelson, 2010).

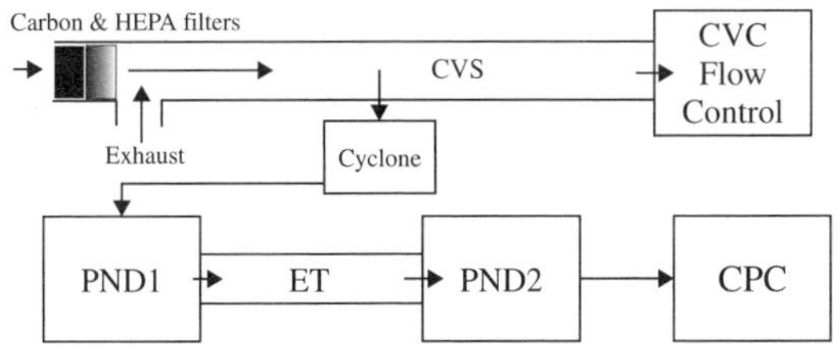

Figure 15.3 *Solid-particle-number measurement scheme. PND, particle number diluter; ET, evaporation tube; CPC, condensation particle counter; CVS, constant volume sampler.*

15.5.4 Sizing Techniques

The scanning mobility particle sizer (SMPS) has historically been the standard instrument used to measure nanoparticle size distributions. However, with a normal scan time of over 1 minute, it is not suitable for rapidly changing aerosol sources such as engines under non-steady-state conditions. Indeed, legislated drive cycles are by their very nature transient.

One of the first instruments to allow a real-time measurement of the particle size distribution from engines was the electrical low-pressure impactor (ELPI; Ahlvik *et al.*, 1998). This consists of a charger and a series of impactor plates. When a particle lands on a plate, a current is detected and the aerodynamic diameter is inferred. While not having as high a spectral or temporal resolution as the electrical mobility devices described later, the range of particle size measurable extends from a few nanometres up to 10 μm, and the use of impactor plates allows the size-segregated collection of sample for off-line testing, for example by chemical analysis.

In recent years, fast particle-mobility spectrometers have become available, which despite compromising on the sensitivity and spectral resolution of an SMPS, offer much faster data rates (up to 10 Hz) and, more importantly, short response times (down to 200 ms). These *differential mobility spectrometers* (Reavell, Hands and Collings, 2002; Johnson *et al.*, 2004) first charge the aerosol with a unipolar diffusion charger, which places a higher level of charge than a bipolar charger. As the name suggests, the charge is (usually) net positive rather than a net neutral distribution. The particles then pass into a classification column, which is similar to a differential mobility analyser (DMA) in that there is a central high-voltage electrode (which here repels the positively charged particles), and a sheath air flow, which carries the particles towards the other end of the column. The particles move in trajectories that depend upon their charge–drag ratio, eventually landing on a series of metal detection rings placed along the inside of the outer wall of the classification column. Each detector is connected to a sensitive electrometer circuit, and when a particle lands, a small current is registered (on the order of a few femtoamperes). A data-inversion algorithm uses a charging model and a model of the classifier to generate size spectral density versus diameter from the measured currents. The data in Figure 15.1 were obtained using such an instrument.

15.5.5 Morphology Determination

Transmission electron microscopy (TEM) is commonly used to study the structure of soot aggregates (Figure 15.4). The primary particles are usually clearly distinguishable and their size can be estimated; the number (N) in an agglomerate scales with the *radius of gyration* (R_g) as:

$$N = k_g \left(\frac{2R_g}{d_p} \right)^{D_f}$$

where D_f is the *fractal dimension*, d_p is the primary particle diameter and k_g is a constant (Mandelbrot, 1982). The radius of gyration is the root-mean-squared distance of the primary particles from the centre of the aggregate. For soot aggregates, D_f is usually just less than 2.0 (e.g. Park, Kittelson and McMurry, 2004).

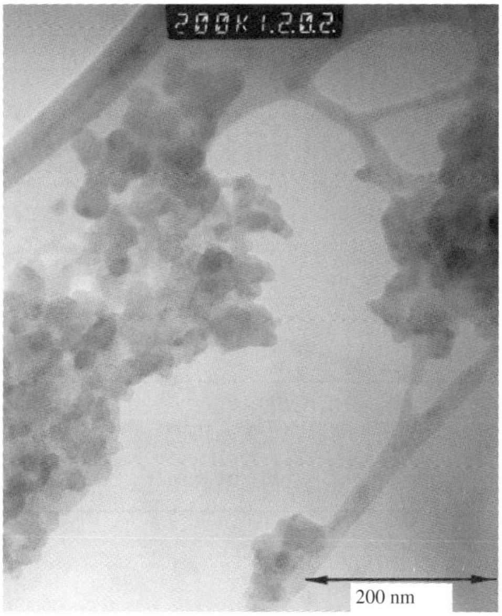

Figure 15.4 *TEM of heavy-duty diesel–soot aggregates; primary particles can clearly be seen. Image courtesy of BP/Dr Peter Harris of Reading University.*

It is possible to express the mass of a particle in terms of a power law dependent upon its diameter:

$$m \propto d_{mo}^{\ D_{fm}}$$

where D_{fm} is the *mass-mobility exponent* (also known as the mobility-diameter-based fractal dimension, which is different to the fractal dimension based on radius of gyration). For spherical particles, $D_{\mathrm{fm}} = 3$, and for agglomerate particles, such as those in the accumulation mode of engine particle emissions, $2 < D_{\mathrm{fm}} < 3$. In order to determine D_{fm} for an aerosol source, it is necessary to measure both the diameter and the mass of the particles. Size selection is usually performed with a DMA and in-series mass selection by an electric-centrifugal particle mass analyser, such as an APM (Park *et al.*, 2003) or a CPMA (Olfert, Symonds and Collings, 2007). Using a CPC as a detector, the peak mass can be determined for a given mobility diameter and particle effective density can be calculated as a function of diameter (Figure 15.5). With this technique, Park *et al.* (2003) found the mass-mobility exponent of heavy-duty soot for various engines to be between 2.33 and 2.41, and Olfert, Symonds and Collings (2007) found that for one Peugeot diesel engine, the mass-mobility exponent ranged from 2.22 to 2.48, and as high as 2.76 at higher load. A high mass-mobility exponent for soot particles usually indicates absorbed volatile material infilling the agglomerate structure; the particle thus becomes 'more spherical'. In the case of the Peugeot engine, the volatile material was thought to be sulfate. Soot from gasoline engines tends to have a higher mass-mobility exponent that that from diesel engines, symptomatic of a higher level of semivolatile material in the exhaust stream.

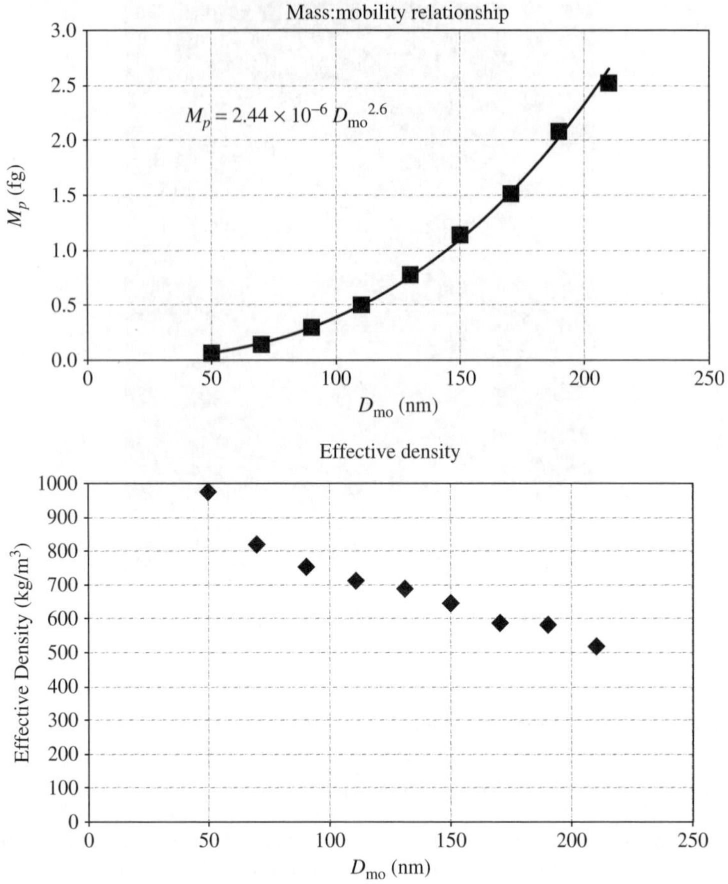

Figure 15.5 *(a) Mass-mobility relationship and (b) effective density of diesel soot, measured with tandem DMA-CPMA (idle condition, $D_{fm} = 2.6$). Data courtesy of Cambustion, used with permission.*

An alternative to a tandem DMA-CPMA experiment or equivalent is to use a measure of aerodynamic diameter in tandem with a DMA such as an ELPI. This technique has been used by Maricq and Xu (2004), for example. As aerodynamic diameter depends upon mass, an effective density can be calculated as a function of mobility diameter by comparing the mobility and aerodynamic diameters, and from this the mass-mobility exponent can be obtained. Maricq and Xu (2004) obtained a mobility-diameter-based fractal dimension for diesel vehicle soot of 2.3.

Once D_{fm} has been determined, it is possible to use spectral instruments to estimate particle mass from a given size spectrum (Kittelson *et al.*, 2004). If a discrete particle size spectrum is used for this calculation then it is prone to excessive noise from the channels representing the largest particles; these have most significance in the calculation as they are the heaviest. One solution to this is to fit a lognormal function to the data (which suppresses spectral noise in the tail of the function) then use Hatch–Choate equations to calculate the diameter of average mass and weight this to calculate the total particle mass (Symonds *et al.*, 2007).

15.6 Amelioration Techniques

15.6.1 Fuel Composition

We have already seen that reducing the sulfur impurities in fuel can reduce the sulfate fraction of engine aerosol dramatically and that organometallic fuel additives can have a soot-suppressing effect (though they can lead to an increase in ash). The combustion quality of diesel fuel is determined by its *cetane* number. This is a measure of the fuel's ignition delay: the delay between ignition and a detectable rise in pressure. On one hand, decreasing the cetane number of the fuel, and thus increasing the ignition delay, allows more time for air entrainment, favouring premixed combustion, and can therefore reduce carbonaceous soot emission (Li, Chippior and Gülder, 1996); on the other, it can lead to increased organic fraction emission due to an increased chance of wall impingement.

In recent years, biofuels such as biodiesel, ethers (such as dimethyl ether, DME) and ethanol have been increasingly used in fuel blends. Biodiesel consists of long-chain ester molecules. Apart from their other environmental and sustainability advantages (though they are not without controversy), the presence of oxygen in these substances leads to a decrease in soot formation during combustion (Yage, Cheung and Huang, 2009), although an increase in volatile particle emission can sometimes be seen (Northrop *et al.*, 2011). It is not currently known exactly why the presence of oxygenates leads to decreased soot production, although it may be as simple as providing more oxygen for combustion (Rakopoulos, Antanopolous and Rakopoulos, 2006).

15.6.2 Control by Engine Design and Calibration

Reducing particulate emissions by engine design and *calibration* (the set-up of the engine operating conditions in the electronic engine management system) is an attractive option for engine manufacturers, as it can often make a considerable difference without incurring the extra cost and complication of aftertreatment systems (e.g. Klindt, 2010).

Fuel pressure and timing are among the most influential calibration parameters that affect particulate emissions. By increasing the fuel pressure, the distance from the injector nozzle to the combustion zone is increased, which increases the amount of air entrained in the spray, making the local air–fuel mixture leaner and reducing the amount of unburnt and partially unburnt fuel, which in turn lowers particulate emissions (Picket and Siebers, 2004). The resulting higher combustion temperatures also lead to a higher rate of in-cylinder soot oxidation. Of course, if the injection pressure is raised too much, fuel can impinge on the cylinder walls or piston, with a resulting *increase* in emissions, including of particulate. The resultant increased cylinder pressure during combustion can also lead to engine damage.

The injection of additional fuel after the main injection event, known as post-injection, can be an effective means of reducing PM emission (Desantes *et al.*, 2011). Splitting the injection means that each injection leads to a less locally rich mixture. However, care must be taken not to post-inject too late and cool the charge, leading to reduced in-cylinder soot oxidation. The relative position and angle of the injector and spark plug in a GDI engine, and whether the fuel is guided by the surface of the piston or localised in the region of the spark plug, can also make a large difference to the quantity and nature of particulate emissions (Price *et al.*, 2006).

Exhaust gas recirculation (EGR) has been introduced to reduce NO_x emissions. A portion of the exhaust gas is recirculated back to the intake manifold (essentially inert gas, which cools the combustion process by lowering the heat capacity and diluting oxygen), thereby reducing the reactions of nitrogen and oxygen favoured at high temperatures which produce NO_x. A side effect of EGR, however, is to increase particulate emissions, due to the decreased availability of oxygen. The trade-off between NO_x and particulate emissions resulting from varying the amount of EGR used presents a difficult challenge to engine calibrators.

15.6.3 Particulate Filters

In recent years, the DPF has provided a very effective solution to the emission of nanoparticles from diesel engines. DPFs consist of a (usually) ceramic substrate placed in a metal container in the exhaust system of the vehicle. The most common substrate construction is the wall flow type (Howitt and Montierth, 1981), in which a grid of channels is formed along the length of the substrate, with alternate channels being blocked off at alternate ends. Exhaust enters via any of the 50% of channels open at the engine end of the DPF, particles are removed as the gas stream passes through the interchannel wall and the particle-reduced gas stream passes into the channels that are open at the tailpipe end. This configuration serves to maximise the surface area available to filtration for a given volume of filter. Common substrate materials include cordierite and silicon carbide. SiC has a higher melting point than cordierite, but it does have a high thermal expansion coefficient, which warrants SiC filters being built in sections (which are cemented to form the overall substrate) in order to allow for thermal expansion.

On a microscale, the walls of a DPF contain many irregular pores. Initially, when a DPF is clean, filtration occurs by deposition of the pore walls (Figure 15.6). After some time, soot gets deposited across the necks of the pores. As these pores fill up, the backpressure across the DPF caused by gas being pumped through it by the engine rapidly increases. Eventually, most of the pores are filled and filtration occurs through a layer of soot on the channel walls – the so-called cake-filtration stage. As more soot arrives, the cake layer increases in thickness and the backpressure continues to rise, although at a slower rate than during the pore-filling stage. The performance of a DPF is normally expressed in terms of its filtration efficiency. When a DPF is clean, at the start of the pore-filling stage, its filtration efficiency can be as low as 50%, but as the DPF begins to fill, its efficiency rapidly increases. At the cake-filtration stage, the efficiency can reach over 99%. This is factored into emissions-testing protocols.

Eventually, the DPF fills with soot to the extent that the backpressure becomes detrimental to the performance of the engine (or more likely, damages the DPF, as discussed later). At this point, it is necessary to *regenerate* the DPF. There are two forms of regeneration, which apply to different DPF systems to varying degrees. In passive regeneration, soot is continuously oxidised at a slow rate by the thermal action of the exhaust stream. Catalytic coatings on the DPF substrate or catalytic fuel additives can be used to make this reaction more favourable at lower temperatures. In active regeneration, an automated process causes

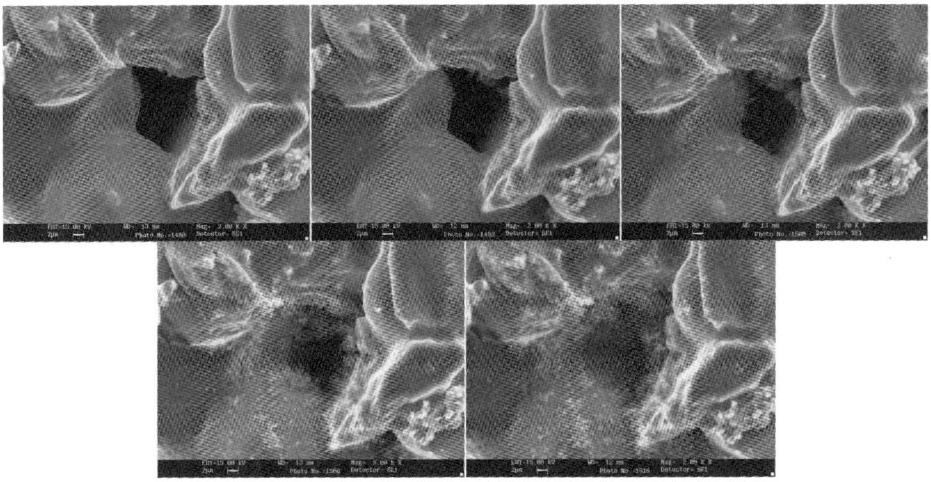

Figure 15.6 Optical microscopy of progressive DPF pore filling. Reproduced with permission from Payne, 2011.

the DPF to be regenerated when the engine management system detects a sufficiently large pressure drop across the DPF. Methods of active regeneration commonly use a change in engine control strategy (e.g. changing injection timing) to cause the exhaust gases to increase in temperature. Other methods include injecting fuel into the exhaust stream and electric heating. There is usually an efficiency and fuel economy penalty to pay during regeneration.

Soot oxidation can occur through reaction with either oxygen or NO_2 (which is usually present in diesel exhaust). One variant of the DPF, the Continuously Regenerating Trap (CRT; Cooper and Thoss, 1989), uses an oxidation catalyst before the filter to oxidise NO to NO_2. Combustion with NO_2 can occur at normal exhaust operating temperatures (~ 250 °C) and this variant thus improves passive regeneration performance.

Active regeneration occurs at temperatures in excess of 600 °C and the oxidation reaction is itself exothermic. The temperatures reached can be in excess of 1000 °C, which can place considerable stress on the ceramic substrate. The maximum backpressure reached before active regeneration is induced may in practice be much less than that which in itself would be tolerated by the engine, in order to prevent the absolute mass of soot accumulated on the filter from reaching a level such that the heat released during regeneration permanently damages the substrate.

It should be remembered that DPFs are most effective at removing solid particles; that is to say, the carbonaceous and ash fractions. Nucleation-mode particles may not normally even be formed until the exhaust system or real-world dilution, so they may pass through a DPF as their gas-phase precursors. It is common for volatile materials to adsorb on to solid particles, meaning that if the solid particles have been removed by the DPF, homogenous nucleation of volatiles passing through the DPF into a particle mode may be promoted. The formation of sulfuric acid in a DOC and its desorption from the DOC and DPF can be made worse under the high-temperature conditions of DPF regeneration, and large quantities of nucleation aerosol have been observed during regeneration (Campbell *et al*., 2006).

The DPF also captures ash particles (including any formed from fuelborne additives intended to catalyse regeneration), but these are not usually removed by regeneration. Although the proportion of ash in the engine exhaust is small, over time this can lead to a gradual reduction in the capacity of the DPF.

The widespread adoption of DPFs has been driven by ever more stringent legislation, such as the PMP project and the Euro 5 and 6 standards for light- and heavy-duty engines in Europe and the US2007 standard for heavy-duty engines in the USA (light-duty diesel vehicles are not yet common there). A number of retrofit programmes are in place globally, particularly for off-road applications (especially vehicles used indoors or in confined spaces). A number of bus fleets have been retrofitted and low-emission zones are becoming widespread in Europe, which mandate the retrofitting of DPFs.

With the advent of the Euro 6 emission standards, which will introduce a particle-number standard for gasoline-engine particle emissions, some manufacturers are considering the use of gasoline particulate filters (GPFs, e.g. Mikulic *et al.*, 2010). Due to the increased temperatures involved in gasoline combustion, continuous thermal passive regeneration tends to be a feature of GPFs. This means that the level of soot filling for optimal filtration efficiency may never be reached. However, as the level of particulate emissions from even direct-injection gasoline engines is much less than that from diesel engines, the amount of attenuation required from a GPF is usually much less than that required by a DPF.

Acknowledgements

The author would like to thank Professor Nick Collings of Cambridge University Engineering Department and Chris Nickolaus, Mark Peckham and Kingsley Reavell of Cambustion for their useful suggestions.

References

Abdul-Khalek, I., Kittelson, D.B. and Brear, F. (2000) Nanoparticle Growth During Dilution and Cooling of Diesel Exhaust: Experimental Investigation and Theoretical Assessment. SAE Technical Paper 2000-01-0515.

Ahlvik, P., Ntziachristos, L., Keskinen, J. and Virtanen, A. (1998) Real Time Measurements of Diesel Particle Size Distribution with an Electrical Low Pressure Impactor. SAE Technical Paper 980410.

Andersson, J., Giechaskiel, B., Muñoz-Bueno, R. *et al.*, (2007) Particle Measurement Programme (PMP): Light-duty Inter-laboratory Correlation Exercise (ILCE_LD) Final Report, Joint Research Centre, European Commission, http://www.unece.org/trans/main/wp29/wp29wgs/wp29grpe/grpeinf54.html (last accessed 8 August 2013).

Andersson, J., Mamakos, A., Giechaskiel, B., Carriero, M. and Martini, G., (2010) Particle Measurement Programme (PMP): Heavy-duty Inter-laboratory Correlation Exercise (ILCE_HD) Final Report, Joint Research Centre, Ispra http://www.unece.org.unecedev.colo.iway.ch/fileadmin/DAM/trans/doc/2009/wp29grpe/PMP-24-02e.pdf (last accessed 15 August 2013).

Burtscher, H. (2005) Physical characterization of particulate emissions from diesel engines: a review. *Journal of Aerosol Science*, **36**, 896–932.

Buzzard, N.A., Clark, N.N. and Guffey, S.E. (2009) Investigation into pedestrian exposure to near-vehicle exhaust emissions. *Environmental Health*, **8**(13). doi: 10.1186/1476-069X-8-13.

Campbell, B., Peckham, M., Symonds, J. *et al.*, (2006) Transient Gaseous and Particulate Emissions Measurements on a Diesel Passenger Car Including a DPF Regeneration Event. SAE Technical Paper 2006-01-1079.

Carpentieri, M., Kumar, P. and Robins, A. (2011) An overview of experimental results and dispersion modelling of nanoparticles in the wake of moving vehicles. *Environmental Science and Technology*, **159**, 685–693.

Cooper, B.J. and Thoss, J.E. (1989) Role of NO in Diesel Particulate Emission Control. SAE Technical Paper 890404.

Dec, J.E. (1997) A Conceptual Model of Di Diesel Combustion Based on Laser-Sheet Imaging. SAE Technical Paper 970873.

Desantes, J.M., Bermúdez, V., García, A. and Linares, W.G. (2011) A comprehensive study of particle size distributions with the use of postinjection strategies in DI diesel engines. *Aerosol Science and Technology*, **45**(10), 1161–1175.

Eastwood, P. (2008) *Particulate Emissions from Vehicles*, SAE International/John Wiley & Sons, Ltd, Chichester.

Ehara, K., Hagwood, C. and Coakley, K.J. (1996) Novel method to classify aerosol particles according to their mass-to-charge ratio – aerosol particle mass analyser. *Journal of Aerosol Science*, **27**, 217–234.

Fortuin, J.P.F., Van Dorland, R., Wauben, W.M.F. and Kelder, H. (1995) Greenhouse effects of aircraft emissions as calculated by a radiative transfer model. *Annales Geophysicae*, **13**(4), 413–418.

Fujiwara, Y., Tosaka, S. and Murayama, T. (1993) Formation Process of SOF in the Combustion Chamber of IDI Diesel Engines. SAE Technical Paper 932799.

Gidney, J.T., Twigg, M.V. and Kittelson, D.B. (2010) Effect of organometallic fuel additives on nanoparticle emissions from a gasoline passenger car. *Environmental Science and Technology*, **44**(7), 2562–2569.

Greeves, G. and Wang, C.H.T. (1982) Origins of Diesel Particulate Mass Emission. SAE Technical Paper 810260.

Hands, T., Nickolaus, C., Symonds, J. and Finch, A. (2010) Real-Time Particle Emissions from 2-Stroke Motorbikes with and without PMP Sampling System. SAE Technical Paper 2010-32-0047.

Heywood, J.B. (1988) *Internal Combustion Engine Fundamentals*, McGraw-Hill International Editions.

Howard, J.B. and Kausch, W.J. (1980) Soot control by fuel additives. *Progress in Energy and Combustion Science*, **6**, 263–276.

Howitt, J.S. and Montierth, M.R. (1981) Cellular Ceramic Diesel Particulate Filter. SAE Technical Paper 810114.

Johnson, T., Caldow, R., Pocher, A. *et al.*, (2004) An Engine Exhaust Particle Sizer Spectrometer for Transient Emission Particle Measurements. SAE Technical Paper 2004-01-1341.

Johnson, B.T., Hargrave, G.K. and Reid, B. (2011) Crankcase Sampling of PM from a Fired and Motored Compression Ignition Engine. SAE Technical Paper 2011-24-0209.

Jung, H., Kittelson, D.B. and Zachariah, M.R. (2003) The Influence of Engine Lubricating Oil on Nanoparticle Emissions and Kinetics of Oxidation. SAE Technical Paper 2003-01-3179.

Kato, S., Takayama, Y., Sato, G.T. *et al.*, (1997) Investigation of Particulate Formation of DI Diesel Engine with Direct Sampling from Combustion Chamber. SAE Technical Paper 972969.

Kawai, T., Goto, Y. and Odaka, M. (2004) Influence of Dilution Process on Engine Exhaust Nano-particles. SAE Technical Paper 2004-01-0963.

Kittelson, D.B. (1998) Engines and nanoparticles: a review. *Journal of Aerosol Science*, **29**(5/6), 575–588.

Kittelson, D., Hands, T., Nickolaus, C. *et al.*, (2004) Mass correlation of engine emissions with spectral instruments. Proceedings of 2004 Japanese SAE Annual Congress. Paper No. 20045462.

Klindt, K. (2010) Reducing the particulate emission numbers in DI gasoline engines. Proceedings of Directions in Engine-Efficiency and Emissions Research (DEER) Conference.

Kumar, P., Robins, A., Vardoulakis, S. and Britter, R. (2010) A review of the characteristics of nanoparticles in the urban atmosphere and the prospects for developing regulatory control. *Atmospheric Environment*, **44**, 5035–5052.

Li, X., Chippior, W.L. and Gülder, Ö. (1996) Effects of Fuel Properties on Exhaust Emissions of a Single Cylinder DI Diesel Engine. SAE Technical Paper 962116.

Liu, Z.G., Vasys, V.N., Dettmann, M.E. *et al.* (2009) Comparison of strategies for the measurement of mass emissions from diesel engines emitting ultra-low levels of particulate matter. *Aerosol Science and Technology*, **43**(11), 1142–1152.

Lyyränen, J., Jokiniemi, J., Kauppinen, E.I. *et al.* (2004) Comparison of different dilution methods for measuring diesel particle emissions. *Aerosol Science and Technology*, **38**, 12–23.

Mandelbrot, B. (1982) *The Fractal Geometry of nature*, Freeman, San Francisco, CA.

Maricq, M.M., Chase, R.E., Podsadlik, D.H. and Vogt, R. (1999) Vehicle Exhaust Particle Size Distributions: A Comparison of Tailpipe and Dilution Tunnel Measurements. SAE Technical Paper 1999-01-1461.

Maricq, M.M. and Xu, N. (2004) The effective density and fractal dimension of soot particles from premixed flames and motor vehicle exhaust. *Journal of Aerosol Science*, **35**, 1251–1274.

Mikulic, I., Koelman, H., Majkowski, S. and Vosejpka, P. (2010) A study about particle filter application on a state-of-the-art homogeneous turbocharged 2l DI gasoline engine. *Aachener Kolloquium Fahrzeug- und Motorentechnik*, **19**, 1–18.

Miller, K.A., Siscovick, D.S., Sheppard, L. *et al.* (2007) Long-term exposure to air pollution and incidence of cardiovascular events in women. *New England Journal of Medicine*, **356**, 447–458.

Northrop, W.A., Madathil, P.V., Bohac, S.V. and Assanis, D.N. (2011) Condensation growth of particulate matter from partially premixed low temperature combustion of biodiesel in a compression ignition engine. *Aerosol Science and Technology*, **45**(1), 26–36.

Olfert, J. and Collings, N. (2005) New method for particle mass classification – The Couette centrifugal particle mass analyser. *Journal of Aerosol Science*, **36**, 1338–1352.

Olfert, J.S., Symonds, J.P.R. and Collings, N. (2007) The effective density and fractal dimension of particles emitted from a light-duty vehicle with diesel oxidation catalyst. *Journal of Aerosol Science*, **38**, 69–82.

Park, K., Cao, F., Kittelson, D.B. and McMurry, P.H. (2003) Relationship between particle mass and mobility for diesel exhaust particles. *Environmental Science and Technology*, **37**, 577–583.

Park, K., Kittelson, D.B. and McMurry, P.H. (2003) A closure study of aerosol mass concentration measurements: comparison of values obtained with filters and by direct measurements of mass distributions. *Atmospheric Environment*, **37**, 1123–1230.

Park, K., Kittelson, D.B. and McMurry, P.H. (2004) Structural properties of diesel exhaust particles measured by transmission electron microscopy (TEM): relationship s to particle mass and mobility. *Aerosol Science and Technology*, **38**, 881–889.

Payne, S. (2011) Experimental studies of diesel particulate filtration. PhD thesis. University of Cambridge.

Petzold, A., Marsh, R., Johnson, M. *et al.* (2011) Evaluation of methods for measuring particulate matter emissions from gas turbines. *Environmental Science and Technology*, **45**, 3562–3568.

Pickett, L.M. and Siebers, D.L. (2004) Soot in diesel fuel jets: effects of ambient temperature, ambient density and injection pressure. *Combustion and Flame*, **138**, 114–135.

Pipho, M.J., Ambs, J.L. and Kittelson, D.B. (1986) In-cylinder Measurements of Particulate Formation in an Indirect Injection Diesel Engine. SAE Technical Paper 860024.

Price, P., Stone, R., Collier, T. and Davies, M. (2006) Particulate Matter and Hydrocarbon Emissions Measurements: Comparing First and Second Generation DISI with PFI Engines in Single Cylinder Optical Engines. SAE Technical Paper 2006-01-1263.

Rakopoulos, C.D., Antanopolous, K.A. and Rakopoulos, D.C. (2006) Multi-zone modelling of diesel engine fuel spray development with vegetable oil, bio-diesel or diesel fuels. *Energy Conversion and Management*, **47**, 1550–1573.

Ramanathan, V. and Carmichael, G. (2008) Global and regional climate changes due to black carbon. *Nature Geoscience*, **1**, 221–227.

Reavell, K., Hands, T. and Collings, N.A. (2002) Fast Response Particulate Spectrometer for Combustion Aerosols. SAE Technical Paper 2002-01-2714.

Schmid, O., Hagen, D.E., Whitefield, P.D. *et al.* (2011) Methodology for particle characterisation in the exhaust flows of gas turbine engines. *Aerosol Science and Technology*, **38**(11), 1108–1122.

Symonds, J.P.R., Reavell, K.S.J., Olfert, J.S. *et al.* (2007) Diesel soot mass calculation in real-time with a differential mobility spectrometer. *Journal of Aerosol Science*, **38**, 52–68.

UNECE R83 (United Nations Economic Commission for Europe), (2011) World Forum for Harmonization of Vehicle Regulations, Regulation Number 83 – Emission of Pollutants According to Engine Fuel Requirements, available from http://www.unece.org/fileadmin/DAM/trans/main/wp29/wp29regs/r083r4e.pdf (last accessed 15 August 2013).

US EPA (United States Environmental Protection Agency) (2002) Health Assessment Document for Diesel Engine Exhaust.

Williams, R.L., Perez, J.M. and Griffing, M.E. (1985) A Review of Sampling Condition Effects Upon Polynuclear Aromatic Hydrocarbons (PNA) from Heavy-duty Diesel Engines. SAE Technical Paper 852081.
Witze, P.O. and Green, R.M. (2005) Comparison of Single and Dual Spray Fuel Injectors during Cold Start of a PFI Spark Ignition Engine Using Visualization of Liquid Fuel Films and Pool Fires. SAE Technical Paper 2005-01-3863.
Yage, D., Cheung, C.S. and Huang, Z. (2009) Comparison of the effect of biodiesel-diesel and ethanol-diesel on the particulate emissions of a direct injection diesel engine. *Aerosol Science and Technology*, **43**(5), 455–465.

16

Movement of Bioaerosols in the Atmosphere and the Consequences for Climate and Microbial Evolution

Cindy E. Morris, Christel Leyronas, and Philippe C. Nicot
INRA, UR407 de Pathologie Végétale, France

16.1 Introduction

The presence of bioaerosols in the atmosphere has a multitude of consequences for human, animal, and plant health, for various biogeochemical and atmospheric processes, and for the conservation and maintenance of buildings and monuments. 'Bioaerosols' refer to particulate aerosol matter of biological origin. The major types of bioaerosols are primary and secondary biological aerosols and biogenic aerosols. Primary biological aerosols are bits of organisms or intact cells. They can be alive or dead. Secondary biological aerosols result from physical or chemical processes (condensation, oxidation, coating, etc.) that modify primary biological aerosols. They are distinct from biogenic aerosols, which are products of metabolism and often undergo secondary chemical processes that lead to their impact on the atmosphere. They include dimethylsulfide and other volatile organic carbons such as methane, for example. Whereas materials from nonbiological sources (such as sulfates, black carbon, nitrates, mineral dust, and sea salt) and organic carbon from forest fires constitute the bulk of atmospheric aerosols (Mahowald *et al.*, 2011), primary biological aerosols can constitute up to 25% of the aerosol particles in the size range 0.2–50.0 μm (Jaenicke, 2005). A very detailed presentation of primary biological aerosols is given by Després *et al.* (2012). Primary biological aerosols comprise various forms of microorganism, including bacterial or algal aggregates and single cells, fungal and bacterial spores,

Aerosol Science: Technology and Applications, First Edition. Edited by Ian Colbeck and Mihalis Lazaridis.

fragments of fungal mycelium, and virus particles. They can also originate from macroorganisms; these include insect parts, pollen grains, fern and moss spores, bits of plant tissue, fragments of animal tissue, and so on.

The presence of microorganisms in the atmosphere was first brought to light over 200 years ago by Spallanzani in the middle of the eighteenth century (Capanna, 1999), and his findings were expanded upon by Pasteur at the end of the nineteenth century (Pasteur, 1890). In the early twentieth century, the horizons of aerobiology were extended by the discovery that certain fungal spores were transported upward to several kilometers above the earth's surface. This was found by the team of E.C. Stakman at the University of Minnesota in 1921, who utilized a variety of spore-sampling devices hung out from US Army fixed-wing planes during 50 flights over the front of a wheat stem rust epidemic as it moved from southern Texas to southern Wyoming (Stakman *et al.*, 1923). These experiments demonstrated for the first time that microorganisms are present in the atmosphere at the altitudes of clouds and beyond, and opened the way for a series of subsequent high-altitude sampling efforts, some involving eminent pilots, including Charles Lindberg. Aerobiology was again taken to new heights when meteorological rockets and balloons were used to sample the atmosphere, revealing the presence of bacteria and fungi beyond the troposphere at altitudes up to 70 km (Imshenetsky, Lysenko, and Kazakov, 1978; Wainwright *et al.*, 2003). Although most aerobiology research effort has focused on the emission and dissemination of pollen, there has been a renewed surge in studies into the microbial component of the primary biological aerosols, as this offers the opportunity to explore simultaneously diverse research themes, including long-distance transport, particle–climate interactions, the atmosphere as a habitat, and the impact of aerial dissemination on the evolutionary history of organisms and its consequence for disease epidemiology. Therefore, this chapter will focus on microorganisms in the atmosphere.

In this chapter we will discuss the origins and transport of microbial populations in the atmosphere and their interactions with atmospheric physical and chemical properties. We will devote particular attention to the methodology used to study these phenomena and the dynamics of the processes revealed. In spite of the past progress made in examining the interaction of microorganisms with the atmosphere, communication and collaboration between the disparate disciplines that contribute to these studies (physics, biology, chemistry, and modeling) is still hampered by the complexity of certain concepts and phenomena. Therefore, in this chapter we attempt to provide a somewhat simplified presentation of some of the particularly complex subjects. This is not meant to belittle the importance of the technical and theoretical foundations underlying these subjects but rather is to allow the reader to construct an intellectual framework for reasoning that can then be enriched with the precise details if and when this becomes necessary.

The sources of microorganisms in outdoor air (plant, soil, and water surfaces in particular) and the ecology of microorganisms in these habitats have been well described. Furthermore, recent reports, reviews, and books have provided considerable information on the diversity and abundance of atmospheric microorganisms – including clouds – and the techniques deployed for these studies (Lacey and West, 2006; Amato *et al.*, 2007; Brodie *et al.*, 2007; Burrows *et al.*, 2009b; Frohlich-Nowoisky *et al.*, 2009; Delort *et al.*, 2010; Bowers *et al.*, 2011; Després *et al.*, 2012). Our story will begin with the launch of microorganisms into the atmosphere from these sources and will cover their ascent and their fate in the atmosphere in the context of the major tropospheric processes that assist in this transport, as illustrated in Figure 16.1. The launch, or emission, is well understood for certain microorganisms, such as fungi, but remains rather elusive for others. The subsequent transport in the earth's boundary layer (either ascension or deposition) is a process whose net outcome can be measured, but such measurements are still challenging at present. The microorganisms that successfully reach the free troposphere are generally those whose aerodynamic properties allow their trajectory to be estimated with the same tools used to predict the dissemination of very light particulate matter or gases. While they are suspended in the atmosphere, microorganisms can play the same four principal roles of inert aerosols that are of major importance to the climate and to air quality (Mahowald *et al.*, 2011). They can reflect light, they can participate in chemical reactions, they can be cloud condensation nuclei (CCN), and they can be ice nuclei (IN). But in contrast with inert aerosols, these interactions can have consequences for the evolutionary history of the living microorganisms that constitute biological aerosols. This chapter will address these interactions with the atmosphere: how microorganisms might affect atmospheric processes and how dissemination impacts life history.

16.2 Emission: Launch into the Atmosphere

The specific mechanisms involved in the emission of microbial particles into the atmosphere from their source can be considered the big black box of aerobiology. Although general processes involved in passive and active emissions from dry and wet surfaces have been described, our ability to predict and quantify these emissions is still part of the realm of speculation. Emission is critical to permitting microorganisms to escape from the laminar boundary layer that surrounds the surfaces on which they grow or are attached and to be propelled into the planetary boundary layer where they can begin their voyage (Figure 16.1).

Although the mechanisms of emission and aerosolization of bacteria have been studied (Burrows *et al.*, 2009a; Simon *et al.*, 2011), the most detailed information about microorganism emission comes from the study of fungi (Ingold, 1965; Meridith, 1973). Like other biological particles, fungal spores can be released into the air by passive mechanisms involving energy input from an external source (wind, agricultural practices such as combining, etc.). This leads to the emission not only of the fungal spores themselves but of any microbial 'hitchhikers'. Fungal spores can often carry other microorganisms on their surfaces, such as when spores of the loose smut of wheat and barley carry spores of the fungus that causes ergot (Cherewick, 1953) or when uredospores of rust carry bacteria (French, Novotny, and Searles, 1964; Doherty and Preece, 1978). But the peculiarity of fungi is that their spores can also be released by active mechanisms, whereby they project them into the atmosphere through complicated mechanisms involving specialized fungal structures and changes in the pressures of tissues due to wetting and drying. Both types of release are influenced by microclimatic parameters (Meridith, 1973; Aylor, 1990). These release mechanisms are essential to the escape of spores and other small biological particles from the laminar boundary layer and to their reaching the turbulent boundary layer, where they can be dispersed widely.

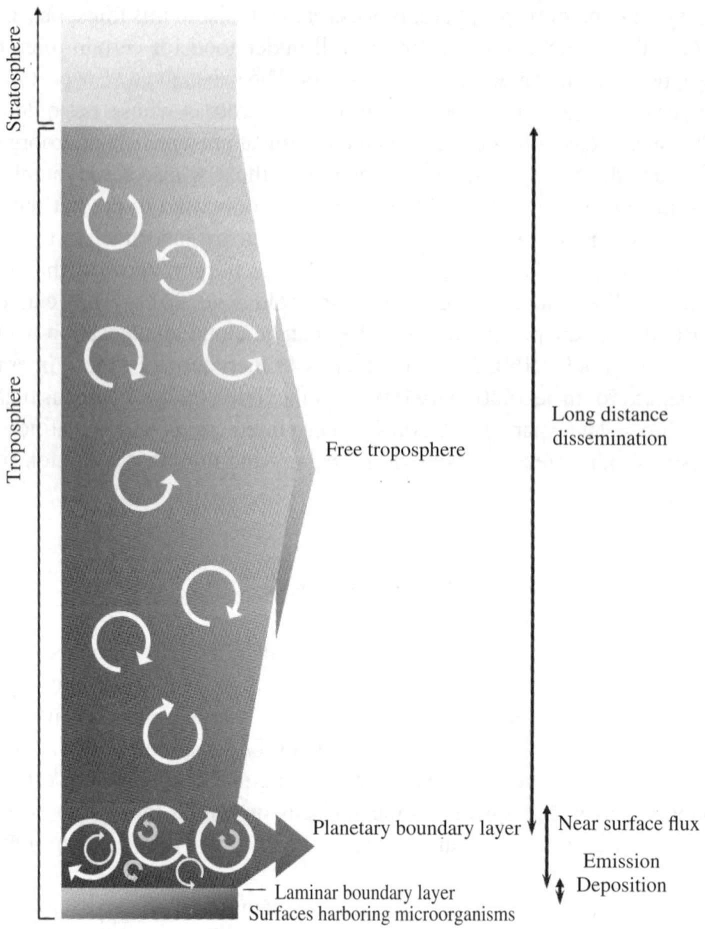

Figure 16.1 *Vertical range and motors of transport of microorganisms in the atmosphere. In order to become part of the aerosols transported in the atmosphere, microbial cells, spores, or particles containing microorganisms are emitted from the surfaces that harbor them (mainly leaves, soil, and water). In some cases, these surfaces are under plant canopies with depths of several centimeters to tens of meters, as in forests. From here, they can move into the earth's planetary boundary layer, which is several kilometers in depth. Movement of biological aerosols in this part of the troposphere can be greatly influenced by eddies caused by thermal convection and topographical features of the earth's surface (indicated as curved arrows in the figure). Particles can ascend via upward convective and turbulent eddies. They descend due to rainfall, turbulent eddies, or when the force of gravity overcomes the upward forces and the resistance of the air. The outcome of these vertical movements is called 'flux': the number of particles passing through a 'window' of a given surface per unit of time in a net upward or downward direction. Particles can also move horizontally in the planetary boundary layer. Vertical flux in the planetary boundary layer (called 'near-surface flux') and horizontal movements are highly influenced by topographical features of the earth's surface and are particularly variable. In contrast, direct capture of biological aerosols in order to describe their movement is more difficult in the free troposphere because of the lower density of biological aerosols compared to that in the planetary boundary layer. Therefore, models of atmospheric circulation are usually used to estimate the movement of biological aerosols, as well as of many other types of aerosol, in this part of the troposphere. Particles can pass the troposphere–stratosphere boundary. However, it is in the troposphere where they can influence processes with subsequent impacts on climate processes and it is there that they are most likely to survive in order to return back to the earth's surface.*

16.2.1 Active Release

Active release is prevalent among fungi in certain species in the Ascomycota and Basidiomycota phyla and has been well described in several reviews (Meridith, 1973; Elbert *et al.*, 2007). In Ascomycota, active release of spores relies on the reaction of turgid cells (asci) to the water supply. Ascospores are usually released after wetting by rain or dew. In Basidiomycota, active discharge of basidiospores is powered by the rapid movement of a droplet of fluid over the spore surface, called the 'surface tension catapult', which induces a swift shift in the center of gravity of the spore responsible for the launch. Spores can be ejected substantial distances depending on the fungal species: 2–300 mm in the case of ascospores and 0.04–1.26 mm in the case of basidiospores (Jones and Harrison, 2004; Fischer *et al.*, 2010).

16.2.2 Passive Release

Turbulent wind movements are very efficient at lofting all sorts of matter into the atmosphere. The remarkable and rare event of fish (Gudger, 1929) or frogs being swept up into the sky and later falling as rain can occur when stormy wind speeds of 160 km/h are attained. That being the case, it is easy to see that gentle breezes of 20 km/h or light air of 5 km/h is enough to lift bacterial cells whose weight is measured in picograms or fungal spores weighing under a microgram. Over land, the aerial parts of plants are considered a principal source of the microorganisms in the atmosphere (Lighthart, 1997), with surfaces covered by vegetation generally being stronger sources of microbial aerosols than is barren or fallow soil (Lindemann *et al.*, 1982; Lighthart and Shaffer, 1994; Lighthart, 1999). The formation of aerosols containing microorganisms from plant surfaces is likely a result of wind stress, which can directly lift the microorganisms, or of secondary impacts caused by the wind stress-induced deformations of leaves. For fungal spores, the minimum wind speed required to remove them varies according to the species (e.g., 0.36–0.5 m s^{-1} for conidia of *Botrytis cinerea* and 0.76–2 m s^{-1} for aeciospores of *Puccinia coronifera*) according to the work of Stepanov, as described by Gregory (1961). Wind gusts are more efficient in spore removal than is streamlined wind (Jarvis, 1962; Aylor, 1990). Removal by wind can be facilitated by the elevated spore-bearing structures produced on the surfaces of plant tissues found in Deuteromycota (conidiophores). These structures can twist as a result of rapid changes in atmospheric humidity, resulting in spore detachment (Fitt, Creighton, and Bainbridge, 1985). Reductions in relative humidity and increased exposure to infrared and visible radiation can also lead to the release of conidia, resulting from the decrease in surface-tension effects (bonding effects) caused by the decrease in moisture on the leaf surface (Jones and Harrison, 2004). The release of fungal spores can also result from the fall of raindrops on diseased plant tissues: the air expelled by the impact of a raindrop on a surface can be enough to lift the spores from that surface; the droplet can also shake the leaf and confer momentum to the spore in that way (McCartney, 1991; Jones and Harrison, 2004).

These passive-release mechanisms have been described for fungi, but it is very likely that they can affect other biological particles as well. The forces of wind, raindrops, and various human activities can lead to the release of algae into the atmosphere from plants, soils, and the surfaces of buildings (Sharma *et al.*, 2007). With plant surfaces in particular, biological particles may be removed by wind (well known for pollen), lifted by raindrops (bacteria), and released when water-binding forces decrease during drying, for example. Drying of leaf surfaces occurs as a result of biological processes or of changing atmospheric conditions. These mechanisms can be compounded by changes in the charge of leaf surfaces during the day, which modify the electrostatic attraction or repulsion of biological particles (Leach, 1987).

Above water surfaces, the creation of aerosols containing microorganisms occurs through bubble bursting. When waves break, their dissipated energy entrains a dense plume of air bubbles in the bulk water. When these bubbles rise to the surface, they break, leading to the formation of tens to hundreds of film and jet drops (Fuentes *et al.*, 2010). Marine aerosols are believed to account for the majority of the global aerosol flux (all types of aerosol considered as a whole) (O'Dowd and de Leeuw, 2007) and their formation can contribute to the presence of biological particles in the atmosphere in remote regions, such as above the central Arctic Ocean (Leck and Bigg, 2005).

16.2.3 Quantifying Emissions

Various devices have been developed to quantify the release of microbial cells or cell fragments and to assess how it can be affected by such factors as wind speed, air temperature, humidity, the growth substrate, the microbial species, and the age of microbial growth (Harrison and Lowe, 1987; Willoquet *et al.*, 1998; Górny *et al.*, 2001; Sivasubramani *et al.*, 2004; Kanaani *et al.*, 2009; Lee *et al.*, 2010). Beyond a minimum threshold of air velocity, most of the available fungal spores might be released within seconds or over a prolonged period, depending on the species and wind speed (Harrison and Lowe, 1987; Willoquet *et al.*, 1998). Even under the most favorable release conditions, however, not all the cells or cell fragments available for release will actually be emitted into the atmosphere. For example, in one experiment less than 10% of the spores of *B. fabae* present on bean leaves and only 6.2% of those of *B. cinerea* were removed by air currents up to 10 m s^{-1} (Harrison and Lowe, 1987). Even smaller release efficiencies (0.05–2.3%) were observed for various fungal species on four building materials taken from mold-problem homes (Sivasubramani *et al.*, 2004). Extrapolating from nonlinear models representing the effect of wind speed on the efficiency of spore release, Willoquet *et al.* (1998) estimated that a wind speed of at least 25 m s^{-1} would be necessary to disperse more than 90% of the spores of *Erysiphe necator* (syn. *Uncinula necator*) present on infected grape leaves. Aerosolization studies showed that the rate of fragmentation of fungal structures during the release phase was also dependent on the species and wind speed (Kanaani *et al.*, 2009; Lee *et al.*, 2010).

The rate of global emission of microorganisms into the atmosphere has also been estimated by inferring the rates needed to ensure observed atmospheric concentrations. Based on this reasoning, the results of simulation models suggest that bacteria-containing particles are emitted at a rate of about 10^{24} particles per year (0.04–1.8 Tg $year^{-1}$ of bacterial cells) (Burrows *et al.*, 2009a). Likewise, rates of emission of larger microbial particles (with diameter 9–60 μm) have been estimated, but for specific regions rather than on a global scale (Wilkinson *et al.*, 2011). Estimates of regional emissions of fungi as a function of land-use type have also been made based on measures of atmospheric concentrations at different heights and on broad assumptions about air circulation. These estimates suggest that rates of emission range from 8 fungal spores m^{-2} s^{-1} over tundra to 2500 over crops (Sesartic and Dallafior, 2011). Estimates of global emissions of fungal spores have also been made using mannitol as a proxy. Mannitol emissions are on the order of 28 Tg year^{-1}, with 25% of this constituting fine-mode ($PM_{2.5}$) aerosols, and about 7% of this fraction representing fungal spores (Heald and Spracklen, 2009).

16.3 Transport in the Earth's Boundary Layer

16.3.1 Motors of Transport

Apart from certain fungi that can expel their spores into the air as described in the previous section, microorganisms do not have their own means of propulsion. It is the movement of air molecules that entrains microbial particles, unless the particles have aerodynamic properties that allow the forces of gravity to overcome this entrainment. In the earth's boundary layer, the fluid behavior of air is highly dependent on local topography and climate. Local microclimate is the result of interactions between the atmosphere and surface properties of the land cover (type of plant and structure of the canopy, irrigation, abundance of bare soil, etc.). These interactions lead to local variations in temperature, relative humidity, and wind speed near the plant canopy. Topographical features such as hedges and mountains act as wind barriers. The influence of land cover and topography on variations in air movements is further compounded by ambient temperature and incident radiation, which have seasonal and diurnal cycles. Due to the complexity and variability of the air movements in the planetary boundary layer, the ability to predict the trajectories of particles is strongly dependent on rigorous, direct measures of near-surface phenomena.

Despite the small size of microorganisms, their transport in the earth's boundary layer is nonetheless influenced by gravity and hence by their aerodynamic properties. An important aerodynamic property is their terminal velocity: the speed at which they fall in still air as a result of the counterbalance of gravity and the drag of the air (Lacey and West, 2006). Another is the aerodynamic diameter: the diameter that a particle would need to have in order to attain its terminal velocity if it were a perfect sphere with a density of 1 g cm^{-3}. The aerodynamic properties of microorganisms depend approximately on their size and mass, with larger, heavier particles such as fungal spores having a greater propensity to be affected by gravity than smaller particles such as bacterial cells. However, it has been reported that it is not possible to accurately estimate aerodynamic properties, and in particular aerodynamic diameter, solely based on physical diameter (Reponen *et al.*, 2001). The density of spores and other microbial particles is an essential component of aerodynamic diameter, but this varies widely, and independently of particle size. The irregular, nonspherical form of a particle can also compound the difficulty in assessing aerodynamic diameter. But in fact few efforts have been made to measure the true aerodynamic properties of microbial particles, such as terminal velocity or aerodynamic diameter. The terminal velocity of uredospores, the form of spore produced by rust fungi that is most frequently disseminated via the atmosphere, is in the range of 1–5 cm s^{-1} (Li *et al.*, 2009). This is comparable to the terminal velocities of a wide variety of fungal spores, which Gregory reported as being in the range of 0.3–1.4 cm s^{-1} (Gregory, 1961). Certain fungal spores have surface features that enhance buoyancy (such as spines that serve as sorts of wings), leading to a remarkable capacity to remain aloft, surpassing that of other spores of similar physical diameter (Stakman and Christensen, 1946). Overall, microbial particles are small enough that the effects of gravity are often masked by the effects of air movements, leading to long residence times in the atmosphere – on the order of 2–15 days (Burrows *et al.*, 2009a).

In addition to gravity, there are two predominant phenomena in the atmosphere that entrain microbial particles, leading to their upward or downward movement: thermal convection cells and turbulent eddies. Convection cells occur when a volume of air is heated. The air expands, becomes less dense than the surrounding air, and thus ascends; when it cools, it becomes denser and sinks. The same phenomenon occurs when water is heated in a saucepan: it starts to move as a result of the temperature differential between the bottom and the top. As the earth's surface is heated by the incident radiation from the sun, the heat is conducted into the atmosphere, leading to a temperature differential. This results in convection, moving the heat upward into the atmosphere. Sensible heat is the amount of energy that is thus exchanged, which plays an important role in the flux of microbial and other particles in the atmosphere. Turbulent eddies are the swirling and reverse current created when air flows encounter an obstacle. The obstacle might be a physical object, the friction between the surface roughness of a landscape and the moving air (wind), or a difference in air pressure.

16.3.2 Quantifying Near-Surface Flux

The flux of microbial particles in the planetary boundary layer is referred to as 'near-surface flux' and concerns essentially vertical movement. Quantitative measures of flux are indispensable to determining the net direction of this movement, thereby allowing the identification of effective sources of microbial aerosols and of the strength of these sources. These measures are also useful in predicting the trajectories and destinies of these aerosols, by providing data for the parameterization of mesoscale dispersion models (tens of meters to several kilometers) or long-distance dissemination models (tens to thousands of kilometers). The presence of microorganisms in the atmosphere and the comparison of their relative abundance at different times, in different geographical locations, and over different types of land cover have been assessed, in the whole of aerobiology literature, almost solely in terms of the concentration of microorganisms in a given volume of air (Jones and Harrison, 2004; Burrows *et al.*, 2009b; Bowers *et al.*, 2011). Although these assessments are useful for describing the diversity of microorganisms in the atmosphere, such measures are merely snapshots of a perpetual movement of these organisms, of uninterrupted comings and goings, of arrivals and departures. Concentrations of microorganisms in the atmosphere can fluctuate rapidly as a result of punctuated changes in atmospheric transportation barriers (Tallapragada, Ross, and Schmale, 2011).

Particle flux (also called 'flux density') is the amount of particles passing through a unit area per unit time (particles m^{-2} s^{-1}). Flux is analogous to concentration multiplied by speed (particles $m^{-3} \times m\ s^{-1} =$ particles m^{-2} s^{-1}). One can visualize this by imagining cars on a highway: the intensity of traffic is generally expressed as the number of vehicles passing a virtual barrier per unit time. This is a flux. A snapshot of cars on the road at any one time will not tell us whether the traffic is moving or in what direction, but it does give us information on the concentration of different brands of cars on the road. Such a snapshot cannot tell the difference between the invasion of a city and an exodus; nor can it distinguish dense traffic from a parking lot.

There have only been two reports of the measurement of near-surface flux of microorganisms in the atmosphere (Lindemann *et al.*, 1982; Lighthart and Shaffer, 1994). This is in contrast to the notable development of the measurement of flux of gases, and in particular of pollutants and greenhouse gases such as O_3 and CO_2, in order to characterize emissions and their sources. Most of the methods currently deployed for the measurement of gas flux, based on the calculation of eddy covariance, involve high-frequency measurements (on the order of 5–20 Hz; that is, 5–20 measurements per second) of gas concentrations and of micrometeorological parameters, including wind speed and temperature (Langford *et al.*, 2009). These methods can be applied to gases present in the atmosphere at high enough concentrations to be detected in the small volumes of air that are collected in these short time intervals. For rare gases and other aerosols (such as certain volatile organics, various stable isotopes, pesticides, pollen, and microbial aerosols), however, the samples of air collected at such high frequencies will generally be devoid of the target chemical or particle. Therefore, the methods commonly used to measure the flux of abundant gases cannot be used for microorganisms, rare gases, or other rare particles. Research on the atmospheric flux of rare chemicals and particles has thus generally been much slower than that for abundant gases.

The two reported attempts to measure microbial flux in the atmosphere (Lindemann *et al.*, 1982; Lighthart and Shaffer, 1994) deployed a gradient method based on micrometeorological and bacterial concentration measurements at two heights. The field set-ups utilized a vertical structure, several meters in height, on which air samplers and meteorological equipment were installed in order to obtain data at two heights for air density, temperature, and wind speed (Lindemann *et al.*, 1982) and for temperature and sensible heat flux (Lighthart and Shaffer, 1994), in addition to the concentrations of microbial populations. The calculations of flux from these measurements involved computing the difference in microbial concentration at the two heights, coupled to estimations of the atmospheric flow conditions between the two heights based on the meteorological data. These latter estimations are based on some very strong assumptions about atmospheric heat transfer, which pertain to very particular landscape conditions where there is sufficient fetch; that is, where the air movement in the local vicinity is sufficiently undisturbed by topographical features (flat, open regions in particular).

An alternative micrometeorological approach has been developed to measure the flux of particles and rare gases. Relaxed eddy accumulation (REA) is a conditional sampling technique in which samples of air are selectively directed into up- or down-draught reservoirs according to the net direction of the vertical wind velocity at the time of sampling (Businger and Oncley, 1990). It has been deployed for studies of diverse atmospheric constituents, including stable isotopes (Bowling, Tans, and Monson, 2001), mercury (Olofsson *et al.*, 2005), halocarbons (Hornsby *et al.*, 2009), and aerosols in general (Gaman *et al.*, 2004; Held *et al.*, 2008). The air from each reservoir is subsequently analyzed and flux is calculated from the difference in concentration between the two reservoirs.

Efforts are underway to develop REA systems adapted to the measurement of flux of microorganisms. The major challenge of this work will most likely be in adapting it to samplers with sufficient throughput to detect microorganisms that are typically present at concentrations of only tens to thousands of particles per cubic meter of air. First, the measurement of flux requires active particle collection at a known flow rate of air, and therefore is not adapted to the use of passive traps. Passive traps, such as open Petri dishes containing agar medium, are useful in measuring rates of deposition (numbers of particles falling on a given surface per time unit) but cannot be used to measure concentrations (see review by Sesartic and Dallafior, 2011, for example). For active samplers, particles are caught in a stream of air and drawn into the sampler, and ultimately captured by impaction on a solid surface (an agar medium or a filter, for example), caused to impinge into a liquid, or trapped in the cyclone of a swirling liquid. These types of sampler are described in Table 16.1. In the currently available active samplers adapted to recovering microbial particles, flow rates are below $1\ m^3\ min^{-1}$. There is also considerable variation in the efficiency of samplers in terms of their capacity to capture small particles. The cut-off size of a sampler is defined as the particle size below which it collects less than 50% of the available particles. The higher the cut-off size, the lower its efficiency in collecting small particles. A sampler's cut-off size depends on such characteristics as the diameter of its nozzle and the jet velocity through the same. In a study of seven impactors, Yao and Mainelis (2006) concluded that most of the bioaerosol samplers tested substantially underestimated bacterial concentration, especially for single bacterial cells with diameters of $0.5-1.0\ \mu m$. On the other hand, most of the samplers appeared to be suitably efficient for the collection of larger fungal spores.

Sampling efficiency is also related to outside air velocity. The balance between outside air velocity and inside jet velocity can affect collection efficiency. Ideally, the velocity of the air entering a nozzle should be equal to the local wind velocity (isokinetic sampling – see review by Sesartic and Dallafior, 2011, for example). With anisokinetic sampling, there may be over- or under-collection of particles, depending on their size (Nicholson, 1995). Efficiency is also linked to the biological characteristics of the microorganisms. During sampling, microorganisms may lose viability as a result of mechanical stress (impaction) and dehydration (filter collection). If quantification is achieved through the growth of microorganisms on a synthetic nutrient medium, cells that cannot grow into colonies on the medium will not be taken into account.

Near-surface flux describes the net vertical movement of particles in the atmosphere. Horizontal movement also occurs in the earth's boundary layer. Studying the relatively short-distance dispersal (mesoscale dispersal) of various plant pathogens (e.g., within and between cultivated fields) has been a major preoccupation of plant pathology. The typical trend of exponential decay in the density of spores from a suspected source and of the incidence of associated disease has been described for many plant pathogens (Gregory, 1961; Zadocks and Bosch, 1994). More recent work has described the specific atmospheric barriers to mesoscale dispersal (Schmale *et al.*, 2012). In general, mesoscale dispersal occurs in the planetary boundary layer. Long-distance dispersal (LDD), on the other hand, is more challenging to assess and can occur both in the planetary boundary layer and beyond into the free troposphere.

Table 16.1 *The major types of sampler used for microbial aerosols, the processes by which they function, and their advantages and disadvantages.*

Sampler type	Throughput of commercially available samplers	Principle of operation	Advantages	Disadvantages
Passive trap	No through-put	Collection by sedimentation on sticky surface	Simple, cheap	Preferential collection of coarse particles due to settling under gravity Concentration cannot be calculated
Solid impaction	Up to 600 l min^{-1}	Particles caught in a jet and collected on a surface	Concentration can be calculated (number of particles per cubic meter of air)	Loss of viability due to mechanical shock of the particle with the surface Analysis is mainly by culture methods or microscopic observation (not suitable for molecular analysis)
Liquid collection (impingement)	Max. 30 l min^{-1}	Particles caught in a jet and directed into a liquid	Concentration can be calculated (number of particles per cubic meter of air) Dilution of liquid if suspension is too concentrated Liquid sample can be analyzed using a variety of methods (culture, molecular, etc.)	Devices made of glass (fragile for routine use) Liquid evaporation limits sampling duration Not suitable for rare bioaerosols due to low throughout
Cyclone	Up to 1 m^3 min^{-1}	Particles sucked by a centrifugal force and collected in liquid	Concentration can be calculated Dilution of liquid if suspension too concentrated	Liquid evaporation limits sampling duration
Collection on filter	Up to 100 l min^{-1}	Particles sucked and caught on a filter	When polycarbonate or cellulose acetate filters are used, analysis by culture or molecular methods is possible Concentration can be calculated	Possible loss of viability due to dehydration of living cells on filter as a result of the air flow

16.4 Long-Distance Transport: From the Boundary Layer into the Free Troposphere

Long-distance transport of microorganisms is an important issue given their long residence time in the atmosphere. Bacterial residence times (particles of 1–3 μm diameter) have been estimated to be 1 to 2 weeks if they are active as CCN or IN (leading to their sequestering). If they are not removed from the atmosphere, they can have residence times of 100 days or more (Burrows *et al.*, 2009a). For larger microbial particles (9–60 μm in diameter), residence time is likely to be on the order of several days to 2 weeks, depending on the site and conditions of emission (Wilkinson *et al.*, 2011). Air masses can traverse oceans and continents in several days, so these residence times suggest the potential for long-distance movement across the planet.

16.4.1 Scale of Horizontal Long-Distance Transport

Long-distance transport has been defined in terms of horizontal movement. For plant pathogens, it is usually called 'long-distance dispersal' and is defined as the transport of viable forms of the pathogen, capable of causing infection, at a distance of 1000 km or more from the source (Nagarajan and Singh, 1990). Knowledge of the identity and localization of the source is an integral part of this definition. However, it should be noted that the term 'dispersal' is often used in spite of an absence of knowledge about the source. Many cases of LDD of microorganisms over continents or across the planet have been reported (Brown and Hovmøller, 2002). Unlike dust plumes such as those monitored via satellites following a massive dust storm, fire, or volcanic eruption (Flentje *et al.*, 2010), microbial aerosols cannot easily be visualized and traced from an intense, well-defined source. Since it is almost impossible to specifically tag the microorganisms in a suspected source, LDD is never measured directly. Corroborating evidence for LDD involves using analytical techniques based on population genetics, the modeling of air mass trajectories, and remote sensing.

Fungi, and the rusts in particular, have been a major model for the study of LDD. Rusts are obligate biotrophs that are unable to grow on synthetic microbiological media and require specific host plants for their development. Early progress in the study of the dissemination of these fungi was based on the distinctive morphology of their spores – a morphology that also contributes to their aerodynamic buoyancy and propensity for flight. The restricted host range of rust fungi and the specific climatic conditions required for epidemics of the diseases they cause were also useful in inferring the LDD of these fungi. By coupling ground surveys to weather and climate analyses and to remote sensing, two important transatlantic rust-spore LDD events were revealed in the 1970s (Nagarajan and Singh, 1990): spores of *Hemileia vastatrix*, the causal agent of coffee-leaf rust, are suspected to have been carried from the Republic of Angola to Brazil and spores of *Puccinia melanocephala*, the causal agent of sugarcane rust, are suspected to have been transported from Cameroon to the Dominican Republic. In both cases, it was not formally proven that the long-distance transport occurred in the atmosphere, but the assumption is based on analysis of detailed meteorological data (Brown and Hovmøller, 2002). The advent of molecular techniques that substantiate the kinship of microorganisms in separate samples has provided additional evidence for LDD. For example, the results of classical epidemiological studies have suggested that spores of *Puccinia striiformis* f. sp. *tritici*, the fungus that causes yellow rust on wheat, frequently migrate in North-West Europe between the UK, Germany, France, and Denmark. Amplified fragment length polymorphism (AFLP) fingerprinting showed that a single clonal population was present in the four countries, up to 1700 km apart. In five cases, specific pathogen clones were dispersed between the UK and Denmark, and on at least two occasions clones were also spread from the UK to Germany and France (Hovmøller, Justesen, and Brown, 2002).

LDD can also be inferred from modeling of air mass trajectories between suspected microorganism sources and destinations of interest. The most widely used model for aerobiological applications is the hybrid single-particle Lagrangian integrated trajectory (HYSPLIT) model (NOAA ARL; http://ready.arl.noaa.gov/HYSPLIT.php). Retro-trajectories of air masses can provide insights into the potential sources of aerosols. Based on trajectories estimated using HYSPLIT, Prospero *et al.* (2005) reported the interhemispheric transport of viable fungi and bacteria from Africa to the Caribbean with soil dust. By coupling air mass trajectories to molecular data on the similarity of populations at the suspected source and at the destination, Jeon *et al.* (2011) showed that dust storms in the Takla Makan and Gobi Deserts across China affected the airborne bacterial community in Seoul, South Korea, in terms of culturable bacterial concentration, genetic structure, and diversity. The results provide evidence that an Asian dust storm can transport bacteria associated with desert dust over several thousand kilometers.

16.4.2 Altitude of Long-Distance Transport

Although microorganisms have been directly observed in the free troposphere and stratosphere (Imshenetsky *et al.*, 1978; Wainwright *et al.*, 2003; Yang *et al.*, 2008) – above cloud height – as well as in the planetary boundary layer, the altitude at which LDD occurs has generally been inferred from data about atmospheric circulation. For example, in order to explain the viability of fungi and bacteria associated with dust from African deserts and collected in Barbados, Prospero *et al.* (2005) suggested that a substantial part of the transport must take place in the planetary boundary layer (below cloud height). At such altitudes, the relative humidity is higher than that above cloud height, minimizing the effects of desiccation. In contrast, air masses arriving in Barbados from the North Atlantic often sink from the middle troposphere, where temperatures and relative humidity can be very low and ultraviolet (UV) intensity very high – factors that could kill microorganisms. These hypotheses, based on the different origins of air, have been corroborated by the microbial viability observed in their samples. The aerodynamic properties of spores of the Asian soybean rust (*Phakopsora pachyrhizi*) have also been taken to suggest transport in the free troposphere. When transported in the planetary boundary layer, the spores of this fungus generally encounter conditions that cause them to sediment rapidly (Krupa *et al.*, 2006), meaning they are unlikely to travel long distances unless they are transported to higher altitudes. In a study by Krupa *et al.* (2006) of the dispersal of Asian soybean rust from southern Texas and the Yucatan Peninsula to the Midwestern United States, these authors identified vertical motions of air that indicated a ventilation of the boundary layer in upwind areas, suggesting the possible injection of uredospores into the free troposphere, where they could be transported for long distances before settling by wet or dry deposition.

16.5 Interaction of Microbial Aerosols with Atmospheric Processes

Microbial cells have surface and metabolic properties that confer a capacity to interact with physical and chemical processes in the atmosphere. Two of their potential roles in atmospheric processes include an action on radiative forcing via light scattering and absorbance (Jaenicke, 2005) and an effect on condensation via their action as CCN (Ariya and Amyot, 2004). These have not been explored as intensely as have the other suspected roles: the potential for microorganisms to alter cloud chemistry and to induce glaciation of clouds and thereby induce precipitation. Due to the relatively low abundance of microbial aerosols compared to mineral and nonbiological aerosols in the atmosphere, their impact on atmospheric cooling and other consequences of radiative forcing are likely to be insignificant and will not be further discussed here. A wide range of particles in the atmosphere can act as CCN by fostering condensation of atmospheric water vapor on their surfaces, a key step in the formation of clouds. Although the chemical surface properties of particles are critical for CCN activity, particle size has an overriding influence on CCN efficiency above certain dimensions. Particles with a diameter of 1 μm or more are well into the range of sizes that are highly efficient at causing condensation, with size being more important in this regard than surface chemistry (Dusek *et al.*, 2006). It thus seems likely that bacterial, fungal, and algal cells will be inherently good CCN even if they might not be as abundant as the nonbiological CCN. Després *et al.* (2012) summarized the available data on the hygroscopic growth of bacteria, fungal spores, pollen, and algae, which confirm the capacity of microbiological particles to condense and accumulate water on their surfaces but show that it does not rival the capacity of atmospheric hygroscopic salts.

There has been intense research on the potential for microorganisms, and bacteria in particular, to contribute to the oxidative reactions and other transformations of organic compounds in the atmosphere (Delort *et al.*, 2010). These reactions are particularly important in the acidification of the atmosphere, in the subsequent formation of secondary aerosols, and in their ensuing impacts on climate, air quality, and geochemical cycles. The bulk of the chemical reactions that occur in the atmosphere are driven by photochemistry and permit transformation of chemicals in dry suspension. However, photochemical reactions come to a halt at night. Furthermore, the wet conditions of cloud droplets would permit the action of microbial enzymes. Hence, under certain conditions (in clouds at night), the metabolic activity of microorganisms suspended in wet aerosol particles has a catalytic potential similar to that of photochemistry (Delort *et al.*, 2010). In addition, as we pointed out in a previous work (Morris *et al.*, 2011), it is likely that microbial cells also participate in nonmetabolic chemical modification of atmospheric chemistry. Molecules can be desorbed from biological surfaces (Cote, Kos, and Ariya, 2008), chemicals can be released during cell lysis, and cells can participate in collision-coalescence processes, for example.

The potential for microorganisms to catalyze the freezing of cloud droplets has received by far the most attention of all the types of interaction that microbial cells can have with the atmosphere. The temperature of spontaneous freezing of pure water is about −37 °C (Murray *et al.*, 2010). At temperatures below 0 °C but above the spontaneous freezing temperature, water does not freeze unless a catalyst acts to help it overcome the thermodynamic barriers to changing from the metastable super-cooled liquid phase to the solid phase. A wide range of particles naturally present in the atmosphere can catalyze freezing at temperatures warmer than the spontaneous freezing temperature of water (Mossop, 1963;

Szyrmer and Zawadzki, 1997). But in the absence of a sufficient quantity of IN active at the temperature of the cloud, cloud droplets are able to super cool, even in clouds approaching temperatures of −37 °C. If they do not freeze, droplets in clouds in mid-latitudes can remain suspended, as there is no effective means for them to aggregate into larger, heavier precipitation-sized ice particles capable of falling toward the ground as rain or snow. Under some conditions, such as heavy rainfalls and tropical storms, sufficiently large cloud droplets can collide efficiently and coalesce to form raindrops directly in the liquid phase. In these cases, freezing is not a critical step in the formation of rain. Because microbial species with ice nucleation activity have been found in clouds (Sands *et al.*, 1982; Amato *et al.*, 2005; Attard *et al.*, 2012), and because they have the potential to catalyze ice formation at relatively warm temperatures where the bulk of the inert atmospheric IN are impotent (at temperatures warmer than −10 °C) (Szyrmer and Zawadzki, 1997; Morris, Georgakapoulos, and Sands, 2004; Després *et al.*, 2012), it is easy to understand that there is considerable excitement around the potential role of biological ice nucleators in this window where the minimum cloud temperature does not fall below −10 °C.

The real-life potential of atmospheric microorganisms to influence rainfall has been inferred from the processes and phenomena just described. This reasoning has been further bolstered by experimental data showing that seeding clouds with ice-nucleation-active bacteria can lead to their glaciation, both for clouds created in laboratory chambers (Möhler *et al.*, 2008) and for naturally-occurring clouds (Ward and DeMott, 1989). These results clearly support the hypothesis that ice-nucleation-active microorganisms can induce freezing of cloud water droplets under conditions typical of mid-latitude clouds. Under these conditions, freezing of droplets is a requisite for their accumulation into drops of sufficient size to fall from clouds as precipitation. However, ice crystals in clouds are also necessary to setting off the process required for lightning. Recent work has deployed regional atmospheric modeling to evaluate the impact of ice-nucleation-active bacteria on the electrification of clouds and the total number of lightning flashes (Gonçalves *et al.*, 2011). This work suggests that under certain conditions, the presence of ice-nucleation-active bacteria in clouds leads to a reduction in precipitation but an increase in the formation of ice, which subsequently contributes to an increased number of lightning flashes.

16.6 Implications of Aerial Transport for Microbial Evolutionary History

Evaluating the importance of aerial transport to habitats that are invaded by airborne microorganisms or to the health of plants and animals suffering from the diseases they cause is an essential preoccupation of epidemiology. The converse issue – the way in which aerial dispersal impacts the evolutionary history of the microorganisms themselves – is also a fundamental problem that is receiving increased attention. It is reasonable to assume that microorganisms capable of surviving the seemingly stressful conditions of intense UV exposure, drying, and rather cold temperatures will have traits that reflect biological adaptation to them. However, the capacity to survive these conditions is part of the complex biology of microorganisms that proliferate in environments very different from the extremes of the atmosphere. As a consequence of the balancing act of the different aspects of the life of a microorganism, it is likely that the functional traits that lead to their survival are much more complex than what can be measured by direct tests of desiccation or UV resistance, for example. We expect that genomic and metabolomic analyses of microorganisms – approaches that integrate disparate functions in the cellular biology of these organisms – will provide novel insight that can move our understanding beyond the current, somewhat piecemeal, perspective. Nevertheless, in the following analysis of how aerobiology impacts the evolution of microorganisms, we will reason with the data currently available.

Obligate biotrophy is a trait that is likely to have an important link to aerial dispersal. Obligately biotrophic organisms need a living host in order to proliferate; they cannot survive on dead materials. Therefore, such organisms must be assured of finding their hosts either by tethering themselves to them (as with certain parasitic plants) or by being able to disperse sufficiently to encounter them across their ecological range. Movement is critically important to obligately biotrophic fungi (such as those responsible for rusts, powdery mildews, and some downy mildews) that attack annual plants or plants such as deciduous trees, whose susceptible tissue has annual cycles. Some of these fungi produce overwintering spores, but the airborne spores are predominant in their life cycle. Tobacco blue mold, caused by *Peronospora tabacina*, spreads northward in the USA during the growing season, along an advancing disease wave front from the southernmost to the northernmost tobacco growing areas, via the production of airborne spores (Aylor, 2003). Progression of wheat stem rust, caused by *Puccina graminis*, from Mexico to Canada through what has been called the Wheat Belt of North America, is also based on the seasonal reintroduction of spores into northern growing areas from the southern latitudes (Aylor, 2003). These are just a couple of the many examples of the dissemination of obligately biotrophic fungi.

The expansion of crops that has taken place since the advent of agriculture from 10 000 to 5000 years ago and the subsequent homogenization of the genotypes of the plants (crops) that cover the earth's surface have likely reinforced the positive natural capacity of such fungi to disseminate aerially. If it were possible to identify specific genetic markers of the capacity for aerial dissemination, it would be exciting to compare modern grain rust fungi with the traces of those that might be captured in natural archives such as ice cores. However, at present we can only surmise, from the rapidity of the emergence of new virulent races of rust fungi, that the homogenization of the crop genetic landscape has reinforced the propensity for aerial dissemination. When a resistance gene is introduced widely in crop plants, if a mutation to virulence happens in a single genetic line of the pathogen that effectively disperses via LDD then previously resistant cultivars will become susceptible across large areas where they are cultivated and the pathogen will thereby increase its ecological range (Brown and Hovmøller, 2002). Such a phenomenon has occurred for wheat yellow leaf rust (*Puccinia striiformis*) in North-West Europe. In the early 1990s, a resistance gene was widely introduced into wheat in order to control this rust. The first rust outbreak on wheat harboring this resistance gene was reported in the UK in 1994, involving one pathogen genotype. In 1997, isolates with this genotype were detected in Denmark and France, and one year later they were found in Germany. In 1995, another virulent genotype of *P. striiformis* was found in England, and 2 years later it was found elsewhere in North-West Europe, at sites up to 1700 km apart. The most likely explanation for the widespread distribution of both virulent genotypes across North-West Europe is wind dispersal of spores from the UK (Brown and Hovmøller, 2002). The introduction of disease-resistance genes in cultivated plants is considered a critical factor in the emergence of new microorganism pathotypes, which develop the means to overcome the resistance. However, it might very well be that the widespread use of these resistance genes is also enhancing aerial dissemination. In order to assess this impact, it would be very interesting to determine whether there is a relationship between the aerodynamic properties of obligate biotrophic fungi, for example, and the spatial distribution of their hosts, particularly among the pathogens of wild and widely cultivated graminaceous species.

For the cases of LDD mentioned earlier (sugar cane and coffee rusts), the capacity to move to new regions where the host was located led to the apparent emergence of the diseases where they had not been seen before. For obligate biotrophs, this invasive capacity would require that a sensitive host species was present in the new region or that the pathogen had sufficient mutation rates to ensure the emergence of variants capable of living on the newly encountered plant species. But for microorganisms capable of saprophytic growth (i.e., most fungi and bacteria) or with large host ranges, the capacity to proliferate successfully in a new region could be achieved through versatility or polyphagy. The plant pathogenic bacterium *Pseudomonas syringae* seems to be affiliated with the water cycle, moving upward in aerosols from plant surfaces into clouds and depositing with snow and rainfall (Morris *et al.*, 2008). Surprisingly, the genotype of this bacterium found in the widest range of environmental substrates is also that with the greatest versatility in terms of the number of different plant species that it can attack (Morris *et al.*, 2010). This is a seeming paradox because one would expect that there would be some fitness cost for being a generalist. However, the overall versatility of this genomic group of *P. syringae* is likely due to multifunctional toxins, as indicated by the unusual abundance of genes for toxins and the scarcity of host-specific virulence factors in this group compared to those in the other genetic groups of this bacterium (Baltrus *et al.*, 2011). In other words, this widely dispersed group of *P. syringae* has a somewhat different evolutionary history to the rest of the genomic groups of *P. syringae*, in that it has accumulated fewer specific virulence factors and more toxins than the other groups, the latter providing wide adaptability at a relatively low cost. Is this related in any way to its being dispersed by the atmosphere? It should be noted that this same genetic group of *P. syringae* also contains the most highly effective ice nucleators in this species as a whole (Morris *et al.*, 2010). Furthermore, based on measures of the frequency of ice-nucleation-active strains of this bacterium from its various substrates and habitats, it has been observed that precipitation, particularly snow, provides the strongest positive selection for ice nucleation activity of all of the substrates studied. These observations led to the proposal that ice nucleation activity, by allowing the bacterium to cause ice formation and subsequently precipitation, is an active form of aerial transport, ensuring the deposition of this bacterium (Morris *et al.*, 2010). If so, cells depositing as rain or snowfall would be deposited in a wide range of environmental contexts. Therefore, adaptability would be necessary to ensuring the survival and proliferation of bacteria benefiting from this means of aerial transport.

References

Amato, P., Menager, M., Sancelme, M. *et al.* (2005) Microbial population in cloud water at the Puy de Dôme: implications for the chemistry of clouds. *Atmospheric Environment*, **39**, 4143–4153.

Amato, P., Parazols, M., Sancelme, M. *et al.* (2007) Microorganisms isolated from the water phase of tropospheric clouds at the Puy de Dôme: major groups and growth abilities at low temperatures. *FEMS Microbiology Ecology*, **59**, 242–254.

Ariya, P.A. and Amyot, M. (2004) New directions: the role of bioaerosols in atmospheric chemistry and physics. *Atmospheric Environment*, **38**, 1231–1232.

Attard, E., Yang, H., Delort, A.-M. *et al.* (2012) Effects of atmospheric conditions on ice nucleation activity of Pseudomonas. *Atmospheric Chemistry and Physics*, **12**, 10667–10677.

Aylor, D.E. (1990) The role of intermittent wind in the dispersal of fungal pathogens. *Annual Review of Phytopathology*, **28**, 73–92.

Aylor, D.E. (2003) Spread of plant disease on a continental scale: role of aerial dispersal of pathogens. *Ecology*, **84**, 1989–1997.

Baltrus, D.A., Nishimura, M.T., Romanchuk, A. *et al.* (2011) Dynamic evolution of pathogenicity revealed by sequencing and comparative genomics of 19 *Pseudomonas syringae* isolates. *PLoS Pathogens*, **7**, e1002132.

Bowers, R.M., McLetchie, S., Knight, R. and Fierer, N. (2011) Spatial variability in airborne bacterial communities across land-use types and their relationship to the bacterial communities of potential source environments. *ISME Journal*, **5**, 601–612.

Bowling, D.R., Tans, P.P. and Monson, R.K. (2001) Partitioning net ecosystem carbon exchange with isotopic fluxes of CO_2. *Global Change Biology*, **7**, 127–145.

Brodie, E.L., DeSantis, T.Z., Parker, J.P.M. *et al.* (2007) Urban aerosols harbor diverse and dynamic bacterial populations. *Proceedings of the National Academy of Sciences of the United States of America*, **104**, 299–304.

Brown, J.K. and Hovmøller, M.S. (2002) Aerial dispersal of pathogens on the global and continental scales and its impact on plant disease. *Science*, **297**, 537–541.

Burrows, S.M., Butler, T., Jöckel, P. *et al.* (2009a) Bacteria in the global atmosphere – Part 2: modelling of emissions and transport between different ecosystems. *Atmospheric Chemistry and Physics*, **9**, 9281–9297.

Burrows, S.M., Elbert, W., Lawrence, M.G. and Pöschl, U. (2009b) Bacteria in the global atmosphere – Part 1: review and synthesis of literature data for different ecosystems. *Atmospheric Chemistry and Physics*, **9**, 9263–9280.

Businger, J.A. and Oncley, S.P. (1990) Flux measurement with conditional sampling. *Journal of Atmospheric and Oceanic Technology*, **7**, 349–352.

Capanna, E. (1999) Lazzaro Spallanzani: at the roots of modern biology. *Journal of Experimental Zoology*, **285**, 178–196.

Cherewick, W.J. (1953) Association of ergot with loose smut of wheat and of barley. *Phytopathology*, **43**, 461–463.

Cote, V., Kos, G. and Ariya, P.A. (2008) Microbial and de novo transformation of dicarboxylic acids by three airborne fungi: *Aspergillus fumigatus*, *Aspergillus flavus*, and *Penicillium chrysogenum*. *Science of the Total Environment*, **390**, 530–537.

Delort, A.-M., Vaïtilingom, M., Amato, P. *et al.* (2010) A short overview of the microbial population in clouds: potential roles in atmospheric chemistry and nucleation processes. *Atmospheric Research*, **98**, 249–260.

Després, V.R., Huffman, J.A., Burrows, S.M. *et al.* (2012) Primary biological aerosol particles in the atmosphere: a review. *Tellus*, **B 64**, 015598. doi: 015510.013402/tellusb.v015564i015590.015598.

Doherty, M.A. and Preece, T.F. (1978) *Bacillus cereus* prevents germination of uredospores of *Puccinia allii* and the development of rust disease of leek, *Allium porrum*, in controlled environments. *Physiological Plant Pathology*, **12**, 123–132.

Dusek, U., Frank, G.P., Hildebrandt, L. *et al.* (2006) Size matters more than chemistry for cloud-nucleating ability of aerosol particles. *Science*, **312**, 1375–1378.

Elbert, W., Taylor, P.E., Andreae, M.O. and Poeschl, U. (2007) Contribution of fungi to primary biogenic aerosols in the atmosphere: wet and dry discharged spores, carbohydrates, and inorganic ions. *Atmospheric Chemistry and Physics*, **7**, 4569–4588.

Fischer, M.W.F., Stolze-Rybczynski, J.L., Cui, Y. and Money, N.P. (2010) How far and how fast can mushroom spores fly? Physical limits on ballistospore size and discharge distance in the Basidiomycota. *Fungal Biology*, **114**, 669–675.

Fitt, B.D.L., Creighton, N.F. and Bainbridge, A. (1985) Role of wind and rain in dispersal of *Botrytis fabae* conidia. *Transactions of the British Mycological Society*, **85**, 307–312.

Flentje, H., Claude, H., Elste, T. *et al.* (2010) The Eyjafjallajökull eruption in April 2010 – detection of volcanic plume using in-situ measurements, ozone sondes and lidar-ceilometer profiles. *Atmospheric Chemistry and Physics*, **10**, 10085–10092.

French, R.C., Novotny, J.D. and Searles, R.B. (1964) Properties of bacteria isolated from wheat stem rust spores. *Phytopathology*, **54**, 970–973.

Frohlich-Nowoisky, J., Pickersgill, D.A., Despres, V.R. and Pöschl, U. (2009) High diversity of fungi in air particulate matter. *Proceedings of the National Academy of Sciences of the United States of America*, **106**, 12814–12819.

Fuentes, E., Coe, H., Green, D. *et al.* (2010) Laboratory-generated primary marine aerosol via bubble-bursting and atomization. *Atmospheric Measurement Techniques*, **3**, 141–162.

Gaman, A., Rannik, Ü., Aalto, P. *et al.* (2004) Relaxed eddy accumulation system for size-resolved aerosol particle flux measurements. *Journal of Atmospheric and Oceanic Technology*, **21**, 933–943.

Gonçalves, F.L.T., Martins, J.A., Albrecht, R.I. *et al.* (2011) Effect of bacterial ice nuclei on the frequency and intensity of lightning activity inferred by the BRAMS model. *Atmospheric Chemistry and Physics Discussions*, **11**, 26143–26171.

Górny, R.L., Reponen, T., Grinshpun, S.A. and Willeke, K. (2001) Source strength of fungal spore aerosolization from moldy building material. *Atmospheric Environment*, **35**, 4853–4862.

Gregory, P.H. (1961) *The Microbiology of the Atmosphere*, Interscience Publishers, Inc., New York.

Gudger, E.W. (1929) I. More rains of fishes. *Journal of Natural History Series*, **10**(3), 1–26.

Harrison, J.G. and Lowe, R. (1987) Wind dispersal of conidia of *Botrytis* spp. pathogenic to *Vicia faba*. *Plant Pathology*, **36**, 5–15.

Heald, C.L. and Spracklen, D.V. (2009) Atmospheric budget of primary biological aerosol particles from fungal spores. *Geophysical Research Letters*, **36**, L09806. doi: 09810.01029/02009GL037493.

Held, A., Patton, E., Rizzo, L. *et al.* (2008) Relaxed eddy accumulation simulations of aerosol number fluxes and potential proxy scalars. *Boundary-Layer Meteorology*, **129**, 451–468.

Hornsby, K.E., Flynn, M.J., Dorsey, J.R. *et al.* (2009) A relaxed eddy accumulation (REA)-GC/MS system for the determination of halocarbon fluxes. *Atmospheric Measurement Techniques*, **2**, 437–448.

Hovmøller, M.S., Justesen, A.F. and Brown, J.K.M. (2002) Clonality and long-distance migration of *Puccinia striiformis* f.sp. *tritici* in north-west Europe. *Plant Pathology*, **51**, 24–32.

Imshenetsky, A.A., Lysenko, S.V. and Kazakov, G.A. (1978) Upper boundary of the biosphere. *Applied and Environmental Microbiology*, **35**, 1–5.

Ingold, C.T. (1965) *Spore Liberation*, Clarendon Press, Oxford.

Jaenicke, R. (2005) Abundance of cellular material and proteins in the atmosphere. *Science*, **308**, 73.

Jarvis, W.R. (1962) The dispersal of spores of *Botrytis cinerea* fr. in a raspberry plantation. *Transactions of the British Mycological Society*, **45**, 549–559.

Jeon, E.M., Kim, H.J., Jung, K. *et al.* (2011) Impact of Asian dust events on airborne bacterial community assessed by molecular analyses. *Atmospheric Environment*, **45**, 4313–4321.

Jones, A.M. and Harrison, R.M. (2004) The effects of meteorological factors on atmospheric bioaerosol concentrations – a review. *Science of the Total Environment*, **326**, 151–180.

Kanaani, H., Hargreaves, M., Ristovski, Z. and Morawska, L. (2009) Fungal spore fragmentation as a function of airflow rates and fungal generation methods. *Atmospheric Environment*, **43**, 3725–3735.

Krupa, S., Bowersox, V., Claybrooke, R. *et al.* (2006) Introduction of soybean rust spores into the Midwestern United States – a case study. *Plant Disease*, **90**, 1254–1259.

Lacey, M.E. and West, J.W. (2006) *The Air Spora*, Springer, Dordrecht.

Langford, B., Davison, B., Nemitz, E. and Hewitt, C.N. (2009) Mixing ratios and eddy covariance flux measurements of volatile organic compounds from an urban canopy (Manchester, UK). *Atmospheric Chemistry and Physics*, **9**, 1971–1987.

Leach, C. (1987) Diurnal electrical potentials of plant leaves under natural conditions. *Environmental and Experimental Botany*, **27**, 419–430.

Leck, C. and Bigg, K. (2005) Biogenic particles in the surface microlayer and overlaying atmosphere in the central Arctic Ocean during summer. *Tellus*, **57B**, 305–316.

Lee, J.H., Hwang, G.B., Jung, J.H. *et al.* (2010) Generation characteristics of fungal spore and fragment bioaerosols by air flow control over fungal cultures. *Journal of Aerosol Science*, **41**, 319–325.

Li, X., Yang, X., Mo, J. and Guo, T. (2009) Estimation of soybean rust uredospore terminal velocity, dry deposition, and the wet deposition associated with rainfall. *European Journal of Plant Pathology*, **123**, 377–386.

Lighthart, B. (1997) The ecology of bacteria in the alfresco atmosphere. *FEMS Microbiology Ecology*, **23**, 263–274.
Lighthart, B. (1999) An hypothesis describing the general temporal and spatial distribution of alfresco bacteria in the earth's atmospheric surface layer. *Atmospheric Environment*, **33**, 611–615.
Lighthart, B. and Shaffer, B.T. (1994) Bacterial flux from chaparral into the atmosphere in mid-summer at a high desert location. *Atmospheric Environment*, **28**, 1267–1274.
Lindemann, J., Constantinidiou, H.A., Barchet, W.R. and Upper, C.D. (1982) Plants as source of airbone bacteria, including ice nucleation-active bacteria. *Applied and Environmental Microbiology*, **44**, 1059–1063.
Mahowald, N., Ward, D.S., Kloster, S. *et al.* (2011) Aerosol impacts on climate and biogeochemistry. *Annual Review of Environment and Resources*, **36**, 45–74.
McCartney, H.A. (1991) Airborne dissemination of plant pathogens. *Journal of Applied Bacteriology*, **70**, 49–59.
Meridith, D.S. (1973) Significance of spore release and dispersal mechanisms in plant disease epidemiology. *Annual Review of Phytopathology*, **11**, 313–342.
Möhler, O., Georgakopoulos, D.G., Morris, C.E. *et al.* (2008) Heterogeneous ice nucleation activity of bacteria: new laboratory experiments at simulated cloud conditions. *Biogeosciences*, **5**, 1425–1435.
Morris, C.E., Georgakapoulos, D. and Sands, D.C. (2004) Ice nucleation active bacteria and their potential role in precipitation. *Journal de Physique Iv, France*, **121**, 87–103.
Morris, C.E., Sands, D.C., Bardin, M. *et al.* (2011) Microbiology and atmospheric processes: research challenges concerning the impact of airborne micro-organisms on the atmosphere and climate. *Biogeosciences*, **8**, 17–25.
Morris, C.E., Sands, D.C., Vanneste, J.L. *et al.* (2010) Inferring the evolutionary history of the plant pathogen *Pseudomonas syringae* from its biogeography in headwaters of rivers in North America. *Europe and New Zealand mBio*, **1**(3), e00107-10–e00107-20.
Morris, C.E., Sands, D.C., Vinatzer, B.A. *et al.* (2008) The life history of the plant pathogen *Pseudomonas syringae* is linked to the water cycle. *ISME Journal*, **2**, 321–334.
Mossop, S.C. (1963) Atmospheric ice nuclei. *Zeitschrift für Angewandte Mathematik und Physik*, **14**, 456–486.
Murray, B.J., Broadley, S.L., Wilson, T.W. *et al.* (2010) Kinetics of the homogeneous freezing of water. *Physical Chemistry Chemical Physics*, **12**, 10380–10387.
Nagarajan, S. and Singh, D.V. (1990) Long-distance dispersion of rust pathogens. *Annual Review of Phytopathology*, **28**, 139–153.
Nicholson, K.W. (1995) Physical aspects of bioaerosol sampling and deposition, in *Bioaerosols Handbook* (eds C.S. Cox and C.M. Wathes), CRC Press, Boca Raton, FL, pp. 27–53.
O'Dowd, C.D. and de Leeuw, G. (2007) Marine aerosol production: a review of the current knowledge. *Philosophical Transactions of the Royal Society*, **365**, 2007–2043.
Olofsson, M., Sommar, J., Ljungström, E. *et al.* (2005) Application of relaxed eddy accumulation technique to quantify Hg0 fluxes over modified soil surfaces. *Water Air and Soil Pollution*, **167**, 331–352.

Pasteur, L. (1890) Mémoire sur les corpsuscules organisés qui existent dans l'atmosphère. *Annales de Chimie et de Physique*, **3**, 5–110.

Prospero, J.M., Blades, E., Mathison, G. and Naidu, R. (2005) Interhemispheric transport of viable fungi and bacteria from Africa to the Caribbean with soil dust. *Aerobiologia*, **21**, 1–19.

Reponen, T., Grinshpun, S.A., Conwell, K.L. *et al.* (2001) Aerodynamic versus physical size of spores: measurement and implication for respiratory deposition. *Grana*, **40**, 119–125.

Sands, D.C., Langhans, V.E., Scharen, A.L. and de Smet, G. (1982) The association between bacteria and rain and possible resultant meteorological implications. *Journal of the Hungarian Meteorological Service*, **86**, 148–152.

Schmale, D.G. III,, Ross, S.D., Fetters, T.L. *et al.* (2012) Isolates of *Fusarium graminearum* collected 40–320 meters above ground level cause *Fusarium* head blight in wheat and produce trichothecene mycotoxins. *Aerobiologia*, **28**, 1–11.

Sesartic, A. and Dallafior, T.N. (2011) Global fungal spore emissions, review and synthesis of literature data. *Biogeosciences*, **8**, 1181–1192.

Sharma, N.K., Rai, A.K., Singh, S. and Brown, R.M. (2007) Airborne algae: their present status and relevance. *Journal of Phycology*, **43**, 615–627.

Simon, X., Duquenne, P., Koehler, V. *et al.* (2011) Aerosolisation of *Escherichia coli* and associated endotoxin using an improved bubbling bioaerosol generator. *Journal of Aerosol Science*, **42**, 517–531.

Sivasubramani, S.K., Niemeier, R.T., Reponen, T. and Grinshpun, S.A. (2004) Fungal spore source strength tester: laboratory evaluation of a new concept. *Science of the Total Environment*, **329**, 75–86.

Stakman, E. and Christensen, C.M. (1946) Aerobiology in relation to plant disease. *The Botanical Review*, **12**, 205–253.

Stakman, E.C., Henry, A.W., Curran, G.C. and Christopher, W.N. (1923) Spores in the upper air. *Journal of Agricultural Research*, **24**, 599–606.

Szyrmer, W. and Zawadzki, I. (1997) Biogenic and anthropogenic sources of ice-forming nuclei: a review. *Bulletin of the American Meteorological Society*, **78**, 209–228.

Tallapragada, P., Ross, S.D. and Schmale, D.G. III, (2011) Lagrangian coherent structures are associated with fluctuations in airborne microbial populations. *Chaos: An Interdisciplinary Journal of Nonlinear Science*, **21**, 033122.

Wainwright, M., Wickramasinghe, N.C., Narlikar, J.V. and Rajaratnam, P. (2003) Microorganisms cultured from stratospheric air samples obtained at 41 km. *FEMS Microbiology Letters*, **218**, 161–165.

Ward, P.J. and DeMott, P.J. (1989) Preliminary experimental evaluation of Snomax snow inducer, *Pseudomonas syringae*, as an artificial ice nucleus for weather modification. *Journal of Weather Modification*, **21**, 9–13.

Wilkinson, D.M., Koumoutsaris, S., Mitchell, E.A.D. and Bey, I. (2011) Modelling the effect of size on the aerial dispersal of microorganisms. *Journal of Biogeography*, **39**, 89–97.

Willoquet, L., Berud, F., Raoux, L. and Clerjeau, M. (1998) Effects of wind, relative humidity, leaf movement and colony age on dispersal of conidia of Uncinula necator, causal agent of grape powdery mildew. *Plant Pathology*, **47**, 234–242.

Yang, Y., Itahashi, S., Yokobori, S.-I. and Yamagishi, A. (2008) UV-resistant bacteria isolated from upper troposphere and lower stratosphere. *Biological Sciences in Space*, **22**, 18–25.

Yao, M. and Mainelis, G. (2006) Investigation of cut-off sizes and collection efficiencies of portable microbial samplers. *Aerosol Science and Technology*, **40**, 595–606.

Zadocks, J.C. and Bosch, F.V. (1994) On the spread of plant disease: a theory on foci. *Annual Review of Phytopathology*, **33**, 503–521.

17

Disinfection of Airborne Organisms by Ultraviolet-C Radiation and Sunlight

Jana S. Kesavan and Jose-Luis Sagripanti
Edgewood Chemical Biological Center, Aberdeen Proving Ground, USA

17.1 Introduction

Bioaerosols are a serious concern to human health because of their potential role in the transmission of infectious diseases during natural epidemics or after intentional release of biological agents through acts of terrorism or warfare. Extensive research has been conducted toward understanding the disinfection of organisms deposited on surfaces, suspended in water, and contaminating food, all of which have been discussed elsewhere (Block, 2001). In contrast, the disinfection of microorganisms in aerosols has received much less attention, mainly because of the difficulties inherent in conducting the experiments. Airborne organisms are difficult to reach with the liquid disinfectants that are commonly used to treat medical devices, foods, and drinking water. Therefore, the primary means of organism inactivation in aerosols is ultraviolet (UV) radiation.

Radiation from sunlight is used as a disinfectant to kill organisms. Radiation of <290 nm wavelength is absorbed by atmospheric ozone before sunlight reaches the earth's surface; therefore, the longer UV wavelengths in sunlight provide the main germicidal agent in the environment (Jagger, 1985; Giese, 1976; Lytle and Sagripanti, 2005). Germicidal UV lights are relatively easy to use indoors; hence, light with 254 nm wavelength is frequently used to disinfect the air and surfaces in health-care and biomedical industrial settings.

Aerosol Science: Technology and Applications, First Edition. Edited by Ian Colbeck and Mihalis Lazaridis.

Organisms appear to be more susceptible to UV radiation when they are suspended in air (King, Kesavan, and Sagripanti, 2011). For this reason, conclusions derived from studies of organisms on liquids or surfaces are generally not transferable to aerosolized organisms. This increased susceptibility is likely a result of the comparatively easier mixing in air, combined with its lower rate of UV absorption. In addition, the aerosolization process itself may weaken or damage an organism, thereby rendering it more susceptible to UV irradiation. The effects of dehydration and oxygenation may also contribute to the increased vulnerability of airborne microbes.

This report provides the background to and an overview of UV radiation, selected organisms of potential interest to aerosol scientists and engineers, and the effects of UV light on aerosolized organisms. The aerosol generation methodology and equipment and the test methodology used at the US Army Edgewood Chemical Biological Center (ECBC) are provided as examples for researchers to compare to their own experimental conditions and as guidance to those newer to this field.

17.2 UV Radiation

Electromagnetic radiation is a fluctuation of electric and magnetic fields in space and is classified according to the frequency of these fluctuations as: radio waves; microwaves; infrared (IR), visible, and UV waves; X-rays; and γ-rays. The basic unit of electromagnetic radiation is the photon, which has no mass or electric charge and travels at the speed of light in a wavelike pattern. UV radiation is a non-ionizing radiation that is emitted by atoms when their electrons descend from higher to lower energy states to yield photons of specific wavelength and energy range. The emission of UV radiation from a mercury gas-filled light is illustrated in Figure 17.1. In this system, two high-voltage electrodes are placed in a mercury-filled chamber. The arc between the electrodes provides the energy that allows the electrons of the mercury atoms to enter a higher energy state, and photons are released as these electrons fall back to their basal low-energy states.

The UV wavelength region is between the visible and the X-ray regions. The lower wavelength of the UV region is 10 nm and the upper wavelength limit is 380 nm (or 400 nm,

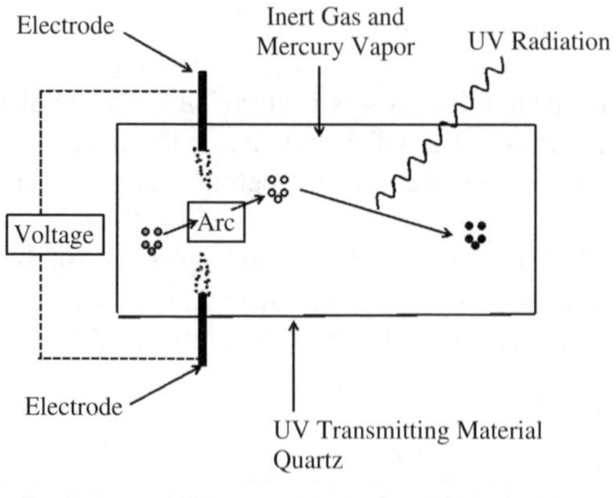

Figure 17.1 *Release of UV radiation.*

Table 17.1 Wavelength ranges of UV regions.

Region	Wavelength (nm)
Far or vacuum UV	10–190
UV-C	190–290
UV-B	290–320
UV-A	320–380

depending on the author), which is the beginning of the range that is visible to human eyes. As shown in Table 17.1, the spectrum of UV light is divided into four regions: far UV or vacuum UV, UV-A, UV-B, and UV-C. These regions have markedly different effects on microorganisms. Far-UV (<190 nm wavelength) radiation from the sun is absorbed by the atmosphere and does not reach the earth. For this reason, far-UV radiation has little relevance to the inactivation of organisms within the environment. Far-UV wavelengths are highly deleterious to all life forms, but they are difficult to produce in the laboratory.

The precise boundaries between UV-A, UV-B, and UV-C often vary among authors and depend on whether the division is based on health effects or on the physical principles of the illumination sources. In general, from a health perspective, UV-B and UV-C wavelengths are defined as germicidal, while UV-A radiation is considered nongermicidal. UV-A radiation includes the wavelengths between 320 and 380 (or 400) nm and is used in tanning lamps. UV-B radiation spans the wavelengths between 290 and 320 nm. UV-C radiation contains the wavelengths between 190 and 290 nm. An alternate definition by the International Commission on Illumination (Vienna, Austria) defines UV-B radiation as having a wavelength between 280 and 315 nm.

The availability of low- and medium-pressure mercury lamps makes UV radiation relatively easy and inexpensive to obtain and use. Low-pressure mercury lamps are referred to as 'monochromatic' in that they produce a narrow band of UV-C radiation at a wavelength of about 254 nm. Medium-pressure mercury lamps are described as 'polychromatic' and produce a broader but flatter spectrum that extends from about 200 to 400 nm.

Organisms are affected by UV intensity (flux) and exposure time, and the multiplication of these is called the UV exposure (fluence). UV meters measure the flux (in watts per square meter), and this is multiplied by exposure time (seconds) to obtain the fluence (in joules per square meter). In addition to direct radiation from the sun or from lamps, reflection from surfaces and particles in the environment is also important. Highly reflective environments such as white sand and fresh snow can significantly increase the UV irradiance in areas shaded from direct illumination and thus disinfect organisms effectively (Ben-David and Sagripanti, 2010).

17.3 Sunlight

The increasing focus on epidemiology, biodefense, and public health has led increasing numbers of scientists to become interested in evaluating organism decontamination via sunlight exposure (Coohill and Sagripanti, 2009). The germicidal effect of solar radiation has been known for many years and was described as early as 1878 (Black and Veatch

Corporation, 2010). The role of sunlight in human epidemics was demonstrated by an increase in hospital admissions for influenza during the burning season in Brazil, when smoke blocked the UV irradiation and reduced the inactivation of viruses in the air (Mims, 2005). Sunlight is known to kill organisms, but conducting germicidal experiments using direct sunlight is challenging. Sunlight is difficult to control and the risk of inadvertent contamination of samples is especially high outdoors.

Many factors affect the amount of UV radiation that reaches the earth's surface, including latitude, altitude, solar zenith angle (the angle between the sun and zenith during the day), the day of the year, and weather conditions (Lytle and Sagripanti, 2005). Environmental parameters such as total column ozone, the presence of clouds, pollutants, dusts, and aerosols, and reflection from the ground also affect radiation levels (Ben-David and Sagripanti, 2010). It is important to note that the atmospheric ozone layer absorbs radiation of wavelengths <290 nm, completely eliminating germicidal UV-C radiation and decreasing the amount of UV-B radiation that reaches the ground. For example, in regions with less atmospheric ozone, more UV-B radiation reaches the ground (Puskeppeleit *et al.*, 1992; Lubin and Jensen, 1995). Additionally, the amount of UV-B radiation that reaches the earth is greater at high altitudes, low latitudes, during summer months, and peaks at solar noon.

Researchers have used many methods to remove portions of the sun's spectrum in order to allow evaluation of the effects of UV-A and UV-B radiation both separately and in combination. For example, a single layer of Saran Wrap (Dow Chemical Company, Midland, MI) allows penetration of UV-A and UV-B radiation but protects samples from environmental contaminants during exposure (Riesenman and Nicholson, 2000). In the same study, a 1.25 cm-thick glass plate was used to block UV-B and allow UV-A radiation to reach the samples. In addition to natural solar radiation, artificial lights that simulate sunlight are also used in experiments. Exposure to artificial light is easier to control during experiments conducted inside the laboratory. Furthermore, medium-pressure mercury arc lamps that emit a UV spectrum in the range of 280–320 nm (peak at 302 nm) wavelength can be used with filters that block wavelengths below 290 nm to yield the UV-B wavelengths (Riesenman and Nicholson, 2000).

Many researchers have measured the UV radiation doses received at different geographical locations. Heisler *et al.* (2004) measured the UV-B dose in Baltimore, MD for a 14 month period. When their measurements were compared to those from two other sites in Maryland, the results indicated that the UV radiation doses were similar at the three sites. The UV-B radiation was greatest during June–July (when the solar elevation angle was also the highest) and depended on the annual stratospheric O_3 cycle. The results of this study indicated that the yearly average daily UV dose was 35 600 J/m^2. The maximum daily dose of 70 000 J/m^2 was observed in summer and the minimum daily dose of 10 000 J/m^2 in winter. On clear days, the maximum total UV-B irradiance in Baltimore ranged from about 3 W/m^2 (in June) to 0.6 W/m^2 (in December).

Hicke *et al.* (2008) studied radiation levels in different parts of the country. The UV-B-radiation-measuring instruments used in their study were designed to replicate the erythemal action spectrum. An erythemal action spectrum is a parameter that describes the relative effectiveness of different wavelengths of energy in producing a skin response. An action spectrum is used as a 'weighting factor' for the UV spectrum, in order to identify the actual biologically effective dose for a given result. Reports from this study indicated that over an 8-year period, the mean annual irradiance for Maine was 0.15 W/m^2 and for New Mexico was 0.35 W/m^2. The mean monthly irradiance for New Mexico was 0.5 W/m^2 in June and 0.18 W/m^2 in January, and for Maine was 0.25 W/m^2 in June and 0.05 W/m^2 in January. This shows that the amount of UV radiation varies throughout the USA and changes throughout the year.

For a given date, the flux does not change, on average, from year to year; therefore, previously published data can be reanalyzed using newly obtained flux measurements. For many sites in the USA, full daily UV-A and UV-B fluence measurements are obtained using broadband sensors or sensors tuned to certain wavelengths. This research is conducted by the UV-B Monitoring and Research Program (UVMRP), a data-collection and research program of the US Department of Agriculture, headquartered at Colorado State University (Fort Collins, CO). The UVMRP Web site (http://UVB.nrel.colostate.edu) includes information on local UV-B and UV-A radiation that has been obtained since 1993.

17.4 Selected Organisms

17.4.1 Bacterial Endospores

Organisms that form spores are among the life forms most resistant to disinfection and inactivation (Block, 2001). Endospores are highly resistant to a variety of stresses, such as toxic chemicals, desiccation, temperature and pressure extremes, and high doses of ionizing or UV radiation.

The high level of resistance of spores is related to their architecture. An illustration of the typical bacterial spore layers, including the exosporium, coats, outer membrane, cortex, germ cell wall, inner membrane, and core, is shown in Figure 17.2. The outermost layer is the exosporium, which is a large, loose-fitting structure composed of proteins, including glycoproteins in some *Bacillus* species. The amount of exosporium varies with the species; for example, *Bacillus subtilis* has a thin layer of exosporium, if any. Inside the exosporium are the spore coats, which are complex structures formed by many proteins in

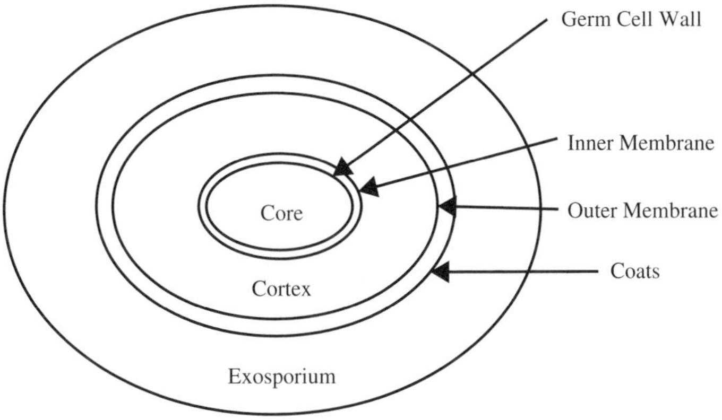

Figure 17.2 *Spore structure. Reproduced with permission from Setlow (2003). Copyright © 2003, Elsevier.*

multiple layers. The outer membrane, which is necessary for spore formation, lies under the spore coats. Located within the outer membrane, the cortex is essential for reduction of the water content in the spore core and for formation of a dormant spore. The cortex is degraded in spore germination, and this degradation is required for spore core expansion and subsequent outgrowth. The germ cell wall (composed of peptidoglycan) under the cortex becomes the cell wall of the outgrowing spore. The inner spore membrane serves as a strong permeability barrier to chemicals that damage spore DNA. The spore core contains DNA, enzymes, ribosomes, and transfer RNA.

Microorganisms use several mechanisms to protect themselves from environmental stresses such as UV radiation:

1. Nucleocapsids and cytoplasm contain UV-radiation-absorbing proteins.
2. The majority of bacterial spores identified at high altitudes contain pigments such as carotenoids and melanins that scavenge the free radicals produced by oxidative damage caused by UV radiation (Moeller *et al.*, 2005).
3. Spores can contain a biochemical pathway that repairs the DNA damage caused by UV radiation.
4. Spores are protected by thick spore protein coats consisting of inner and outer coat layers.
5. The core's low water content keeps DNA in a specific conformation (the A form) that is more resistant to UV radiation (Setlow, 2006).

Bacillus endospores are the best characterized species among spore-forming bacteria and are used as a model system for the study of the resistance of bacterial endospores to environmental extremes. Studies have shown that *B. anthracis* Sterne spores and *B. subtilis* spores have identical UV inactivation kinetics; therefore, *B. subtilis* strains can reliably be used as a biodosimetry model for the UV inactivation of *B. anthracis* spores (Nicholson and Galeano, 2003; Menetrez *et al.*, 2006). However, various researchers have identified different inactivation kinetics for the same organisms. This may be due to a lack of consistent exposure techniques between laboratories, varying spore concentrations in the liquid samples (mixed or not mixed), the quality and state of the spores used in the studies, or the proportion of vegetative bacteria in the samples. The differences identified between some species pertain to plasmid content, proteins, and the presence and thickness of spore coats.

17.4.2 Vegetative Bacteria

Vegetative bacteria continuously grow and reproduce. Unlike endospores, vegetative bacteria lack spore coats and an exosporium; however, they have a cell wall that is composed of peptidoglycan (protein and sugar). Vegetative bacteria can be classified as either Gram positive or Gram negative, depending on whether the cells retain crystal violet dye during the Gram-staining protocol. Gram-positive bacteria have a thick peptidoglycan layer within their walls, which also contain modified alcohols called teichoic acids (Kratz, 2005). Some bacteria, such as *Mycobacterium tuberculosis*, have a waxy outer layer that is resistant to disinfectants. Gram-negative bacteria have a thin peptidoglycan layer within their walls and an additional outside layer called the 'outer membrane'. The outer membrane is composed of phospholipids, lipopolysaccharides, lipoproteins, and channel proteins. Some bacteria do not have Gram-positive or Gram-negative cell walls. The nuclear material of vegetative bacteria is the target site for damage by UV radiation; it is located inside the fluid-filled cytoplasm, which is itself within the plasma membrane.

17.4.3 Viruses

Viruses, which are among the smallest infectious agents, exist in the boundary between living organisms and chemical entities. Viruses are composed of nucleic acids surrounded by a protein coat called the 'capsid'. Different viruses can infect and then replicate within a variety of host cells. Some viruses have an additional layer surrounding the capsid, called the 'envelope', that is made up of phospholipids, proteins, and carbohydrates. The envelope is taken from the host as the virus exits the host cell. The nucleic acids in a virus can be DNA or RNA and may be double- or single-stranded, and some viruses contain both DNA and RNA.

Many viruses (more than 90) can cause disease in humans (Fields and Knipe, 1990; Knipe and Howley, 2001). Because of their apparent fragility and sensitivity to salts, solvents, and pressure, viruses might be expected to be more sensitive than bacteria to other environmental effects. However, recent observations and experimental measurements indicate that viruses can reach further into the environment than was previously expected. For example, an information leak from the former Soviet Union reported an accidental infection of naval personnel 11 miles offshore from a smallpox testing site (Tucker and Zilinskas, 2003), which indicated that viruses can survive for long time periods and over great distances outdoors.

Lytle and Sagripanti (2005) reviewed published data on virus sensitivity to UV light and devised an approach for estimating the survival of various viruses after their release at any location and at any time of the year. Ben-David and Sagripanti (2010) modeled the use of solar UV radiation to inactivate viruses aerosolized in the atmosphere. They estimated that a full day of sun exposure would on average decrease the infectivity of UV-sensitive viruses by 3 $\log_{10}$. Similar calculations were used by public agencies to estimate viral persistence in the environment during recent influenza epidemics (Sagripanti and Lytle, 2007).

Although mammalian viruses are the causative agents of human epidemics, few have been used in UV experiments and other survival studies, as a result of safety concerns. Bacteriophages are often studied as surrogates to infectious viruses; however, phages differ greatly in morphology, composition, and UV resistance when compared with mammalian viruses of interest in public health and biodefense (Lytle and Sagripanti, 2005). For these reasons, bacteriophages are not useful surrogates for the study of the sensitivity of mammalian viruses to UV light.

17.5 Effects of UV Light on Aerosolized Organisms

17.5.1 Cell Damage Caused By UV Radiation

DNA is an organism's most sensitive target, and UV light chemically modifies the genetic material of microorganisms (Lytle and Sagripanti, 2005). Unsaturated bonds present in biological molecules, such as coenzymes, hormones, and electron carriers, may also be susceptible to UV damage.

Nucleic acids have absorption peaks at 200 and 265 nm. The 200 nm radiation is absorbed by molecules of ribose and phosphate in the DNA backbone, and the 265 nm radiation is absorbed by the nucleotide bases thymine, adenine, cytosine, and guanine (Kowalski, 2009). Thus, among the UV radiation regions, UV-C radiation (254 nm, close to the maximal absorption of DNA) inactivates organisms most effectively, UV-A radiation (320–380 nm) inactivates organisms least effectively, and UV-B radiation (290–320 nm) falls in the middle. Sensitivity to UV light correlates inversely with the size (or molecular weight) of the genome, such that smaller DNA molecules are more severely affected by UV radiation (Lytle and Sagripanti, 2005).

UV radiation can cause crosslinks between two adjacent thymine bases. The result is the formation of stable thymine dimers that inhibit replication and inactivate bacteria and DNA viruses (Masschelein and Rice, 2002). In addition, thymine forms crosslinks with thymine in adjacent DNA strands and proteins, inactivating the DNA and negatively affecting reproduction. Crosslinks can also occur between cytosine and guanine, but more energy is required for this due to the three hydrogen bonds between them. Other dimers formed after UV exposure include cytosine dimers, cytosine–thymine dimers, uracil dimers, uracil–thymine dimers, and uracil–cytosine dimers. Dimer formation occurs mainly following UV-A exposure, whereas lesions of other types occur as a result of UV-B and UV-C exposure. Fewer dimers are formed in the hydrated state, as it is less compact and the nucleotides are relatively far apart (Schreier *et al.*, 2007).

Uracil takes the place of thymine in RNA. Uracil dimers are formed following the UV exposure of RNA, but not to the extent that thymine dimers are formed in DNA. RNA also covalently links with proteins in response to UV irradiation. The pyrimidine-pyrimidone (6-4) series of lesions (called 6-4 photoproducts) also occur as a result of UV exposure.

Because it has more structural stability, double-strand RNA is more stable after UV exposure than is single-strand RNA (Rauth, 1965; Becker and Wang, 1989). Parallel findings in DNA viruses indicate that double-strand DNA viruses are also more resistant than single-strand DNA viruses (Tseng and Li, 2007). In addition, DNA viruses are more resistant than RNA viruses with the same number of strands (double or single).

Like nucleotide crosslinks, nucleotides can also bind with water in response to UV exposure, in a process called photohydration. Photohydration reactions cause cytosine and uracil to bond with water molecules; however, this does not occur with thymine. Unlike DNA and RNA damage, photohydration is independent of wavelength (Cerutti *et al.*, 1969; Remsen, Miller, and Cerutti, 1970).

17.5.2 Photorepair

Many environmental bacteria live in balance between UV damage and repair. The repair mechanisms start immediately after sunlight irradiation and continue into the night, with the organisms needing to be fully repaired by the next morning. The repair that occurs during the night is called dark repair. The use of light energy for repair after UV damage is called photorepair, photorecovery, or photoreactivation (Peccia and Hernandez, 2001; Xue and Nicholson, 1996). Photorepair activities show an initial peak in the UV-A region (320–380 nm). In addition, visible light (380–750 nm) is also used by cells undergoing photorepair. The degree of photoreactivation of vegetative bacteria depends on the particular species and strain. Photorepair does not occur in the spore stage, but does occur during and after spore germination.

Three primary repair processes are used by organisms in response to UV radiation. In the first process, called the 'excision repair process' or the 'cut-and-patch mechanism', a series of cellular enzymes replace thymidine dimers with undimerized thymine. In the second, certain spores and vegetative bacteria use the photons in visible and UV light to reverse DNA damage by monomerizing dimers. In the third, cells are rescued from severe DNA damage: polymerization around DNA damage is followed by excision of the damaged areas and normal DNA synthesis. However, this process can lead to mutation, due to the addition of extra bases (Coohill and Sagripanti, 2008; Hanawalt, 1989). Photorepair processes are optimized at different temperatures, depending on the enzymes used by particular cells. Humidity also affects the photorepair process; exposure to high-relative-humidity (RH) environments generally supports photoreactivation by facilitating overall cell function.

Repair intensity depends on the amount of damage sustained and the biological organization of the microorganism. The repair mechanism involves the production of enzyme–substrate complexes at the DNA lesion site, where light energy is absorbed for the repair. Photorepair fixes the damage that results from narrow-band UV-C-radiation (254 nm) exposure. Radiation produced by broad-spectrum UV lamps damages the enzymes and other microbial constituents, and photorepair processes are unable to repair this damage.

17.5.3 Typical Survival Curve for UV Exposure

The kill effect of UV radiation can be quantified by several parameters, such as the doses required to kill 90 and 99% of the present organisms (D_{90} and D_{99}, respectively) and the dose required for 37% survival of the organisms (D_{37} or e^{-1}). D_{90} and D_{99} are often indicated as $F_{-1\log}$ and $F_{-2\log}$, respectively (King, Kesavan, and Sagripanti, 2011). D_{37} corresponds to the amount of radiation required to produce, on average, one lethal hit per organism (Lytle and Sagripanti, 2005).

A curve describing the survival of organisms following UV exposure has been modeled to incorporate different complexities, such as how the organism is damaged and how it repairs the damage. The disinfection curve shown in Figure 17.3 reflects a simple exponential

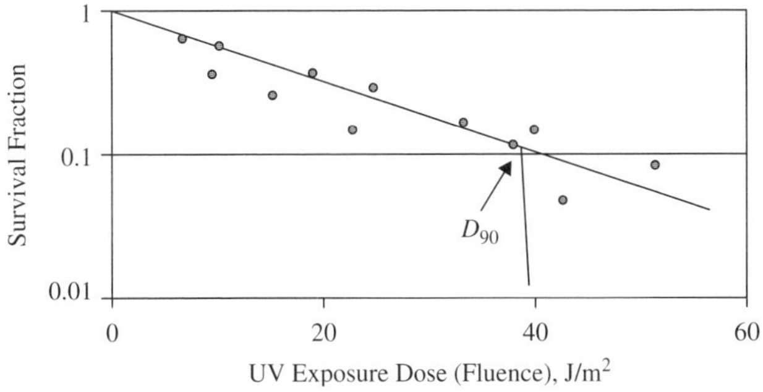

Figure 17.3 *Example of a disinfection or decay curve that shows first-order decay. Adapted with permission from Kowalski, 2009. Copyright © 2009, Springer-Verlag Berlin Heidelberg.*

(Equation 17.1). This is a first-order decay model that provides relatively accurate values for D_{90} and D_{99}. Higher values for rate constants (or slope) indicate fast disinfection, and lower values indicate slow disinfection:

$$S = e^{-kD} \tag{17.1}$$

where S is the survival fraction (the number of organisms surviving radiation divided by the number of organisms before irradiation), k is the UV rate constant (generally in m^2/J), and D is the dose or fluence (generally in J/m^2).

Often, a tiny fraction (sometimes <1%, but as high as 40–50% when protected by media or pigments) of the organisms exhibits a higher level of resistance and a significantly lower rate constant. In this case, a fast decay is observed, which is attributable to the susceptible population; this is followed by a slower decay that corresponds to the apparently resistant population. If the resistant population is on the order of 1% or less, it can be identified graphically after the D_{99} value has been reached. In this two-stage decay curve (shown in Figure 17.4), survival is described by:

$$S = (1 - f)e^{-Dk_1} + fe^{-Dk_2} \tag{17.2}$$

where f is the resistant fraction, k_1 is the first-stage rate constant, and k_2 is the second-stage rate constant.

The early part of the curve can also have a shoulder region, where there is no permanent damage. The bacteria can repair the damage caused by low levels of UV exposure. This has been experimentally shown by a lack of change in culturable bacteria concentrations with increasing exposure to certain levels of UV radiation (reviewed in Coohill and Sagripanti, 2008). The length of the shoulder region reflects the ability of the organisms to withstand and repair damage, and is dependent upon the particular organism and the assay sensitivity. The shoulder region is followed by the first-stage linear region on a $\log_{10}$-linear graph, as

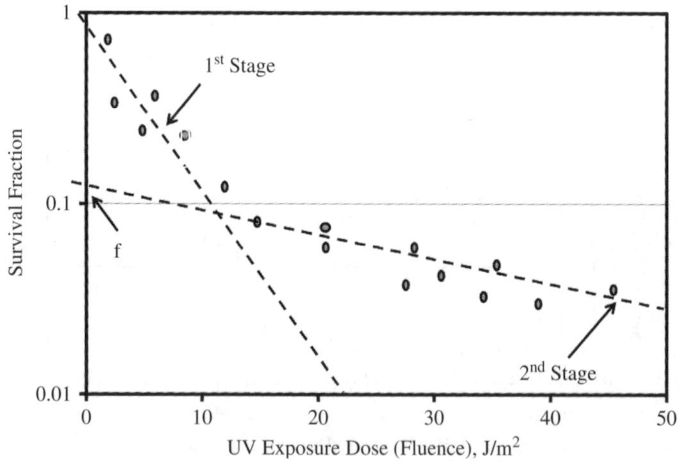

Figure 17.4 Example of a disinfection curve that shows first- and second-stage decays. Adapted with permission from Kowalski, 2009. Copyright © 2009, Springer-Verlag Berlin Heidelberg.

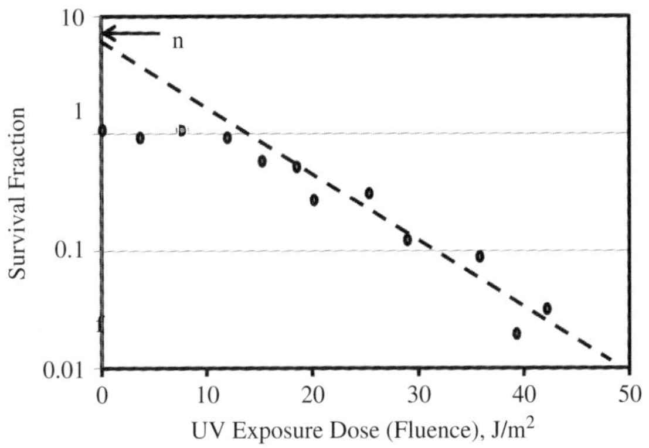

Figure 17.5 *Example of a disinfection or decay curve that shows shoulder and first-stage decay regions. Adapted with permission from Kowalski, 2009. Copyright © 2009, Springer-Verlag Berlin Heidelberg.*

shown in Figure 17.5 (Coohill and Sagripanti, 2008). The multihit model (Kowalski, 2009) shows the survival curve that includes the shoulder and linear portion as:

$$S(t) = 1-(1-e^{-kD})^n \tag{17.3}$$

where k is the rate constant (the slope of the linear portion of the curve) and n is the shoulder constant, which is equal to the y-axis intercept. Equations used to describe microbial decay that include a shoulder followed by two (fast and slow) rate decays also exist, but they are not shown here.

As shown in Figure 17.4, the decay curves can end in a tail that indicates the presence of a resistant population. There can also be other reasons for the appearance of tailing. For example, tailing often occurs in the decay curves of experiments conducted with bacteria on agar surfaces. Agar surfaces have microscopic pits and fissures. Bacteria inside these microcrevices are exposed to less or no UV radiation, and therefore appear to be resistant. If 1 bacterium in 1000 were shielded then the linear portion of the decay curve would end at the $-3 \log_{10}$ level and tailing would be evident after that point. For bacteria exposed on filter paper, the resulting decay curves are determined by the filter paper structure and moisture content. To reduce these surface effects, some researchers have irradiated bacteria while stirring them in dilute solutions in order to expose all of them to comparable levels of UV radiation (Moeller *et al.*, 2005). For these experiments, the medium must be transparent, and cell density must be sufficiently low that no shadow effects occur between the cells in the center of the container and those closer to the container walls.

17.5.4 The UV Rate Constant

When organisms are exposed to UV light, they decay as a function of dose (fluence), where the slope of the logarithmic decay curve is defined as the rate constant, k. The rate constant is affected by RH, temperature, DNA conformation (A or B), irradiance level, photoreactivation, type of UV lamp, dose range, culture medium and method, and the specific strain of organism. As discussed in the previous section, the resulting decay curves have different shapes for different organisms, with the linear portions resulting in different rate constants. It is difficult to make comparisons between organisms that have different shapes of decay. Therefore, the D_{90} value is frequently used to compare different organisms, and a rate constant (known as the 'UV susceptibility', with units of square meters per joule) is calculated as:

$$D_{90} = \frac{-\ln(1-0.9)}{k} = \frac{2.3}{k} \tag{17.4}$$

This equation does not describe the actual shape of the decay curve, which may have a shoulder and one or two linear portions; therefore, D_{90} is used more often to compare studies or organisms. The dose needed to reduce the microbial population to 10% of the original amount is indicated as T_1, and the fluence that reduces the microbial load by 4 $\log_{10}$ is indicated as T_4. Alternatively, a parameter derived from quantum mechanics and statistical damage can be used to characterize microbial sensitivity to UV radiation. The D_{37} value is equal to the reciprocal of the slope and corresponds to the UV fluence that produces, on average, one lethal hit to the organisms, resulting in a survival rate of 37%. D_{37} can be calculated by dividing the fluence that inactivates a 1 $\log_{10}$ virus load (as obtained from the linear portion of the graphs) by 2.3 (the natural logarithmic base). A lower value for D_{37} indicates a higher sensitivity to inactivation by UV radiation.

17.5.5 RH and Temperature Effects

The effect of RH level on microbial susceptibility to UV radiation has been studied for many years, but a clear consensus has not been achieved. As mentioned in the previous section, DNA exists in A and B conformations. The conformations change in response to variations in RH level. The A conformation occurs during low RH conditions, when water molecules are desorbed from DNA as a result of the hydrophobic nature of the tighter internal channels. During high RH conditions, the DNA hydrates into the B conformation. Most bacteria tend to exhibit decreased susceptibility to UV when exposed to high RH levels, but some respond with the opposite effect or no effect at all, and viruses respond with mixed effects. In studies of *Escherichia coli*, *Mycobacterium parafortuitum*, and *Staphylococcus epidermis*, low susceptibility was identified during exposure to high RH conditions (Kowalski, 2009). Xu *et al.* (2005) reported similar results in experiments in which *M. parafortuitum* cells were aerosolized to determine the effectiveness of upper-room-air UV germicidal irradiation. McDevitt *et al.* (2007) showed that the susceptibility of vaccinia virus increased with decreasing RH. In addition, Tseng and Li (2007) evaluated the survival fraction of four viruses at 55 and 85% RH levels, and the results indicated that susceptibility was increased at lower RH levels. Similar effects were observed in a study by Lin and Li (2002), who found that microorganism susceptibility at 80% RH was lower than at 50% RH. The organisms used in this test were *E. coli* cells, *B. subtilis* spores, *Candida famata* var. *flareri* cells, and *Penicillium citrinum* spores. A few studies have indicated that susceptibility is high for organisms in high RH environments. For example, for some strains of *Serratia marcescens*, disinfection increases at higher RH levels (Kowalski, 2009).

Kowalski (2009) reported that the D_{90} value for inactivating organisms in high-RH air was fivefold higher than for organisms in low-RH air. In that same report, the D_{90} value for inactivating organisms in water was twofold higher than that for organisms in high-RH air.

In any case, the apparently contradictory reports regarding changes attributable to RH describe only modest effects on the susceptibility of organisms to UV radiation.

For example, normal environmental RH changes (50–70%) only affected the survival of the influenza virus by 9%, while extreme RH changes (15–90%) affected the virus 12.5-fold (Hemmes, Winkler, and Kool, 1960). An infected patient's nasal secretions can contain 10^7 viral particles per milliliter (Couch, 1995), and an increase in RH may reduce the viable particle concentration to 10^6 particles per milliliter. This reduction is not significant when considered in the context of preventing disease transmission. In comparison, the effects of solar radiation or indoor UV lamps on the inactivation of microorganisms both outdoors and indoors are substantial.

Temperature also influences the effects of UV radiation. Higher temperatures can cause structural changes in cell-membrane phospholipids, proteins, and DNA. Ko, Melvin, and Burge (2002) reported that UV effectiveness was significantly increased when temperature was increased from 4 to 25 °C. UV radiation is likely to cause mutagenic effects at high temperatures, where DNA becomes more flexible, due to a more active metabolism.

17.5.6 Bacterial Clusters

The majority of the previous studies intended to establish the effects of germicidal UV lights or sunlight were conducted using single spores on surfaces and in air (King, Kesavan, and Sagripanti, 2011; Foarde *et al.*, 2006; reviewed in Coohill and Sagripanti, 2009). However, the outer layers of spores in a cluster might protect the organisms in the center. The killing effect of UV light on all the organisms in the cluster may be underestimated by data obtained from single-spore dispersion tests.

Clumping also shields bacteria in the center from UV effects and can therefore affect survival curves. On the basis of previous calculations for irradiation with 254 nm wavelength, 61% of the incident light will be transmitted through one organism (Coohill, 1986). Similarly, two organisms shielding a third organism will only allow 37% of the incident UV light to reach the latter. Therefore, clumping can protect the bacteria in the center from UV light and thereby cause the tailing observed in the decay curves from some experiments.

17.6 Disinfection of Rooms Using UV-C Radiation

UV lamps generating radiation of 254 nm wavelength (UV-C) have been proposed as a supplemental germicidal control in health-care facilities (TB Infection-Control Guidelines Work Group, 1994) because of the associated low costs, ease of application, and potential efficiency. UV-C lamps can be used in enclosed mechanical ventilation system ducts and locally recirculating units, as well as in open configurations in rooms.

In several studies (Rutala, Gergen, and Weber, 2010; Xu *et al.*, 2003), a reduction in bacterial spore and vegetative cell numbers occurred in response to upper-room-air UV irradiation. As expected, UV-C lights installed to irradiate upper-room air were reported to be more effective at inactivating vegetative bacteria than bacterial spores (Miller and Macher, 2000). In one of the first studies to demonstrate the benefits of germicidal UV lights, the number of influenza infections in a veterans' hospital was shown to be reduced when compared to infections in nearby, nonirradiated hospital rooms (McLean, 1956). The US Centers for Disease Control and Prevention (CDC) recommends that hospital isolation rooms be ventilated with at least 12 clean, outside-air changes each hour (Siegel *et al.*, 2007). In addition, the general rule for the number of UV lightbulbs in a room is that one 30 W fixture should be used for every 18 m^2 of floor area or for every seven people in the room. In a study by Xu *et al.* (2005) that included fourfold higher UV wattage than is recommended by CDC guidelines, it was shown that UV irradiation that is uniformly distributed throughout a room is more effective than UV radiation of nonuniform distribution.

In a study conducted by Menzies *et al.* (2003), it was shown that UV-C lights installed in office ventilation systems reduced microbial and endotoxin concentrations on irradiated surfaces within the ventilation systems by 99% (2 $\log_{10}$) and also reduced respiratory and musculoskeletal symptoms in office personnel. Ko, Melvin, and Burge (2002) showed that UV germicidal irradiation in an upper room significantly reduced culturable bacteria concentrations in a lower room. A study conducted in a test chamber (King, Kesavan, and Sagripanti, 2011) showed that bacterium culturability was reduced in response to UV-C-radiation exposure. These findings indicate that UV-C irradiation can contribute to the control of contagious disease.

17.7 Sunlight Exposure Studies

In previous studies, both the sun and artificial light sources designed to simulate the sunlight spectrum (Munakata, 1999; Munakata *et al.*, 2000a,2000b) were used to expose organisms to direct and indirect radiation. The effects of both direct and indirect radiation must be considered in order to completely explain the effects of UV radiation on organisms.

Analysis of sunlight-exposure studies indicates that the UV sensitivity of *B. anthracis* Sterne spores is similar to that of *B. subtilis* spores (strain SN624, ATCC 6633), which suggests that *B. subtilis* can be used as a surrogate for *B. anthracis* Sterne in UV decontamination studies (Nicholson and Galeano, 2003). Also, a carotene-like pigment found in *Bacillus atrophaeus* DSM 675 spores at high altitudes affects the resistance of the spores to environmental UV radiation (Moeller *et al.*, 2005).

UV doses measured by the US Department of Agriculture were used to determine virus kill rates from sunlight exposure (Sagripanti and Lytle, 2007; Lytle and Sagripanti, 2005). The flux at each wavelength in the UV-B region was multiplied by an appropriate normalized action spectrum value that was based on the 254 nm kill effect. It was determined that the germicidal activity of sunlight peaks at 300–305 nm, with the solar flux between these wavelengths accounting for 75% of the total effective germicidal solar flux (Sagripanti and Lytle, 2007). Thus, a UV dosimeter that is sensitive between 295 and 305 nm provides a usable correlation between solar-light dose and germicidal effect (Coohill and Sagripanti, 2009).

For reasons of convenience, artificial sunlight is preferred in testing environmental solar effects (US Environmental Protection Agency, 2010). Beebe *et al.* (1962) exposed *B. anthracis* Ames to 600 $\mu W/cm^2$ (1 W/m^2) of artificial sunlight using General Electric (Schenectady, NY) type RS sunlamps, with wavelengths <305 nm removed. Mice were exposed to nonirradiated and irradiated *B. anthracis* spores in order to determine the effect of simulated solar UV on virulence.

In addition to such experiments, mathematical models that predict virus inactivation by solar radiation have been developed by Lytle and Sagripanti (2005) and Ben-David and Sagripanti (2010). These authors have predicted the inactivation of smallpox virus, as well as of Ebola, Marburg, Crimean-Congo, Junin, and other hemorrhagic viruses and of Venezuelan equine encephalitis in various parts of the world. The use of such calculations allows researchers to predict the survival of organisms in outdoor settings while avoiding the dangers of working with infectious agents.

17.8 Testing Considerations

In many of the previous studies on the effects of UV radiation, not all variables were controlled and not all of the information necessary to duplicate the experiments and confirm the findings was reported. Variables that should be reported in UV-C- and solar-exposure studies include temperature, RH, moisture state of the particle (wet or dry), surface conditions (if microbes are deposited on surfaces), dosimetry, environmental conditions (date, time of day, altitude, latitude), organism stage (exponential or stationary phase), amount of debris present, and the ratio of vegetative to spore bacteria, as well as any other relevant chemical or biological conditions (Coohill and Sagripanti, 2009). Biological parameters such as the ability of the organism to repair, its lifecycle stage, sample temperature, and the UV spectrum of the radiation are among the most critical variables to specify if reproducibility is intended.

Previous studies have shown that the type of microbial damage that occurs following solar UV irradiation depends on wavelength. For example, the biocidal effectiveness of the 300 nm wavelength is 100-fold higher than that of the 320 nm wavelength. In laboratory-based experiments, a relatively narrow range of wavelengths is generally used to mimic solar radiation. A disadvantage of using the narrow-range exposure method is that the solar spectrum has been reported to produce biological effects that can differ by an order of magnitude from those of laboratory-based, monochromatic radiation studies. In addition, UV-A and visible radiation can support photorepair of the damage caused by other wavelengths, and this effect would not be observed in the exposure of microbes to a narrow range of wavelengths.

Sunlight exposure time and duration are important parameters for predicting radiation wavelength and dose to organisms. The greatest solar damage occurs at solar noon, with less damage occurring on either side of noon due as a result of the decrease in UV-B flux. Therefore, sun exposure time must be recorded and dosimetry must be accurately measured.

Unexplained differences between the effects of artificial sources and sunlight on microbial inactivation have been suggested. The formation of different photoproducts has been reported in response to exposure to UV-A, UV-B, and UV-C radiation (Nicholson, Setlow, and Setlow, 2002). Therefore, extreme care must be exercised when selecting exposure wavelengths and matching the proper dosimeter with a wavelength or spectrum response that accurately reflects the dose received by the organisms during testing.

Three distinct methods are used by researchers to determine the effects of UV radiation on microorganisms: on surfaces, in liquids, and in aerosols. Often, the disinfection of organisms on surfaces is studied by exposing organisms on agar plates; this method only allows organisms to be exposed to UV light on one side. Alternatively, organisms in liquid or air suspensions can be exposed to UV light from all directions. When organisms are exposed to an identical radiation source under any of the three physical conditions (surface, liquid, or aerosol) different fluencies are required to attain comparable damage.

The aim of many studies is to determine the sensitivity of bacterial spores to UV. The quality and state of the spores in a sample affect their response to UV irradiation. The ratio of vegetative bacteria to spores in a bacterial preparation strongly affects the response of that preparation to UV irradiation, as spores are 5- to 10-fold more resistant to UV radiation than are vegetative bacteria of the same species (Coohill and Sagripanti, 2008). The physical condition of the sample also affects inactivation: organisms in liquid or on agar plates have layers of water around them, which can offer a level of protection that is not experienced by organisms on dry surfaces or suspended in air.

During solar exposure, spore inactivation is independent of temperature; however, vegetative bacteria are more resistant at higher temperatures, because vegetative cells, unlike spores, can undergo photoreactivation to repair the suffered damage. When it is desirable to reduce or eliminate the photoreactivation, vegetative bacteria are exposed to sunlight at or near 0 °C. Furthermore, spores can remain in the dormant state for many weeks after exposure but vegetative bacteria must be cultured immediately. Repair-deficient mutants of vegetative bacteria are used in some studies to determine inactivation kinetics, but these tests occur rapidly and require shorter exposure times that are difficult to measure accurately.

17.8.1 Test Methodology in Our Laboratory

We have developed a methodology for determining the disinfection efficiencies of UV-C radiation and simulated sunlight on organisms suspended in air. We have tested single spores and spore clusters of different sizes. A similar methodology was used to study the effect of UV light on vegetative bacteria. Organisms deposited on membrane filters and agar surfaces were used to determine the decontamination effectiveness against organisms on surfaces. Chambers filled with bioaerosols were used for aerosol testing.

17.8.1.1 Equipment Used During Testing

For the aerosol disinfection studies, a Collison nebulizer (BGI, Inc., Waltham, MA) was used to generate particles containing single organisms. For aerosol testing, a Sono-Tek aerosol generator (Milton, NY) was used to produce bioaerosol clusters. Each instrument and test chamber is briefly described in the following sections. Examples of single spores and clusters generated during testing are shown in Figure 17.6.

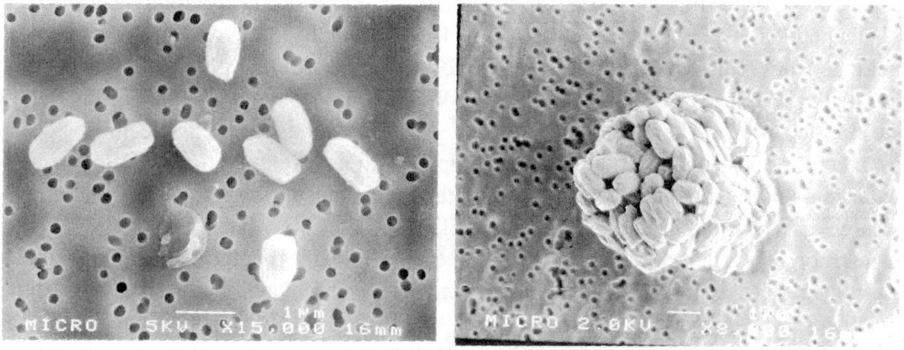

(a) (b)

Figure 17.6 *(a) Single spore. (b) Approximately 4 µm clustered bacterial spores.*

Figure 17.7 *A Collison nebulizer. See plate section for colour version.*

17.8.1.2 Collison Nebulizer

Collison nebulizers (Figure 17.7) with 4, 6, 24, and 36 jets were used at ECBC to generate single-spore particles for aerosol testing. Test organisms were added to deionized water, and this solution was used in the nebulizer. To minimize the possibility of doublet and triplet organisms occurring in each aerosol particle, relatively low organism concentrations were used. The Collison nebulizer was connected to a compressed-air source. The air exited at high velocity from small holes inside the nebulizer. The low pressure created in the exit region caused the water-organism solution to be drawn from the bottom of the nebulizer through a second tube (in accordance with the Bernoulli effect). The liquid exited the tube as a thin filament. As it accelerated in the airstream, the liquid filament stretched until it broke into droplets. The spray stream was directed on to a wall. Larger droplets impacted the wall and were removed from the air.

17.8.1.3 Sono-Tek Aerosol Generator

Sono-Tek ultrasonic atomizing nozzles, like the one shown in Figure 17.8, were used to produce droplets of the liquid and material mixtures suspended in liquids, which were

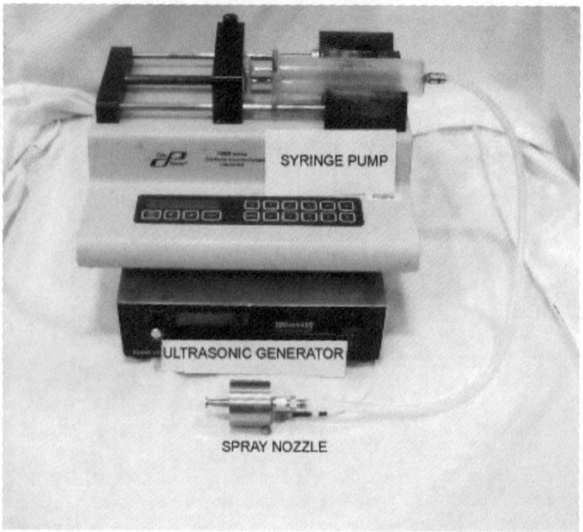

Figure 17.8 Sono-Tek aerosol generators. See plate section for colour version.

then dried to yield a dry aerosol of residual material particles. In biological experiments, the aerosolized mixture was composed of organisms suspended in water. The final particle size was determined from the material concentration in the feed liquid and the characteristic size of the Sono-Tek generated droplets. The rate of aerosol generation was dependent upon the rate at which the fluid mixture was fed to the Sono-Tek nozzle. Five models of Sono-Tek aerosol generator are available for the production of particles with mean diameters ranging from 23 to 70 μm (at 120–25 kHz operating frequencies) and flow rates from a few microliters per second to about 6 gallons per hour. We used the 25 kHz model, fed by a peristaltic pump, and the 120 kHz microbore model, fed by a syringe pump. These were used to create controllable aerosol concentrations of about 500 to 50 000 particles per liter of air inside a chamber.

17.8.1.4 Aerosol Exposure Test Methodology

Figure 17.9 is an illustration of one of our aerosol exposure chamber, which has dimensions of $36 \times 17 \times 14$ inches and a volume of $0.14\ m^3$. Single-spore particles were generated using the Collison nebulizer and bioclusters were generated using the Sono-Tek ultrasonic atomizing nozzles. After the aerosol was generated, the chamber air was mixed to produce a uniform aerosol concentration. A baseline air sample was obtained on gel filters before UV exposure occurred. The organisms were then exposed to UV light for different lengths of time. Air samples were collected on gel filters at the end of each exposure. Control samples were collected in a similar fashion, but without the UV light exposures (while the UV light was turned off). The amount of kill due to UV exposure was determined by comparing the baseline samples with the UV light-exposed and control samples (King, Kesavan, and Sagripanti, 2011).

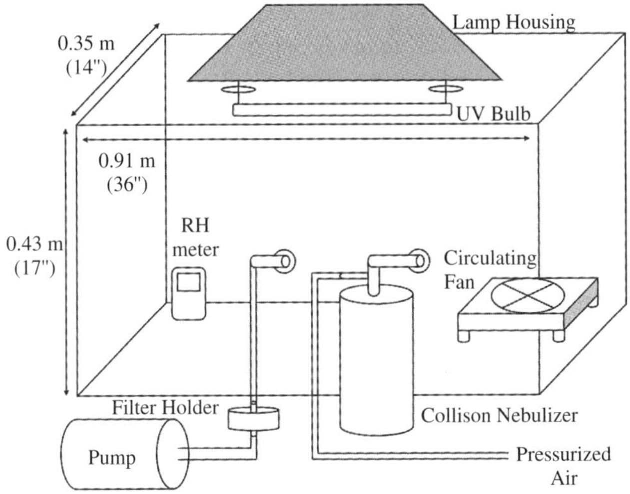

Figure 17.9 *Aerosol exposure chamber. First published in King, Kesavan, and Sagripanti (2011).*

17.9 Discussion

The unintentional release of bioaerosols from infected patients is of concern to public health and the large scale intentional or accidental release of aerosolized microorganisms is of concern to biodefense. Dilution and directed ventilation reduce risk by decreasing microbial concentration and removing organisms, but these methods do not kill germs. Air-filtration technology allows the number of infectious organisms in a room to be reduced, with high-efficiency particulate air (HEPA) filtration removing >99.99% of the airborne particles that arrive at the filter media, but implementation of HEPA filtration often requires costly engineering modifications. Among the air- and surface-disinfection technologies available, the one with the widest potential application may be UV-C disinfection, as it is relatively easy to use and is highly effective. In addition, UV-C lamps can be installed in buildings for a fraction of the cost of HEPA filtration systems. As indicated by the research summarized in this report, care must be taken to obtain accurate measures of microbial sensitivity to UV radiation. However, once accurate microbial sensitivity measurements are available, researchers and engineers can use the data in a wide range of applications. The use of UV-C radiation to inactivate germs suspended in aerosols or deposited on to surfaces should allow for the development of precise and effective engineering solutions for the protection of health and life. Understanding the effect of natural sunlight on airborne germs should assist in understanding and thus predicting microbial dissemination and its role in natural and artificial epidemics.

References

Becker, M.M. and Wang, Z. (1989) B–A transitions within a 5 S ribosomal RNA gene are highly sequence-specific. *Journal of Biological Chemistry*, **264**(7), 4163–4167.

Beebe, J.M., Dorsey, E.L., Guse, D.G., and Hunt, G.R. (1962) Stability and Virulence Relationships of Air-Borne *Bacillus anthracis* Spores Under Stress of Light and Humidity. Technical Memorandum 18, US Army Biological Laboratories, Fort Detrick, MD.

Ben-David, A. and Sagripanti, J.-L. (2010) A model for inactivation of microbes suspended in the atmosphere by solar ultraviolet radiation. *Photochemistry and Photobiology*, **86**, 895–908.

Black and Veatch Corporation (2010) *White's Handbook of Chlorination and Alternative Disinfectants*, 5th edn, John Wiley & Sons, Inc, Hoboken, NJ.

Block, S.S. (2001) *Disinfection, Sterilization, and Preservation*, 5th edn, Lippincott Williams & Wilkins, Philadelphia, PA.

Cerutti, P.A., Miller, N., Pleiss, M.G. *et al.* (1969) Photohydration of uridine in the RNA of coliphage R17 I. Reductive assay for uridine photohydration. *Biochemistry*, **64**, 731–738.

Coohill, T.P. (1986) Virus cell interactions as probes for vacuum ultraviolet radiation damage and repair. *Photochemistry and Photobiology*, **44**, 359–363.

Coohill, T.P. and Sagripanti, J.-L. (2008) Overview of the Inactivation by 254 nm ultraviolet radiation of bacteria with particular relevance of biodefense. *Photochemistry and Photobiology*, **84**, 1084–1090.

Coohill, T.P. and Sagripanti, J.-L. (2009) Review: bacterial inactivation by solar ultraviolet radiation compared with sensitivity to 254 nm radiation. *Photochemistry and Photobiology*, **85**, 1043–1052.

Couch, R.B. (1995) Orthomyxoviruses, in *Medical Microbiology*, 4th edn, University of Texas Medical Branch, Galveston, TX, pp. 1–22.

Fields, B.N. and Knipe, D.M. (eds) (1990) *Fields Virology*, 2nd edn, Raven Press, New York.

Knipe, D.M. and Howley, P.M. (2001) *Fields Virology*, 4th edn, Lippincott Williams and Wilkins, Philadelphia, PA.

Foarde, K., Franke, D., Webber, T. *et al.* (2006) Technology Evaluation Report: Biological Inactivation Efficiency by HVAC In-Duct Ultraviolet Light Systems, US Environmental Protection Agency, Office of Research and Development, National Homeland Security Research Center.

Giese, A.C. (1976) *Living with Our Sun's Ultraviolet Rays*, Plenum Press, New York.

Hanawalt, P.C. (1989) Concepts and models for DNA repair: from *Escherichia coli* to mammalian cells. *Environmental and Molecular Mutagenesis*, **14**(16), 90–98.

Heisler, G.M., Grant, R.H., Gao, W. and Slusser, J.R. (2004) Solar ultraviolet-B radiation in urban environments: the case of Baltimore Maryland. *Photochemistry and Photobiology*, **80**(3), 422–428.

Hemmes, J.H., Winkler, K.C. and Kool, S.M. (1960) Virus survival as a seasonal factor in influenza and poliomyelitis. *Nature*, **188**, 430–431.

Hicke, J.A., Slusser, J., Lantz, K. and Pascual, F.G. (2008) Trends and interannual variability in surface UVB radiation over 8 to 11 years observed across the United States. *Journal of Geophysical Research*, **113**, D21302.

Jagger, J. (1985) *Solar-UV Actions on Living Cells*, Praeger, New York.

King, B., Kesavan, J. and Sagripanti, J.-L. (2011) Germicidal UV sensitivity of bacteria in aerosols and on contaminated surfaces. *Aerosol Science and Technology*, **45**, 645–653.

Ko, G., Melvin, F.W. and Burge, H.A. (2002) The characterization of upper-room ultraviolet germicidal irradiation in inactivating airborne microorganisms. *Environmental Health Perspectives*, **110**(1), 95–101.

Kowalski, W. (2009) *Ultraviolet Germicidal Irradiation Handbook: UVGI for Air and Surface Disinfection*, Springer, New York.

Kratz, R.F. (2005) *Barron's E-Z Microbiology*, Barron's Educational Series, New York.

Lin, C. and Li, C. (2002) Control effectiveness of ultraviolet germicidal irradiation on bioaerosols. *Aerosol Science and Technology*, **36**(4), 474–478.

Lubin, D. and Jensen, E.H. (1995) Effects of clouds and stratospheric ozone depletion on ultraviolet radiation trends. *Nature*, **377**, 710–713.

Lytle, D.C. and Sagripanti, J.-L. (2005) Predicted inactivation of viruses of relevance to biodefense by solar radiation. *Journal of Virology*, **79**(22), 14244–14252.

Masschelein, W.J. and Rice, R. (2002) *Ultraviolet Light in Water and Wastewater Sanitation*, CRC Press, Boca Raton, FL.

McDevitt, J.J., Lai, K.M., Rudnick, S.N. *et al.* (2007) Characterization of UVC light sensitivity of vaccinia virus. *Applied and Environmental Microbiology*, **73**(18), 5760–5766.

McLean, R.L. (1956) Comments on reducing influenza epidemics among hospitalized veterans by UV irradiation of droplets in the air. *American Review of Respiratory Disease*, **83**(Suppl), 36–38.

Menetrez, M.Y., Foarde, K.K., Webber, T.D. *et al.* (2006) Efficiency of UV irradiation on eight species of *Bacillus*. *Journal of Environmental Engineering and Science*, **5**, 329–334.

Menzies, D., Popa, J., Hanley, J.A. *et al.* (2003) Effect of ultraviolet germicidal lights installed in office ventilation systems on workers' health and wellbeing: double-blind multiple crossover trial. *Lancet*, **362**, 1785–1791.

Miller, S.L. and Macher, J.M. (2000) Evaluation of a methodology for quantifying the effect of room air ultraviolet germicidal irradiation on airborne bacteria. *Aerosol Science and Technology*, **33**, 274–295.

Mims, F.M. (2005) Avian influenza and UV-B blocked by biomass smoke. *Environmental Health Perspectives*, **113**(12), A806–A807.

Moeller, R., Horneck, G., Facius, R. and Stackebrandt, E. (2005) Role of pigmentation in protecting *Bacillus* sp endospores against environmental UV radiation. *FEMS Microbiology Ecology*, **51**, 231–236.

Munakata, N. (1999) Comparative measurements of solar UV radiation with spore dosimetry at three European and two Japanese sites. *Journal of Photochemistry and Photobiology B*, **53**, 7–11.

Munakata, N., Kazadzis, S., Bais, A.F. *et al.* (2000a) Comparisons of spore dosimetry and spectral photometry of solar-UV radiation at four sites in Japan and Europe. *Photochemistry and Photobiology*, **72**(6), 739–745.

Munakata, N., Makita, K., Bolsee, D. *et al.* (2000b) Spore dosimetry of solar UV radiation: applications to monitoring of daily irradiance and personal exposure. *Advances in Space Research*, **26**(12), 1995–2003.

Nicholson, W.L. and Galeano, B. (2003) UV resistance of *Bacillus anthracis* spores revisited: validation of *Bacillus subtilis* spores as UV surrogates for spores of *B. anthracis* sterne. *Applied and Environmental Microbiology*, **69**(2), 1327–1330.

Nicholson, W.L., Setlow, B. and Setlow, P. (2002) UV photochemistry of DNA in vitro and in *Bacillus subtilis* spores at earth ambient and low atmospheric pressure: implications for spore survival on other planets or moons in the solar system. *Astrobiology*, **2**, 417–425.

Peccia, J. and Hernandez, M. (2001) Photoreactivation in airborne *Mycobacterium parafortuitum. Applied and Environmental Microbiology*, **67**(9), 4225–4232.

Puskeppeleit, M., Quintern, L.E., Naggar, S.E. *et al.* (1992) Long-term dosimetry of solar UV radiation in Antarctica with spores of *Bacillus subtilis. Applied and Environmental Microbiology*, **58**(8), 2355–2359.

Rauth, A.M. (1965) The physical state of viral nucleic acid and the sensitivity of viruses to ultraviolet light. *Biophysical Journal*, **5**, 257–273.

Remsen, J.F., Miller, N. and Cerutti, P.A. (1970) Photohydration of uridine in the RNA of coliphage R17 II. The relationship between ultraviolet inactivation and uridine photohydration. *Proceedings of the National academy of Sciences of the United States of America*, **65**(2), 460–466.

Riesenman, P.J. and Nicholson, W.L. (2000) Role of the spore coat layers in *Bacillus subtilis* spore resistance of hydrogen peroxide, artificial UV-C, UV-B, and solar UV radiation. *Applied and Environmental Microbiology*, **66**(2), 620–626.

Rutala, W.A., Gergen, M.F. and Weber, D.J. (2010) Room decontamination with UV radiation. *Infection Control and Hospital Epidemiology*, **31**(10), 1025–1029.

Sagripanti, J.-L. and Lytle, C.D. (2007) Inactivation of influenza virus by solar radiation. *Photochemistry and Photobiology*, **83**, 1278–1282.

Schreier, W.J., Schrader, T.E., Koller, F.O. *et al.* (2007) Thymine dimerization in DNA is an ultrafast photoreaction. *Science*, **315**(5812), 625–629.

Setlow, P. (2003) Spore germination. *Current Opinion in Microbiology*, **6**, 550–556.

Setlow, P. (2006) Spores of *Bacillus subtilis*: their resistance to and killing by radiation, heat, and chemicals. *Journal of Applied Microbiology*, **101**, 514–525.

Siegal, J.D., Rhinehart, E., Jacckson, M., and Chiarello, L.: the Healthcare Infection Control Practices Advisory Committee (2007) Guideline for Isolation Precautions: Preventing Transmission of Infectious Agents in Healthcare Settings, http://www.cdc.gov/hicpac/2007IP/2007isolationPrecautions.html (last accessed August 12, 2013).

TB Infection-Control Guidelines Work Group (1994) Guidelines for preventing the transmission of *Mycobacterium* tuberculosis in health-care facilities. *Morbidity and Mortality Weekly Report*, **43**(RR-13), 1–132.

Tseng, C. and Li, C. (2007) Inactivation of viruses on surfaces by ultraviolet germicidal irradiation. *Journal of Occupational and Environmental Hygiene*, **4**, 400–405.

Tucker, J.B. and Zilinskas, R.A. (2003) The 1971 smallpox outbreak in the Soviet city of Aralsk: implications for variola virus as a bioterrorist threat. *Critical Reviews in Microbiology*, **29**, 81–95.

US Environmental Protection Agency (2010) Investigation of Simulated Sunlight in the Inactivation of *B. anthracis* and *B. subtilis* on Outdoor Materials, Report EPA/600/R-10/048, Office of Research and Development, National Homeland Security Research Center, Washington, DC.

Xu, P., Kujundzic, E., Peccia, J. *et al.* (2005) Impact of environmental factors on efficacy of upper-room air ultraviolet germicidal irradiation for inactivation airborne mycobacteria. *Environmental Science and Technology*, **39**, 9656–9664.

Xu, P., Peccia, J., Fabian, P. *et al.* (2003) Efficacy of ultraviolet germicidal irradiation of upper-room air in inactivating airborne bacterial spores and mycobacteria in full-scale studies. *Atmospheric Environment*, **37**, 405–419.

Xue, Y. and Nicholson, W.L. (1996) The two major spore dna repair pathways, nucleotide excision repair and spore photoproduct lyase, are sufficient for the resistance of *Bacillus subtilis* spores to artificial UV-C and UV-B but not to solar radiation. *Applied and Environmental Microbiology*, **62**(7), 2221–2227.

18

Radioactive Aerosols: Tracers of Atmospheric Processes

Katsumi Hirose
Department of Materials and Life Sciences, Faculty of Science and Technology, Sophia University, Japan

18.1 Introduction

Radionuclides emitted in the atmosphere, which have gaseous forms and contain fine particles, turn into radioactive aerosols through gas–particle conversion and coagulation. The interaction of these radioactive aerosols with the environment has been studied in terms of radiological effects and atmospheric tracers (Chamberlain, 1991; Papastefanou, 2008). Human activities have an effect on the atmospheric environment via the increasing emission of anthropogenic pollutants, including natural and anthropogenic radionuclides. Public interest in radioactive aerosols began in the mid 1950s, when global fallout of fission products from nuclear weapons tests was first observed. The thermonuclear tests at Bikini Atoll (Bravo Test) in 1954 had radiological effects on human health in the form of fallout of radioactive ash. Nuclear reactor accidents such as Windscale, Three Mile Island, Chernobyl, and Fukushima have caused the atmospheric emission of large amounts of radioactive aerosols. Natural radionuclides, typically radon and its decay products, also pose a potential hazard to human health, via their exposure to uranium and other mines. In order to assess the radiological effects of anthropogenic and natural radionuclides, it is important to elucidate the environmental behaviors of radioactive aerosols.

Radioactive aerosols have been applied to tracers of atmospheric processes, including source tracking and transport, wet and dry deposition, residence time, and others (Junge, 1963; Warneck, 1988; Chamberlain, 1991). Natural radionuclides, especially

Aerosol Science: Technology and Applications, First Edition. Edited by Ian Colbeck and Mihalis Lazaridis.

radon progenies and ^{7}Be, have been extensively used in elucidating aerosol atmospheric processes (Papastefanou, 2008; Baskaran, 2011). The anthropogenic radionuclides found in aerosols are powerful transient tracers that allow a better understanding of aerosol atmospheric behaviors, especially the dispersion of radioactive clouds from nuclear accidents, by enabling their sources, including emission processes, to be identified. In this chapter, we examine the atmospheric processes of radioactive aerosols.

18.2 Origin of Radioactive Aerosols

18.2.1 Natural Radionuclides

Natural radioactive aerosols can be classified in the following categories: (i) radioactive aerosols associated with radioactive nuclides of cosmogenic origin, such as ^{7}Be, ^{22}Na, ^{32}P, and ^{35}S; (ii) aerosols associated with radon and thoron decay products; and (iii) terrestrial materials such as ^{40}K, uranium, and thorium.

Typical cosmogenic radionuclides found in aerosols include nuclides with relatively short half-lives: ^{7}Be (53.3 days), ^{32}P (14.3 days), ^{33}P (25.3 days), and ^{35}S (84.7 days); and nuclides with longer half-lives: ^{10}Be (1.5×10^{6} years), ^{22}Na (2.6 years), ^{26}Al (7.17×10^{5} years), ^{32}Si (280 years), and ^{36}Cl (3.01×10^{5} years). Other cosmogenic radionuclides, which primarily exist in gaseous forms, are found in the atmosphere: ^{3}H (12.5 years), ^{14}C (5730 years), and ^{39}Ar (269 years). These are produced in the stratosphere and upper troposphere by the interaction of cosmic-ray particles with atmospheric components. Most are formed by the spallation processes of light atmospheric nuclei, such as nitrogen, oxygen, and carbon, or of heavier nuclei, such as sodium, phosphorus, sulfur, potassium, and calcium (NCRP, 1987). The global average production rates and concentrations of cosmogenic radionuclides in the atmosphere are summarized in Table 18.1. Among the cosmogenic radionuclides, ^{7}Be is extensively used as an atmospheric tracer because of its high production rate (8.1×10^{-2} atoms cm^{-2} s^{-1}) and a gamma emitter (477.6 keV) that can easily be detected and measured in atmospheric samples. Its production rate depends on cosmic ray intensity, which is inversely related to

Table 18.1 *Production rates and concentrations of cosmogenic radionuclides in the atmosphere (UNSCEAR, 2000).*

Radionuclides	Production rate		Global inventory (PBq)	Concentrations in troposphere (mBq m^{-3})
	Per unit area (atoms m^{-2} s^{-1})	Annual amount (PBq yr^{-1})		
^{7}Be	810	1960	413	12.5
^{10}Be	450	6.4×10^{-5}	230	0.15
^{22}Na	0.86	0.12	0.44	0.0021
^{26}Al	1.4	0.1×10^{-5}	0.71	1.5×10^{-8}
^{32}Si	1.6	8.7×10^{-4}	0.82	2.5×10^{-5}
^{32}P	8.1	73	4.1	0.27
^{33}P	6.8	35	3.5	0.15
^{35}S	14	21	7.1	0.16
^{36}Cl	11	1.3×10^{-5}	5.6	9.3×10^{-8}

variation in sunspots. Production rates of ^{7}Be vary by approximately 25% with the 11-year solar cycle. The other radionuclides referred to in this chapter are mostly beta-emitters, with low production rates in the atmosphere and, therefore, very low concentrations in the troposphere. For ^{7}Be-bearing aerosols, activity median aerodynamic diameters (AMADs) are within the range 0.33 to 1.18 μm (Bondietti and Brantley, 1986; Papastefanou and Ioannidou, 1995; Winkler *et al.*, 1998; Yu and Lee, 2002; Porstendörfer and Gründel, 2003). It is noteworthy that there is a significant contribution from resuspension of deposited radionuclides on land surface to long-lived cosmogenic radionuclides.

The gaseous radon (^{222}Rn, half-life: 3.82 days) and thoron (^{220}Rn, half-life: 55.6 seconds) in the atmosphere are produced as decay products of uranium and thorium, respectively, which are emanated from soil and rocks, with lesser amounts of radon released from the ocean. Radon is therefore a tracer of continental air mass. Radon isotopes emitted from land surface are diffused and advected in the atmosphere and produce progenies (polonium, radioactive bismuth, and radioactive lead) by radioactive decay. The short-lived progenies of ^{222}Rn are ^{218}Po (half-life: 3.05 minutes), ^{214}Pb (26.8 minutes), and ^{214}Bi (19.7 minutes), whereas most of the radionuclides of the ^{220}Rn decay chain are very short-lived (less than 1 minute), except ^{212}Pb (10.64 hours) and ^{212}Bi (3.05 minutes). A long-lived radionuclide in the ^{222}Rn decay chain is ^{210}Pb (22.2 years). Pb-210 is also a tracer of continental air mass (Turekian and Cochran, 1981). Although most of the atmospheric ^{210}Pb is an airborne radon progeny, aerosol ^{210}Pb contains contributions from other sources, such as resuspension of deposited ^{210}Pb and soil ^{210}Pb, and biomass burning (Hirose *et al.*, 2011). The progenies of ^{210}Pb – that is, ^{210}Bi (5.012 days) and ^{210}Po (138.4 days) – are also important to elucidating the atmospheric behaviors of aerosols. Radon progenies, which are formed as positive ions, attach submicrometer particles – so-called 'condensation nuclei' – in the air. In surface air, the AMADs of ^{210}Pb-bearing particles range from 0.28 to 0.77 μm (Sanak, Gaudry, and Lambert, 1981; Winkler *et al.*, 1998; Porstendörfer and Gründel, 2003).

Major isotopes of uranium (^{238}U, 4.47×10^{9} years) and thorium (^{232}Th, 1.41×10^{10} years) in airborne dust and rainwater are of lithogenic origin; uranium mine operation sites are common sources of uranium- and thorium-bearing particles (Martin, 2003). Therefore, uranium and thorium in rainwater and airborne dust are primarily tracers of soil dust and contaminated soil particles resulting from human activities. On the other hand, significant amounts of uranium and thorium are also released to the atmosphere by industrial activities such as coal burning. Fly ash particles are considerably enriched in several radionuclides (Tadmor, 1986; Kolb, 1989). In addition to the major isotopes of uranium and thorium, several long-lived isotopes of uranium- and thorium-series have also been observed in rainwater and airborne particles; for uranium: ^{235}U (half-life: 7.04×10^{8} years) and ^{234}U (2.45×10^{5} years), a progeny of ^{234}Th (24.1 days) belonging to the U-238 series; and for thorium: ^{230}Th (7.54×10^{4} years), a progeny of ^{234}U belonging to the U-238 series, and ^{228}Th (1.91 years), a progeny of ^{228}Ra (5.75 years) belonging to the Th-232 series. Anthropogenic sources, such as ^{235}U released from atmospheric burn-up of nuclear satellites (Krey *et al.*, 1979) and nuclear explosions of ^{235}U (Matsunami and Mamuro, 1975), cause perturbations of isotope ratios of uranium, whereas natural fractionation processes in the course of erosion of rocks can occur between progenies of uranium and thorium; for example, ^{238}U–^{234}U and ^{232}Th–^{228}Th. Therefore, the isotope ratios of uranium and thorium can provide important information on sources of airborne particles and transport processes.

18.2.2 Anthropogenic Radionuclides

The anthropogenic radionuclides in aerosols consist of fission products (typically, ^{137}Cs (half-life: 30.17 years), ^{90}Sr (28.79 years), ^{89}Sr (50.53 days), and ^{131}I (8.02 days)) and fissile materials such as plutonium. Anthropogenic radionuclide-bearing aerosols have been produced by atmospheric nuclear weapons testing, satellite burn-up, nuclear reactor accidents, and other sources (UNSCEAR, 2000).

Anthropogenic radionuclides were first injected into the atmosphere by atmospheric nuclear explosions at Alamogardo (New Mexico, USA) and Hiroshima and Nagasaki (Japan) in 1945. During the period of 1945–1980, 541 atmospheric nuclear weapons tests were conducted by the USA, the USSR, China, France, and the UK (UNSCEAR, 2000). For example, the Nevada test site in the USA was the location for 86 atmospheric nuclear tests from 1951 to 1962. Anthropogenic radionuclides released from atmospheric nuclear tests, serious nuclear reactor accidents, and other sources (e.g., atmospheric burn-up of nuclear satellites) are spread across the globe and are almost globally detectable in the environment (UNSCEAR, 2000). The dominant source of anthropogenic radionuclides in the environment is atmospheric nuclear weapons testing. Peaks in the annual deposition of anthropogenic radionuclides occurred in 1963 in northern hemisphere stations, just after the 1961–1962 large-scale atmospheric nuclear testing conducted by the USA and the USSR; the maximum deposition in southern hemisphere stations appeared in 1964. After the Limited Test Ban Treaty in 1963, atmospheric nuclear tests were conducted by China and France. Until the early 1980s, the atmospheric levels of anthropogenic radionuclides were dominantly supported by Chinese and French nuclear explosions.

Other dominant sources of anthropogenic radionuclides in the atmosphere are nuclear reactor and nuclear facility accidents; Windscale (UK) in 1957, Three Mile Island (USA) in 1979, Chernobyl (former USSR) in 1986, and Fukushima Daiichi NPP (nuclear power plant) (Japan) in 2011 are recorded as major nuclear reactor accidents. Especially large amounts of radionuclides were emitted to the atmosphere from the Chernobyl and Fukushima accidents. Table 18.2 shows the estimated releases of ^{90}Sr, ^{131}I, ^{137}Cs, ^{134}Cs (half-life: 2.06 years), and ^{144}Ce (284.9 days) from the Nevada tests, the thermonuclear tests (H-bomb tests), the 1957 Windscale accident, a 1957 accident at a separation plant

Table 18.2 *Release of fission products in weapons tests and accidents.*

Source	^{90}Sr (PBq)	^{131}I (PBq)	^{137}Cs (PBq)	^{134}Cs (PBq)	^{144}Ce (PBq)
Nevada atmospheric tests (total yield 1 MT)	4	4×10^3	6	–	2×10^2
All H-tests (total yield 200 MT)	8×10^2	8×10^5	1×10^3	–	4×10^4
Windscale accident	0.3×10^{-3}	0.9	0.08	–	~0.01
Chemical explosion in Urals	4	–	0.03	–	50
Three Mile Island accident	–	0.6×10^{-3}	–	–	–
Chernobyl accident	8	1.8×10^3	86	50	170
Fukushima accident	0.14	140	15	18	0.1×10^{-3}

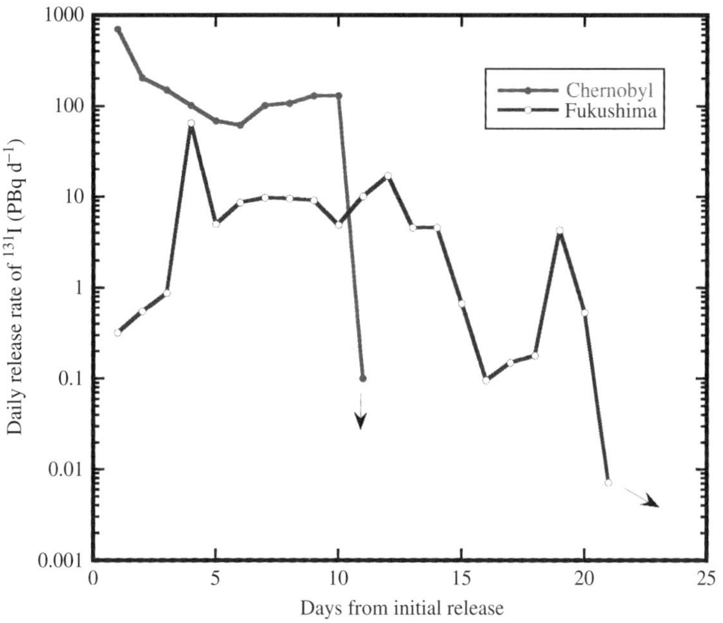

Figure 18.1 *Temporal changes in daily ^{131}I emission rates from the nuclear reactors at Chernobyl and Fukushima Daiichi NPP. See plate section for colour version.*

in the Urals, the 1986 Chernobyl accident, and the 2011 Fukushima accident. Figure 18.1 shows ^{131}I emission histories from the Chernobyl and Fukushima Daiichi NPP accidents. Although the release of ^{131}I from the Chernobyl accident was about half of that from the Nevada tests, ^{131}I from Chernobyl affected the human thyroid as a result of its short period of emission (less than 10 days); the release of ^{131}I from the thermonuclear explosions was about 3 orders of magnitude greater than that at Chernobyl, but only a small fraction reached ground because most decayed in the stratosphere.

Transuranics (typically plutonium isotopes and americium) and ^{235}U, which are fissile materials and products of neutron reaction, are released in the atmosphere by atmospheric nuclear tests, satellite accidents, launched nuclear batteries and reactors, and nuclear reactor accidents. With atmospheric nuclear explosions and the reentry of nuclear satellites, most radionuclides are emitted into the atmosphere as fine particles. Larger particles, including plutonium and ^{235}U, are released in nuclear reactor accidents such as that at Chernobyl. About 15 PBq of ^{239,240}Pu and 0.3 PBq of ^{238}Pu have been released into the atmosphere by atmospheric nuclear tests. 0.63 PBq of ^{238}Pu was injected into the upper atmosphere by the burn-up of the US satellite SNAP-9A in 1964. The reentries of the Soviet satellites Cosmos 954 in January 1978 and Cosmos 1402 in February 1984 (Krey *et al.*, 1979; Sakuragi, Meason, and Kuroda, 1983) dispersed 50 kg of fine particles of ^{235}U. In the 1950s, 20 kg of uranium was emitted into the atmosphere from the Windscale reactor. Following the Chernobyl accident, about 6–8 tonnes of UO_2 and 0.035, 0.03, 0.042, and ~6 PBq of ^{238}Pu, ^{239}Pu, ^{240}Pu, and ^{241}Pu, respectively, were released into the atmosphere (Victorova and Garger, 1990; UNSCEAR, 2000).

18.3 Tracers of Atmospheric Processes

18.3.1 Transport of Radioactive Aerosols

Anthropogenic radionuclides originating from atmospheric nuclear explosions are injected into the stratosphere and upper troposphere; the injection altitude is dependent on the scale and height of the explosion (Reiter and Bauer, 1975); most thermonuclear explosions conducted in the troposphere inject radionuclides into the stratosphere. The stratosphere is of scientific interest because of the possibility of tracing stratosphere transport. The stratosphere aerosol layer was first described by Junge and Manson (1961). Anthropogenic radionuclides produced by a nuclear explosion immediately attach to submicrometer aerosols, although several fission products are initially present in gaseous forms (e.g., ^{137}Xe (half-life: 3.818 minutes) → ^{137}Cs). Stratospheric sampling of anthropogenic radionuclides has revealed that anthropogenic radioactivity derived from atmospheric nuclear testing persists in the stratosphere for years and that radionuclides are associated with particles of below 0.02 μm radius at distances above 27 km from the surface of the earth and with particles of nearly 0.1 μm radius at between 21 km and the tropopause (Martell, 1966). The change in the particle size distribution of radionuclide-bearing particles reflects the change of natural aerosols in stratosphere; a sulfate aerosol layer with submicrometer radius exists at around 21 km altitude (Junge and Manson, 1961). For troposphere circulation, trade-wind circulation in the lower latitude has been successfully traced based on the dispersion of radionuclides emitted from the nuclear weapon tests at or near the earth's surface close to the equator in 1952 and 1954 (Machta, List, and Huber, 1956). The westerly circulation in the midlatitude has been traced by using radioactivity measurements and an air trajectory analysis of the Chinese nuclear explosion in 1965 (Kuroda, Miyake, and Nemoto, 1965).

Nuclear reactor accidents are a typical example of lower-atmospheric injection of radioactive aerosols. Large amounts of anthropogenic radionuclides were injected into the lower troposphere from the Chernobyl accident (from April 28 to May 6, 1986). As a result, several anthropogenic radionuclides were observed in northern hemisphere air and rainwater, mainly in the Belarus, Ukraine, and Scandinavian regions. Most of the radioactivity reaching Western Europe was carried on aerosols with a diameter of less than 2 μm (Jost *et al.*, 1986; Winkelmann *et al.*, 1987). On the other hand, larger particles, which bear a higher proportion of refractory elements, such as plutonium isotopes, reached Lithuania and Scandinavia (Devell *et al.*, 1986; Persson, Rodhe, and de Geer, 1987). The Chernobyl radioactivity was observed in the air at remote sites. Levels of ^{137}Cs and ^{134}Cs (half-life: 2.046 years) in surface air and rainwater collected at Tsukuba, Japan in May 1986, as well as of volatile short-lived radionuclides (^{131}I (8.02 days), ^{103}Ru (39.26 days), and others), increased markedly over the previous month (Aoyama *et al.*, 1986), whereas plutonium did not show any marked increase in Japan (Hirose and Sugimura, 1990), although higher depositions of the Chernobyl-derived transuranics were observed in Eastern Europe (IAEA, 2006). After the 2011 Great East Japan Earthquake, a severe accident occurred at the Fukushima Daiichi NPP and huge amounts of radionuclides were released into the environment. Figure 18.2 shows the temporal variation of ^{137}Cs activity concentration in surface air in the central part of Japan at this time. The Fukushima-derived ^{137}Cs concentration in surface air varied greatly, affected by such factors as the change in the pathway of radioactive plume and the change in emission intensity at the accidental site.

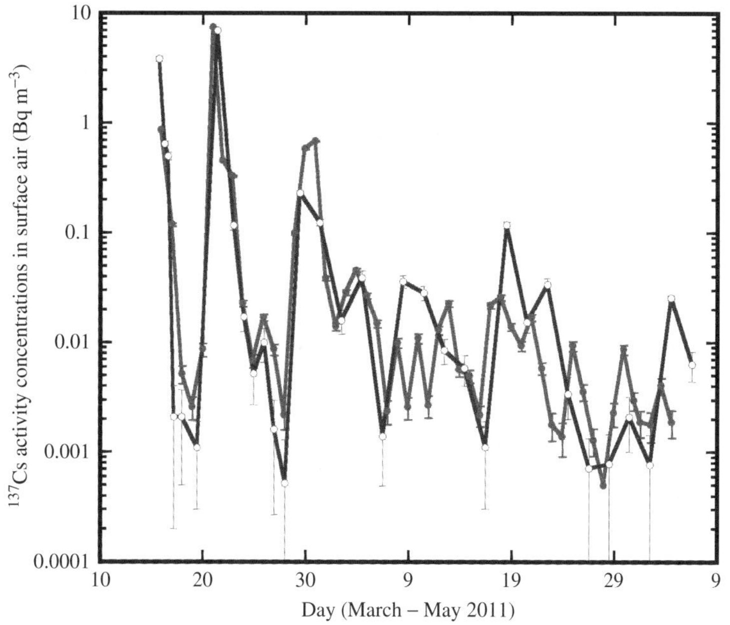

Figure 18.2 *Temporal changes in Fukushima-derived* ^{137}*Cs concentrations in surface air at Tsukuba and Inage, Japan. Open circle: Tsukuba, closed circle: Inage. Data cited from Amano* et al. *(2012) and Doi* et al. *(2013). See plate section for colour version.*

Radioactive aerosols have been used for the verification of model simulation studies. Numerical model simulation has developed to predict the dispersion of radionuclides emitted in the atmosphere by nuclear plant accidents (Imai *et al.*, 1985; Terada and Chino, 2008). It is essential to determine the 3D atmospheric distribution and deposition of radioactive aerosols by environmental monitoring in order to assess radiological effects for the public and the environment, and numerical modeling provides an important tool for estimating the local, regional, and global dispersion of accidentally released radionuclides. After the Chernobyl accident in 1986, numerical models were used to study the distribution of radionuclides across Europe (e.g., Hass *et al.*, 1990; Brandt, Christensen, and Frohn, 2002) and globally (Anspaugh, Catlin, and Goldman, 1988). The dispersion of the Fukushima-derived radionuclides in 2011 has been extensively studied using numerical model simulation:

- (i) Simulation models were used to estimate total amounts and emission histories of radionuclides emitted from the Fukushima Daiichi NPP based on monitoring data (Chino *et al.*, 2011; Stohl *et al.*, 2011).
- (ii) Numerical models were used to simulate the dispersion of the radioactive cloud; typically, temporal change in the dispersion of the radioactive plume at global scale are calculated by diffusion and advection models using meteorological data (Takemura *et al.*, 2011). The numerical model, including back- and forward-trajectory analysis, was able to reproduce when the Fukushima-derived radioactive cloud reached North America, Europe, and Asia (Hernández-Ceballos *et al.*, 2012; Lujaniené

et al., 2012). Regional-scale model simulation (Huh, Hsu, and Lin, 2012) revealed that the Fukushima-derived radioactive plume was predominantly transported toward the south-west under phases of north-easterly winds in the first week, April 6–7, 2011.

- (iii) Numerical models were applied to reproduce the spatial distribution of deposition of the Fukushima-derived radionuclides; models roughly reproduced the observed temporal and spatial distributions of radioactive deposition at the regional scale, although the flow of the radioactive cloud was restricted by land topography, due to surface emission (Morino, Ohara, and Nishizawa, 2011; Yasunari *et al.*, 2011).

18.3.2 Dry Deposition

Dry deposition is an important pathway for the removal of radioactive aerosols. Theoretical and experimental studies of particle and gas dry deposition processes have been carried out and reviewed by many researchers (Chamberlain, 1953; Slinn, 1978; Sehmel, 1980; Underwood, 2001). Dry deposition, which depends on land-surface conditions (such as grass, plat plain, etc.) and on meteorological conditions, is controlled by gravitational settling, Brownian diffusion, and compaction, as is the physicochemical form of radioactive aerosols. Radioactive aerosols are an important tool by which to gain a better understanding of the dry deposition of aerosols in the environment.

The dry deposition velocity has been applied in order to evaluate aerosol transfer to the earth's surface. The dry deposition velocity, $V_{d,r}$, is defined as the ratio of dry deposition flux to land-to-surface air concentration:

$$V_{d,r} = F_r / C_{air,r} \tag{18.1}$$

where F_r and $C_{air,r}$ are the deposition flux of radioactive aerosols ($\mathrm{Bq\,m^{-2}\,s^{-1}}$) and the surface air concentration of radioactive aerosols ($\mathrm{Bq\,m^{-3}}$), respectively. The dry deposition velocity is a function of particle size and wind velocity; it increases with decreasing particle diameter at particle sizes less than 0.1 μm, which can be controlled by the Brownian diffusion; it exhibits a minimum in the particle size range of 0.1–1.0 μm; and it increases with increasing particle diameter at particle sizes above 1 μm, where impaction and gravitational settling govern the shape of the increasing curve. The rate of gravitational settling of soil particles is calculated to be around $1 \times 10^{-3}\ \mathrm{m\,s^{-1}}$ for a mean particle size of 7 μm and a particle density of about $1.5\ \mathrm{g\,cm^{-3}}$, which is the minimum value, since the dry deposition velocity increases as a result of friction velocity and surface conditions (Sehmel, 1980).

The dry deposition velocities of radioactive aerosols have been determined by field observations. They depend on the type of radionuclide and isotope; for ^{7}Be-bearing particles, the dry deposition velocity varies from 1.0×10^{-3} to $3.4 \times 10^{-2}\ \mathrm{m\,s^{-1}}$ (Chamberlain, 1953; Small, 1960; Peirson *et al.*, 1973; Young and Silker, 1980; Turekian, Benninger, and Dion, 1983; Todd *et al.*, 1989; Papastefanou *et al.*, 1995; Rosner, Hotzl, and Winkler, 1997; McNeary and Baskaran, 2003); for ^{210}Pb-bearing particles, it varies from 0.7×10^{-2} to $1.1 \times 10^{-2}\ \mathrm{m\,s^{-1}}$ (Turekian *et al.*, 1983; Todd *et al.*, 1989; McNeary and Baskaran, 2003); and for ^{137}Cs-bearing particles, it varies from 3.8×10^{-4} to $6.3 \times 10^{-2}\ \mathrm{m\,s^{-1}}$ (Aoyama, Hirose, and Takatani, 1992; Papastefanou *et al.*, 1995; Rosner *et al.*, 1997). Amano *et al.* (2012) have determined the dry deposition velocities of the Fukushima-derived ^{131}I and ^{137}Cs during the period of March 14–17 to be around 2×10^{-3} to $3 \times 10^{-3}\ \mathrm{m\,s^{-1}}$ for ^{134}Cs and ^{137}Cs and 1×10^{-3} to $3 \times 10^{-3}\ \mathrm{m\,s^{-1}}$ for ^{131}I. The dry deposition velocities of thorium isotopes ranged from 3.8×10^{-3} to 4.0×10^{-2} m s^{-1} (Hirose, 2000) and from 2.0×10^{-3} to 6.4×10^{-2} m s^{-1} over the same period (Crecelius *et al.*, 1978).

18.3.3 Wet Deposition

The major pathway of deposition of the natural radioactivity in the atmosphere is precipitation scavenging. The wet deposition velocity ($V_{wet,R}$) of the radionuclides is a useful tool for elucidating the wet removal processes of radionuclides (as is dry deposition velocity, which is calculated from the wet deposition flux (D_R: Bq m^{-2} s^{-1}) and surface air concentration ($C_{a,R}$: Bq m^{-3}) of the radionuclides using the equation $V_{wet,R}=D_R/C_{a,R}$). The wet deposition velocities of radioactive aerosols vary considerably and depend on the type of radionuclide and rainfall event (Hirose, 2000). Washout ratios are usually used to describe the wet removal processes of atmospheric pollutants (Englemann, 1971; Barrie, 1985). The wet deposition velocity can be calculated from:

$$V_{wet,R} = W\, I_R \tag{18.2}$$

where I_R is rainfall rate (mm h^{-1}) and W is the washout ratio (or scavenging ratio). Washout ratio is given by a ratio of the concentration in bulk precipitation (C_{rain}: Bq m^{-3}) to the concentration in surface air (C_{air}: Bq m^{-3}): C_{rain}/C_{air}. It must be noted that as another definition of washout ratio, W′ is equal to $\rho C_{rain}/C_{air}$, where C_{rain} and C_{air} are the concentration in bulk precipitation (Bq kg^{-1}) and surface air (Bq m^{-3}) and ρ is the density of air at standard conditions (1.2 kg m^{-3} at 20 °C and 760 mm Hg) (McNeary and Baskaran, 2003). The following relationship is established between different defined washout ratios: $W=10^3W'/\rho$.

The washout ratios of natural and anthropogenic radionuclides for individual rainfall events are usually on the order of 10^6; the washout ratios of ^{232}Th, ^{230}Th, and ^{228}Th range from 0.13×10^6 to 3.2×10^6, from 0.087×10^6 to 1.9×10^6, and from 0.16×10^6 to 4.6×10^6, respectively (Hirose, 2000); Harvey and Matthews (1989) give a value of 0.92×10^6 for ^{7}Be, while McNeary and Baskaran (2003) give values of from 0.18×10^6 to 1.7×10^6 with an average of 0.79×10^6 for ^{7}Be and from 0.046×10^6 to 1.9×10^6 with an average of 0.53×10^6 for ^{210}Pb. For anthropogenic radionuclides (1×10^6) derived from atmospheric nuclear tests (Slinn, 1978) and the Chernobyl radioactivity, ^{137}Cs: 0.19×10^6 to 0.53×10^6, ^{103}Ru: 0.16×10^6 to 0.37×10^6, and ^{90}Sr: 0.50×10^6 to 1.4×10^6 (Hirose, Takatani, and Aoyama, 1993).

The washout ratio of atmospheric aerosols depends on meteorological, physical, and chemical factors. From a theoretical standpoint (Slinn, 1978; Harvey and Matthews, 1989), the weak dependence of washout ratio on the rainfall rate is predicted to be:

$$W = a I_R^{-b} \tag{18.3}$$

where a and b are constants. For thorium isotopes, the washout ratios have been correlated with rainfall rate; values of a and b for ^{232}Th, ^{230}Th, and ^{228}Th are estimated to be 0.67×10^6 and 0.69, 0.43×10^6 and 0.72, and 0.88×10^6 and 0.81, respectively (Hirose, 2000). The washout ratio of the Chernobyl ^{90}Sr, whose AMAD was about 1 μm, seems to have been dependent on the rainfall rate, with b = 0.22 (Hirose *et al.*, 1993). On the other hand, the washout ratios for ^{137}Cs and ^{103}Ru in the Chernobyl radioactivity were nearly independent of the rainfall rate. The constancy of the washout ratio has been observed for ^{7}Be (Harvey and Matthews, 1989). It is noteworthy that ^{137}Cs and ^{103}Ru in the Chernobyl radioactivity and cosmogenic ^{7}Be attach to primarily submicrometer particles. A theoretical model suggests that b is expected to be in the range of 1/4 (for frontal storms) to 1/2 (for convective storms) (Harvey and Matthews, 1989). However, these findings suggest that the b values of chemical components, reflecting different physical and chemical types of aerosols, may be related to the particle sizes of corresponding chemical component-bearing particles instead of to rainfall type; in other words, b values seem to increase with particle size.

18.3.4 Resuspension

Resuspension is an important process in sustaining a level of anthropogenic radioactive aerosols in the surface air. The radionuclides deposited on to ground and/or vegetation are adsorbed on to fine organic or mineral particles. Some meteorological conditions, such as dry and strong wind, can blow off fragments of dried soil and vegetation. Initially, the resuspension process was concerned with the inhalation of resuspended radioactivity, especially plutonium, in areas contaminated by nuclear tests and accidental release from nuclear facilities. In order to characterize the resuspension of radioactive aerosols, the terms 'resuspension factor', k_r, and resuspension ratio, Λ, have been introduced, which are defined as the ratio of radioactivity concentration in the air (C_{air}: Bq m^{-3}) to the radioactivity deposited on the ground (D_s: Bq m^{-2}) and the ratio of vertical flux (Q: Bq m^{-2} s^{-1}) to the radioactivity deposited on the ground (D_s: Bq m^{-2}), respectively.

$$k_r = C_{air}/D_s \tag{18.4}$$

$$\Lambda = Q/D_s \tag{18.5}$$

The resuspension factor is a measure of the resuspension phenomenon. Resuspension factors have been calculated for aerosols associated with ^{7}Be and ^{137}Cs. For the ^{7}Be-bearing particles, the resuspension factor varies from 1.4×10^{-4} to 4.2×10^{-4} m^{-1} (average 2.3×10^{-4}; Papastefanou *et al.*, 1995); for the ^{137}Cs-bearing particles, from 10^{-8} to 1.2×10^{-4} m^{-1} (Stewart, 1966; Garland and Cambray, 1988; Papastefanou *et al.*, 1995), showing large variability. The lower resuspension factor (a mean value of 3×10^{-10} m^{-1}) was deduced by Shinn, Homan, and Gray (1983) from measurements of aerosol Pu over a bare field near the Savannah River Processing Plant. The resuspension factor depends on meteorological factors such as wind velocity and humidity, land conditions (climatologically factors, industrial and agricultural activities), and the chemical and physical properties of radionuclide-bearing particles.

Since the 1990s, the anthropogenic radionuclides observed in surface air and rainwater across the globe, except in the Chernobyl fallout, are considered to derive from the resuspension of radionuclides deposited on the land surface (Nicholson, 1988; Rosner and Winkler, 2001; Arimoto, Webb, and Conley, 2005; Karlsson *et al.*, 2008). More than 10 years since the cessation of atmospheric nuclear testing, the atmospheric deposition of ^{137}Cs, ^{90}Sr, and ^{239,240}Pu derived from nuclear tests is negligible. It is likely that the anthropogenic radionuclides and their activity ratios in surface air and rainwater reflect their redistribution processes on the land surface. In particular, regional transport of soil dust from the desert and arid region (e.g., the Saharan dust in Europe and the Asian dust in East Asia) is an important factor affecting resuspension of anthropogenic radionuclides (Igarashi *et al.*, 2001, 2003, 2009; Lee, Pham, and Povinec, 2002; Hirose *et al.*, 2003, 2004; Fujiwara *et al.*, 2007; Masson *et al.*, 2010). Although the levels of dominant anthropogenic

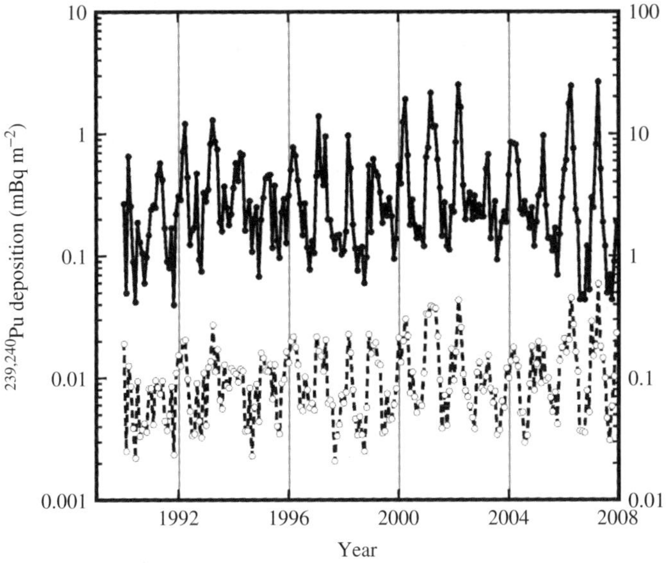

Figure 18.3 *Temporal variations of the monthly ^{239,240}Pu deposition and corresponding specific activity observed at Tsukuba, Japan. Solid line: monthly ^{239,240}Pu deposition; dotted line: ^{239,240}Pu specific activity.*

radionuclides in surface air in the 1990s and 2000s were supported by resuspension, atmospheric behavior differs among the anthropogenic radionuclides. For example, the ^{90}Sr and ^{137}Cs deposition fluxes observed in Japan since 1990 have been decreasing slowly, whereas ^{239,240}Pu deposition has been almost constant since 1985, although there is interannual variability (Figure 18.3). The monthly anthropogenic radionuclide depositions in East Asia exhibit seasonal changes, with a maximum in spring (Igarashi *et al.*, 2001, 2003; Hirose *et al.*, 2003; Hirose, Igarashi, and Aoyama, 2008); of ^{137}Cs, ^{90}Sr, and ^{239,240}Pu, ^{239,240}Pu deposition shows the most typical seasonal change (Hirose *et al.*, 2003, 2008). The annual and seasonal changes in ^{239,240}Pu deposition coincide with the occurrence of the Kosa (Asian dust) event observed in Japan in the spring (typically, March and April) (Hirose, Igarashi, and Aoyama, 2007). These findings suggest that the major origin of resuspension of the anthropogenic radioactive aerosols in Japan in the 1990s and 2000s is aeolian dust produced in the East Asian deserts and arid areas (Igarashi *et al.*, 2001; Hirose *et al.*, 2003).

Natural radionuclides, especially lithogenic elements, are a useful tool for elucidating resuspension processes. Most of the thorium in aerosols originates from soil particles (Hirose, 2000). The variation of thorium in surface air therefore provides a key to solving factors controlling resuspension. The temporal variation of ^{232}Th deposition in the 1990s and 2000s shows a marked increase in spring, coinciding with Asian dust (Kosa) events. Residual materials in deposition samples consist of soil dust, fly ash, sea salt, and others. The soil dust in residual materials comprises local soil and long-range transported soil particles: so-called 'Kosa'. The specific activities of thorium in deposition samples collected at Tsukuba range from 1.5 to 23.0 mBq g^{-1} (Hirose, Kikawada, and Igarashi, 2012); the thorium concentration in surface soil at Tsukuba is 17 mBq g^{-1} on average (range: 4.4–32.0 mBq g^{-1}, unpublished data). The ^{232}Th specific activities in deposition are approximately lower than the mean ^{232}Th concentration in surface soil because soil particles occupy part of the residual materials in deposition. The ^{230}Th/^{232}Th activity ratios vary according to sources, with the ratio high in local cultivated fields, due to fertilization, and low in arid and desert soils; they can therefore be used to differentiate between locally and remotely derived ^{232}Th (Hirose *et al.*, 2010). The ^{230}Th/^{232}Th activity ratios in aerosol and deposition samples show large variability, with high ratios occurring in early spring. These high ^{230}Th/^{232}Th ratios can be attributed to local dust storms, which cause resuspension of soils from cultivated fields. The ^{230}Th/^{232}Th ratios in deposition also allow us to separate locally and remotely derived ^{232}Th fractions. The results reveal that both locally and remotely derived ^{232}Th depositions show seasonal variations with maxima in spring, although the remotely derived fraction is dominant over the locally derived one. The locally derived ^{232}Th deposition shows a peak in early spring, which can be attributed to local dust-storm events. The ^{232}Th deposition maximum later in spring is attributable to the remotely derived fraction, corresponding to the Kosa events. Thorium isotopes in aerosols clearly show a coexistence of two components of resuspension: locally and remotely derived.

18.3.5 Other Processes

In addition to resuspension, biomass burning (e.g., forest fires) is another potential source of several anthropogenic radionuclides in the air and rainwater (e.g., in highly contaminated areas near the Chernobyl accident site), although the radioactive aerosols emitted by biomass burning depend on the chemical properties of the radionuclides (^{137}Cs $\gg$ Pu) (Lujanienė, Aninkevicius, and Lujanas, 2009). In fact, sporadic high ^{137}Cs concentrations in surface air were observed near the region heavily contaminated by the Chernobyl fallout (Grabowska *et al.*, 2003).

18.3.6 Application of Multitracers

Use of radionuclides with different physical and chemical properties as multitracers is effective in solving complicated atmospheric aerosol processes. Typically, many kinds of radionuclide, with different physical and chemical properties, are emitted into the atmosphere by nuclear reactor accidents. Compositions of the Chernobyl-derived radionuclides in air samples provide important information on the atmospheric processes of radioactive aerosols, as do reactor conditions at the site of the accident. The ^{134}Cs/^{137}Cs activity ratio in surface air for the Chernobyl accident, equal to 0.5, was constant during the period from May to June 1986, whereas the ^{131}I/^{137}Cs activity ratio observed at Tsukuba, Japan increased exponentially, and the ^{103}Ru/^{137}Cs activity ratio at Chernobyl showed higher values from May 20 to June 10 (Figure 18.4). Table 18.3 shows the relative surface air concentrations of the fission products and plutonium to ^{137}Cs at different monitoring stations. The differences in the atmospheric behaviors of different Chernobyl-derived radionuclides are attributable to the differences in their chemical properties (e.g., gas–particle exchange for radioiodine) and the method of release of the radioactive plume (larger amounts of ^{103}Ru were released in the secondary stage than in the initial explosion at Chernobyl). Table 18.3 reveals that there was a progressive reduction in the relative concentrations of the refractory fission products (^{95}Zr and ^{141}Ce; boiling points: 4400 and 2900 °C)

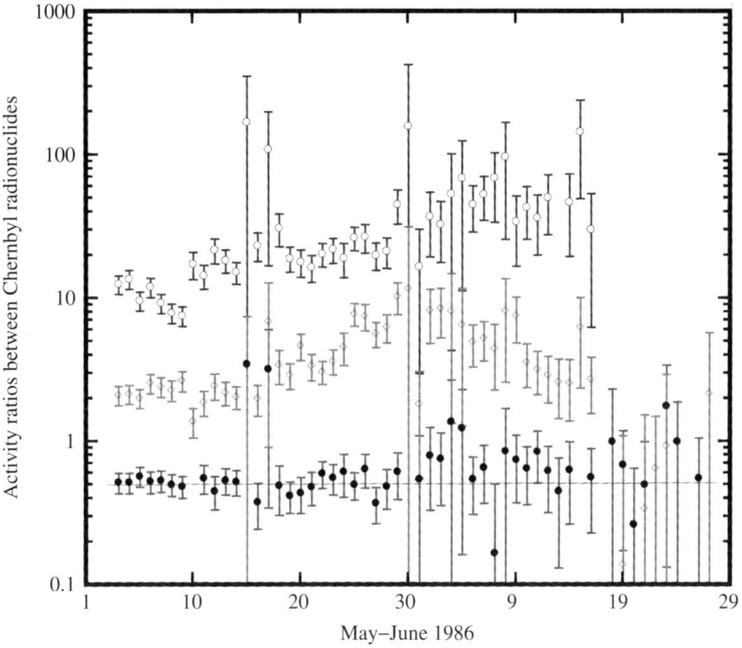

Figure 18.4 *Temporal variations of $^{131}I/^{137}Cs$ (open circle), $^{103}Ru/^{137}Cs$ (open rhombic), and $^{134}Cs/^{137}Cs$ (closed circle) activity ratios in surface air samples observed at Tsukuba, Japan. See plate section for colour version.*

Table 18.3 *Relative concentrations of the Chernobyl radionuclides in surface air.*

	Near Chernobyl	Vartyshevka (140 km SE) (1)	Baltic region (2)	Munich (3)	Harwell (4)	New York (5)	Tsukuba (6)
^{137}Cs	1	1	1	1	1	1	1
^{131}I	10	5	19	14	12	7	
^{90}Sr	–	–	0.02	0.01	–	–	0.001
^{95}Zr	3	1	0.2	0.02	–	0.01	–
^{103}Ru	6.8	–	0.4	1.3	1.7	0.3	2
^{140}Ba	2.5	3.2	1	0.7	0.6	0.2	–
^{141}Ce	3.7	–	0.1	0.02	0.01	0.02	–
$^{239,240}Pu$	–	–	2×10^{-5}	2×10^{-5}	–	–	2.8×10^{-6}

Reference: (1) USSR State Committee (1986); (2) Median of results for various locations from Krey *et al.* (1986), Devell *et al.* (1986), Cambray *et al.* (1987), and Aarkrog (1988); (3) Winkelmann *et al.* (1987); (4) Cambray *et al.* (1987); (5) Larsen *et al.* (1986); (6) Hirose (1995).

and plutonium (3228 °C) with increasing distance from Chernobyl. Although ruthenium is also refractory, its oxide, RuO_4, is relatively volatile. In fact, Chernobyl-derived ^{103}Ru was contained in submicrometer particles (0.35–0.65 μm) (Aoyama *et al.*, 1992). Ru, therefore, was more persistent in long-range transport. The $^{90}Sr/^{137}Cs$ ratio within the 80 km zone was 0.3, and a similar ratio was found by Aarkrog (1988) in a sample of soil from Kiev. For remote sites, the $^{90}Sr/^{137}Cs$ ratio in deposition near Munich, Germany and at Tsukuba, Japan, was about 0.01 (Winkelmann *et al.*, 1987; Hirose, 1995). These findings suggest that the effect of the Chernobyl radioactivity on the total deposition depended on the type of radionuclide. In order to elucidate factors controlling the differences in the atmospheric behaviors between different radionuclides, the relative contribution, F, of the Chernobyl-derived radionuclides to the Chernobyl-derived ^{137}Cs must be introduced; $F = (D_R/D_{Cs}) \cdot (I_R/I_{Cs})^{-1}$, where D_R and D_{Cs} are the depositions of the Chernobyl-derived radionuclide and ^{137}Cs, respectively, and I_R and I_{Cs} are the total amounts released of the Chernobyl-derived radionuclide and ^{137}Cs, respectively. The relative contributions, F, of the Chernobyl-derived ^{90}Sr and $^{239,240}Pu$ were only 1.5 and 0.18%, respectively – much smaller than the contribution of the Chernobyl-derived ^{137}Cs (Hirose, 1995). The difference in deposition behavior between the Chernobyl-derived radionuclides is mainly attributable to the difference in size between the radionuclide-bearing particles (size order: $^{137}Cs < ^{90}Sr < ^{239,240}Pu$); larger particles are more easily removed from the atmosphere by wet and dry deposition (Aoyama *et al.*, 1992; Hirose *et al.*, 1993, 2001; Hirose, 1995). Approximately 60% of the ^{95}Zr and $^{141,144}Ce$ emitted in the Chernobyl accident was deposited within 80 km, whereas 27% of the radiocesium and radioiodine was deposited in this zone. For the 2011 Fukushima accident, the dominant detected radionuclides in the air were ^{133}I, ^{132}I, ^{131}I, ^{134}Cs, ^{136}Cs, ^{137}Cs, ^{132}Te, and ^{129m}Te (Amano *et al.*, 2012; Doi *et al.*, 2013). Considerable amounts of refractory radionuclides were not detected in surface air samples in Japan. When radioactive decay is corrected at the end of the Fukushima Daiichi NPP incident, the ratios between isotopes are almost constant during the period March to April 2011: $^{133}I/^{131}I$, $^{134}Cs/^{137}Cs$, $^{136}Cs/^{137}Cs$, and $^{129m}Te/^{132}Te$ were 1.3, 1.1, 0.21, and 0.092, respectively. On the other hand, there were larger variations between elements; decay-corrected $^{131}I/^{137}Cs$, decay-corrected $^{132}Te/^{137}Cs$, and decay-corrected $^{99m}Tc/^{137}Cs$ were in the ranges 8 to 1.0×10^3, 4 to 65, and 0.16 to 262, respectively. Large variations of $^{131}I/^{137}Cs$, $^{132}Te/^{137}Cs$, and $^{99m}Tc/^{137}Cs$ suggest that atmospheric emission of the Fukushima-derived radionuclides and the removal of these radionuclides from the atmosphere were greatly affected by the chemical and physical properties of elements. In fact, significant amounts of ^{131}I and ^{133}I existed in gaseous form, whereas radiocesium attached to relatively larger particles (>1 μm).

18.3.7 Atmospheric Residence Time of Radioactive Aerosols

Classical compartment models (Krey and Krajewski, 1970) are more effective than computer model simulation in providing an improved conceptual understanding of natural processes; corresponding atmospheric residence times are important values for constraining the timescale of aerosols that exist in the compartment. Land-based sampling of anthropogenic radionuclides can provide important information on the stratospheric motion of aerosols. Long-range monitoring of ^{90}Sr, ^{137}Cs, $^{239,240}Pu$, and ^{238}Pu reveals that their temporal variations reflect the global air motion of radionuclide-bearing particles in stratosphere (Hirose *et al.*, 1987). Long-term measurements of anthropogenic radioactivity in surface aerosols and deposition allow us to have information on the transport processes and residence times of aerosols in the upper troposphere and the stratosphere (Reiter and Bauer, 1975; Katsuragi, 1983; Hirose *et al.*, 1987). Long-term observation of atmospheric radionuclides in the northern hemisphere mid-latitude region indicates that the annual deposition of radioactive debris from the thermonuclear tests varies with apparent stratospheric residence times of 0.5–1.7 years, which suggests that three layers with different timescales – that is, the upper stratosphere, the lower stratosphere (below 21 km), and the active mixing and exchange (AME) layer (just above the troposphere) – exist in the stratosphere. The half-transport rates from the upper-to-lower-stratosphere compartment, from the lower-stratosphere-and-AME-layer compartment, and from the AME layer to the troposphere, are 0.5, 0.7, and 0.3 years, respectively (Hirose *et al.*, 1987). The lower stratosphere with the longest half-residence time (HRT) corresponds to the stratospheric sulfate layer.

The Chernobyl ^{137}Cs concentration in surface air (Figure 18.5) in Japan decreased exponentially, which apparently corresponds to a half-life of 6.5 days. The geometric mean of weekly measurements of ^{137}Cs at Chilton, Gibraltar, Tromso, and Hong Kong reveals that the ^{137}Cs concentration in surface air declines with a half-life of 6.3 days (Cambray *et al.*, 1987). On the other hand, long-term measurements of the Chernobyl-derived ^{137}Cs deposition provides an apparent HRT of about 25 days (Aoyama, 1999). Longer-term measurements near Munich have shown that continuing surface air concentrations of ^{137}Cs decline with a half-life of 230 days (Hötzl, Rosner, and Winkler, 1989). This longer half-life of ^{137}Cs is attributed to resuspension of locally deposited ^{137}Cs from the monitoring site. Although most of the Chernobyl radioactivity was injected in the lower troposphere, small amounts of the Chernobyl ^{137}Cs (about 0.5% of total release) were transported into the stratosphere (Jaworowski and Kownacka, 1988; Aoyama, Hirose, and Sugimura, 1991; Aoyama, 1999). For the Fukushima accident, the maximum ^{137}Cs deposition occurred in March 2011 and exponentially decreased with a half-life of around 12 days during the period of March to June 2011 (Figure 18.6) (Hirose, 2012). Longer-term monitoring of

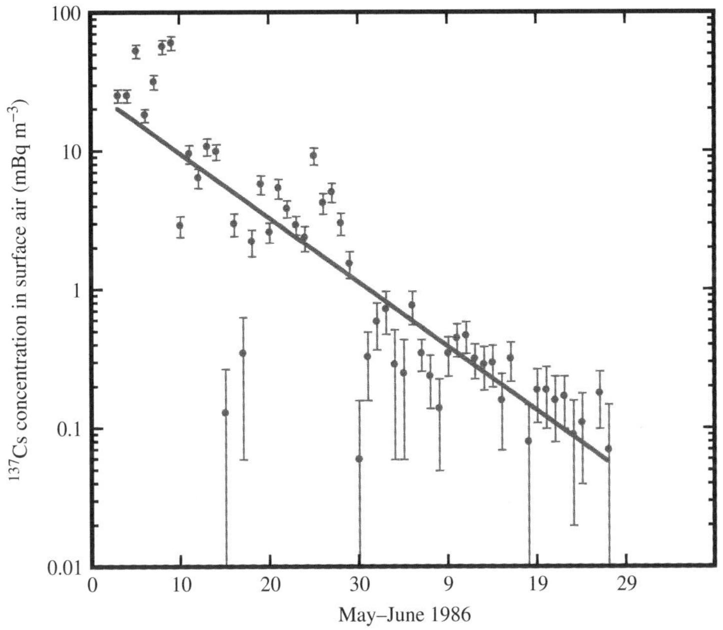

Figure 18.5 *Temporal variation of the Chernobyl-derived ^{137}Cs concentration in surface air at Tsukuba, Japan.*

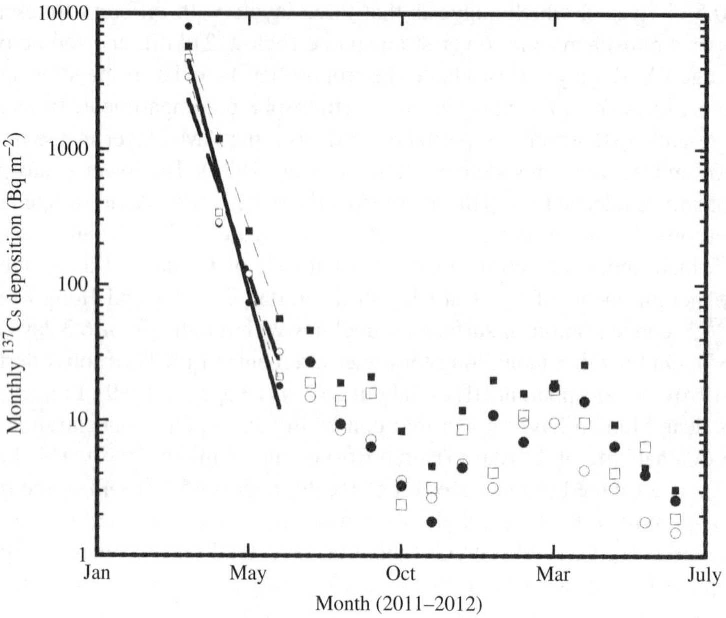

Figure 18.6 *Temporal variation in monthly deposition of the Fukushima-derived ^{137}Cs observed at monitoring stations in the Kanto Plain. Closed circle: Tokyo; open circle: Chigasaki; closed square: Utsunomiya; open square: Takasaki.*

the monthly ^{137}Cs deposition revealed that there was a longer half-life (50 days) during the period of July to October and that a small second peak in ^{137}Cs deposition occurred in winter and spring of 2012, which might be attributed to the resuspension of locally deposited ^{137}Cs (Hirose, 2013).

Within natural radionuclides, ^{7}Be (like Rn progenies) is extensively used as a tool for solving the atmospheric behaviors of aerosols. Be-7 produced in the upper atmosphere is initially attached to Aitken particles (0.015 μm), which increase their size by coagulation in the troposphere. The residence time of tropospheric aerosols has been estimated from the particle diameter in surface air and the particle growth rate, which is estimated at 0.004–0.005 μm h^{-1} (McMurry and Wilson, 1982). The estimated residence time for ^{7}Be-bearing particles ranges from 2.6 to 11.8 days (Papastefanou and Ioannidou, 1995; Winkler *et al.*, 1998; Yu and Lee, 2002). The implication of the residence time determined from the particle growth rate and particle diameter is rigorously different from that for the tropospheric residence time deduced from the mean concentration of radionuclide in the troposphere air column and the corresponding total depositional flux out of the air column (Ehhalt, 1973). In a 3D chemical tracer model, the tropospheric residence time of ^{7}Be is 21 days (Koch, Jacob, and Graustein, 1996).

Atmospheric ^{210}Pb is important as an input function on land and on sea, and also as a tracer of the atmospheric processes of aerosols (Baskaran, Coleman, and Santschi, 1993). The atmospheric residence time estimated for ^{210}Pb over the west-central USA is 8 days, based on the radioactive equilibrium of ^{210}Pb and its short-lived progenies (Moore, Poet, and Martell, 1973). Lambert *et al.* (1982) estimated a global mean aerosol residence time of 6.5 days by using the atmospheric inventories of ^{222}Rn and ^{210}Pb extrapolated from observations and computing the ^{210}Pb deposition sink to balance ^{222}Rn decay. The residence times of tropospheric aerosols inferred from a global 3D simulation of ^{210}Pb was 5–10 days, depending on season and latitude (Balkanski *et al.*, 1993). Koch *et al.* (1996), using a 3D chemical tracer model similar to that of Balkanski *et al.* (1993), calculated a tropospheric residence time of 9 days for ^{210}Pb-bearing particles at different latitudes from 80° S to 80° N.

18.4 Tracer of Environmental Change

Long-term variation of anthropogenic radionuclides in aerosols and deposition reflects changes in the earth's surface, including migration of radionuclides across the land surface by erosion, caused by increasing human activities and resultant climate change. The temporal variations in the monthly depositions and corresponding specific activities of anthropogenic radionuclides in depositions (activity in residue, in Bq g^{-1}) are shown in Figure 18.3 (Hirose *et al.*, 2003, 2008). The $^{239,240}Pu$ specific activities, which were high in the dust season (January to May) and low in the non-dust season (Figure 18.3), showed no overall decrease during the period 1990–2007. Compared to other years, high $^{239,240}Pu$ specific activities occurred in the dust seasons (March or April) of 2000, 2001, 2002, 2006, and 2007. Plutonium isotope ($^{240}Pu/^{239}Pu$) ratios (Hirose *et al.*, 2004) suggest that the plutonium in the recent deposition samples originated from nuclear weapons testing. The ^{137}Cs specific activities, which, like those of plutonium, were high in the dust season and low in the non-dust season (Figure 18.7a), gradually decreased during the period 1990–2006 (Hirose *et al.*, 2008). The apparent HRTs of the ^{137}Cs specific activity in the dust and non-dust seasons are calculated to have been 30 ± 10 and 16 ± 4 years, respectively, during the period 1992–2006, taking into account the effect of the Chernobyl ^{137}Cs. The apparent HRT of the ^{137}Cs specific activity in the dust season almost coincides with the radioactive half-life of ^{137}Cs, suggesting that ^{137}Cs did not migrate into the surface soils of the East Asian continent, whereas the apparent HRT of the ^{137}Cs specific activity in the non-dust season is consistent with that in Japanese surface soils (14.4 years) (Igarashi *et al.*, 2003), which means that ^{137}Cs in local soils gradually migrated due to surface water flow. The ^{90}Sr specific activities, which showed no difference between the dust and non-dust seasons (in contrast to ^{137}Cs and plutonium), gradually decreased (Figure 18.7b). The apparent HRT of the ^{90}Sr specific activity has been calculated to be 12 ± 2 years during the period 1990–2005, which coincides with previous estimates of HRTs in annual depositions (10 years) and surface soil (11 years) (Igarashi *et al.*, 2003). This finding suggests that in spite of the environmental conditions of the sources of ^{90}Sr, it gradually migrates from the soil surface layer by such processes as downward movement with percolating water flow in the soil column and incorporation from small soil particles on to larger ones. A change in deposition signals, such as the $^{239,240}Pu$ specific activities and $^{137}Cs/^{90}Sr$ ratios, was observed in the 2000s (Hirose *et al.*, 2008; Igarashi *et al.*, 2009); these findings suggest that this change reflects recent desertification in the North China Plain, north-eastern China, and the Korean Peninsula, resulting from overcropping and climate change. This set of

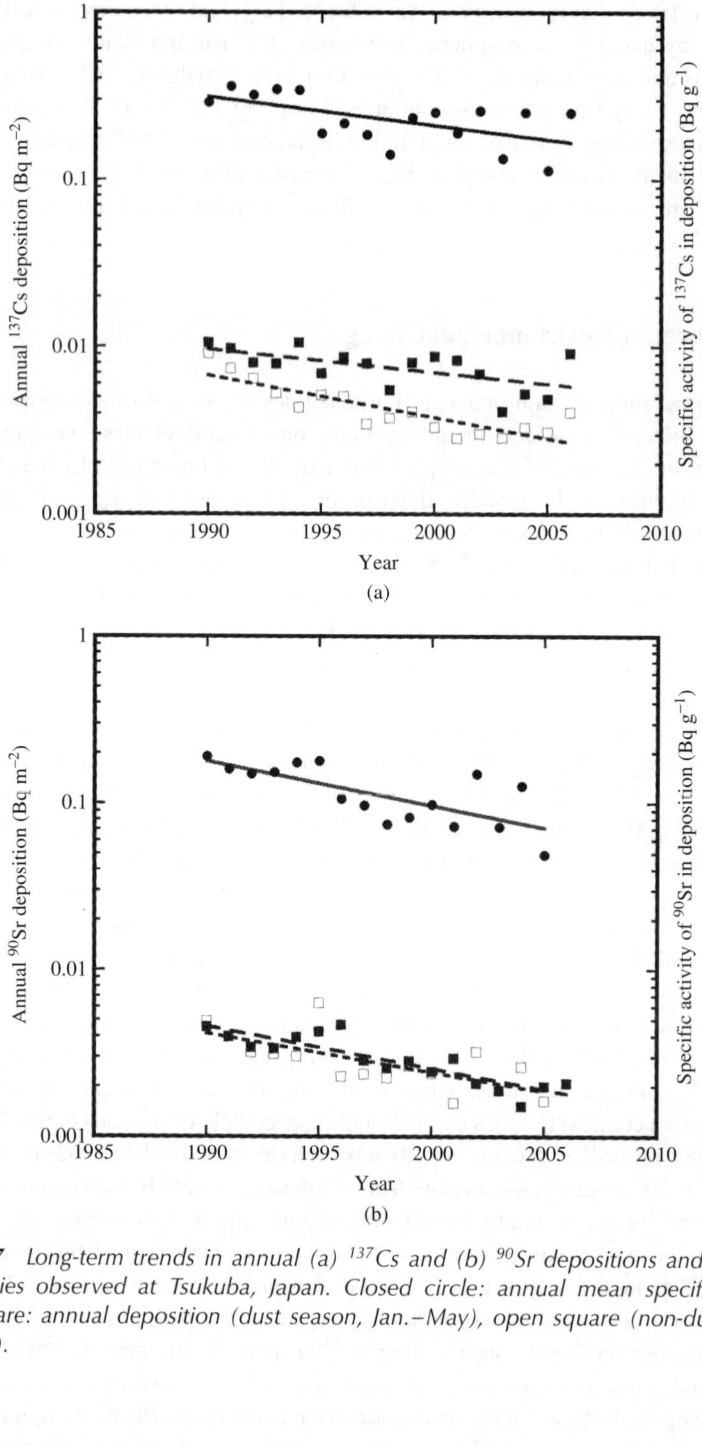

Figure 18.7 *Long-term trends in annual (a) ^{137}Cs and (b) ^{90}Sr depositions and their specific activities observed at Tsukuba, Japan. Closed circle: annual mean specific activity, closed square: annual deposition (dust season, Jan.–May), open square (non-dust season, June–Dec.).*

anthropogenic radionuclides in deposition is a useful tool for monitoring ongoing environmental changes in the terrestrial environment.

Natural radionuclides in aerosols show long-term variability due to changes of sources and sink, which is accompanied by increasing human activities and resulting climate change. Typically, ^{7}Be in aerosols and deposition shows a roughly 11-year cycle, coinciding with solar activities. Although aerosol thorium is of terrestrial origin, ^{232}Th deposition exhibits long-term variation (Hirose *et al.*, 2012). The annual ^{232}Th deposition showed an increasing trend during the period 1990–2007, as shown in Figure 18.8. In contrast to the monthly ^{232}Th deposition in spring, the temporal variation of the annual ^{232}Th deposition did not correspond to that of annual frequency of the Kosa events observed in Japan. Figure 18.8 indicates that the remotely derived annual ^{232}Th deposition exhibits an increasing trend with an increasing rate of 2% per year for the period 1990–2007, although the correlation of the increasing trend is weak ($r = 0.45$). On the other hand, the locally derived annual ^{232}Th deposition statistically shows no trend ($r = 0.37$). This result reveals that most of the increase in the annual ^{232}Th deposition is attributable to remotely derived ^{232}Th. Time series data on the annual mean specific activity of ^{232}Th (Figure 18.7) range from 3.1 to 11.2 Bq g^{-1}, with an average of 7.4 Bq g^{-1}, showing an increasing trend with an increasing rate of 3.4% per year ($r = 0.68$) from 1990 to 2007. It is difficult to explain the increasing trend of the annual mean specific activity of ^{232}Th using the trend in mineral dust load, because the mineral dust load seems not to have a long-term trend. This implies that the contribution of ^{232}Th-enriched dust increased during the period 1990–2007. Therefore, a possible cause of the increasing ^{232}Th deposition is considered to be an increment of ^{232}Th enriched dust rather than of mineral dust load.

Coal burning is a potentially important source of atmospheric thorium in East Asia, since coal consumption in Japan and China increased from 0.026×10^{15} to 0.084×10^{15} g C and

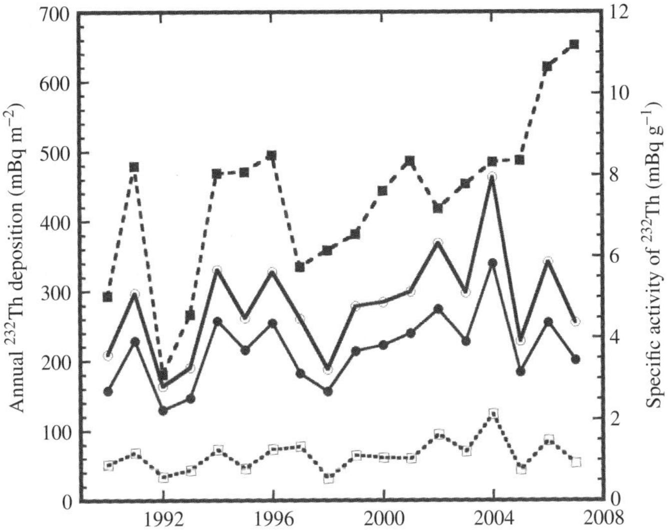

Figure 18.8 *Long-term trends in annual ^{232}Th deposition and its specific activity observed at Tsukuba, Japan. Closed square: annual mean specific activity, open circle: annual deposition, closed circle: remotely derived deposition, open square: locally derived deposition.*

from 0.9×10^{15} to 2.2×10^{15} g C, respectively, during the period 1990–2005 as a result of the increasing activities of coal power plants in that region. Although the major part of coal combustion-derived fly ash is retained in the electrostatic precipitators (Papastefanou, 2010), 1–3% fly ash escapes into the atmosphere as fine particles. About 0.8×10^{12} and 0.02×10^{12} Bq of thorium are estimated to have been released into the atmosphere as a result of coal burning by China in 2007 (2.6×10^{15} g C) and by Japan in 2005, respectively, assuming a thorium content of 0.03 Bq g^{-1} in coal (Tadmor, 1986; Lu, Jia, and Wang, 2006) and that 1% of the thorium in coal is released to the atmosphere from coal combustions. Compared to soil, fly ash is enriched with respect to ^{232}Th (Tadmor, 1986; Lu *et al.*, 2006; Papastefanou, 2010). This suggests the hypothesis that the increasing coal combustion from thermal power plants affects the long-term trend of thorium deposition at Tsukuba.

Interannual change in ^{210}Pb deposition has been recognized (Beks, Eisma, and van der Plicht, 1998; Baskaran, 2011). Since East Asia is downstream of the Asian outflow, including of anthropogenic pollutants and Asian dust, studies on the depositional behavior of ^{210}Pb in East Asia might be useful in tracking changes in Asian outflow. Su and Huh (2003) observed an anomalously high ^{210}Pb deposition in 1998, which corresponded to an El Niño event. Long-term variability in the ^{210}Pb deposition in East Asia is closely related to that in the amount of precipitation, especially in winter, which reflects climate change. The ^{210}Pb concentration in rainwater also shows long-term variability (Hirose *et al.*, 2011). However, it is not clear why an enhanced ^{210}Pb concentration in rainwater in East Asia should occur as a result of the El Niño event. To explain interannual change in the ^{210}Pb deposition, Beks *et al.* (1998) examined the possibility of there being additional sources of atmospheric ^{210}Pb or ^{222}Rn apart from the ^{222}Rn exhalation from the land surface. Volcanic eruption and industrial emission have been mentioned as candidates for being additional sources of atmospheric ^{210}Pb. However, estimation of the global contribution from volcanic sources has suggested that volcanic emissions of ^{210}Pb (and local industrial emissions) play only a minor role compared with ^{222}Rn exhalation (Beks *et al.*, 1998). On the other hand, the industrial emission of ^{210}Pb from Chinese burning is estimated to be 0.2–0.6 PBq $year^{-1}$, which corresponds to about 0.4–1.2% of the ^{210}Pb originating from global ^{222}Rn exhalation (Hirose *et al.*, 2011). Although this estimate suggests a minor contribution to global ^{222}Rn exhalation, an ongoing increase in the Chinese emission of pollutants will seriously affect the environmental radioactivity in East Asia. Biomass burning, which is enhanced in El Niño and recorded a maximum in the 1998 El Niño (van der Werf *et al.*, 2006), affects global atmospheric chemistry (Crutzen and Andreae, 1990) and is another candidate additional source of atmospheric ^{210}Pb. The ^{210}Pb release from biomass burning is calculated to be approximately 1.6 PBq $year^{-1}$ (Hirose *et al.*, 2011), corresponding to about 3% of the ^{210}Pb originating from global ^{222}Rn exhalation. As with other causes, the variation in the ^{210}Pb concentration in rainwater may derive from interannual changes in continental naked areas in winter, which can act as ^{210}Pb sources, and/or in ^{210}Pb transport from source areas, both resulting from climate change.

18.5 Conclusion

Natural and anthropogenic radionuclides can be effectively used as atmospheric tracers. The temporal and spatial distributions of radioactive aerosols and their deposition provide knowledge concerning the atmospheric behaviors of aerosols and existing environmental change resulting from increasing human activities.

In March 2011, the Fukushima NPP accident following a big earthquake and resulting tsunami in East Asia led to large amounts of radioactive aerosols and gases being released into the environment, causing serious public concern over the radiological risk. However, the Fukushima accident also provided a rare opportunity to obtain new information concerning the atmospheric behaviors of radioactive aerosols, such as verification of aerosol dispersion and deposition models. Study of radioactive aerosols, including development of measurements, is therefore an up-to-date topic, although radioactive aerosol studies have a history going back more than 60 years.

References

Aarkrog, A. (1988) The radiological impact of the Chernobyl debris compared with that from nuclear weapons fallout. *Journal of Environmental Radioactivity*, **6**, 151–162.

Amano, H., Akiyama, M., Chunlei, B. *et al.* (2012) Radiation measurements in the Chiba Metropolitan area and radiological aspects of fallout from the Fukushima Daiichi Nuclear Power Plants accident. *Journal of Environmental Radioactivity*, **111**, 42–52.

Anspaugh, L.R., Catlin, R.J. and Goldman, M. (1988) The global impact of the Chernobyl reactor accident. *Science*, **242**, 1513–1519.

Aoyama, M. (1999) Geochemical studies on behavior of anthropogenic radionuclides in the atmosphere, PhD thesis, Kanazawa University.

Aoyama, M., Hirose, K. and Sugimura, Y. (1991) The temporal variation of stratospheric fallout derived from the Chernobyl accident. *Journal of Environmental Radioactivity*, **13**, 103–115.

Aoyama, M., Hirose, K., Suzuki, Y. *et al.* (1986) High level radioactive nuclides in Japan in May. *Nature*, **321**, 819–820.

Aoyama, M., Hirose, K. and Takatani, S. (1992) Particle size dependent dry deposition velocity of the Chernobyl radioactivity, in *Precipitation Scavenging and Atmospheric Exchange Processes; Fifth International Conference*, Vol. **3** (eds S.E. Schwartz and W.G.N. Slinn), Hemisphere, Washington, DC, pp. 1581–1593.

Arimoto, R., Webb, J.L. and Conley, M. (2005) Radioactive contamination of atmospheric dust over southeastern New Mexico. *Atmospheric Environment*, **39**, 4745–4754.

Balkanski, Y.J., Jacob, D.J., Gardner, G.M. *et al.* (1993) Transport and residence time of tropospheric aerosols inferred from a global three-dimensional simulation of ^{210}Pb. *Journal of Geophysical Research*, **98**, 20573–20586.

Barrie, L.A. (1985) Scavenging ratios, wet deposition, and in-cloud oxidation: an application to the oxides of sulphur and nitrogen. *Journal of Geophysical Research*, **90**, 5789–5799.

Baskaran, M. (2011) Po-210 and Pb-210 as atmospheric tracers and global atmospheric Pb-210 fallout: a review. *Journal of Environmental Radioactivity*, **102**, 500–513.

Baskaran, M., Coleman, C.H. and Santschi, P.H. (1993) Atmospheric depositional fluxes of ^{7}Be and ^{210}Pb at Galveston and College Station, Texas. *Journal of Geophysical Research*, **98**, 20555–20571.

Beks, J.P., Eisma, D. and van der Plicht, J. (1998) A record of atmospheric ^{210}Pb deposition in The Netherlands. *Science of the Total Environment*, **222**, 35–44.

Bondietti, E.A. and Brantley, J.N. (1986) Characteristics of Chernobyl radioactivity in Tennessee. *Nature*, **322**, 313–314.

Brandt, J., Christensen, J.H. and Frohn, L.M. (2002) Modelling transport and deposition of caesium and iodine from the Chernobyl accident using the DREAM model. *Atmospheric Chemistry and Physics*, **2**, 397–417.

Cambray, R.S., Cawse, P.A., Garland, J.A. *et al.* (1987) Observations on radioactivity from the Chernobyl accident. *Nuclear Energy*, **26**, 77–101.

Chamberlain, A.C. (1953) Experiments on the deposition of iodine-131 vapour onto surfaces from an airstream. *Philosophical Magazine*, **44**, 1145–1153.

Chamberlain, A.C. (1991) *Radioactive Aerosols*, Cambridge Environmental Chemistry Series, Vol. **3**, Cambridge University Press.

Chino, M., Nakayama, H., Nagai, H. *et al.* (2011) Preliminary estimation of released amounts of ^{131}I and ^{137}Cs accidentally discharged from the Fukushima Daiichi nuclear power plant into the atmosphere. *Journal of Nuclear Science and Technology*, **48**, 1129–1134.

Crecelius, E.A., Robertson, D.E., Abel, K.H. *et al.* (1978) Atmospheric deposition of ^{7}Be and other elements on the Washington coast, in *Pacific Northwest Laboratory Annual Report for 1977 to the DOE Assistant Secretary for Environment: Ecological Sciences, PNL-2500 PT-2*, Pacific Northwest Laboratory, Battelle, pp. 7.25–7.26.

Crutzen, P.J. and Andreae, M.O. (1990) Biomass burning in the tropics: impact on atmospheric chemistry and biogeochemical cycles. *Science*, **250**, 1669–1678.

Devell, L., Tovedal, H., Bergström, V. *et al.* (1986) Initial observations of fallout from the reactor accident at Chernobyl. *Nature*, **321**, 192–193.

Doi, T., Masumoto, K., Toyoda, A. *et al.* (2013) Anthropogenic radionuclides in the atmosphere observed at Tsukuba: characteristics of the radionuclides derived from Fukushima. *Journal of Environmental Radioactivity***122**, 55–62.

Ehhalt, D.H. (1973) Turnover times of ^{137}Cs and HTO in the troposphere and removal rates of natural particles and water vapor. *Journal of Geophysical Research*, **78**, 7076–7086.

Englemann, R.J. (1971) Scavenging prediction using ratios of concentrations in air and precipitation. *Journal of Applied Meteorology*, **10**, 493–497.

Fujiwara, H., Fukuyama, T., Shirato, Y. *et al.* (2007) Deposition of atmospheric ^{137}Cs in Japan associated with the Asian dust event of March 2002. *Science of the Total Environment*, **384**, 306–315.

Garland, J.A. and Cambray, R.S. (1988) Deposition, resuspension and the long-term variation of airborne radioactivity from Chernobyl. Proceedings of 6th Symposium International de Radioecologie de Caderache, Centre d'Etudie Nucleaire de Cadarache, France, tome 1, pp. B26–B31.

Grabowska, S., Mietelski, J.W., Kozak, K. and Gaca, P. (2003) Gamma emitters on micro-Becquerel activity level in air at Krakow (Poland). *Journal of Atmospheric Chemistry*, **46**, 103–116.

Harvey, M.J. and Matthews, K.M. (1989) ^{7}Be deposition in a high-rainfall area of New Zealand. *Journal of Atmospheric Chemistry*, **8**, 299–306.

Hass, H., Memmesheimer, M., Geis, H. *et al.* (1990) Simulation of the Chernobyl radioactive cloud over Europe using the EURAD model. *Atmospheric Environment*, **24A**, 673–692.

Hernández-Ceballos, M.A., Hong, G.H., Lozano, R.L. *et al.* (2012) Tracking the complete revolution of surface westerlies over Northern Hemisphere using radionuclides emitted from Fukushima. *Science of the Total Environment*, **438**, 80–85.

Hirose, K. (1995) Geochemical studies in the Chernobyl radioactivity in environmental samples. *Journal of Radioanalytical and Nuclear Chemistry – Articles*, **197**, 331–342.

Hirose, K. (2000) Dry and wet deposition behaviors of thorium isotopes. *Journal of Aerosol Research, Japan*, **15**, 256–263.

Hirose, K. (2012) 2011 Fukushima Daiichi nuclear power plant accident: summary of regional radioactivity deposition monitoring results. *Journal of Environmental Radioactivity*, **111**, 13–17.

Hirose, K. (2013) Temporal variation of monthly ^{137}Cs deposition observed in Japan: effects of the Fukushima Daiichi nuclear power plant accident, *Applied Radiation and Isotopes*, **66**, 1675–1678.

Hirose, K., Aoyama, M., Katsuragi, Y. and Sugimura, Y. (1987) Annual deposition of Sr-90, Cs-137 and Pu-239,240 from the 1961–1980 nuclear explosions: a simple model. *Journal of the Meteorological Society of Japan*, **65**, 259–277.

Hirose, K., Igarashi, Y. and Aoyama, M. (2007) Recent trends of plutonium fallout observed in Japan: comparison with natural lithogenic radionuclides, thorium isotopes. *Journal of Radioanalytical and Nuclear Chemistry*, **273**, 115–118.

Hirose, K., Igarashi, Y. and Aoyama, M. (2008) Analysis of 50 years records of atmospheric deposition of long-lived radionuclides in Japan. *Applied Radiation and Isotopes*, **66**, 1675–1678.

Hirose, K., Igarashi, Y., Aoyama, M. and Inomata, Y. (2010) Depositional behaviors of plutonium and thorium at Tsukuba and Mt. Haruna in Japan indicate the sources of atmospheric dust. *Journal of Environmental Radioactivity*, **101**, 106–112.

Hirose, K., Igarashi, Y., Aoyama, M. *et al.* (2003) Recent trends of plutonium fallout observed in Japan: plutonium as a proxy for desertification. *Journal of Environmental Monitoring*, **5**, 1–7.

Hirose, K., Igarashi, Y., Aoyama, M. and Miyao, T. (2001) Long-term trends of plutonium fallout observed in Japan, in *Plutonium in the Environment* (ed A. Kudo), Elsevier Science, pp. 251–266.

Hirose, K., Kikawada, Y., Doi, T. *et al.* (2011) ^{210}Pb deposition in the Far East Asia: controlling factors of its spatial and temporal variations. *Journal of Environmental Radioactivity*, **102**, 514–519.

Hirose, K., Kikawada, Y. and Igarashi, Y. (2012) Temporal variation and provenance of thorium deposition observed at Tsukuba, Japan. *Journal of Environmental Radioactivity*, **108**, 24–28.

Hirose, K., Kim, C.K., Kim, C.S. *et al.* (2004) Plutonium deposition observed in Daejeon. Korea: wet and dry depositions of plutonium. *Science of the Total Environment*, **332**, 243–252.

Hirose, K. and Sugimura, Y. (1990) Plutonium isotopes in the surface air in Japan: effect of Chernobyl accident. *Journal of Radioanalytical and Nuclear Chemistry – Articles*, **138**, 127–138.

Hirose, K., Takatani, S. and Aoyama, M. (1993) Wet deposition of long-lived radionuclides derived from the Chernobyl accident. *Journal of Atmospheric Chemistry*, **17**, 61–73.

Hötzl, H., Rosner, G. and Winkler, R. (1989) Long-term behaviour of Chernobyl fallout in air and precipitation. *Journal of Environmental Radioactivity*, **10**, 157–171.

Huh, C.A., Hsu, S.-C. and Lin, C.-Y. (2012) Fukushima-derived fission nuclides monitored around Taiwan: free tropospheric versus boundary layer transport. *Earth Planetary Science Letter*, **319–320**, 9–14.

International Atomic Energy Agency (IAEA) (2006) *Experimental Consequences of Chernobyl Accident and their Mediation: Twenty Years of Experience. Report of the Chernobyl Forum Expert Group 'Environment'*, Radiological Assessment Reports Series, IAEA, Vienna.

Igarashi, Y., Aoyama, M., Hirose, K. *et al.* (2003) Resuspension: decadal monitoring time series of the anthropogenic radioactivity deposition in Japan. *Journal of Radiation Research*, **44**, 319–328.

Igarashi, Y., Aoyama, M., Hirose, K. *et al.* (2001) Is it possible to use ^{90}Sr and ^{137}Cs as tracers for the aeolian transport? *Water Air and Soil Pollution*, **130**, 349–354.

Igarashi, Y., Inomata, Y., Aoyama, M. *et al.* (2009) Possible change in Asian dust source suggested by atmospheric anthropogenic radionuclides during the 2000s. *Atmospheric Environment*, **43**, 2971–2980.

Imai, K., Chino, M., Ishikawa, H. *et al.* (1985) *SPEEDI:* A Computer Code System for the Real-time Prediction of Radiation Dose to the Public Due to An Accidental Release, JAERI 1297, Japan Atomic Research Institute.

Jaworowski, Z. and Kownacka, L. (1988) Tropospheric and stratospheric distributions of radioactive iodine and caesium after the Chernobyl accident. *Journal of Environmental Radioactivity*, **6**, 145–150.

Jost, D.T., Gaggeler, H.W., Baltensparger, V. *et al.* (1986) Chernobyl fallout in size-fractionated aerosol. *Nature*, **324**, 22–23.

Junge, C.E. (1963) *Air Chemistry and Radioactivity*, Academic Press, New York.

Junge, C.E. and Manson, J.E. (1961) Stratospheric aerosol studies. *Journal of Geophysical Research*, **66**, 2163–2182.

Karlsson, L., Hernandez, F., Rodríguez, S. *et al.* (2008) Using ^{137}Cs and ^{40}K to identify natural Saharan dust contributions to PM_{10} concentrations and air quality impairment in the Canary Islands. *Atmospheric Environment*, **42**, 7034–7042.

Katsuragi, Y. (1983) A study of ^{90}Sr fallout in Japan. *Papers in Meteorology and Geophysics*, **33**, 277–291.

Koch, D.M., Jacob, D.J. and Graustein, W.C. (1996) Vertical transport of tropospheric aerosols as indicated by ^{7}Be and ^{210}Pb in a chemical tracer model. *Journal of Geophysical Research*, **101**, 18651–18666.

Kolb, W. (1989) Seasonal fluctuations of the uranium and thorium contents of aerosols in ground-level air. *Journal of Environmental Radioactivity*, **9**, 61–75.

Krey, P.W., Klusek, C.S., Sanderson, C. *et al.* (1986) Radiochemical characterization of Chernobyl fallout in Europe, in Report EML 460, in (ed H.L. Volchok), US Department of Energy, pp. 155–218.

Krey, P.W. and Krajewski, B.T. (1970) Comparison of atmospheric transport model calculations with observations of radioactive debris. *Journal of Geophysical Research*, **75**, 2901–2908.

Krey, P.W., Leifer, R., Benson, W.K. *et al.* (1979) Atmospheric burn-up of the Cosmos-954 reactor. *Science*, **205**, 583–585.

Kuroda, P.K., Miyake, Y. and Nemoto, J. (1965) Strontium isotopes: global circulation after the Chinese nuclear explosion of 14 May 1965. *Science*, **150**, 1289–1290.

Lambert, G., Polian, G., Sanak, J. *et al.* (1982) Cycle du radon et de ses descendants: application a l'étude des èchanges troposphère-stratosphère. *Annales de Geophysique*, **38**, 497–531.

Larsen, R.J., Sanderson, C., Rivera, W. and Zamichieli, M. (1986) The characterization of radionuclides in North American and Hawaiian surface air and deposition following the Chernobyl accident, in Report EML 460, in (ed H.L. Volchok), US Department of Energy, pp. 1–140.

Lee, S.H., Pham, M.K. and Povinec, P.P. (2002) Radionuclide variations in the air over Monaco. *Journal of Radioanalytical and Nuclear Chemistry*, **254**, 445–453.

Lu, X., Jia, X. and Wang, F. (2006) Natural radioactivity of coal and its by-products in the Baoji coal-fire power plant, China. *Current Sciences*, **91**, 1508–1511.

Lujaniené, G., Aninkevicius, V. and Lujanas, V. (2009) Artificial radionuclides in the atmosphere over Lithuania. *Journal of Environmental Radioactivity*, **100**, 108–119.

Lujaniené, G., Bycenkiené, S., Povinec, P.P. and Gera, M. (2012) Radionuclides from the Fukushima accident in the air over Lithuania: measurement and modeling approaches. *Journal of Environmental Radioactivity*. doi: 10.1016/j.jenvrad.2011.12.004

Machta, L., List, R.J. and Huber, L.F. (1956) World-wide travel of atomic debris. *Science*, **124**, 474.

Martell, E.A. (1966) The size distribution and interaction radioactive and natural aerosols in the stratosphere. *Tellus*, **18**, 486–498.

Martin, P. (2003) Uranium and thorium series radionuclides in rainwater over several tropical storms. *Journal of Environmental Radioactivity*, **65**, 1–18.

Masson, O., Piga, D., Gurriaran, R. and D'Amico, D. (2010) Impact of an exceptional Saharan dust outbreak in France: PM_{10} and artificial radionuclides concentrations in air and in dust deposit. *Atmospheric Environment*, **44**, 2478–2486.

Matsunami, T. and Mamuro, T. (1975) Study of uranium deposition by basin method. *Annual Report of Radiation Center Osaka Prefecture*, **16**, 22–24.

McMurry, P.H. and Wilson, J.C. (1982) Growth laws for the formation of secondary ambient aerosols: Implications for chemical conversion mechanisms. *Atmospheric Environment*, **16**, 121–134.

McNeary, D. and Baskaran, M. (2003) Depositional characteristics of ^{7}Be and ^{210}Pb in southeastern Michigan. *Journal of Geophysical Research*, **108**(D7), 4201. doi: 10.1029/2002JD003021

Moore, H.E., Poet, S.E. and Martell, E.A. (1973) ^{222}Rn, ^{210}Pb, ^{210}Bi, and ^{210}Po profiles and aerosol residence times versus altitude. *Journal of Geophysical Research*, **78**, 7065–7075.

Morino, Y., Ohara, T. and Nishizawa, M. (2011) Atmospheric behavior, deposition, and budget of radioactive materials from the Fukushima Daiichi nuclear power plant in March 2011. *Geophysical Research Letters*, **38**, L00G11.

National Council on Radiation Protection and Measurements (NCRP) (1987) Exposure of the Population in the United States and Canada from Natural Background Radiation, Report No. 94, NCRP.

Nicholson, K.W. (1988) A review of particle resuspension. *Atmospheric Environment*, **22**, 2639–2651.

Papastefanou, C. (2008) Radioactive Aerosols, in The *Radioactivity in the Environment Series*, Vol. **12** (ed M.S. Baxter), Elsevier.

Papastefanou, C. (2010) Escaping radioactivity from coal-fired plants (CPPs) due to coal burning and associated hazards: a review. *Journal of Environmental Radioactivity*, **101**, 191–200.

Papastefanou, C. and Ioannidou, A. (1995) Aerodynamic size association of ^{7}Be in ambient aerosols. *Journal of Environmental Radioactivity*, **26**, 273–282.

Papastefanou, C., Ioannidou, A., Stoulos, S. and Manolopoulou, M. (1995) Atmospheric deposition of cosmogenic ^{7}Be and ^{137}Cs from fallout of the Chernobyl accident. *Science of the Total Environment*, **170**, 151–156.

Peirson, D.H., Cawse, P.A., Salmon, L. and Cambray, R.S. (1973) Trace elements in the atmospheric environment. *Nature*, **241**, 252–256.

Persson, C., Rodhe, H. and de Geer, L.E. (1987) The Chernobyl accident – a meteorological analysis of how radionuclides reached and were deposited in Sweden. *Ambio*, **16**, 20–31.

Porstendörfer, J. and Gründel, M. (2003) Comparison of the activity size distribution of the radionuclide aerosols in outdoor air, in *Book of Abstracts, Dresden Symposium on Radiation Protection, Dresden, Germany*, Dresden University of Technology, Dresden, p. 24.

Reiter, E.R. and Bauer, E. (1975) *Residence Times of Atmospheric Pollutant*, CIAP Monograph, Vol. **1**, US Department of Transportation, Washington, DC.

Rosner, G., Hotzl, H. and Winkler, R. (1997) Long-term behaviour of plutonium in air, and deposition and the role of resuspension in semi-rural environment in Germany. *Science of the Total Environment*, **196**, 255–261.

Rosner, G. and Winkler, R. (2001) Long-term variation (1986–1998) of post-Chernobyl ^{90}Sr, ^{137}Cs, ^{238}Pu and ^{239,240}Pu concentrations in air, depositions to ground, resuspension factors and resuspension rates in south Germany. *Science of the Total Environment*, **273**, 11–25.

Sakuragi, Y., Meason, J.L. and Kuroda, P.K. (1983) Uranium and plutonium isotopes in the atmosphere. *Journal of Geophysical Research*, **88**, 3718–3724.

Sanak, J., Gaudry, A. and Lambert, G. (1981) Size distribution of ^{210}Pb aerosols over oceans. *Geophysical Research Letters*, **8**, 1067–1069.

Sehmel, G.A. (1980) Particle and gas dry deposition: a review. *Atmospheric Environment*, **14**, 983–1011.

Shinn, J.H., Homan, D.N. and Gray, D.D. (1983) Plutonium aerosol fluxes and pulmonary exposure rates during re-suspension from bare soils near a chemical separation facility, in *Precipitation Scavenging, Dry Deposition and Re-suspension* (eds H.R. Pruppacher, R.G. Semonin and W.G.N. Slinn), Elsevier, Amsterdam, pp. 1131–1143.

Slinn, W.G.N. (1978) Parametrizations for resuspension and for wet and dry deposition of particles and gases for use in radiation dose calculations. *Nuclear Safety*, **19**, 205–219.

Small, S.H. (1960) Wet and dry deposition of fallout materials at Kjeller. *Tellus*, **12**, 308–314.

Stewart, K. (1966) The resuspension of particulate material from surface, in *Surface Contamination* (ed B.R. Fish), Pergamon, Oxford, pp. 63–74.

Stohl, A., Seibert, R., Wotawa, G. *et al.* (2011) Xenon-133 and casesium-137 releases into the atmosphere from the Fukushima Dai-ichi nuclear power plant: determination of the source term, atmospheric dispersion, and deposition. *Atmospheric Chemistry and Physics Discussions*, **11**, 28319–28394.

Su, C.-C. and Huh, C.-A. (2003) Factors controlling atmospheric fluxes of ^{7}Be and ^{210}Pb in the northern Taiwan. *Geophysical Research Letters*, **30**, 2018.

Tadmor, J. (1986) Radioactivity from coal-fired plants: a review. *Journal of Environmental Radioactivity*, **4**, 177–204.

Takemura, T., Nakamura, H., Takigawa, M. *et al.* (2011) A numerical simulation of global transport of atmospheric particles emitted from the Fukushima Daiichi Nuclear Power Plant. *Sora*, **7**, 101–104.

Terada, H. and Chino, M. (2008) Development of an atmospheric dispersion model for accidental discharge of radionuclides with the function of simultaneous prediction for multiple domains and its evaluation by application to the Chernobyl nuclear accident. *Journal of Nuclear Science and Technology*, **45**, 920–931.

Todd, J.F., Wong, G.T.F., Olsen, C.R. and Larsen, I.L. (1989) Atmospheric depositional characteristics of beryllium-7 and lead-210 along southeastern Virginia coast. *Journal of Geophysical Research*, **94**, 11106–11116.

Turekian, K.K., Benninger, L.K. and Dion, E.P. (1983) ^{7}Be and ^{210}Pb total deposition fluxes at New Haven, Connecticut and at Bermuda. *Journal of Geophysical Research*, **88**, 5411–5415.

Turekian, K.K. and Cochran, J.K. (1981) ^{210}Pb in surface air at Enewetak and the Asian dust to the Pacific. *Nature*, **292**, 522–524.

Underwood, B.Y. (2001) *Review of Deposition Velocity and Washout Coefficient*, AEA Technology, Harwell.

UNSCEAR (2000) *Sources and Effects of Ionizing Radiation: Sources*, Vol. **1**, United Nations, New York, p. 654.

USSR State Committee on the Utilisation of Atomic Energy (1986) The Accident at the Chernobyl Nuclear Power Plant and its Consequences, Safety Series No. 75-INSAG-1. IAEA, Vienna.

Victorova, N.V. and Garger, E.K. (1990) Biological monitoring of the deposition and transport of radioactive aerosol particles in the Chernobyl NEP zone of influence, Proceedings of the CEC Seminar on Comparative Assessment of the Environmental Impact of Radionuclides Released During Three Major Nuclear Accidents, Kyshtym, Windscale, Chernobyl, EUR 13574, pp. 223–236.

Warneck, P. (1988) *Chemistry of Natural Atmosphere*, Academic Press, San Diego, CA, pp. 360–373.

van der Werf, G.R., Randerson, J.T., Giglio, L. *et al.* (2006) Interannual variability in global biomass burning emissions from 1997 to 2004. *Atmospheric Chemistry and Physics Discussions*, **6**, 3423–3441.

Winkelmann, I., Endrulat, H.-J., Fouasnon, S. *et al.* (1987) *Radioactivity Measurements in the Federal Republic of Germany after the Chernobyl Accident*, Institut für Strahlenhygiene des Bundesgesundheitamtes, Neuherberg.

Winkler, R., Dietl, F., Franck, G. and Tschiersch, J. (1998) Temporal variation of ^{7}Be and ^{210}Pb size distribution in ambient aerosol. *Atmospheric Environment*, **32**, 983–991.

Yasunari, T.J., Stohl, A., Hayano, R.S. *et al.* (2011) Ceaium-137 deposition and contamination of Japanese soils due to the Fukushima nuclear accident. *Proceedings of the National Academy of Sciences of the United States of America*, **108**, 19530–19534.

Young, J.A. and Silker, W.B. (1980) Aerosol deposition velocities on the Pacific and Atlantic Oceans calculated from ^{7}Be measurements. *Earth Planetary Science Letter*, **50**, 92–104.

Yu, K.N. and Lee, L.Y.L. (2002) Measurement of atmospheric ^{7}Be properties using high efficiency gamma spectroscopy. *Applied Radiation and Isotopes*, **57**, 941–946.

Index

Aerosol Science: Technology and Applications, First Edition. Edited by Ian Colbeck and Mihalis Lazaridis.